The Guanine-Nucleotide Binding Proteins

Common Structural and Functional Properties

NATO ASI Series

Advanced Science Institutes Series

A series presenting the results of activities sponsored by the NATO Science Committee, which aims at the dissemination of advanced scientific and technological knowledge, with a view to strengthening links between scientific communities.

The series is published by an international board of publishers in conjunction with the NATO Scientific Affairs Division

A	**Life Sciences**	Plenum Publishing Corporation
B	**Physics**	New York and London
C	**Mathematical and Physical Sciences**	Kluwer Academic Publishers Dordrecht, Boston, and London
D	**Behavioral and Social Sciences**	
E	**Applied Sciences**	
F	**Computer and Systems Sciences**	Springer-Verlag
G	**Ecological Sciences**	Berlin, Heidelberg, New York, London,
H	**Cell Biology**	Paris, and Tokyo

Recent Volumes in this Series

Volume 160—Early Influences Shaping the Individual
edited by Spyros Doxiadis

Volume 161—Research in Congenital Hypothyroidism
edited by F. Delange, D. A. Fisher, and D. Glinoer

Volume 162—Nematode Identification and Expert System Technology
edited by Renaud Fortuner

Volume 163—Leishmaniasis: The Current Status and New Strategies for Control
edited by D. T. Hart

Volume 164—Cochlear Mechanisms: Structure, Function, and Models
edited by J. P. Wilson and D. T. Kemp

Volume 165—The Guanine-Nucleotide Binding Proteins: Common Structural and Functional Properties
edited by L. Bosch, B. Kraal, and A. Parmeggiani

Volume 166—Vascular Dynamics: Physiological Perspectives
edited by N. Westerhof and D. R. Gross

Series A: Life Sciences

The Guanine-Nucleotide Binding Proteins

Common Structural and Functional Properties

Edited by

L. Bosch and B. Kraal

Leiden University
Leiden, The Netherlands

and

A. Parmeggiani

Ecole Polytechnique
Palaiseau, France

Springer Science+Business Media, LLC

Proceedings of an EMBO-NATO-CEC Advanced Research Workshop
on the Guanine-Nucleotide Binding Proteins:
Common Structural and Functional Properties,
held August 6-11, 1988,
in Renesse, The Netherlands

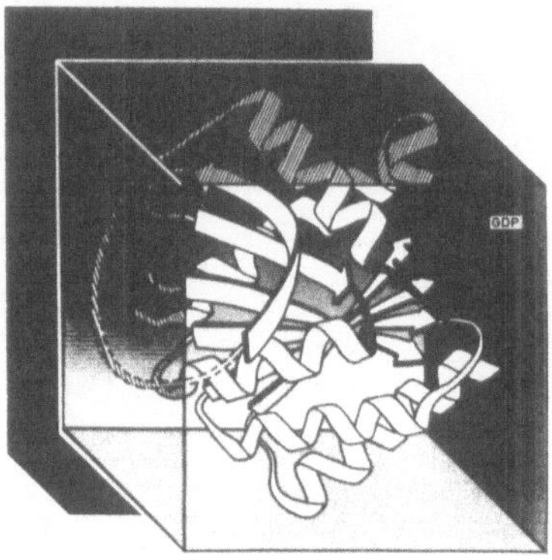

Library of Congress Cataloging in Publication Data

EMBO-NATO-CEC Advanced Research Workshop on the Guanine-Nucleotide Binding Proteins: Common Structural and Functional Properties (1988: Renesse, Netherlands)

The guanine-nucleotide binding proteins: common structural and functional properties / edited by L. Bosch and B. Kraal and A. Parmeggiani.

p. cm.—(NATO ASI series. Series A, Life sciences; v. 165)

"Proceedings of an EMBO-NATO-CEC Advanced Research Workshop on the Guanine-Nucleotide Binding Proteins: Common Structural and Functional Properties, held August 6-11, 1988 in Renesse, The Netherlands"—T.p. verso.

"Published in cooperation with NATO Scientific Affairs Division."

Includes bibliographical references and indexes.

ISBN 978-1-4757-2039-6 ISBN 978-1-4757-2037-2 (eBook)

DOI 10.1007/978-1-4757-2037-2

1. G proteins—Structure-activity relationships—Congresses. I. Bosch, L. II. Kraal, B. III. Parmeggiani, A. IV. North Atlantic Treaty Organization. Scientific Affairs Division. V. Title. VI. Series.

QP552.G16E52 1988 89-4038

574.19′245—dc20 CIP

Originally published by Plenum Press, New York in 1989.

Softcover reprint of the hardcover 1st edition 1989

PREFACE

This volume contains the proceedings of the EMBO-NATO-CEC Advanced Research Workshop on "Guanine-nucleotide binding proteins. Common structural and functional properties", which was held in Renesse, The Netherlands, August 6-11, 1988.

The transmission of information is one of the most important processes in cellular life and involves the most diverse physiological functions. The cellular membrane, as the obligatory target for external signals, harbours complex pathways transducing the signals from the receptors of the external stimuli to the cytoplasmic effector. Heterotrimeric proteins are fundamental components of these pathways. Other proteins that are monomeric may be found associated with the membrane or in soluble form in the cytoplasm, and can also function in signal transduction. Intracellular transmission of signals may proceed in an analogous fashion, protein synthesis being a well-known example.
It is one of the most remarkable and puzzling observations of recent years that all of these proteins share common properties, both functionally and structurally. Their primary structures show pronounced similarities, in most cases concentrated in the NH_2-terminal portion of the molecule. They all bind guanine nucleotides (hence the general name of G-proteins) and are GTPases, a crucial enzymatic activity, since it converts the active complex induced by GTP into the inactive one induced by GDP. Consensus sequences have been identified as responsable for interacting with the different parts of the guanine nucleotide.
To emphasize the importance of this class of proteins in the pathology of mammals, it is sufficient to mention that alteration of cell proliferation, such as oncogenic transformation, is associated with the mutation of the *ras* product p21, a guanine-nucleotide binding protein.

For many years there has been little, if any, interaction between the groups of scientists involved in the study of these various protein families. The exponential increase of knowledge of their functioning and the growing notion of mechanistic similarities have led to the organization of the above-mentioned workshop. Its major goal was to bring together a number of specialists, actively engaged in the study of one or more of the major families of these proteins, in order to stimulate the exchange of information and to coordinate the efforts.
For four and a half day, similarities and diversities of these proteins were discussed, taking advantage of the detailed analysis of two model proteins: the prokaryotic elongation factor Tu and the *ras* protein p21. They are the only members of these families that have been crystallized, which has led to the elucidation of their three-dimensional structures.

The workshop was highly favoured by the relaxing atmosphere in a comfortable environment far away from the tumult of daily life, and the salted sea breeze of Zeeland. It could not have been organized without the generous support from the European Molecular Biology Organization, from the NATO Scientific Affairs Division, and from the Commission of the European Communities. Financial contributions from the Royal Netherlands Academy of Arts and Sciences, Gist Brocades NV, Rhône-Poulenc Santé, and Du Pont de Nemours (Nederland) BV are also gratefully acknowledged.

We have the feeling that everyone participated intensively and that the exchange of information was useful and exciting. We hope that any of the stimulating propositions will lead to further investigations and the future development of the field. If so, this will be the most appreciated reward for the effort of the organizers.

L. Bosch,
B. Kraal,
A. Parmeggiani

CONTENTS

STRUCTURE AND FUNCTION OF TUBULINS

THREE-DIMENSIONAL STRUCTURES AND CONSENSUS SEQUENCE ELEMENTS

NEW STRUCTURAL DATA ON ELONGATION FACTOR-Tu:GDP BASED ON X-RAY CRYSTALLOGRAPHY

J. Nyborg and T. la Cour

Department of Biostructural Chemistry
University of Aarhus
Langelandsgade 140
DK-8000 Aarhus C
Denmark

INTRODUCTION

The GDP binding proteins (G-proteins) have attracted much attention in recent years for many good reasons. One good reason is that some of these represent a biomolecular switching mechanism with a possibility of amplification of extracellular signals. Most (if not all) have a specific GDP- binding domain of about 200 residues. This domain can be in an "on-state" complexed with GTP. This state is transformed to the "off-state" by hydrolysis of GTP to GDP. The G-proteins in their "on-state" influence the reactions of other proteins to amplify the signal.

For the protein factors in the biosynthesis of proteins, the "on-state" binds aminoacylated transfer RNA (aa-tRNA), which in turn binds to the ribosome. EF-Tu from *E. coli,* is one of the best characterized factors, and there now exist two independently determined models for the structure of Domain I (Jurnak, 1985; la Cour *et al.*, 1985).

A model for the structure of Elongation Factor Tu (EF-Tu) in its complex with GDP containing all three domains is proposed here (see Figures 1 and 2). The data are from X-ray diffraction experiments on proteolytically cleaved material in crystals diffracting to a maximum resolution of 2.6 Å. The model is based on phases to 2.9 Å resolution and has been subjected to a few cycles of least squares refinement. The fragment containing residues 45 to 58 is not found in the crystals.

In previous communications from this laboratory, we have described EF-Tu as consisting of three domains. In order to have a more logical nomenclature, we have renamed the domains II and III so that their names represent their sequential order along the polypeptide rather than the historical naming reflecting their spatial relation with respect to domain I. Domain I (residues 1-44, 59-199), or the G-domain has an α/β type structure with a central core of six β-strands surrounded by α-helices. It is the GDP/GTP-binding domain

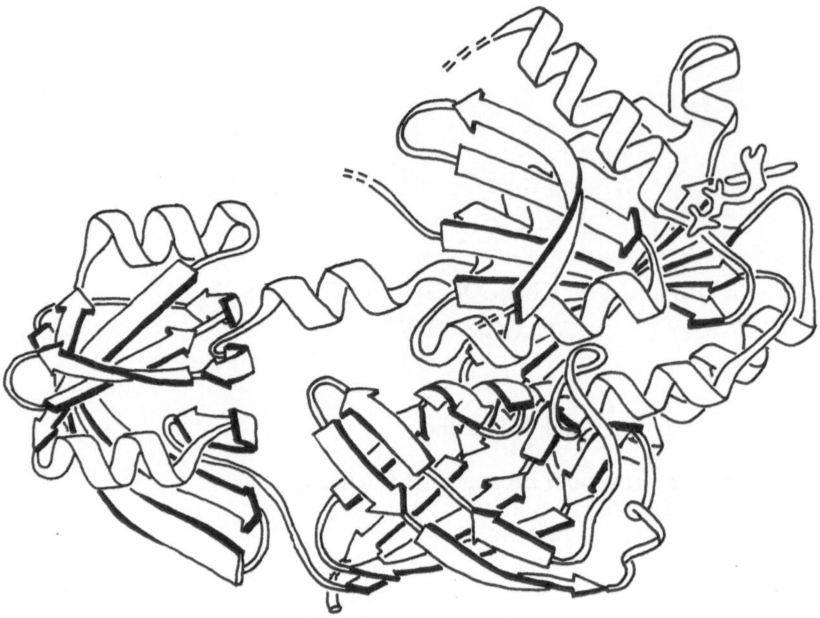

Figure 1. "Side view" of EF-Tu:GDP. At the top right is seen domain I with GDP. At the left is domain II and at the bottom right domain III.

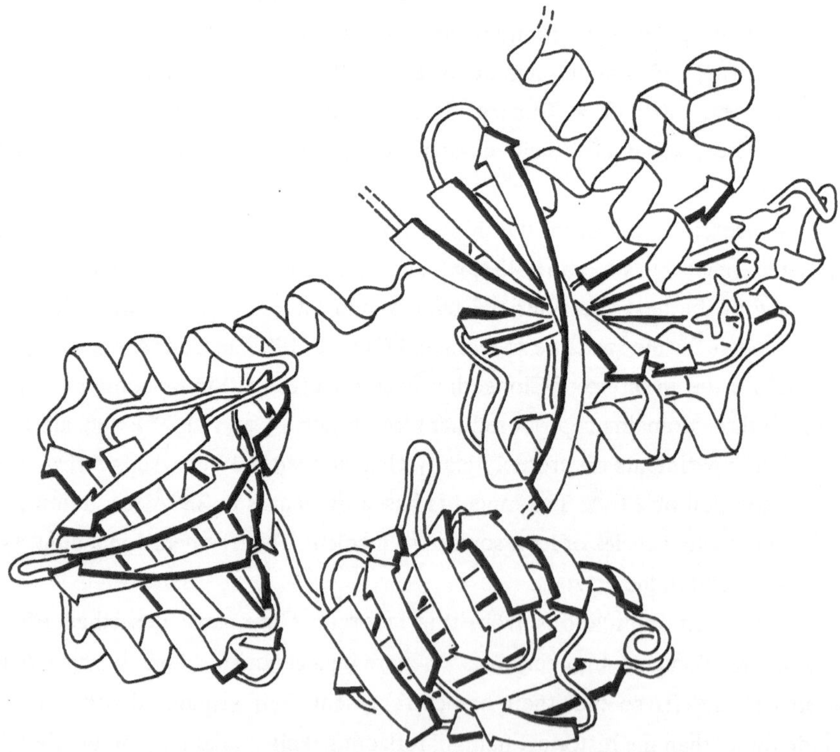

Figure 2. "Top view" of EF-Tu:GDP. Domains are in the same relative orientation as in Figure 1. Arrows are β-strands and cylindrical ribbons are α-helices.

Table 1. Residue assignment to secondary structure elements

Residue	Assignment	Comments
1-10		not seen in density
10-17	β–strand, Ia	well defined
18-23	loop	well defined
24-40	α-helix, IA	well defined at the N-terminal end
41-44		not seen in density
59		not seen in density
60-70	β-strand Ib	well defined
71-73	loop	well defined
74-81	β-strand, Ic	well defined
82-84	loop	well defined
85-93	α-helix, IB	well defined
94-99	loop	not well defined, exists possibly in two conformations
100-106	β-strand, Id	well defined
107-112	loop	well defined
113-125	α-helix, IC	well defined
126-128	loop	well defined
129-135	β-strand, Ie	well defined
136-142	loop	well defined
143-160	α-helix, ID	well defined
161-168	loop	well defined
169-172	β-strand, If	well defined
173	loop	well defined
174-179	α-helix, IE	well defined at places
180-182	loop	no good densities
183-199	α-helix, IF	well defined
200-202	loop, Bridge	well defined
203-208	α-helix, Bridge	ill defined
209-210	loop	ill defined
211-221	β-strand, IIa	well defined
222-224	loop	not seen in density
225-229	β-strand, IIb	well defined
230-239	loop	well defined
240-247	β-strand, IIc	well defined
248-249	loop	well defined
250-255	β-strand, IId	well defined
256-259	loop	less well defined
260-267	α-helix, IIA	ill defined
268-271	loop	less well defined
272-278	β-strand, IIe	well defined
279-280	loop	less well defined
281-286	loop	ill defined, perhaps helical
287-290	loop	less well defined
291-294	β-strand, IIf	well defined
295-300	loop, Bridge	well defined
301-311	β-strand, IIIa	well defined
312-321	loop	less well defined
322-330	β-strand, IIIb	well defined
331-332	loop	well defined
333-341	β-strand, IIIc	well defined
342-354	loop	at places less well defined
355-364	β-strand, IIId	well defined
365-371	loop	less well defined
372-378	β-strand, IIIe	well defined
379-380	loop	less well defined
381-391	β-strand, IIIf	well defined
392-393	loop	less well defined

where GTP hydrolysis to GDP takes place and hence the primary function of the molecule, the control of the recognition of aa-tRNA. We shall here present a model of the rest of the molecule containing the C-terminal residues 200-393. These are evenly distributed between the domains II (residues 210-295) and III (residues 301-393). The residues 200-209 form the only covalent link between domains I and II while the residues 296-300 form the only covalent link between domains II and III.

The two domains II and III both have a classical anti-parallel β-structure consisting of six β-stands. Due to the right-handed twist of the sheets, they are wrapped into cylinders with a barrel shape and therefore belongs to a common class of tertiary structures called anti-parallel β-barrels.

METHOD

The structure of EF-Tu has been found from X-ray crystallographic data from three different crystal forms (Type I, orthorhombic; Type II, orthorhombic; Type III tetragonal). The phase determination has been done for Type I crystals using Hg, Pt and Pb derivatives, thus producing an initial low resolution structure (Morikawa *et al.*, 1978). Data are collected to a maximum resolution of 2.6 Å using film methods (Arndt and Wonacott, 1977) Experimental phases have been determined using standard methods (Blundell and Johnson, 1976).

A note of warning is perhaps necessary here. The quality of the electron density map at initial stages and at medium resolution, i.e. 2.9 Å, is not high. Normally, one is able to trace part of the main chain and to judge side chains by size. A lot of additional information needs to be included in the interpretation of the map. The information used here includes: The amino acid sequence, positions of heavy atoms, secondary structure predictions, biochemical data related to sequence, sequence homology to related proteins, knowledge of folding patterns in other proteins and modelling many alternative tracings.

Problems are often found at the surface of a protein. Internal β-strands can be well defined but loops between them hard to see. If there is not much doubt about the assignment of residues to two consecutive β-strands, then one can safely postulate that they are connected somehow. Often also α-helices at the surface are badly defined and are seen as only half helical turns.

The residue assignments to secondary structure elements and comments on the density found in Table 1 illustrates the problems. Note the nomenclature for β-strands and α-helices where the roman number is the number of a domain, small letters are strands and capital letters are helices.

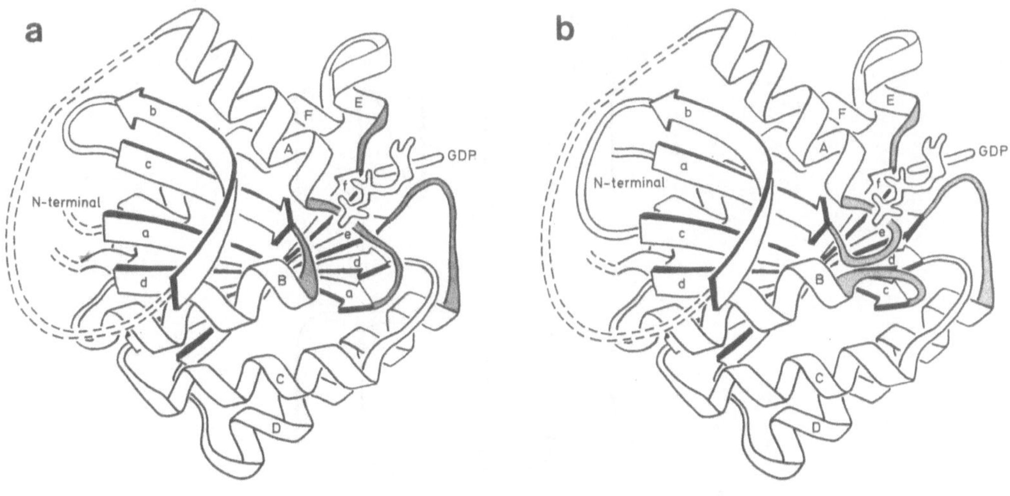

Figure 3. a) The G-domain (domain I) of EF-Tu. The position of GDP is shown. The four GDP binding loops are shown in grey. b) The G-domain of EF-Tu with the β-strand order of p21.

DOMAIN I

Two models for the structure of domain I of EF-Tu have recently been reported (Jurnak, 1985; la Cour *et al.*, 1985). The two models are in all main details similar.

The GDP cofactor binds to the surface of domain I which has a typical α/β type structure (Levitt and Chotia, 1976). The central core of the domain consists of a twisted β-sheet made up of five parallel β-strands and one anti-parallel β-strand. The connections between the parallel strands contain six α-helices, as shown in Figure 3a. The present assignment of amino acid residues to the secondary structure elements of the domain is given in Table 1. A schematic diagram of the topology of the central β-sheet is shown in Figure 4. This shows a folding arrangement of β-strands which can be described as -2, +1, +2x, +1x, +1x (Richardson, 1981). The GDP-binding site is located at the carboxy end of the β-sheet, such as found in other nucleotide-binding proteins (Brändén, 1980). The amino acid residues involved in the interaction between the nucleotide and the protein are found in four loops connecting β-strands with α-helices, see Figure 5. Sequence comparisons between EF-Tu and other GDP/GTP-binding proteins (Halliday, 1984; Leberman and Egner, 1984; Möller and Amons, 1985; Dever *et al.*, 1987) have led to consensus sequences for all these proteins in three of the four loops. The three consensus sequences are

Ala/Gly-X-X-X-X-Gly-Lys-Thr/Ser
Asp-X-X-Gly
Asn-Lys-X-Asp

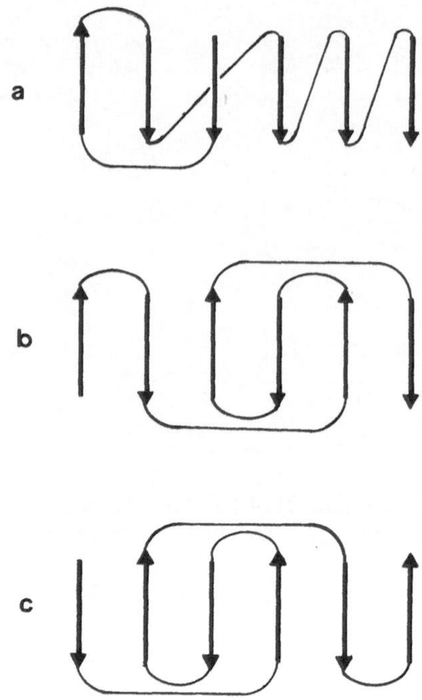

Figure 4. Folding topology diagram of a: domain I, b: domain II, c: domain III.

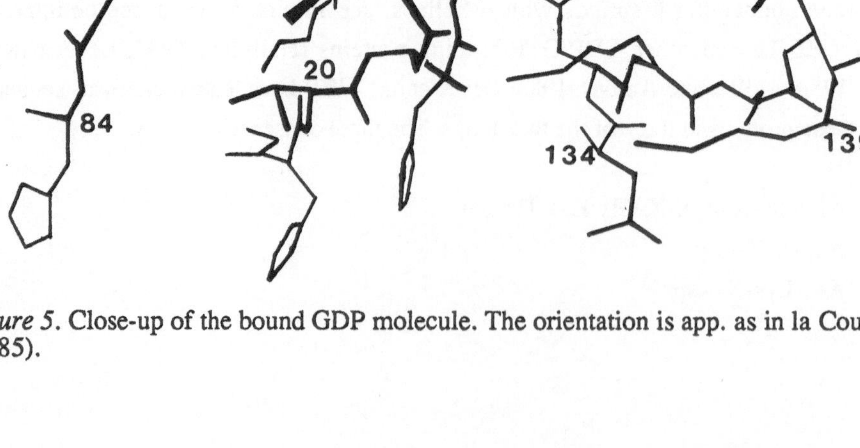

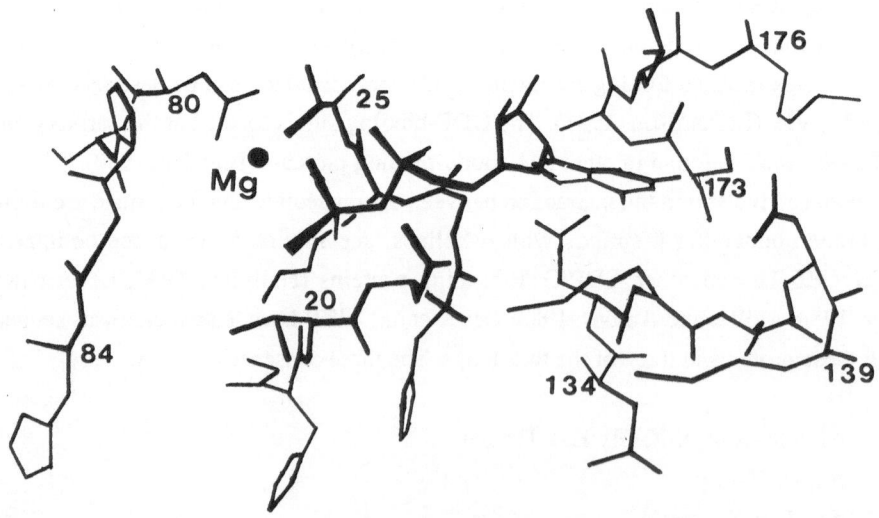

Figure 5. Close-up of the bound GDP molecule. The orientation is app. as in la Cour *et al.* (1985).

The first two sequences are those of the loops between strand Ia and helix IA and between strand Ic and helix IB, respectively, and together define the phosphoryl binding site. The third sequence is for the loop between strand Ie and helix ID and defines the guanine specificity.

This and other sequence homologies have been the basis for the postulation of a model of the G-domain of p21 (McCormick *et al.*, 1985) and of Initiation Factor 2 (Cenatiempo *et al.*, 1987).

Most of the conserved amino acid residues seem to play important structural roles in the binding of GDP. The phosphoryl group is bound to the loop from strand Ia to helix IA, where most of the main chain amides are pointing towards the β-phosphate. Such an arrangement is found in many nucleotide-binding proteins, where the N-terminal end of a helix apparently forms the right electrostatic environment for binding negatively charged groups (Hol *et al.*, 1978). In most cases, the loop is very short (one residue long) and therefore requires a Gly-residue at the C-terminal end of the β-strand (see Schultz and Schirmer, 1979). In EF-Tu, the loop is rather long and flexible and contains residues Gly(18)-His-Val-Asp-His-Gly-Lys-Thr(25). In the model, the positions of Gly(18) and Gly(23) let the β-strand and the α-helix come sufficiently close together. The Lys(24) residue can be built to have its positively charged side chain amino group close to the β-phosphate of GDP. The β-phosphate thus seems to be held in position by the main chain amides of the loop, the dipole of the helix and the positive charge on Lys(24).

In the density a strong peak is found between the β-phosphate, the C-terminal end of strand Ic and the N-terminal end of helix IA. It is approximately 3.8 Å away from the centre of the β–phosphate. This is believed to be the position of a Mg^{2+} ion. Residue Thr(25) can be built to have its hydroxyl group and residue Asp(80) can be built to have its acidic group pointing towards the Mg^{2+} position. The Mg^{2+} ion thus seems to be held in position by the negative charges on the β–phosphate and of Asp(80), and possibly having some of many oxygen atoms in this area (including some postulated waters) as ligands in an octahedral complex. This is in good agreement with a Mg^{2+}-O distance in such a complex of about 2.1 Å (Nyborg and Danielsen, 1970; Hanic *et al.*, 1971).

The role of Gly(83) is obscure. It is positioned at the beginning of helix IB. From model building of GTP into the GDP binding site, it is apparent that the γ-phosphate cannot be accommodated into the model. It is conceivable that movement of the loop between strand Ic and helix IB will influence the position of helix IB which - because it forms part of the interface between domain I and domain III - in turn could influence the relative position of these two domains. This could then constitute part of the GTP conformation of EF-Tu and signal the "on-state" of the G-domain.

The ribose ring is positioned with its "endo-side" against the protein. It is in the 2'-endo conformation, which leaves the 2'-and 3'-hydroxyls exposed to the solvent, in agreement with observations on modifications or substitutions of these groups (Miller and Weissbach, 1977; Suck and Kabsch, 1981; Sprinzl, 1987).This also places the guanine in the anti-conformation. The ribose apparently is not held in position by any direct interaction with the proteins.

The guanine ring is buried in a hydrophobic cavity defined by residues Gly(23), Leu(175), Lys(136) and Met(139). In the bottom of the cavity is found Asn (135), which can be built to interact with the O(6) of the guanine. Found near the O(6) of guanine is also the hydroxyl of Ser(173). At one side of the entrance to the cavity Asp(138) is located. It can be built (but in a strained conformation) to interact with N(1) and/or N(2) of guanine. There is no apparent interaction between the amino group of Lys(136) and GDP. The electron density of this residue is, however, unusually clear for the otherwise long and flexible side chain, indicating that it it held in position perhaps by interacting with the phosphoryl binding loop. It lacks reactivity towards modification by ethyl acetamide (Antonsson and Leberman, 1984).

The structures of the two models of the G-domain of EF-Tu (Jurnak, 1985; la Cour *et al.*, 1985) are very similar. Of course they will end up being identical within the limits of error.

Recently the structure of p21 has been reported (DeVos *et al.*, 1988). Most of this molecule constitutes a G-domain with a possible membrane bound sequence at the C-terminal. Surprisingly, there is a profound difference in the strand order of p21 and EF-Tu. The folding arrangement of the β-strands of p21 can be described as -1, +2, +1x, +1x, +1x. In Figure 3b is shown a model of EF-Tu with the strand order of that of p21 to highlight the main difference between the two structures. In p21, the loop between helix IA (which is shorter than in EF-Tu) and strand Ib (also shorter) folds over the phosphoryl group of GDP. In p21, the loop between strand Ic and strand Id does not contain a complete helix IB but only a single helical turn. The main difference, however, is the positions of the loops between strand Ia and helix IA and of the loop between strand Ic and helix IB. These are two of the loops with conserved sequences. One of them still forms the binding pocket for the phosphoryl group, i.e. the loop from strand Ia to helix IA. The other loop from strand Ic to helix IB can no longer be part of the Mg^{2+} binding site, unless Mg^{2+} and the phosphoryl are placed differently (Jurnak, 1988).

DOMAIN II

The topology diagram of domain II is shown in Figure 4 and the order of the strands is +1, +3, -1,-1, +3, giving it a typical Greek key connection. After entering the domain at residue 210, the first strand IIa extends to residue 221 (see Table 1). After a sharp turn, the next strand IIb comes back and follows strand IIa until residue 229. A long loop directs the strand on top of the barrel to residue 239 where a bend leads into strand IIc. This strand continues to residue 248 and turns back in an anti-parallel fashion to strand IId which at residue 256 goes into a large strand running on top of the domain. In this strand, the residues have helical conformation but are rather badly defined due to non-specific side chain interactions. A turn at residues 268-270 leads into strand IIe which via a long irregular loop leads into strand IIf exiting the domain at residue 295.

DOMAIN III

From domain II, the extended chain leads into domain III. The topology diagram of this domain is shown in Figure 4 and the strand order is also a typical Greek key being +3, -1, -1, +3, +1. The chain enters the domain into strand IIIa at residue 301 (see Table 1). At residue 312, the loop forming the lid on the barrel begins, and after having spanned the barrel it reenters at residue 322. This strand IIIb continues in a rather irregular way until residue 330 where it turns back into strand IIIc which extends to residue 342. Here the second loop forming the barrel lid starts and the chain runs partly with a helical conformation and loop into strand IIId at residue 355. This long strand ends at residue 364 and the bottom of the barrel is formed by a loop which at residue 372 goes into strand IIIe. This strand makes a sharp β-bend at residues 379-380 and continues in strand IIIf until the C-terminus.

DOMAIN CONTACTS

Except for the covalent links (residues 200-209) between domains I and II and between domains II and III (residues 296-300), the contacts between the domains are made by side chain interactions. The contacts between domains II and III consist of a hydrophobic volume confined by strands IIa and IIf in domain II and the loop between strands IIIb and IIIc together with the end of loop IIId. Also the location of oppositely charged side chains in the two domains bordering the hydrophobic interface suggests salt bridges.

The contact between domain I and III is mainly created via a large surface made up of strands IIIa, IIIe and IIIf on which helices IB and IC are fixed. In this way, domain III acts as a spacer between domains I and II and, as we shall see later, could be the pivot around which the spatial reorientations of domains I and II with respect to each other could take place.

CORRELATION BETWEEN THE STRUCTURE AND BIOCHEMICAL DATA

A great number of biochemical and mutational data lend support to our model and invite to speculation on the mode of function of complex formation between EF-Tu and aa-tRNA. But it should be emphasized that the structure we are describing is the EF-Tu:GDP complex which not only does not recognize aa-tRNA but also is slightly modified by the excision of residues 45-58.

Summarizing the biochemical data on the protection of residues when EF-Tu:GTP is complexed to aa-tRNA, it is striking that these residues reside in domains I and II but not III.

In labelling experiments of lysines, Antonsson and Leberman (1984) found enhanced reactivity of Lys (208) and Lys(390) upon complex formation whereas lysines 2, 4, 237, 248, 263 and 282 showed reduced reactivity. Similarly, Lys(208) and Lys(237) have been shown to crosslink to 3' oxidized tRNA in the presence of kirromycin or ribosomes (van Noort *et al.*, 1984). None of the remaining four lysines in domain III seem to be affected by ternary complex formation.

Photooxidation of histidines has been shown by Jonak *et al.* (1984) to take place at residues 66, 118, 319 and 364 together with either 75, 78 or 84, whereas His(301) is protected against oxidation. Upon ternary complex formation, His(66) and His(118) become protected by the aa-tRNA, whereas no protection is detected of His(319). Photooxidation also takes place at Met(349) and Met(351) and one of Met(91), Met (98) or Met(112), showing these residues to be exposed. Ternary complex formation protects the exposed Met in domain I, whereas Met(349) and Met(351) are still susceptible towards oxidation.

Two kirromycin resistant mutants with mutations in domains II and III have been characterized (Duisterwinkel *et al.*, 1984). In the mutant EF-Tu B_O, Gly(222) is changed into an Asp. Since this mutant is fully functional in the protein biosynthesis, position 222 must be in a region where the replacement Gly to Asp does not alter the chain conformation. This is in agreement with our model where position 222 is located in the flexible loop between strands IIa and IIb.

The other kirromycin resistant mutant EF-Tu A_R becomes resistant by a lowered affinity towards kirromycin. The mutation consists of a replacement of Ala(375) to either Val or Thr. Ala(375) is positioned on strand IIIe on the surface of domain III in the void space between domains I and II. Since one of the properties of this mutant is its higher affinity towards GTP than the wild type, one could imagine that the space between domains I and II is widened when EF-Tu is in its GTP-binding conformation. The substitution of a more bulky side chain than Ala will thereby stabilize the GTP conformation.

A further support for the idea of tRNA interaction taking place mainly on domains I and II is obtained by crosslinking kirromycin to EF-Tu in the presence of aa-tRNA. This crosslinking has been observed to take place at Lys(357) in domain III (van Noort *et al.*, 1984).

The structure of EF-Tu in its complex with GDP presents a picture of a molecule partitioned in three distinct domains. The connections between domains I and II are limited to a single polypeptide chain and through non-covalent forces via domain III. Observations on, for example, tritium exchange (Prinz and Miller, 1973) have demonstrated that in EF-Tu in its GTP conformation, approximately 14 amide hydrogen atoms/molecule become exposed to the solvent and therefore are exchangeable. These hydrogen atoms could belong to the polypeptide chains which form the interface regions between domains making them inaccessible to solvent molecules in the GDP conformation.

We have earlier hypothesized (la Cour *et al.*, 1985) that in order to accommodate a γ-phosphate in the GDP binding pocket, a movement of the loop containing residues 80-84 is necessary. In effect, this will influence the position or orientation of helix IB which is in contact with domain III. In turn, a change of position/orientation of this domain could influence the position/orientation of domain II. The net effect would be that tRNA recognizing regions on domains I and II could be properly oriented with respect to each other and in this way provide the necessary free energy of interaction.

ACKNOWLEDGEMENT

We would like to thank Prof. B.F.C. Clark for his constant interest in the results of our structural work. We would also like to thank the Danish Natural Science Research Council and the Danish Biotechnology Research Programme for their financial support.

REFERENCES

B. Antonsson and R. Leberman (1984) "Modification of Amino Groups in EF-Tu.GTP and the Termary Complex EF-Tu.GTP. Valyl-tRNA^{Val}", Eur. J. Biochem., 141, 483-487.

U.W. Arndt and A.J. Wonacott, Eds. (1977) "The Rotation Method in Crystallography", North-Holland Publishing Company, Amsterdam

T.L. Blundell and L. N. Johnson (1976) "Protein Crystallography", Academic Press, London.

C.-I. Brändén (1980) "Relation between structure and function of α/β-proteins". Qt. Rev. Biophys., 13, 317-338.

Y. Cenatiempo, F. Deville, J. Dondon, M. Grunberg-Manago, C. Sacerdot, J.W.B. Hershey, H.F. Hansen, H.U. Petersen, B.F.C. Clark, M. Kjeldgaard, T.F.M. la Cour, K.K. Mortensen and J. Nyborg (1987) "The Protein Synthesis Initiation Factor 2 G-Domain. Study of a Functionally Active C-Terminal 65-Kilodalton Fragment of IF2 from *Escherichia coli* ", Biochem., 26, 5070-5076.

T.F.M. la Cour, J. Nyborg, S. Thirup and B.F.C. Clark (1985) "Structural details of the binding of guanosine diphosphate to elongation factor Tu from *E. coli* as studied by X-ray crystallography", EMBO J., 4, 2385-2388.

T.E. Dever, M.J. Glymias and W.C. Merrick (1987) "The GTP-Binding Domain: Three Consensus sequence Elements with Distinct Spacing", P.N.A.S., 84, 1814-1818.

F.J. Duisterwinkel, B. Kraal, J.M. de Graaf, A. Talens, L. Bosch, G.W.M. Swart, A. Parmeggiani, T.F.M. la Cour, J. Nyborg and B.F.C. Clark (1984) "Specific alterations of the EF-Tu polypeptide chain considered in the light of its three-dimensional structure", EMBO J., 3, 113-120.

K.R. Halliday (1984) "Regional Homology in GTP-binding Proto-Oncogone Products and Elongation Factors", J. Cycl. Nucl. Prot. Phos. Res., 9, 435-448.

F. Hanic, O. Lindquist, J. Nyborg and A. Zedler (1971) "The Crystal Structure of MgO. $3B_2O_3.5H_2O$)", Coll. Czech. Chem. Comm., 36, 3678-3701.

W.G.J. Hol, P.T. van Duijnen and H.J.C. Berendsen (1978) "The α–helix dipole and the properties of proteins", Nature, 273, 443-446.

J. Jonak, T.E. Petersen, B. Meloun and I. Rychlik (1984) "Histidine residues in elongation factor EF-Tu from *Escherichia coli* protected by aminoacyl-tRNA against photo-oxidation" Eur. J. Biochem., 144, 295-303.

F. Jurnak (1985) "Structure of the GDP Domain of EF-Tu and Location of the Amino Acids Homologous to *ras* Oncogene Proteins", Science, 230, 32-36.

F. Jurnak (1988) "The three-dimensional structure of c-H-*ras* p21: Implications for oncogene and G protein studies", TIBS, 13, 195-198.

R. Leberman and U. Egner (1984) "Homologies in the primary structure of GTP-binding proteins: the nucleotide-binding site of EF-Tu and p21", EMBO J., 3, 339-341.

M. Levitt and C. Chotia (1976), "Structural Patterns in Globular Proteins", Nature, 261, 552-557.

F. McCormick, B.F.C. Clark, T. F. M. la Cour, M. Kjeldgaard, L. Nørskov-Lauritsen and J. Nyborg (1985), "A Model for the Tertiary structure of p21, the Product of the *ras* Oncogene", Science, 230, 78-82.

D.L. Miller and H. Weissbach (1977), "The interactions of Elongation Factor Tu", in "Nucleic Acid-Protein Recognition", 409-440, Academic Press, New York.

K. Morikawa, T.F.M. la Cour, J. Nyborg, K.M. Rasmussen, D. L. Miller and B. F. C. Clark (1978), "High Resolution X-ray Crystallographic Analysis of a Modified Form of the Elongation Factor Tu: Guanosine Disphosphate Complex", J. Mol. Biol., 125, 325-338.

W. Möller and R. Amons (1985) "Phosphate-binding sequences in nucleotide-binding proteins", FEBS Lett., 186, 1-7.

J.M. van Noort, B. Kraal, L. Bosch, T.F.M. la Cour, J. Nyborg and B.F.C. Clark (1984) "Crosslinking of tRNA at two diffrent sites of the elongation factor Tu", P.N.A.S. 81, 3969-3972.

J. Nyborg and J. Danielsen (1970), "Direct Determination of the Crystal Structure of Bis (dichlorophosphate) bis (phosphorylchloride) Magnesium: Mg $(PO_2Cl_2)_2$ $(POCl_3)_2$", Acta Chem. Scand., 24, 59-71.

M.P. Prinz and D.L. Miller (1973) "Evidence for Conformational Changes in Elongation Factor Tu Induced by GTP and GDP", Biochem. Biophys. Res. Commun., 53, 149-156.

J.S. Richardson (1981), "The Anatomy and Taxonomy of Protein Structure", Adv. Protein. Chem., 34, 167-339.

G.E. Schulz and R.H. Schirmer (1979), "Principles of Protein Structure", Springer-Verlag, Heidelberg.

M. Sprinzl (1987) "Affinity labelling of the GDP/GTP binding site in *Thermus thermophilus* Elongation Factor Tu", Personal communication.

D. Suck and W. Kabsch (1981), "X-ray Determination of the GDP-binding site of *Escherihia coli* Elongation Factor Tu by Substitution with ppGpp", FEBS Lett., 126, 120-122.

A. M. de Vos, L. Tong, M. V. Milburn, P.M. Matias, J. Jancarik, S. Noguchi, S. Nishimura, K. Miura, E. Ohtsuka and S.-H. Kim (1988) "Three-Dimensional Structure of an Oncogene Protein: Catalytic Domain of Human c-H-*ras* p21", Science, 239, 888-893.

PROGRESS ON THE THREE-DIMENSIONAL STRUCTURAL DETERMINATION OF TRYPSIN-MODIFIED EF-TU-GDP

Frances Jurnak, Michelle Nelson, Marilyn Yoder, Susan Heffron and Suet Miu

Department of Biochemistry
University of California
Riverside, California 92521

INTRODUCTION

Elongation factor (EF-)Tu is a cytoplasmic protein whose primary function is to recognize and transport aminoacyl tRNAs to the ribosome during protein synthesis (for review, see [1,2]). In order to carry out its function, EF-Tu binds to different ligands, including GDP, GTP, EF-Ts, aminoacyl-tRNA and ribosomal proteins, during each elongation cycle. Biochemical studies have indicated that EF-Tu undergoes a series of discrete conformational changes as the protein changes ligands. The long-term objective of the crystallographic studies is to determine the atomic details of the conformational changes during the elongation cycle by X-ray diffraction techniques.

The objective of the present report is to communicate the status of the three-dimensional structural studies of a trypsin-modified form of *Escherichia coli* EF-Tu-GDP. A brief overview of the biochemical and crystallographic methods that have been used in the structural determination as well as a tentative description of the entire polypeptide backbone will be given. The GDP binding domain will be described in more detail and will be compared to the three-dimensional structure of the *ras* p21 protein. Finally, research on a new EF-Tu-antibiotic complex will be presented and its significance discussed.

METHODS

Trypsin-modified EF-Tu Crystals

For reasons related to crystal size and order, the X-ray diffraction studies have focused on a form of EF-Tu which has been treated with trypsin[3]. Biochemical analysis has shown that, under the conditions of the experiment, trypsin partially or completely cleaves the arginine residues at position 7, 44 and 58 and that subsequent repurification removes a 14 amino acid fragment, Ala 45-Arg 58. Activity assays demonstrate that the trypsin treatment reduces the affinity of the protein for aminoacyl-tRNA by a factor of 1000[4], slightly affects the EF-Ts-catalyzed exchange of guanine nucleotides, but does not alter the GDP or GTP binding properties of the protein. Crystals of the trypsin-modified form of EF-Tu have been grown to very large sizes and belong to one of three space groups: $P4_32_12$, $C222_1$ or $P2_12_12_1$[5-7]. The crystal polymorphism could not be controlled initially, causing technical difficulties in the X-ray analysis, but studies have now shown that the presence or absence of a $H_2PO_4^{2-}$ determines the specific space group in which the protein crystallizes[8]. Although the crystal polymorphism was initially a detriment to the crystallographic studies, it is now considered an advantage for the molecular averaging methods now being used to solve the three-dimensional structure.

X-ray Diffraction Data Collection

All X-ray diffraction data were collected at the Multiwire Area Detector Facility in Xoung Nguyen-huu's laboratory at the University of California, San Diego. To date, the data shown in Table 1 have been collected. As indicated by the agreement factor, R_{SYM}, between multiply-recorded reflections, the data sets collected are very good, a quality which is important in obtaining an interpretable electron density map. The initial structural report on the GDP binding domain of EF-Tu was based on the phasing information obtained from one Pt and one Hg derivative at a resolution of 2.7 Å. Since the initial report, two additional Hg derivatives have been collected and included. The data have also been extended to a resolution of 2.6 Å for three of the derivatives and to 2.0 Å for the native set.

Crystallographic Techniques

Because the structural interpretation can only be as good as the phasing information used to calculate the electron density maps, a brief description of the different types of calculations used in the crystallographic determination of the EF-Tu structure is given.

Initially, the standard crystallographic technique of multiple isomorphous replacement (MIR) was applied using one Pt and one Hg derivative but the resulting electron density map was not interpretable. To improve the electron

Table 1. X-ray Diffraction Data for Trypsin-modified EF-Tu-GDP

Space Group	Type	Resolution	Reflections	R_{SYM}	Sites
$C222_1$	Native	2.0 Å	191,724	5.5%	–
	Pt	2.4 Å	115,495	6.6%	4
	Hg 1	2.7 Å	72,211	4.7%	6
	Hg 2	2.6 Å	107,599	5.9%	4
	Hg 3	2.6 Å	82,115	6.4%	2
$P2_12_12_1$	Native	2.7 Å	80,710	7.5%	–
	Pt	3.3 Å	57,465	9.4%	2
	Hg	2.8 Å	73,842	6.7%	4

Table 2. Refined Heavy Atom Derivative Parameters

Compound	Occupancy	Atomic Coordinates X	Y	Z	Temperature Factor
Pt	0.0737	0.2412	0.4118	0.3360	62.70
	0.8730	0.4056	0.2603	0.0890	60.00
	0.6722	0.2316	0.2673	0.3995	68.08
	0.6921	0.2595	0.2676	0.1504	96.66
HG 1	0.5382	0.1494	0.0630	0.0433	33.40
	0.5594	0.4357	0.1527	0.2933	33.27
	0.4912	0.2968	0.2779	0.0255	77.99
	0.0604	0.2195	0.3017	0.2745	75.68
	0.1722	0.0202	0.2886	0.1698	10.99
		0.2301	0.0184	0.4216	46.65
HG 2	0.3106	0.4355	0.1539	0.2936	32.80
	0.2936	0.1490	0.0624	0.0434	28.50
	0.2257	0.2221	0.2948	0.2665	144.96
	0.1210	0.2948	0.2760	0.0217	73.92
HG 3	0.1085	0.2349	0.2738	0.2737	18.77
	0.1145	0.3063	0.2623	0.0256	33.64

density image, additional phasing information was obtained using an image enhancement technique developed by B.C. Wang (9). The combination of these two techniques resulted in a partially interpretable map in the region of the GDP binding domain. Amino acid residues 8-190 could be assigned to the electron density unambiguously, but the polypeptide chain corresponding to residues 191-393 could not be traced clearly. Consequently, the three-dimensional structure of only the GDP binding domain of EF-Tu was reported (10).

To complete the structural determination of the entire molecule, diffraction data for additional heavy atom derivatives were collected and the resolution of the calculations was extended to 2.6 Å. The current polypeptide tracing is based on phasing information derived from the application of MIR methods using the atomic coordinates of the four heavy atom derivatives shown in Table 2 and the final refinement and phasing statistics shown in Table 3. Because several ambiguities yet remain in the assignment of the amino acids to the electron density, the tracing is only considered to be an approximation of the final structure.

As a result of the difficulties in finding unique heavy atom derivatives which bind in an isomorphous manner to EF-Tu, it is doubtful that phasing information sufficient to solve the structure unambiguously will be forthcoming from additional applications of the MIR method. Therefore, to improve the electron density maps further, a number of different types of crystallographic techniques have been applied. These techniques include density modification methods, calculation of phases from a known partial

Table 3. Final Refinement and Phasing Statistics for $C222_1$ Form of EF-Tu-GDP

2.6 Å Resolution					Total
Midpoint of Shell (Å)	7.66	4.27	3.30	2.78	
Centric Figure of Merit	0.73	0.53	0.52	0.42	0.57
Observations	228	214	180	146	1536
Overall Figure of Merit	0.79	0.59	0.49	0.37	0.52
Observations	1596	2793	3200	3292	21763
Overall R Factor*					
Pt Derivative	0.60	0.73	0.73	0.76	0.68
Hg Derivative 1	0.49	0.63	0.64	0.63	0.58
Hg Derivative 2	0.55	0.66	0.64	0.64	0.62
Hg Derivative 3	0.74	0.76	0.73	0.74	0.74

$$*R = \left[\sum_{hkl} (F_{CALC} - F_{OBS})^2 / (F_{OBS})^2 \right]$$

model and molecular averaging methods. To date, most of the techniques applied only resulted in a minor improvement in the clarity of the electron density map. However, encouraging results have recently been obtained from the application of real space molecular averaging methods to the two copies of the EF-Tu molecule in the asymmetric unit of the $C222_1$ crystal form. In the near future, molecular averaging methods will be extended to include copies of the protein molecule in space groups $P2_12_12_1$ and $P4_32_12$ of trypsin-modified EF-Tu in a manner similar to that used to solve the structure of the major histocompatibility antigen, HLA-A2[11]. Ultimately, a total of five unique copies of EF-Tu will be averaged across three space groups. From the preliminary molecular averaging results, it is clear that the current polypeptide tracing contains minor errors in a few connections between the major secondary structural elements.

Procedures which utilize the image enhancement procedure of B.C. Wang are no longer used in the crystallographic analysis because more recent research has raised questions about the introduction of artifacts into the electron density by these methods. Also, results from secondary structural predictions, which are relatively inaccurate, are not used in order to avoid introducing a bias into the polypeptide chain interpretation.

RESULTS

The crystallographic results are based on an interpretation of a 2.6 Å resolution electron density map phased by MIR methods using a unique Pt derivative and three Hg derivatives which share common substitution sites. At this stage of a crystallographic analysis, the polypeptide backbone can usually be traced and the orientation of the amino acid side chains deduced. The side chains are typically modeled into the electron density using known geometries of the amino acids, but the position of each atom in the side chain in a particular structure is not accurately established until the crystallographic refinement calculations have been carried out at higher resolution. Therefore, all intramolecular interactions described below should be considered as tentative.

The protein is structured into three domains. The first domain, comprising amino acid residues 1-200, is the GDP binding domain and is shown in Figure 1. The topology of this domain is that of a six-stranded, mostly antiparallel, central beta sheet which is surrounded on both sides by several helices. The amino terminal beta strand is the central or third strand of the sheet, a feature which is found in many nucleotide binding proteins. The GDP is located at the carboxy terminal edge of the beta sheet, with one edge of the ligand exposed to solvent. The phosphate groups are located at the amino terminus of a long alpha helix, a mode of binding which is universally found in all nucleotide binding proteins. The phosphate groups are nestled into a surface loop which is bounded by two glycines, Gly 18 and Gly 23, found in the invariant Gly-4X-Gly-Lys sequence. Because glycines exhibit a broad range of ϕ and ψ dihedral angles, these glycines are likely to play a key role in releasing the ligand during the GDP/GTP exchange process. At this stage of the crystallographic analysis, the invariant lysine, Lys 24, appears to interact with the beta phosphate group; the adjacent semi-invariant threonine, Thr 25, appears to interact with the alpha phosphate group.

The invariant Asp-2X-Gly region is located at the carboxy terminal edge of the second beta strand of the sheet. The side chain of the invariant aspartic acid, Asp 80, appears to be oriented toward the Mg^{2+} in a manner which suggests an electrostatic interaction linking the ligand to the protein, a common but not universal feature in nucleotide binding proteins. The invariant glycine, Gly 83, participates in a typical Pro-Gly turn at the end of the beta strand and is in a position to play a key role in any conformational changes that occur as the protein changes ligands throughout the elongation cycle. At this stage of the crystallographic analysis, the ribose ring appears to have the C-2′ *endo* conformation, with the hydroxyl groups oriented outwards, toward the solvent region. The 2′-OH group appears to interact with the side chain of Lys 176 which is located on a neighboring loop.

Figure 1. Stereodiagram of the GDP binding domain of trypsin-modified EF-Tu. The key invariant amino acids found in all G proteins are indicated by labels and side chains where appropriate. These include Gly 18, Gly 23, Lys 24, Thr 25, Asp 80, Gly 83, Asn 135, Lys 136, Asp 138 and Lys 176.

In comparison to other nucleotide ligands in known protein structures, the position of the guanine ring is quite unusual. Unlike other base rings, the guanine moiety is not buried deep within the protein. Instead, one edge of the guanine ring is exposed to solvent, in a manner which suggests a mechanism for the exchange of nucleotides. The ring substituents which give guanine its unique character appear to interact with residues found in the invariant Asn-Lys-X-Asp pattern found in all G proteins. The keto oxygen at position 6 of the guanine ring appears to interact with Asn 136 and the amino group at position 2 appears to interact, albeit weakly, with Asp 138. The invariant lysine, Lys 137, extends over one side of the guanine ring, apparently making contact with the oxygen in the ribose ring[12]. A second lysine, Lys 176, extends over the other side of the guanine ring and appears to interact with the 2′-OH group on the ribose. This latter lysine is found on a surface loop which borders one side of the ligand and may contribute other direct or indirect interactions to the guanine ring.

The current polypeptide tracing of the entire protein is shown in Figure 2. Because some errors are known to exist in the polypeptide loop connections in the carboxy terminal region of the protein, the remaining portions of the model will not be described in detail. Domain 2 extends from residues 208-290 and Domain 3, from residues 295-393. The tracing indicates that Domain 2 and Domain 3 are each folded into a topology called a beta barrel. Unlike the present Aarhus model, there appears to be very little helical structure present.

One of the greatest difficulties in solving the structure has been in tracing the polypeptide chain (191-207) that connects Domain 1 to Domain 2. In the initial electron density maps, the region appeared to be quite weak, thereby indicating that the conformation of the polypep-tide chain is somewhat flexible. In the most recent electron density maps, this region has become clearer, thereby permitting the interpretation shown in Figure 2. The correct placement of this region of the polypeptide chain is particularly important in order to determine the relative position of Domains 2 and 3 with respect to Domain 1. The observed flexibility of the chain is suggestive of a hinge mechanism whereby only a slight movement of the connecting strand is sufficient to cause a large shift in the position of Domain 3 relative to Domain 1. Such a mechanism could serve as means of communicating small conformational changes in the GDP binding domain over long intramolecular distances to the remaining domains and could be a key element in the overall conformational changes known to occur in the protein during the elongation cycle.

The relative position of Domains 2 and 3 with respect to Domain 1 gives rise to a very unusual packing that is not found in other protein structures, that is, there appears to be a large cavity in the center of the protein. The presence of this cavity was first suggested by R. Leberman and colleagues[13], based upon their interpretation of a 5 Å electron density map and later, by the Aarhus group in their crystallographic studies[14]. However, it is only recently that our

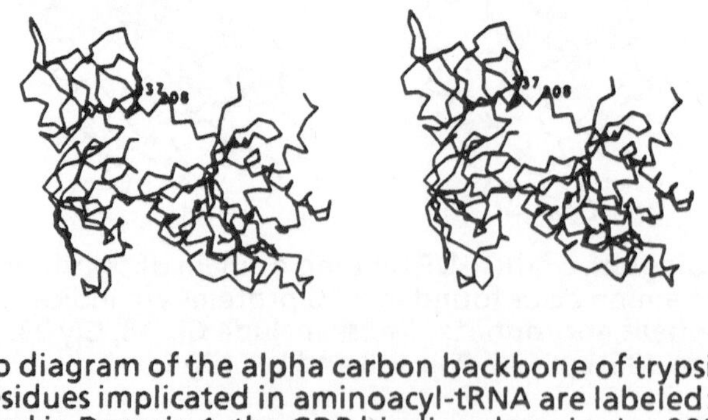

Figure 2. Stereo diagram of the alpha carbon backbone of trypsin-modified EF-Tu-GDP. The residues implicated in aminoacyl-tRNA are labeled: His 66 and Cys 81 are located in Domain 1, the GDP binding domain; Lys 208 and Lys 237 are located in Domain 2.

electron density maps could unambiguously confirm the cavity. The presence of such a large cavity is quite unusual in protein structures and is likely to be of interest to those scientists who study intramolecular protein packing and interactions.

Finally, the last feature which deserves mention is the position of four amino acids which have been implicated in the interaction between EF-Tu and aminoacyl-tRNA. These include His 66 which can be cross-linked to the amino epsilon group of lysyl-tRNA[15]; Cys 81, which when bound to mercurial reagents, is inactive in forming the ternary complex[16]; and the pair of lysines, 208 and 237, which have been crosslinked to the acceptor stem of aminoacyl-tRNA[17]. Although these positions are too few to suggest something other than a very highly speculative model for aminoacyl-tRNA binding, they do, at least, indicate the general region of interaction between the protein and aminoacyl-tRNA. It is not too surprising, given the relative sizes of the two macromolecules, to find that the aminoacyl-tRNA will likely form specific interactions with EF-Tu over a rather large surface area of the protein.

DISCUSSION

A number of investigators have now noted the amino acid sequence similarity between EF-Tu and other G proteins which are activated by a GDP/GTP exchange mechanism[18-20]. One of the most complete studies was that of Halliday who compared the first 187 amino acids of EF-Tu with H-*ras* p21[19]. She found five noncontiguous regions which shared amino acid sequence homology. Three of these regions are characterized by a discrete pattern of invariant amino acids and consequently, can easily be discerned in other GDP binding proteins. The location of these regions at or near the GDP binding site was identified by high resolution X-ray diffraction studies of EF-Tu[10,21]. The first region, characterized by a Gly-4X-Gly-Lys pattern, contains the loop into which the phosphate groups of GDP are nestled. The second region is characterized by an Asp-2X-Gly pattern. The side chain of the invariant Asp appears to interact with the Mg^{2+} ion which bridges the GDP to the protein. The third region of similarity contains the Asn-Lys-X-Asp pattern. Asn and Asp appear to interact with substituents on the guanine ring and thus confer specificity for GDP on the protein.

The other two regions of amino acid sequence similarity, initially identified by Halliday, do not contain a discrete set of invariant amino acids and therefore, are not easily identified in other G protein sequences. Nevertheless, these regions are near to the GDP binding site in EF-Tu and appear to contribute important interactions. One region, including amino acid residues 99-107 in EF-Tu, is the fourth beta strand in the central beta sheet. The second region, from 154-178, contains the sixth beta strand which ends in a loop over one side of the guanine ring. This loop contains a lysine residue, Lys 176, which passes over the guanine ring and appears to interact with the 2′OH group on the ribose ring.

The sequence similarities and identities noted by Halliday represent 40% of the amino acids of the GDP binding domain. If one considers secondary structural predictions for *ras* p21 and compares it to the actual structure of EF-Tu, then the percentage of amino acids with predicted structural homology rises to 75-80%. As a result of these predictions, three-dimensional models of the *ras* p21 were built based on the atomic coordinates of the GDP binding domain of EF-Tu[21,22]. Because the three-dimensional structure of a genetically engineered form of the H-*ras* p21 has also been solved by X-ray crystallographic methods[23], it is now possible to compare the two protein structures in order to deter-mine the validity of the various prediction schemes. From a visual comparison only[24], it is clear that the EF-Tu and the H-*ras* p21 structures agree quite well in the latter portion of the structure, that is, the region from approximately 100-200 in EF-Tu, which is folded into a mononucleotide fold. Within this region, there exist two obvious differences: residues 175-180 form a small helix in EF-Tu whereas none exists in H-*ras* p21 and secondly, the loop containing residues 137-145 in EF-Tu, appears to differ in conformation from that in H-*ras* p21. This loop contains the invariant Asp which interacts with the amino group substituent on position 2 of the guanine ring. Such differences in loop conformation and the orientation of specific interacting residues with the guanine ring may contribute to the observed differences in the guanine nucleotide binding properties between EF-Tu and H-*ras* p21.

In contrast to the carboxy terminal region of the GDP binding domain, there are striking differences in the amino terminal portion of the structure. Spatially, the beta strands of the central sheet appear to be nearly superimposable, yet the assigned amino acids do not correlate well with the homology predictions of Halliday or others. For instance, the invariant Asp-2X-Gly, found in all G proteins, is assigned to different spatial regions in both structures. In EF-Tu, the region containing the invariant Asp-2X-Gly sequence is located at the carboxy terminal end of the second beta strand. The invariant Gly and the preceding Pro participate in a typical Pro-Gly turn. In H-*ras* p21, the Asp-2X-Gly region has been assigned to the amino terminus of the third beta strand. Also, in contrast to EF-Tu, the side chain of the invariant Asp does not appear to interact with the metal ion, either Mg^{2+} or Ca^{2+}, in the H-*ras* p21 structure. As a result of the relative difference in the amino acid assignments, the topology of the two structures differ in the placement of two polypeptide connections between the three beta strands.

If the spatial assignment of the amino acids were temporarily ignored, then the overall structure of the polypeptide backbone would indeed be very similar, corresponding to approximately 77% spatial similarity. Considering only the backbone, the dissimilarities in the amino terminal region would be relatively small: Helix A (residues 24-40) in EF-Tu is longer than the comparable one in H-*ras* p21 by two turns and Helix B (residues 83-93) in EF-Tu is analogous to a coiled, but not helical, structural element in H-*ras* p21. Also, the amino acid sequence from Ala 45-Arg 58 is missing in the trypsin-modified form of EF-Tu but, in H-*ras* p21, this region forms a loop which contacts one side of the GDP ligand. This latter loop contributes a Phe residue (Phe 28) which is perpendicular to the guanine ring. Although the missing segment in EF-Tu could conceivably form a similar type of loop in the intact protein, the nearest phenylalanine residue in EF-Tu would appear to be too far to interact with the guanine ring. Instead, Lys 176 in EF-Tu extends over this side of the guanine ring and appears to interact with the 2′-OH group on the ribose ring. The side chain of the comparable lysine residue, Lys 145 in H-*ras* p21, is oriented away from the guanine ring[12]. Such differences in the amino acid interaction with the guanine ring may account for the differences in the guanine nucleotide binding properties of EF-Tu and H-*ras* p21.

Four out of the five amino acid regions predicted to be similar by Halliday are located in approximately the same spatial regions in both the EF-Tu and the H-*ras* p21 structures. The fifth, that containing the Asp-2X-Gly invariant amino acids, has a different spatial location in each of the structures. Is this difference meaningful? Until both structures have been refined to a resolution of 2.2 Å, the question will be difficult to answer definitively. Should the differences remain after the crystallographic refinement process has been completed for both structures, then it would very unusual, indeed, to find two proteins which share spatially similar tertiary structures but have homologous amino acid sequences assigned to different structural regions. In fact, there are no other known homologous proteins where similar amino acid sequences, characterized by invariant amino acids, are found in different spatial locations in the tertiary structures. It is our opinion that the difference is not real but occurs because a mistake has been made in one or both structures in the assignment of the amino acids in this region. Such mistakes do occur in crystallographic analyses, not as a result of electron density map interpretation, but as a result of errors in the crystallographic phases used to calculate the electron density maps. Difficulties with finding independent and isomorphous heavy atom derivatives plagued both structures. Therefore, the investigators had to rely on alternate methods of deriving phasing information. These methods, such as the B.C. Wang image enhancement procedure used by our group for EF-Tu as well as by the Kim group for *ras* p21, are relatively new techniques and are not fully understood in terms of the type of artifacts that could be introduced into the electron density calculations by them. In order to circumvent these potential artifacts in the EF-Tu structure, the image enhancement technique of B.C. Wang has been eliminated from all calculations and the structure of the GDP binding domain is being redetermined by the real space molecular averaging methods discussed previously. These latter methods have a long record of producing reliable structural results, most notably in the virus structures[25]. It is hoped that these methods will reaffirm the assignment of the Asp-2X-Gly region in EF-Tu.

Finally, it must be reiterated that it is quite unusual for a relatively small monomer of 43,000 daltons to bind to so many different types of ligands as does EF-Tu during the elongation cycle of protein synthesis. It would be even more surprising to find that the *ras* p21 protein, with half the amino acids, could carry out all of the analogous functions ascribed to EF-Tu or signal transduction G proteins. To date, *ras* p21 appears to be most similar to the GDP binding domain of EF-Tu. In mutant studies, in which the GDP binding domain of EF-Tu has been cloned, isolated and characterized, the domain behaves more like *ras* p21 in its guanine nucleotide binding properties[26]. Furthermore, of all the EF-Tu ligands, only the ribosome displays a clear effect enhancing the GTPase activity of the G domain, suggesting that Domains 2 and 3 of the intact protein are involved in the interactions with aminoacyl-tRNA and EF-Ts[26]. Therefore, it seems highly unlikely, from a structural viewpoint alone, that the *ras* p21 can carry out all the analogous functions ascribed to EF-Tu or to signal transduction G proteins. The structure is too small to interact with as many macromolecular ligands as EF-Tu or the other G proteins.

FUTURE DIRECTIONS

Should the molecular averaging methods fail to produce a clearly interpretable electron density map of the complete EF-Tu molecule, the laboratory is taking another approach to solve the entire structure and to resolve the question of the correct placement of the invariant Asp-2X-Gly region. Recently, large crystals of a trypsin-modified EF-Tu-tetracycline complex have been grown. Although tetracycline is not known to interact with EF-Tu directly, the crystals could only be grown in the presence of specific stoichiometric ratios of EF-Tu to tetracycline. Moreover, the morphology of the pale yellow crystals is entirely different from that of any other trypsin-modified form of EF-Tu. The preliminary X-ray analysis also indicates a difference: the crystals belong to space group $P2_1$ and contain 6 molecules per asymmetric unit. There is no other known crystal form of EF-Tu which belongs to this space group. The crystals have very favorable properties that which make them particularly suited for X-ray diffraction studies, including the ability to bind isomorphously to different types of heavy atom compounds. It is anticipated that the structure of the complex crystals can be solved by MIR methods only. Should these methods not produce an interpretable electron density map, real space molecular averaging methods will be applied. With the large number of multiple copies of the protein-tetracycline complexes per asymmetric unit, it is expected that molecular averaging methods will work well.

Although tetracycline is generally believed to inhibit procaryotic protein synthesis by interacting with ribosomal proteins, a close scrutiny of the original data[27-29] indicates that an alternate interpretation, that of tetracycline binding to EF-Tu and blocking its interaction with the ribosome, is possible. Biochemical studies are underway to determine if tetracycline affects any of the known functions of EF-Tu. Our preliminary biochemical results indicate that tetracycline does not interfere with GDP, GTP or aminoacyl-tRNA binding. Assays to determine the effects on EF-Ts or ribosomal binding are underway.

ACKNOWLEDGEMENTS

We wish to thank Xoung Nguyen-huu at the University of California, San Diego for the use of the Multiwire Area Detector Facility and Mark Saper and Don Wiley at Harvard University for providing us with advice and the molecular averaging programs. We also wish to gratefully acknowledge the financial support of the United States Public Health Service (GM 26895) throughout this work.

References

1. D. L. Miller, D. L., and H. Weissbach, Factors involved in the transfer of aminoacyl-tRNA to the ribosome, in: "Molecular Mechanisms of Protein Biosynthesis," S. Pestka and H. Weissbach, eds., Academic Press, New York (1977).
2. Kaziro, Y., The role of guanosine 5'-triphosphate in polypeptide chain elongation, Biochim. Biophys. Acta 505, 95-127 (1978).
3. F. Jurnak, A. McPherson, A. Wang, and A. Rich, Biochemical and structural studies of the tetragonal crystalline modification of the *Escherichia coli* elongation factor, Tu, J. Biol. Chem. 255:6751-6757 (1980).
4. E. Masuda, A. Louie, F. and Jurnak, Effect of trypsin modifications of elongation factor, Tu, on the equilibrium between Tu-GTP and aminoacyl-tRNA, J. Biol. Chem. 260:8702-8705 (1985).
5. F. Jurnak, D. L. Miller, and A. Rich, Preliminary X-ray diffraction data for tetragonal crystals of trypsinized *E. coli* elongation factor, J. Mol. Biol. 115:103-110 (1977).
6. D. Sneden, D. L. Miller, S. H. Kim, and A. Rich, Preliminary X-ray analysis of the crystalline complex between polypeptide chain elongation factor, Tu, and GDP, Nature (London) 241:530 (1973).
7. W. H. Gast, W. Kabsch, A. Wittinghofer, and R. Leberman, Crystals of a large tryptic peptide (fragment A) of elongation factor EF-Tu from *Escherichia coli*, FEBS Letters 74:88-90 (1977).
8. Jurnak, F., Induction of elongation factor Tu-GDP crystal polymorphism by polyethylene glycol contaminants, J. Mol. Biol. 185:215-217 (1985).
9. B. C. Wang, Resolution of phase ambiguity in macromolecular crystallography, in: "Methods in Enzymology," H.W. Wyckoff, C.H.W. Hirs, and S.N. Timasheff, eds., Academic Press, New York (1985).
10. F. A. Jurnak, Structure of the GDP domain of EF-Tu and location of the amino acids homologous to *ras* oncogene proteins, Science 230:32-36 (1985).
11. P. J. Bjorkman, M. A. Saper, B. Samraoui, W. S. Bennett, J.L. Strominger, and D.C. Wiley, Structure of the human class I histocompatibility antigen, HLA-A2, Nature 329:506-512 (1987).
12. A. M. de Vos, personal communication.
13. Kabsch, W., Gast, W. N., Schulz, G. E. and Leberman, R., Low resolution structure of partially trypsin-degraded polypeptide elongation factor, EF-Tu, from *Escherichia coli*, J. Mol. Biol. 117:999-1012 (1977).

14. B. F. C. Clark, T. F. M. LaCour, J. Fontecilla-Camps, K. Morikawa, K. M. Nielsen, J. Nyborg, and J. R. Rubin, 1982, Three-dimensional structural elements of bacterial elongation factor Tu complexed to GDP, in: "Cell Function and Differentiation," Part C, Alan R. Liss, Inc., New York.
15. L. K. Duffy, L. Gerber, A. E. Johnson and D. L. Miller, Identification of a histidine residue near the aminoacyl transfer ribonucleic acid binding site of elongation factor Tu, Biochemistry 20:4663-4666 (1981).
16. J. Jonik, J. Smrt, A. Holij, and I. Ryclik, Interaction of *Escherichia coli* EF-Tu-GTP and EF-Tu-GDP with analogues of 3′ terminus of aminoacyl-tRNA, Eur. J. Biochem. 105:315-320 (1978).
17. J. M. Van Noort, B. Kraal, T. F .M. LaCour, J. Nyborg, and B. F. C. Clark, Cross-linking of tRNA at two different sites of the elongation factor, Tu, Proc. Natl. Acad. Sci. USA 81:3969-3972 (1984).
18. R. Leberman, and U. Egner, Homologies in the primary structure of GTP-binding proteins: The nucleotide-binding site of EF-Tu and p21, The EMBO Journal 3:339 (1984).
19. K. Halliday, Regional homology in GTP-binding proto-oncogene products and elongation factors, J. Cyclic Nucleotide Prot. Phosphoryl. Res. 9:435 (1984).
20. T. E. Dever, M. J. Glynias, and W. C. Merrick, GTP-binding domain: Three consensus sequence elements with distinct spacing, Proc. Natl. Acad. Sci. USA, 84:1814-1818 (1987).
21. F. McCormick, B. F. C. Clark, T. F. M. la Cour, M. Kjeldgaard, L. Norskov-Lauritsen, and J. Nyborg, A model for the tertiary structure of p21, the product of the *ras* oncogene, Science 230:78-82 (1985).
22. I. S. Sigal, G. M. Smith, F. Jurnak, J. D. Marsico-Ahern, J. S. D'allonzo, E. M. Scolnick, and J. B. Gibbs, Molecular approaches towards an anti-*ras* drug, Anti-Cancer Drug Design 2:107-115 (1987).
23. A.M. de Vos, L. Tong, M. V. Milburn, P. M. Matias, J. Jancarik, S. Noguchi, S. Nishimura, K. Miura, E. Ohtsuka, and S-H. Kim, Three-dimensional structure of an oncogene protein: Catalytic domain of human C-H-*ras* p21, Science 239:888-893 (1988).
24. F. Jurnak, The three-dimensional structure of C-H-*ras* p21: Implications for oncogene and G protein studies, Trends in Biochem. Sci. 13:195-198 (1988).
25. F. Jurnak and A. McPherson, eds., "Biological Macromolecules and Assemblies: Virus Structures," Vol. 1, John Wiley & Sons, New York (1984).
26. A. Parmeggiani, G. W. M. Swart, K. K. Mortensen, M. Jensen, B. F. C. Clark, L. Dente, and R. Cortese, Properties of a genetically engineered G domain of elongation factor Tu, Proc. Natl. Acad. Sci. USA 84:3141-3145 (1987).
27. K. H. Nierhau, and H. G. Wittman, Ribosomal function and its inhibition by antibiotics in procaryotes, Naturwissenschaften 67:234-250 (1980).
28. G. Suarez and D. Nathans, Inhibition of aminoacyl-tRNA binding to ribosomes by tetracycline, Biochem. Biophys. Res. Commun. 18:743-750 (1965).
29. T. Tritton, Ribosome-tetracycline interactions, Biochemistry 16:4133-4138 (1977).

THREE-DIMENSIONAL STRUCTURE OF *ras* p21 PROTEINS

A. M. de Vos, L. Tong, M. V. Milburn, P. M. Matias, and S.-H. Kim

Department of Chemistry
University of California
Berkeley, CA 94720

INTRODUCTION

ras genes (review in Barbacid, 1987) have been found in a large number of eukaryotic organisms, from *Saccharomyces* and *Drosophila* to chicken, rat, and man. Moreover, *ras* gene products of various species show a very high degree of homology: even between proteins from yeast and man there is approximately 54% identity between corresponding amino acids, and *ras* proteins from chicken and man differ only in three amino acids. Such evolutionary conservation implies an important cellular function for these proteins, and they have indeed been implicated in playing a crucial role in cell proliferation and terminal differentiation. Based on these observations and on biochemical similarities with G-proteins, it is thought that *ras* proteins participate as signal transducers at the beginning of the cascade of reactions leading to various essential cellular processes.

The importance of *ras* proteins for the normal functioning of cells is indicated by the finding that *ras* genes belong to the small group of protooncogenes. Inappropriate expression of a protooncogene is an important step in tumorigenesis (Nishimura and Sekyia, 1987), and *ras* oncogenes are the most frequently identified oncogenes in human tumors. The mechanism of activation of mammalian *ras* genes is usually a single base change at one of a few critical positions in their coding sequence, resulting in a single amino acid substitution in the protein (Tabin et al., 1982; Reddy et al., 1982; Taparowski et al., 1982). The activating mutation most commonly found in human tumors is the substitution of Gly-12 by valine; the critical nature of residue 12 is further indicated by the fact that substitution by any residue other than proline makes the protein transformation-competent (Seeburg et al., 1984). Other critical amino acid positions are at residues 13, 15, 59, 61, 63, 116, and 119 (Barbacid, 1987).

In the human genome there are three distinct *ras* genes: c–H–*ras*, c–K–*ras*, and N–*ras*. The proteins encoded by these genes, p21 proteins, have a molecular weight of 21 kDa, and contain 189 amino acids. They have virtually identical amino acid sequences up to residue 165, but the sequence between 165 and 185 is highly variable among different p21 proteins. Like the α-subunit of G-proteins, they bind GDP and GTP (Scolnick et al., 1979), and hydrolize GTP; activated p21 proteins have reduced GTPase activity (Gibbs et al., 1984; McGrath et al., 1984; Sweet et al., 1984). A consensus sequence proposed for GTP-binding proteins (McCormick et al., 1985; Dever et al., 1987), consisting of the elements GXXXXGK,

DXXG, and NKXD (where X can be any amino acid), is found in p21. Attachment to the cytoplasmic side of the cell membrane through the carboxy-terminal residues is necessary for biological activity (Willumsen et al., 1984), but nucleotide binding and GTPase activity are not affected by deletion of the carboxy-terminal 23 residues (Willumsen et al., 1985; Lacal et al., 1986). In analogy to G-proteins, *ras* proteins are thought to accomplish their transduction function by acting as a conformational switch: the GDP-bound conformation is the "signal off" state, whereas the GTP-bound conformation can interact with effector molecules, sending out a second message.

Here, we present the structure of the "catalytic domain" (residues 1-171) of c–H–*ras* p21, with GDP bound to it, and compare the structure of the normal protein to that of the transforming Val-12 mutant. Work on the crystallization of the GTP-bound form is in progress.

EXPERIMENTAL PROCEDURES

"Catalytic domains" of both normal p21 and the Val-12 mutant were prepared as described previously (Miura et al., 1986; Jancarik et al., 1988). Single crystals of these proteins were grown with the sitting-drop technique from solutions containing 10 mg/ml protein, 0.1 M $CaCl_2$, 0.075 M Hepes buffer (pH=7.5), 0.5 mM EDTA, 0.5 mM dithiotreitol, and 0.005% *n*-octyl glucoside. The precipitating agent was 30% polyethylene glycol 400. The crystals of both variants have space group $P6_522$, with a=83.2 Å, and c=105.1 Å. There is one molecule per asymmetric unit, and the crystals have a solvent content of about 54% (Jancarik et al., 1988). The crystals diffract quite well, and are relatively stable under exposure to X-rays. Data was collected for native crystals and five heavy atom derivatives at the Stanford Synchrotron Radiation Laboratory, Palo Alto, California, each data set from one single crystal. The structure of the normal protein was solved at 2.7 Å resolution (De Vos et al., 1988), using a combination of multiple isomorphous replacement, solvent flattening, phase extension, and partial model combination. A difference Fourier map between the normal protein and the Val-12 mutant clearly showed the Val-12 side chain, but did not reveal any major structural differences between the two variants. Subsequently, both structures were refined independently using the TNT-package (Tronrud et al., 1987). Current R-factors at 2.2 Å resolution range from 22-24%.

OVERALL THREE-DIMENSIONAL STRUCTURE

A schematic drawing of the structure of the "catalytic domain" of p21 is shown in Figure 1a. The structure contains a six-stranded β-sheet, four α-helices, and nine connecting loops. The central part of the molecule is the β-sheet, with strand $\beta2$ running anti-parallel to the other five strands (Figure 1b). This β-sheet consists of two smaller sheets, each composed of three strands with only a short stretch of hydrogen-bonding between them. Packing against the convex side of the sheet are helices $\alpha2$ and $\alpha3$, while the remaining two helices, $\alpha1$ and $\alpha4$, pack against the concave face. The axis of helix $\alpha1$ is approximately perpendicular to the axes of the other three. The last two turns of the long carboxy-terminal helix, $\alpha4$, form a notable protrusion from the otherwise globular shape of the protein. The topology of the β-α-β-α-β motif in the carboxy-terminal half of the molecule is identical to the so-called nucleotide-binding motif characterized for ATP-binding proteins (Rao and Rossmann, 1973), but the way GDP binds to p21 (see below) is very different from that found in other nucleotide-binding proteins. Three of the loops (L3, L6, and L8 in Figure 2) are at the side of the carboxy-terminus, whereas the GDP molecule and five other loops (L1, L2, L5, L7, and L9) are located at the opposite side of the molecule. The long loop L4 spans the distance between these sides, and forms the connection between the two three-stranded sub-sheets.

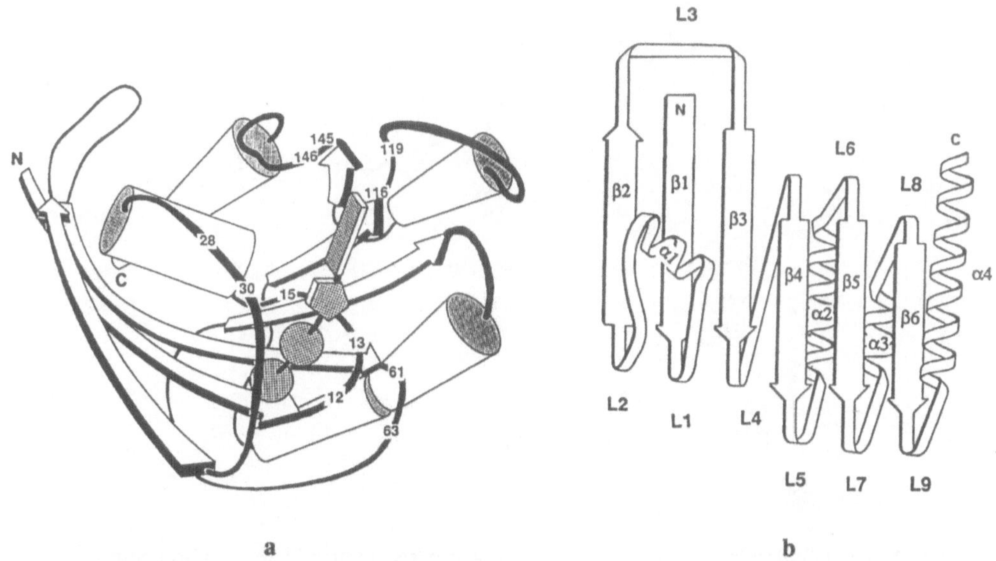

Fig. 1. (a) Backbone structure of the catalytic domain of human c–H-*ras* p21. β-strands are drawn as flattened arrows, α-helices as solid cylinders, and guanine base, ribose, and phosphates as stippled rectangle, pentagon, and spheres, respectively. Several important amino acids are numbered. (b) Topological diagram of the catalytic domain of p21, numbering the structural elements from the amino-terminus.

GDP-BINDING POCKET

In contrast to other nucleotide-binding proteins, the GDP molecule binds to p21 near the surface of the molecule. It fits snugly in a pocket formed by loops L1, L2, L7, and L9, leaving the "top" of the base and the ribose and α-phosphate accessible to solvent. Residues 116, 117, 119, and 120 of loop L7 and residues 145 to 147 of L9 form one side of the cleft for the guanine base, residue 28 the other side (Figure 2). The third element of the consensus sequence proposed for GTP-binding proteins is represented by residues 116-119 as NKCD. At the current stage of refinement, Asn-116 appears to hydrogen-bond to N7 and O6 of guanine, Asp-119 to N1 and N2, and the side chain of Lys-117 runs parallel to the guanine base and reaches the endocyclic ribose oxygen. Other interactions to the guanine ring appear to be made by Ser-145 and the backbone NH of Ala-146 to O6, whereas the hydroxyl groups of the ribose are close to Asp-30. The α-phosphate does not appear to form any interactions with the protein, but the β-phosphate is tightly bound by residues in loop L1. This loop is part of a characteristic strand-loop-helix motif, with the amino terminal end of the helix pointing towards the phosphates. It contains the first element of the consensus sequence as GAGGVGK, residues 10-16. Apart from Lys-16, these residues have no side chains that can hydrogen-bond with the phosphate group. Instead, the β-phosphate is held tightly by the main-chain NH groups of residues 12 to 16. A metal ion, probably Mg^{2+}, is coordinated to two oxygens of the β-phosphate and to the side chain of Ser-17. This divalent cation and the side chain of Lys-16, together with the dipole effect of helix α1, neutralize the negative charges on the phosphates. The second consensus element appears in the form of DTAG for residues 57-60 in p21. These residues are not in contact with GDP, but would be close to the γ-phosphate of GTP (see below).

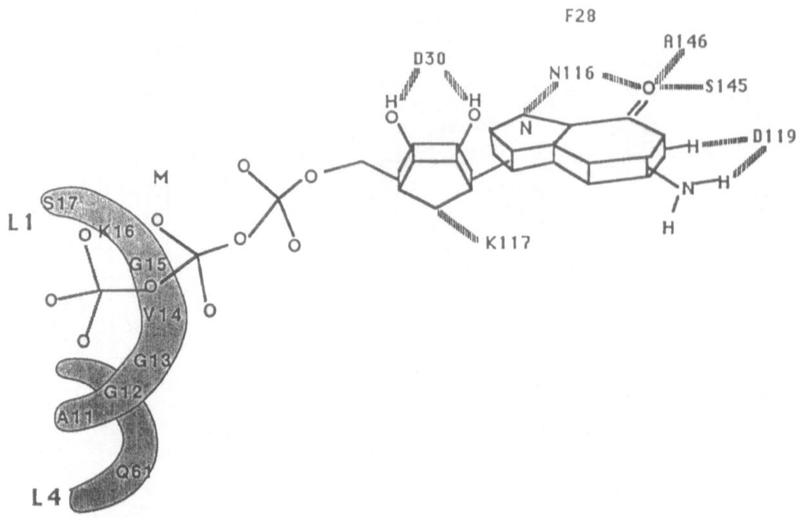

Fig. 2. Nucleotide-binding pocket of p21, showing schematically the environment of the bound GDP. Interactions with the guanine base and the ribose are indicated by shaded lines. Hydrogen bonds between loop L1 and the β-phosphate are not shown. M stands for a metal cation.

DISCUSSION

Comparison with the GTP-binding domain of EF-Tu. The structure of the nucleotide binding domain of elongation factor Tu (EF-Tu-G) was solved a few years ago (Jurnak, 1985; La Cour et al., 1985). The overall structure is surprisingly similar to that of the "catalytic domain" of p21: it, too, has a central six-stranded β-sheet, with α-helices packing against it, and GDP and a metal ion occupy similar positions in the structure. The topology of the carboxy-terminal domain is very similar to that of p21, but there are some important differences in the amino-terminal half of the molecule. Besides the presence of an extra α-helix and the absence of a trypsin-cleaved fragment (41-58), the N-terminus of EF-Tu is situated on a different strand (our β3), resulting in a different conformation of the loop binding the β-phosphate. Models proposed for p21 on the basis of the structure of EF-Tu-G (McCormick et al., 1985; Jurnak, 1985) inherit the same difficulties. Details of the differences will become clear after completion of the crystallographic refinement of both structures at reasonably high resolution.

Two-domain structure. Two domains can be recognized in the structure of the "catalytic domain" of p21, dividing the β-sheet between them: the amino-terminal half of the molecule, consisting of the first three β-strands and the first α-helix, together with the connecting loops, and the carboxy-terminal domain containing the nucleotide-binding motif. This separation of the structure in two parts is supported by the distribution of the temperature factors of the residues in each domain: the residues in the amino-terminal half of the molecule have higher mobility than those in the carboxy-terminal half. The higher mobility of the amino-terminal domain may have a functional significance related to its GTPase activity and putative effector binding function. Moreover, the conformational change that the molecule is thought to undergo upon GTP binding may involve changes in the relative orientation of the two domains (Kim et al., 1988).

Membrane attachment. The long α-helix at the carboxy-terminal end of our structure suggests that there will be a separation of the carboxy-terminal residues in a separate domain in the intact p21 molecule. The residues missing in our structure may interact with the membrane-bound receptor, and the α-helix could transmit signals to the "catalytic domain". Other candidates for interaction with the receptor are the loops at the "membrane side" of the molecule, i.e. loops L3 and L8. The guanine nucleotide is bound at the "cytoplasmic side" of p21. The finding that GDP binds at the surface of the molecule and is rather solvent accessible can be rationalized by assuming this to be necessary for easy exchange. Loops on this side of the molecule are exposed to the cytoplasm and are good candidates for interaction with target proteins. In fact, loop L2 has been identified as candidate for effector binding (Sigal et al., 1986) and as the region interacting with GTPase stimulating protein (Adari et al., 1988; Cales et al., 1988). The structure shows that many of the side chains in this loop are well exposed and pointing into the solvent, explaining the observation that mutations in this region have no effect on GDP/GTP binding and GTPase activity (Sigal et al., 1986), and supporting the evidence for interaction of this loop with other proteins.

Catalytic site of GTP hydrolysis. The β-phosphate of GDP is held rigidly in a particular orientation by the main-chain NH-groups of loop L1, and the metal ion that is bound to the side chain of Ser-17 (Figure 3). A possible mechanism of GTPase catalysis is based on the observation that the likely direction for the phospho-diester bond between the β- and γ-phosphate of GTP is such that it would be straddled by loop L1. The rigid fixation of the β-phosphate would then facilitate attack on its phosphorus atom by a water molecule in an SN2 reaction, releasing the γ-phosphate. An alternative mechanism, similar to that suggested for EF-G (Webb and Eccleston, 1981) and EF-Tu (Eccleston and Webb, 1982), would involve attack on the γ-phosphate. In this case, fixation of the β-phosphate would be opportune to reduce the freedom of the γ-phosphate without impairing its ease of release. Whatever the actual mechanism may be, the position of the side chain of Ala-59 close to the site where the γ-phosphate of GTP would be bound, explains the autophosphorylation of Thr-59 of viral protein (Shih et al., 1982). Moreover, phosphorylation of this residue would prevent the γ-phosphate from leaving, thus presumably freezing the protein in its GTP-bound conformation, explaining the oncogenic activation of Thr-59 mutants.

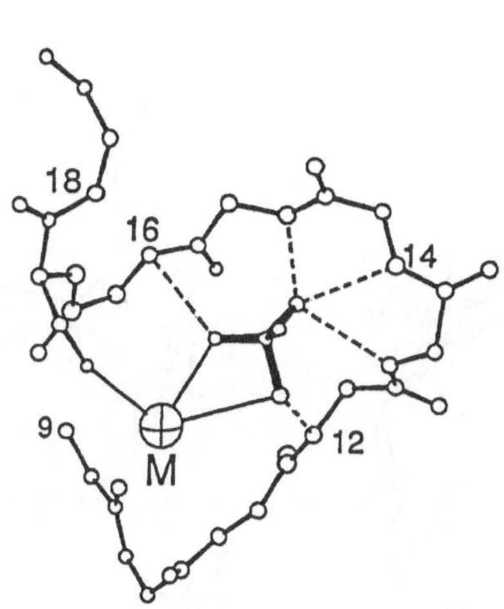

Fig. 3. Phosphate binding by loop L1. The backbone NH-groups of residues 12-16 form hydrogen bonds to the β-phosphate of GDP. A divalent metal ion, M, is bound to the side chain of Ser-17 and to two phosphate oxygens (side chains other than Ser-17 not shown for clarity).

Correlation between oncogenic activation and nucleotide binding. All activating point mutations so far known are situated near the GDP-binding pocket of p21. Three of these, at residues 12, 13, and 15, are mutations in loop L1, the loop binding the β-phosphate of GDP. Residues 59, 61, and 63 are located in loop L4, a loop that is in close contact with loop L1. The last two critical positions, at 116 and 119, are part of loop L7, which binds to the guanine base. Oncogenic activation of p21 by mutation in residues 12, 13, or 15 may be due to the reduction of GTPase activity resulting from conformational changes in the catalytic loop L1; and/or to the enhancement of GDP/GTP exchange rate by the reduction in phosphate binding strength of this loop. The structure of the Val-12 mutant supports these hypotheses (see below). Mutations at residues 116 or 119 could reduce the guanine binding strength, again resulting in an increase in the GDP/GTP exchange rate. Residues 59, 61, and 63 are not in direct contact with GDP, but would be close to the γ-phosphate of GTP; furthermore, they are in direct contact with the catalytic loop L1. Thus, mutations in these residues could affect GTP binding and/or GTPase activity, either by direct contacts with GTP, or by effecting conformational changes in loop L1. In all these cases, the concentration of GTP-bound p21 would be increased relative to the normal situation, thus prolonging the "signal on" state of the protein and inducing transformation.

Differences between normal p21 and the Val-12 mutant. The overall structure of the transforming Val-12 mutant is very similar to that of the normal protein. The significant differences between the two structures are localized in loop L1, around the mutation site. Compared to the tight binding of the β-phosphate in the normal protein, loop L1 has opened up in the case of the Val-12 mutant (Figure 4). The main-chain NH groups of residues 12 and 13 are now too far from the β-phosphate to hydrogen-bond to it. Thus, the β-phosphate is no longer fixed as rigidly as in the normal protein, explaining the drastic reduction in GTPase activity and hence the transforming potential of this mutant. Preliminary results of the crystallographic refinement of a Leu-61 mutant suggest that the same mechanism may hold in this case, presumably through the close contacts between this residue and loop L1.

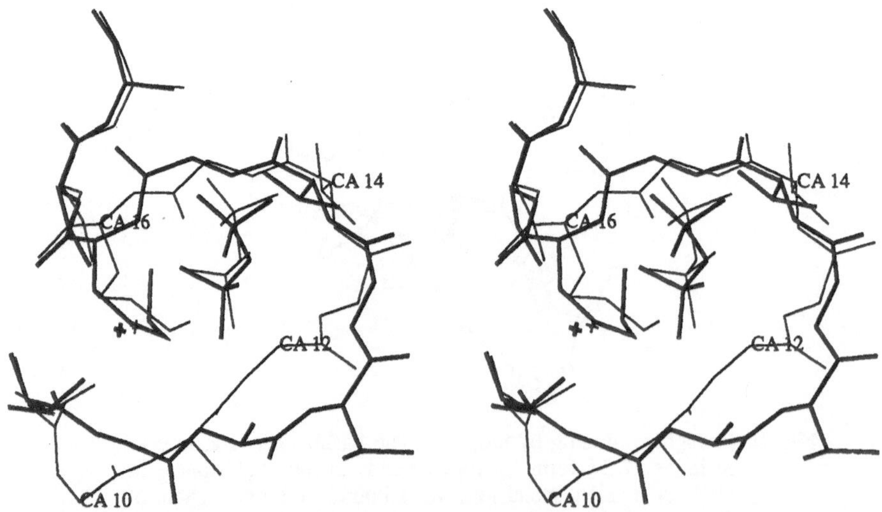

Fig. 4. Stereo drawing showing the differences in phosphate-binding loop L1 between normal p21 and the Val-12 mutant. Thin lines are for normal p21.

REFERENCES

Adari, H., Lowy, D. R., Willumsen, B. M., Der, C. J., and McCormick, F., 1988, Guanosine triphosphate activating protein (GAP) interacts with the p21 *ras* effector binding domain, *Science*, 240:518.

Barbacid, M., 1987, *ras* Genes, *Ann. Rev. Biochem.*, 56:779.

Cales, C., Hancock, J. F., Marshall, C. J., and Hall, A., 1988, The cytoplasmic protein GAP is implicated as the target for regulation by the *ras* gene product, *Nature*, 332:548.

Dever, T. E., Glynias, M. J., and Merrick, W. C., 1987, GTP-binding domain: three consensus sequence elements with distinct spacing, *Proc. Natl. Acad. Sci. USA*, 84:1814.

De Vos, A. M., Tong, L., Milburn, M. V., Matias, P. M., Jancarik, J., Noguchi, S., Nishimura, S., Miura, K., Ohtsuka, E., and Kim, S.-H., 1988, Three-dimensional structure of an oncogene protein: catalytic domain of human c–H–*ras* p21, *Science*, 239:888.

Eccleston, J. F. and Webb, M. R., 1982, Characterization of the GTPase reaction of elongation factor Tu, *J. Biol. Chem.*, 257:5046.

Gibbs, J. B., Sigal, I. S., Poe, M., and Scolnick, E. M., 1984, Intrinsic GTPase activity distinguishes normal and oncogenic *ras* p21 molecules, *Proc. Natl. Acad. Sci. USA*, 81:5704.

Jancarik, J., De Vos, A. M., Kim, S.-H., Miura, K., Ohtsuka, E., Noguchi, S., and Nishimura, S., 1988, Crystallization of human c–H–*ras* oncogene products, *J. Mol. Biol.*, 200:205.

Jurnak, F., 1985, Structure of the GDP domain of EF-Tu and location of the amino acids homologous to *ras* oncogene proteins, *Science*, 230:32.

Kim, S.-H., De Vos, A. M., Tong, L., Milburn, M. V., Matias, P. M., Jancarik, J., Ohtsuka, E., and Nishimura, S., 1988, *Ras* oncogene proteins: three-dimensional structures, functional implications, and a model for signal transduction, *Cold Spring Harbor Symp.*, in press.

Lacal, J. C., Anderson, P. S., and Aaronson, S., A., 1986, Deletion mutants of Harvey *ras* p21 protein reveal the absolute requirement of at least two distant regions for GTP-binding and transforming activities, *EMBO J.*, 5:679.

McGrath, J. P., Capon, D. J., Goeddel, D. V., and Levinson, A. D., 1984, Comparative properties of normal and activated human *ras* p21 protein, *Nature*, 310:644.

La Cour, T. F. M., Nyborg, J., Thirup, S., and Clark, B. F. C., 1985, Structural details of the binding of guanosine diphosphate to elongation factor Tu from *E. coli* as studied by X-ray crystallography, *EMBO J.*, 4:2385.

McCormick, F., Clark, B. F. C., La Cour, T. F. M., Kjeldgaard, M., Norskov-Lauritsen, L., and Nyborg, J., 1985, A model for the tertiary structure of p21, the product of the *ras* oncogene, *Science*, 230:78.

Miura, K., Inoue, Y., Nakamori, H., Iwai, S., Ohtsuka, E., Ikehora, M., Noguchi, S., and Nishimura, S., 1986, Synthesis and expression of a synthetic gene for the activated human c–Ha–*ras* protein, *Jpn. J. Canc. Res. (Gann)*, 77:45.

Nishimura, S. and Sekiya, T., 1987, Human cancer and cellular oncogenes, *Biochem. J.*, 243:313.

Rao, S. T. and Rossmann, M. G., 1973, Comparison of super-secondary structures in proteins, *J. Mol. Biol.*, 76:241.

Reddy, E. P., Reynolds, R. K., Santos, E., and Barbacid, M., 1982, A point mutation is responsible for the acquisition of transforming properties by the T24 human bladder carcinoma oncogene, *Nature*, 300:149.

Scolnick, E. M., Papageorge, A. G., and Shih, T. Y., 1979, Guanine nucleotide-binding activity and an assay for *src* protein of rat-derived murine sarcoma viruses, *Proc. Natl. Acad. Sci. USA*, 76:5355.

Seeburg, P. H., Colby, W. W., Capon, D. J., Goeddel, D. V., and Levinson, A. D., 1984, Biological properties of human c–Ha–*ras*1 genes mutated at codon 12, *Nature*, 312:71.

Shih, T. Y., Stokes, P. E., Smythers, G. W., Dhar, D., and Oroszlan, S., 1982, Characterization of the phosphorylation sites and the surrounding amino acid sequences of the p21 transforming proteins coded for by the Harvey and Kirsten strains of murine sarcoma viruses, *J. Biol. Chem.*, 257:11767.

Sigal, I. S., Gibbs, J. B., D'Alonzo, J. S., and Scolnick, E. M., 1986, Identification of effector residues and a neutralizing epitope of Ha–*ras*-encoded p21, *Proc. Natl. Acad. Sci. USA*, 83:4725.

Sweet, R. W., Yokoyama, S., Kamata, T., Feramisco, J. R., Rosenberg, M., and Gross, M., 1984, The product of *ras* is a GTPase and the T24 oncogenic mutant is deficient in this activity, *Nature*, 311:273.

Tabin, C. J., Bradley, S. M., Bargmann, C. I., Weinberg, R. A., Papageorge, A. G., Scolnick, E. M., Dhar, R., Lowy, D. R., and Chang, E. H., 1982, Mechanism of activation of a human oncogene, *Nature*, 300:143.

Taparowski, E., Suard, Y., Fasano, O., Shimizu, K., Goldfarb, M., and Wigler, M., 1982, Activation of the T24 bladder carcinoma transforming gene is linked to a single amino acid change, *Nature*, 300:762.

Tronrud, D. E., Ten Eyck, L. F., and Matthews, B. W., 1987, An efficient general-purpose least-squares program for macromolecular structures, *Acta Cryst.*, A43:489.

Webb, M. R. and Eccleston, J. F., 1981, The stereochemical course of the ribosome-dependent GTPase reaction of elongation factor G from *Escherichia coli*, *J. Biol. Chem.*, 256:7734.

Willumsen, B. M., Christensen, A., Hubbert, N. L., Papageorge, A., and Lowy, D. R., 1984, The p21 *ras* C-terminus is required for transformation and membrane association, *Nature*, 310:583.

Willumsen, B. M., Papageorge, A., Hubbert, N. L., Bekesi, E., Kung, H.-F., and Lowy, D. R., 1985, Transforming p21 *ras* protein: flexibility in the major variable region linking the catalytic and membrane-anchoring domains, *EMBO J.*, 4:2893.

THE GTP-BINDING DOMAIN REVISITED

Thomas E. Dever and William C. Merrick

Department of Biochemistry
School of Medicine
Case Western Reserve University
Cleveland, Ohio 44106

GTP-BINDING PROTEINS

Recently we explored the similarity of several GTP-binding domains using principals developed from newly reported structural features of EF-Tu (1-4). In this study, it was concluded that many GTP-binding proteins contain the following consensus sequence elements (GXXXXGK, DXXG, and NKXD) and that these elements were spaced in one of two patterns: the distance between the first and second consensus elements was either 40 to 80 amino acids or about 150 amino acids while the second and third consensus elements were usually 40 to 80 amino acids apart. An updated presentation of GTP-binding proteins which conform to this general scheme is given in Tables 1, 2, and 3 although additional sequences of the three major families have not been included (EF-Tu and EF-1α; RAS, YP2, and rho; G proteins). Each of these individual families will be the subject of more extensive study in other chapters in this book. In addition, there will be no discussion of α or β tubulin, both GTP-binding proteins, as neither of these proteins contains the three consensus elements.

The first consensus element of GTP-binding proteins is shown in Table 1. While each of the ten different families has its own rather conserved sequence, only two sequences would appear to be conserved in more than a single family. The most conserved sequence is G/A-H-V-D-H/S-G-K which is represented in the various protein synthesis factors and Lep A (a total of four different families). The

TABLE 1 COMPONENTS OF THE GTP/GDP BINDING SITE

PHOSPHORYL BINDING SEQUENCES

CONSENSUS SEQUENCE	Gly X X X X Gly Lys		REFERENCES
EF-Tu , E. coli	Gly His Val Asp His Gly Lys	(18-24)	(5)
EF-Tu, Euglena chloro.	Gly His Val Asp His Gly Lys	(18-24)	(6)
EF-Tu , yeast mito.	Gly His Val Asp His Gly Lys	(55-61)	(7)
EF-1α, yeast	Gly His Val Asp Ser Gly Lys	(14-20)	(8,9)
EF-1α, A. salina	Gly His Val Asp Ser Gly Lys	(14-20)	(10)
EF-1α, human	Gly His Val Asp Ser Gly Lys	(14-20)	(11)
EF-G, E. coli	Ala His Ile Asp Ala Gly Lys	(16-22)	(12)
EF-2, hampster	Ala His Val Asp His Gly Lys	(26-32)	(13)
LepA, E. coli	Ala His Ile Asp His Gly Lys	(11-17)	(14)
IF-2, E. coli	Gly His Val Asp His Gly Lys	(398-404)	(15)
RAS 1, yeast	Gly Gly Gly Gly Val Gly Lys	(17-23)	(16,17)
RAS 2, yeast	Gly Gly Gly Gly Val Gly Lys	(17-23)	(17,18)
YP2, yeast	Gly Asn Ser Gly Val Gly Lys	(15-21)	(19)
H-ras, N-ras, K-ras, human	Gly Ala Gly Gly Val Gly Lys	(10-16)	(20-22)
p29 ras, rat	Gly Ala Arg Gly Val Gly Lys	(69-75)	(23)
v-ras, mouse	Gly Ala Lys Gly Val Gly Lys	(10-16)	(24)
v-ras H, mouse	Gly Ala Arg Gly Val Gly Lys	(10-16)	(25)
v-ras K, mouse	Gly Ala Ser Gly Val Gly Lys	(10-16)	(26)
rho, Aplysia	Gly Asp Gly Ala Cys Gly Lys	(12-18)	(27)
rho, human	not determined		(27)
PEPCK, chicken	Gly Asn Ser Leu Leu Gly Lys	(237-243)	(28)
PEPCK, chicken mito.	Gly Asp Ser Leu Leu Gly Lys	(same)	(29)
PEPCK, rat liver	Gly Asn Ser Leu Leu Gly Lys	(237-243)	(30)
PEPCK, D. melanogaster	Gly Asn Ser Leu Leu Gly Lys	(263-269)	(31)
GTP:AMP phosphotransferase bovine	Gly Ala Pro Gly Ser Gly Lys	(12-18)	(32)
Guanyltransferase, reovirus	Gly Ala Ala Ala Ala Gly Lys	(893-899)	(33)
Transducin α, bovine	Gly Ala Gly Glu Ser Gly Lys	(36-42)	(34,35)
G_s protein, bovine adrenal	Gly Ala Gly Glu Ser Gly Lys	(47-53)	(36)
G_s protein, rat brain	Gly Ala Gly Glu Ser Gly Lys	(47-53)	(37)
G_o protein, rat brain	- - - - - Gly Lys		(37)
G_i protein, rat brain	Gly Ala Gly Glu Ser Gly Lys	(40-46)	(37)

second most conserved element, G-A-G-E-S-G-K, is found in transducin α and the G proteins. Where there are numerous members to a given family, there is usually a high degree of homology in this region even though there may be considerable difference in the rest of the protein sequence reflecting divergence between the various species.

The second consensus sequence, D-X-X-G, is the most poorly defined of the three elements due to the generally high (greater than 5%) content of aspartic acid and glycine in most proteins which often results in the appearance of this element strictly due to chance. None-the-less, in most of the GTP binding proteins this sequence element only appears once and as noted for the first consensus sequence, within given families the sequences are highly similar as can be seen in Table 2. In several instances, two of these consensus elements are found: in PEPCK from D. melonogaster; in transducin α; and in several of the G proteins. Based upon the position of the consensus element in PEPCK relative to the other three family members, the "correct" element would be the D-S-Q-G-V sequence at residues 344-348. In a similar fashion, using the unique sequence in G_s of D-V-G-G-Q, it is inferred that this sequence in transducin α, G_o and G_i is the "correct" consensus element. In each case, the choice of the the D-X-X-G element results in this element being about 65 amino acid residues from the third consensus element. It should be noted that both of the first two consensus elements also tend to show up in proteins that bind either ATP or nucleic acids (1, 38, 39).

The final consensus element is responsible for guanine specificity in the GTP-binding domain. Based upon the x-ray crystallographic data (2,4), this sequence, N-K-X-D, is thought to provide specificity by an interaction of N with the keto group of guanine, D interacting with the amino group of guanine and K providing a hydrophobic pocket. For known GTP-binding proteins, this sequence seems to be the least conserved as can be seen in Table 3. For example, while the protein synthesis factors and Lep A had similar first and second consensus elements (considering the non-specified amino acids), for these same proteins C, E, V, M, and I are present as the unspecified, single amino acid. This enhanced variation is also seen in the RAS family proteins and to a lesser extent in the G proteins. In addition, there is also greater variation in what would be the three amino acids in the consensus element. There are two possible reasons for this. First, in the absence of crystal structures for several GTP-binding proteins, it is possible that alternate sequences or structures may provide guanine specificity. Secondly, where examined, several of the GTP-binding proteins are capable of using other nucleotides (i.e. ITP or XTP, see below). In addition, either here or at one of the other consensus sequence

TABLE 2

PHOSPHORYL BINDING SEQUENCES

CONSENSUS SEQUENCE	Asp X X Gly	
EF-Tu, E. Coli	Asp Cys Pro Gly His	(80-84)
EF-Tu, Euglena chloro.	Asp Cys Pro Gly His	(80-84)
EF-Tu, yeast mito.	Asp Cys Pro Gly His	(117-121)
EF-1α, yeast	Asp Ala Pro Gly His	(91-95)
EF-1α, A. salina	Asp Ala Pro Gly His	(91-95)
EF-1α, human	Asp Ala Pro Gly His	(91-95)
EF-G, E. coli	Asp Thr Pro Gly His	(87-91)
EF-2, hampster	Asp Ser Pro Gly His	(104-108)
LepA, E. coli	Asp Thr Pro Gly His	(77-81)
IF-2, E. coli	Asp Thr Pro Gly His	(444-448)
RAS 1, yeast	Asp Thr Ala Gly Gln	(64-68)
RAS 2, yeast	Asp Thr Ala Gly Gln	(64-68)
YP2, yeast	Asp Thr Ala Gly Gln	(63-67)
H-ras, N-ras, K-ras, human	Asp Thr Ala Gly Gln	(57-61)
p29 ras, rat	Asp Thr Ala Gly Gln	(116-120)
v-ras, mouse	Asp Thr Ala Gly Gln	(57-61)
v-ras H, mouse	Asp Thr Thr Gly Gln	(57-61)
v-ras K, mouse	Asp Thr Thr Gly Gln	(57-61)
rho, Aplysia	Asp Thr Ala Gly Gln	(59-63)
rho, human	Asp Thr Ala Gly Gln	(same)
PEPCK, chicken	Asp Glu Leu Gly Asn	(318-322)
PEPCK, chicken mito.	Asp Asp Glu Gly Arg	(same)
PEPCK, rat liver	Asp Ala Gln Gly Asn	(318-322)
PEPCK, D. melanogaster	Asp Pro Lys Gly Val	(298-302)
	Asp Ser Gln Gly Val	(344-348)
GTP:AMP phosphotransferase bovine	Asp Leu Thr Gly Glu	(150-154)
Guanyltransferase reovirus	Asp Ile His Gly Leu	(1022-1026)
Transducin α, bovine	Asp Ser Ala Gly Tyr	(146-150)
	Asp Val Gly Gly Gln	(196-200)
G_s protein, bovine adrenal	Asp Val Gly Gly Gln	(223-227)
G_s protein, rat brain	Asp Val Gly Gly Gln	(223-227)
G_o protein, rat brain	Asp Thr Leu Gly Val	
	Asp Val Gly Gly Gln	
G_i protein, rat brain	Asp Leu Ser Gly Val	(123-127)
	Asp Val Gly Gly Gln	(201-205)

TABLE 3

GUANINE SPECIFICITY BINDING SITE

CONSENSUS SEQUENCE	Asn Lys X Asp	
EF-Tu, E. coli	Asn Lys Cys Asp	(135-138)
EF-Tu, Euglena chloro.	Asn Lys Glu Asp	(135-138)
EF-Tu, yeast mito.	Asn Lys Val Asp	(172-175)
EF-1α, yeast	Asn Lys Met Asp	(153-156)
EF-1α, A. salina	Asn Lys Met Asp	(153-156)
EF-1α, human	Asn Lys Met Asp	(153-156)
EF-G, E. coli	Asn Lys Met Asp	(141-144)
EF-2, hampster	Asn Lys Met Asp	(158-161)
LepA, E. coli	Asn Lys Ile Asp	(131-134)
IF-2, E. coli	Asn Lys Ile Asp	(498-501)
RAS 1, yeast	Asn Lys Leu Asp	(123-126)
RAS 2, yeast	Asn Lys Ser Asp	(123-126)
YP2, yeast	Asn Lys Cys Asp	(121-124)
H-ras, N-ras, K-ras, human	Asn Lys Cys Asp	(116-119)
p29 ras, rat	Asn Lys Cys Asp	(175-178)
v-ras, mouse	Asn Lys Cys Asp	(116-119)
v-ras H, mouse	Asn Lys Cys Asp	(116-119)
v-ras K, mouse	Asn Lys Cys Asp	(116-119)
rho, Aplysia	Asn Lys Lys Asp	(117-120)
rho, human	Asn Lys Lys Asp	(same)
PEPCK, chicken	Asn Lys Asp Trp	(388-391)
PEPCK, chicken mito.	Gly Lys Pro Trp	(same)
PEPCK, rat liver	Asn Lys Glu Trp	(388-391)
PEPCK, D. melanogaster	Gly Lys Pro Trp	(414-417)
GTP:AMP phosphotransferase bovine	Asn Lys Ile Trp	(200-203)
Guanyltransferase, reovirus	Thr Lys Gly Glu	(1083-1086)
Transducin α, bovine	Asn Lys Lys Asp	(265-268)
G_s protein, bovine adrenal	Asn Lys Gln Asp	(292-295)
G_s protein, rat brain	Asn Lys Gln Asp	(292-295)
G_o protein, rat brain	Asn Lys Lys Asp	
G_i protein, rat brain	Asn Lys Lys Asp	(270-273)

positions, some of the variation may also correlate with differences in the binding constants for GTP which roughly vary from 3 x 10^{-4} to 10^{-7} M. At present, there is too little quantitative and crystallographic data to draw an accurate conclusion.

OTHER GTP-BINDING PROTEINS ?

If in fact the three consensus elements are present in most GTP-binding proteins, then one should be able to use these elements to screen data bases or analyze recently identified sequences for possible identification of additional GTP-binding proteins. In this fashion, the Lep A protein was tentitively identified as a GTP-binding protein and subsequent direct testing confirmed this (14). Additional proteins have also been identified by such sequence comparisons and they are listed in Table 4 (1,40-42). As noted previously, protein 2C shares extensive

TABLE 4 POSSIBLE GTP BINDING PROTEINS

				REFERENCE
Protein 2C, FMDV	GKSGQGK (110-116)	DDLG (160-163)	NKLD (243-246)	(40)
gst 1, yeast	GHVDAGK (267-273)	DAPG (344-347)	NKMD (406-409)	(41)
tetM, U. urealytium	AHVDAGK (10-16)	DTPG (74-77)	NKID (128-131)	(42)

FMDV = foot and mouth disease virus

homology with other retroviral proteins which map to the same genetic locus (1), however, the other proteins lack any element of the third consensus sequence. To date, the prediction that protein 2C from foot and mouth disease virus would bind GTP and that the other viral protein 2Cs would bind some nucleotide has not (to the authors' knowledge) been directly tested. The second protein in Table 4, yeast gst 1, not only contains the three consensus sequence elements, but in addition shares extensive homology (38%) with EF-1α(41). This protein is required for the G_1 to S transition in the cell cycle of yeast, however the exact mode of action of the protein and whether it binds GTP are currently unknown. The third protein, TetM, is one of the gene products responsible for tetracycline resistence in U. urealyticum. As for the above proteins, the mode of action of the TetM gene product and its ability to bind GTP remain to be determined.

SPACING OF THE THREE CONSENSUS SEQUENCE ELEMENTS

The spacing between the three consensus sequence elements is shown in Table 5 which includes the 10 families known to bind GTP and the 3 presumptive GTP-binding proteins from Table 4. By analysis of

TABLE 5

SPACING OF CONSENSUS GROUPS IN GTP BINDING PROTEINS

FAMILY	SPACE ONE	SPACE TWO
IF-2	40	50
RAS, YP2, rho	41-42	54-54
Protein 2C	44	80
tet M	58	51
EF-Tu, EF-1α	56-71	51-61
Lep A	60	50
EF-G, EF-2	65-72	50-53
gst 1	71	61
PEPCK	75	66
Guanyl transferase	123	57
GTP:AMP phosphotransferase	142	46
Transducin α	154	65
G proteins	165-170	65

previous spacings, a suggestion was made that perhaps two different spacings might be found in GTP-binding proteins. A closer look indicates that perhaps three different spacings may be noted. The first group is characterized by the very short spacing between the first and second consensus sequence elements, 40 to 44 amino acids. The second group is characterized by a 56 to 75 amino acid spacing between the first and second consensus elements, and a 51 to 66 amino acid spacing between the second and third consensus sequence elements. The third group is characterized by a very long distance between the first and second consensus, from 123 to 170 amino acids with average spacing (46 to 65 residues) between the last two elements. While these three different spacings may prove useful in characterizing the GTP-binding domain of different proteins, caution is urged. As noted above, by sequence analysis within the three consensus sequence elements, one observes extensive homology between the protein synthesis factors and the Lep A protein. However, by analysis of spacing between consensus sequence elements, IF-2 falls in the RAS and protein 2C families. Thus, one might ask which grouping is more correct. Unfortunately, only a crystallographic structure would answer this question; as an opinion however, the authors tend to place more weight on the similarities in amino acid sequence.

Along similar lines, a final observation can be made. A number of the GTP-binding proteins appear not to use the energy released with the hydrolysis of GTP either to transfer the gamma phosphate to another substrate or in some manner towards the synthesis of some compound. Rather, the function of these proteins appears to be to undergo a conformational change upon the hydolysis of GTP and in so doing transduce a signal of biological significance (i.e. IF-2, EF-Tu, EF-G, transducin α, or the G proteins). It is possible that this might be inferred by amino acid sequence homology (i.e. the protein synthesis factors), but certainly not by spacing of the consensus sequence elements as "authentic" enzymes fall into two of the three spacing groups. If in fact sequence homology can be used to infer function (enzyme vs signal transducer), one would predict that the yeast protein gst 1, the U. urealytium TetM gene product and the E. coli Lep A protein would be involved in signal transduction (not be enzymes proper) based upon their similarity in sequence to the consensus sequence elements of the protein synthesis factors.

SEQUENCE SPECIFICITY

As noted above, the greatest difference observed in the three consensus sequence elements was in the last element, N-K-X-D, which by x-ray crystallographic analysis is implicated in direct interaction with the substituent groups on the purine ring. Recently, by the use of site-directed mutagenesis, a change of asparagine for aspartic acid in the third consensus element caused the nucleotide specificity for the mutated EF-Tu to be changed from GTP to XTP, a result supporting the crystallographic analysis (44). Presented in Table 6 are some additional proteins which do not conform to the third consensus element, perhaps notably all enzymes. PEPCK and GTP:AMP phosphotransferase are characterized by two different features compared to the normal GTP-binding proteins: first, the K_D for GTP is about 30 μM; second, both ITP and GTP serve equally well as substrates(1). One would assume the use of both nucleotides is the result of the change of tryptophan for aspartic acid in the consensus sequence. The replacement of glycine for asparagine in the consensus sequence of the D. melanogaster and chicken mitochondrial PEPCK seems unusual. To date, there has been no report on any kinetic differences between the mitochondrial or cytoplasmic forms of chicken liver PEPCK, nor differences in substrate specificity (29, 45). To the authors, it would seem unusual that the loss of a hydrogen bond donor and replacement by a much smaller amino acid side chain might not be detectable with the appropriate substrate.

TABLE 6

PURINE SPECIFICITY

PROTEIN	ASN keto group	LYS hydrophobic side chain	X	ASP amino group
PEPCK (rat, chicken-cytosolic) GTP: AMP phosphotransferase				TRP ITP or GTP
EF-Tu (site directed mutant)				ASN XTP
PEPCK (D. melanogaster) PEPCK (chicken, mitochondrial)	GLY			TRP ITP or GTP
GUANYLTRANSFERASE (reovirus)	THR			GLU GTP

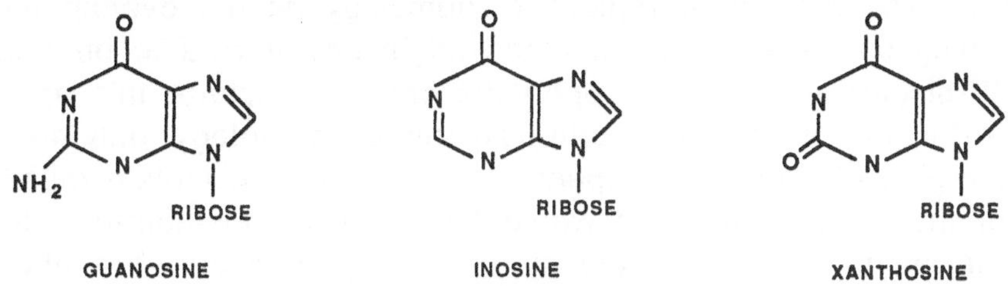

Similarly, the enzyme guanyltransferase has two non-consensus amino acids. The conservative change of glutamic acid for aspartic acid would presumedly continue to provide an interaction with the ring amino group of guanine. The threonine replacement for asparagine could be acceptable assuming the hydroxyl hydrogen can interact with the keto group. It is possible that the two amino acids changed are in part to compensate for one another, the glutamic acid side chain extending the interactive group by one carbon and the threonine side chain hydroxyl being one carbon shorter thus producing a slight twisting of the guanine ring relative to the position it would take in the normal GTP-binding protein. There is also another possible source of confusion for this enzyme. The specificity of the enzyme is for GTP (as the capping nucleotide) and for an RNA molecule which ends in GTP (i.e. pppGpXpXp...)(33,46). At present it is not known which "GTP-binding site" might be represented by the apparent consensus binding domain, but preliminary data would seem to indicate that a region close to the amino terminus is responsible for the binding of the capping GTP(46).

JUST A GTP-BINDING DOMAIN ?

For much of what has been presented above, one might conclude that the three consensus sequence elements could define the inner boundaries of the GTP-binding domain with other portions of the molecule providing additional substrate binding sites. The example presented in Table 7 can be taken as evidence that this is not necessarily true. Two phosphotransferases have been aligned, one which binds GTP and AMP (bovine, 32) and one which binds ATP and AMP (porcine,47). Based upon a similarity in catalytic mechanism and in substrates, we reasoned that one might expect to see regions of sequence homolgy for the nucleotide triphosphate-binding site and for the AMP-binding site. The sequences of homology are indicated by a box around both sequences while the three GTP-binding consensus sequences are indicated by a single boxed line. Four regions of sequence homology are noted. The first region would correspond to the first consensus sequence element which is similar to the glycine rich region that has been observed in numerous ATP-binding proteins (1,38,39). The next three regions of homology do not overlap the other GTP-binding consensus elements and might be inferred to be related to the AMP-binding site. Based upon the crystal structure, this appears to be true for the third and fourth regions of homology only(48). The second region of homology appears to be on the nucleotide triphosphate "half" of the protein and thus would seem unable to interact with AMP. The first region of homology may be shared between the phosphoryl moieties of AMP and the nucleotide triphosphate. However, one must enter some uncertainty as to how similar these two enzyme structures will be as the guanine specificity sequence of N-K-I-W would appear near the carboxy-terminus, an area associated with AMP binding in the ATP:AMP phosphotransferase. For all of this, one should not miss the point that within what would normally be the GTP-binding domain, a second domain is located. Thus, for all its small size, one need not think of the smaller GTP-binding proteins as not having another possible binding site, an idea which might be comforting for proteins such as the RAS family for which no function is yet known and a major inference is that they are signal transducers.

CONCLUSIONS

The material presented above represents a rather modest effort on the part of the authors compared to the considerable effort put forth by a large number of scientists to make available the data that is sorted together in Tables 1-7. To date, the only productive use of this type of study has been the determination that Lep A is a GTP-binding protein(14). Part of the limited, proven usefulness of such studies reflects the recentness of the discoveries and a lack of time to fully develop or test some of the predictions made. However, it is also possible that too great a rush is being made to generate "computer"

TABLE 7

*	G A S A R L L R A I M G A P G S
**	M E E K L K K S K I I - F V V G G P G S
	G K G T V S S R I T K H F E L K H L S S
	G K G T Q C E K I V Q K Y G Y T H L S T
	G D L L R D N M L R G T E I G - V L A K
	G D L L R A E V S S G S A R G K M L S E
	T F I D Q G K L I P D D V M T R L V L H
	I M E K G Q L V P L E T V L D M L R D A
	E L K N L - T Q Y N W L L D G F P R T L
	M V A K V D T S K G F L I D G Y P R E V
	P Q A E A L D R A Y Q I D T V I N L N V
	K Q G E E F E R K I G Q P T L L - - - -
	P F E V I K Q R L T A R W I H P G S G R
	- - - - - - - - - - - - - - - - - - - -
	V Y N I E F N P P K T M G I D D L T G E
	L Y - V D - A G P E T M T K R L L K R G
	P L V Q R E D D R P E T V V K R L K A Y
	E T S G R V D D N E E T I K K P L E T Y
	E A Q T E P V L E Y Y R K K G V L E T F
	Y K A T E P V I A F Y E K R G I V R K V
	S G T E T N K I W P H V Y A F L Q T K L
	N A E G S V D D V F S Q V C T H L D T L
	P Q R S Q E T S V T P
	K

* BOVINE GTP:AMP PHOSPHOTRANSFERASE   ** PORCINE ATP:AMP PHOSPHOTRANSFERASE

answers and consequently forced (and inappropriate) conclusions are drawn. It is difficult to let slide the noted similarities in proteins that bind GTP at a time when there seems to be more emphasis on GTP-binding proteins being involved in regulatory events. Time and considerable work (especially crystallographic work) will be required to determine whether the intriguing GTP-binding consensus elements do in fact represent the correct contacts in the appropriate crystal structures. Hopefully, some of the observations made here will be of value (and with luck be correct). However, as the comparison of the two phosphotransferases points out, paper chemistry/biophysics can be wrong. Therefore, go humble into the good night.

ABBREVIATIONS

IF, initiation factor (prokaryotic); EF, elongation factor; PEPCK, phosphoenolpyruvate carboxykinase; protein family, in the context of this paper, a family of proteins represents proteins of similar function (prokaryotic or eukaryotic) or when function is unknown, a high degree of amino acid sequence homology.

ACKNOWLEDGEMENTS

Supported in part by a Health Training Grant in Metabolism AM 07319 and NIH grant GM 36467.

REFERENCES

1. Dever, T.E., Glynias, M.J., and Merrick, W.C. (1987) Proc. Natl. Acad. Sci. USA 84, 1814-1818.
2. la Cour, T.F.M., Nyborg, J., Thirup, S. and Clark, B.F.C. (1985) EMBO J. 4, 2385-2388.
3. McCormick, F., Clark, B.F.C., la Cour, T.F.M., Kjeldgaard, M., Norskov-Lauritsen, L. and Nyborg, J. (1985) Science 230, 78-82.
4. Jurnak, F. (1985) Science 230, 32-36.
5. Arai, K., Clark, B.F.C., Duffy, L., Jones, M.D., Kaziro, Y., Laursen, R.A., L'Italien, J., Miller, D.L., Nagarkatti, S., Nakamura, S., Nielsen, K.M., Petersen, T.E., Takahashi, K. and Wade, M. (1980) Proc. Natl. Acad. Sci. USA 77, 1326-1330.
6. Montandon, P. and Stutz, E. (1983) Nucleic Acids Res. 11, 5877-5892.
7. Nagata, S., Tsunetsugu-Yokota, Y., Naito, A. and Kaziro, Y. (1983) Proc. Natl. Acad. Sci. USA 80, 6192-6196.
8. Nagata, S., Nagashima, K., Tsunetsugu-Yokota, Y., Fujimura, K.,

Miyazaki, M. and Kaziro, Y. (1984) EMBO J. 3, 1825-1830.
9. Cottrelle, P., Thiele, D., Price, V.L., Memet, S., Micouin, J.Y., Marck, C., Buhler, J.M., Sentenac, A. and Fromagoet, P. (1985) J. Biol. Chem. 260, 3090-3096.
10. Van Hemert, F.J., Amons, R., Pluijms, W.J.M., Van Ormondt, H. and Moller, W. (1984) EMBO J. 3, 1109-1113.
11. Brands, J.H.G.M., Maassen, J.A., Van Hemert, F.J., Amons, R. and Moller, W. (1986) Eur. J. Biochem. 155, 167-171.
12. Ovchinnikov, Y.A., Alakhov, Y.B., Bundulis, Y.P., Bundule, M.A., Dovgas, N.V., Kozlov, V.P., Motuz, L.P. and Vinokurov, L.M. (1982) FEBS Lett. 139, 130-135.
13. Kohno, K., Uchida, T., Ohkubo, H., Nakanishi, S., Nakanishi, T., Fukui, T., Ohtsuka, E., Ikehara, M. and Okada, Y. (1986) Proc. Natl. Acad. Sci. USA 83, 4978-4982.
14. March P.E. and Inouye, M. (1985) Proc. Natl. Acad. Sci. USA 82, 7500-7504.
15. Sacerdot, C., Dessen, P., Hershey, J.W.B., Plumbridge, J.A. and Grunberg-Manago, M. (1984) Proc. Natl. Acad. Sci. USA 81, 7787-7791.
16. DeFoe-Jones, D., Scolnick, E.M., Koller, R. and Dhar, R. (1983) Nature 306, 707-709.
17. Powers, S., Kataoka, T., Fasano, O., Goldfarb, M., Strathern, J., Broach, J. and Wigler, M. (1984) Cell 36, 607-612.
18. Dahr, R., Nieto, A., Koller, R., DeFoe-Jones, D. and Scolnick, E. M. (1984) Nucleic Acids Res. 12, 3611-3618.
19. Gallwitz, D., Donath, C. and Sander, C. (1983) Nature 30, 704-707.
20. Capon, D.J., Chen, E.Y., Levinson, A.D., Seeburg, P.H. and Goeddel, D.V. (1983) Nature 302, 33-37.
21. Taparowsky, E., Shimizu, K., Goldfarb, M. and Wigler, M. (1983) Cell 34, 581-586.
22. Shimizu, K., Birnbaum, D., Ruley, M., Fasano, O., Suard, Y., Edlund, L., Taparowsky, E., Goldfarb, M. and Wigler, M. (1983) Nature 304, 497-500.
23. Rasheed, S., Norman, G.L. and Heidecker, G. (1983) Science 221, 155-157.
24. Reddy, E.P., Lipman, D., Andersen, P.R., Tronick, S.R. and Aaronson, S.A. (1985) J. Virol. 53, 984-987.
25. Dhar, R., Ellis, R.W., Shih, T.Y., Oroszlan, S., Shapiro, B., Maizel, J., Lowy, D. and Scolnick, E. (1982) Science 217, 934-937.
26. Tsuchida, N., Ryder, T. and Ohtsubo, E. (1982) Science 217, 937-939.
27. Madaule, P. and Axel, R. (1985) Cell 41, 31-40.
28. Cook, J.S., Weldon, S.L., Garcia-Ruiz, J.P., Hod Y. and Hanson, R.W. (1986) Proc. Natl. Acad. Sci. USA 83, 7583-7587.
29. Weldon, S. L. & Hanson, R.W., personal communication.
30. Beale, L.G., Chrapkiewicz, N.B., Scoble, H.A., Metz, R.J., Quick, D.G., Noble, R.L., Donelson, J.E., Biemann, K. and Granner, D.K. (1985) J. Biol. Chem. 260, 10748-10760.

31. Gundelfinger, E.D., Hermans-Borgmeyer, I., Grenningloh, G. and Zopf, D. (1987) Nuc. Acids Res. 15, 6745.
32. Tomasselli, A.G., Frank, R. and Schiltz, E. (1986) FEBS Lett. 202, 303-308.
33. Seliger, L.S., Zheng, K. and Shatkin, A.J. (1987) J. Biol. Chem. 262, 16289-16293.
34. Yatsunami, K. and Khorana, H.G. (1985) Proc. Natl. Acad. Sci. USA 82, 4316-4320.
35. Tanabe, T., Nukada, T., Nishikawa,Y., Sugimoto, K., Suzuki, H., Takahashi, H., Noda, M., Haga, T., Ichiyama, A., Kangawa, K.,Minamino, N., Matsuo, H. and Numa, S. (1985) Nature 315,242-245.
36. Robishaw, J.D., Russell, D.W., Harris, B.A., Smigel, M.D. and Gilman, A.G. (1986) Proc. Natl. Acad. Sci. USA 83, 1251-1255.
37. Itoh, H., Kozasa, T., Nagata, S., Nakamura, S., Katada, T., Ui, M., Iwai, S., Ohtsuka, E., Kawasaki, H., Suzuki, K. and Kaziro, Y. (1986) Proc. Natl. Acad. Sci. USA 83, 3776-3780.
38. Fry, D.C., Kuby, S.A. and Mildvan, A.S. (1986) Proc. Natl. Acad. Sci. U.S.A. 83, 907-911.
39. Moller, W. and Amons, R. (1985) FEBS Lett. 186, 1-7.
40. Carroll, A.R., Rowlands, D.J. and Clarke, B.E. (1984) Nucleic Acids Res. 12, 2461-2472.
41. Kikuchi, Y., Shimatake, H. and Kikuchi, A. (1988) EMBO J. 7, 1175-1182.
42. Sanchez-Pescador, R., Brown, J.T., Roberts, M. and Urdea, M.S. (1988) Nuc. Acids Res. 16, 1216-1217.
43. Sanchez-Pescador, R., Brown, J.T., Roberts, M. and Urdea, M.S. (1988) Nuc. Acids Res. 16, 1218.
44. Hwang, Y. and Miller, D.L. (1987) J. Biol. Chem. 262, 13081-13085.
45. T. Nowak, personal communication.
46. A. Shatkin, personal communication.
47. Heil, A., Muller, G., Noda, L.H., Pinder, T., Schirmer, R.H., Schirmer, I. and Von Zabern, I. (1974) Eur. J. Biochem. 43, 131-144.
48. Pai, E.F., Sachsenheimer, W., Schirmer, R.H., and Schulz, G.E. (1977) J. Mol. Biol. 114, 37-45.

STRUCTURE, FUNCTION AND GENETICS
OF TRANSLATIONAL FACTORS

NOVEL MUTANTS OF EF-Tu

Diarmaid Hughes and C.G. Kurland

Department of Molecular Biology
The Biomedical Center, Box 590
S-751 24 Uppsala, Sweden

INTRODUCTION

The process of translation in bacteria is familiar in its broad outlines. However precise knowledge of many of its reactions, their kinetics and the strategies underlying them is still lacking. During the elongation cycle, EF-Tu, in a ternary complex with GTP and aminoacyl-tRNA, plays a role in mediating the interaction between the aminoacyl-tRNA and the ribosome. The elongation cycle in bacteria such as E.coli operates at a rate of up to 20 amino acids per second and with an error frequency in the range of 10^{-3} - 10^{-4} per codon. How EF-Tu contributes to this level of efficiency is unclear. We are studying wild-type and mutant derivatives of EF-Tu in an attempt to define in more detail the elongation cycle and the role of ternary complex in translation. The cycle of elongation may be formally considered to comprise two stages; the selection of a correct aminoacyl-tRNA on a codon-programmed ribosome, and the selection of a new codon involving movement of the selected tRNA relative to the ribosome. The selection of the correct aminoacyl-tRNA species involves a reversible initial selection (I) followed by hydrolysis of GTP on the ternary complex. The low accuracy of this initial selection is increased by one or more kinetic proofreading steps (F) resulting either in the successful formation of a peptide bond or the ejection of the aminoacyl-tRNA from the ribosome (Thompson and Stone, 1977; Ruusala et al, 1982). How EF-Tu influences the initial selection and whether it has any effect on proofreading are questions we wish to address. In addition we have results suggesting that EF-Tu can also influence reading frame selection (Hughes et al, 1987). There are other outstanding questions related to EF-Tu. We would like to understand why the regeneration of EF-Tu.GTP from EF-Tu.GDP is dependent on a separate enzyme, EF-Ts, and why some bacteria carry two genes coding for EF-Tu.

Two genes encoding EF-Tu have been mapped in Salmonella typhimurium (Hughes, 1986). These genes, tufA (minute 71-72) and tufB (minute 88-89) are in similar chromosomal locations to the two tuf genes in the closely related E.coli. We have isolated mutations in each of the tuf genes by selection for resistance to the antibiotic kirromycin. Kirromycin acts by preventing the release of EF-Tu.GDP from the ribosome during elongation, following GTP hydrolysis (Wolf et al, 1977), thus blocking the ribosome. Initial investigations of kirromycin resistant tuf mutations in both E.coli and S.typhimurium showed that some of them caused suppression of a frameshift mutation, trpE91 (Hughes, 1984). This suggested that tuf mutations isolated by selection for kirromycin resistance might provide useful information on the process and accuracy of translation.

ERROR-PRONE MUTANTS OF EF-Tu

Nonsense codon readthrough

Two alleles, tufA8 and tufB103 (chosen initially because they suppress the frameshift mutation trpE91) have been studied by us as examples of error enhancing EF-Tu mutations. We find that these mutations cause suppression of nonsense and frameshift mutations. With a set of 15 nonsense mutations in the lacI part of a lacIZ fusion we have used B-galactosidase activity as a measure of the level of readthrough into lacZ. Strains carrying the EF-Tu mutations tufA8 and tufB103 cause significant increases in readthrough of two out of three UGA mutations (up to 13-fold) and of three out of six UAG mutations (up to 5.5-fold). No significant increase in readthrough was detected at any of six UAA sites tested (Hughes, 1987). In addition seven out of nine UGA mutations tested in the his operon are suppressed (Hughes et al, 1987). These results show that nonsense codon readthrough is stimulated by tufA8 and tufB103 but is specific to certain mutations. Thus no UAA readthrough has been detected and only some UGA and UAA sites are affected. The reasons for this specificity are unknown at present. Also unknown is whether mutant EF-Tu stimulates errors of nonsense readthrough simply by increasing the probability of a missense error event, or whether it might have a more direct effect on the release factor responsible for termination. Nonsense readthrough also occurs in strains carrying only one mutant tuf gene, with the other tuf gene wild-type. In such strains the level of readthrough is intermediate between the base level in the wild-type strains and the level measured in strains carrying both mutations tufA8 and tufB103. Indeed the data show that the level of nonsense readthrough increases approximately additively when both tuf mutations are present. This suggests that tufA8 and tufB103 act independently of one another to cause nonsense readthrough.

Frameshift suppression

We have asked whether mutations in EF-Tu can influence reading frame maintainance. Over thirty frameshift mutations of plus and minus sign in the trp and his operons and the lacI part of a lacIZ fusion were tested for suppression by tufA8 and tufB103. The minus one mutation trpE91, one of four minus one mutations tested in lacI and two plus one mutations in hisD are suppressed (Hughes et al, 1987). Thus frameshifting into both the plus and minus reading frame is enhanced by mutant EF-Tu but is apparently limited to certain frameshift mutations. The efficiency of frameshift suppression is greatest in strains carrying both tufA8 and tufB103, but also occurs in strains carrying one mutant and one wild-type tuf gene. One novel possibility arising from these results is that EF-Tu may have a role in choosing or maintaining the correct reading frame. Alternatively the observed frameshifting may be a consequence of an increase in the level of missense errors, that is, errors of aminoacyl-tRNA selection. If this is shown to be the case it would support the proposal of Kurland (1979) that mismatched codon-anticodon interactions could increase the probability of reading frame errors.

Inactivation of the tufA gene

We have established that each of the mutations tufA8 and tufB103 can increase the level of translational errors in vivo. We wished next to study this phenomenon in vitro using purified mutant and wild-type EF-Tu. Because S.typhimurium carries two genes coding for EF-Tu we considered that analysis of our in vitro data might be complicated if the tufA8 and tufB103 mutations differed significantly in their effects on translation, despite having apparently similar in vivo growth and error characteristics. Thus we attempted to inactivate each of the tuf genes by transposon insertion. We have used a Mu phage derivative MudJ because of its relatively random pattern of insertion and its stability once inserted. The strategy was to use strains carrying one wild-type and one mutant tuf gene, select random MudJ insertions and then screen for kirromycin resistance. Kirromycin resistance could arise because of a spontaneous mutation in the wild-type gene or because of inactivation of that gene by MudJ insertion. In this way we have isolated two independent mutants with MudJ apparently inserted in the tufA gene, leaving tufB as the only active tuf gene. We have not yet succeeded in isolating a similar MudJ inactivation of tufB. Vijgenboom and Bosch (1987) have reported the Mu inactivation of tufB in E.coli and their inability to make a similar inactivation in tufA. They suggest that tufA may be essential for growth despite the apparent physical and functional similarities between the tufA and tufB gene products. We do not know whether this hypothetical situation is reversed in S.typhimurium with the tufB gene essential for some as yet unknown reason. However as the tufB gene in S.typhimurium mutates to kirromycin resistance at a ten-fold higher frequency than tufA (Hughes, 1986) it is possible that we have not yet screened enough potential mutants. An indication that tufB in S.typhimurium may not be an essential gene comes from a novel class of temperature sensitive mutations we have isolated in each of the tuf genes (these are discussed later). The evidence that we have inactivated tufA with MudJ is genetic, enzymatic and physical. An antibiotic resistance within MudJ shows 100% linkage with the kirromycin resistance phenotype of the tufB gene and is also linked to the tufA region of the chromosome. EF-Tu purified from tufB wild-type or tufB103 strains carrying MudJ in tufA has an in vitro error characteristic of pure wild-type or pure error-prone EF-Tu (discussed below). Finally, Southern blotting of chromosomal DNA with overlapping probes of cloned E.coli tufA and fus genes limit the region of the insert tested to an EcoR1-Kpn1 fragment covering the extreme 3' end of fus, the short region between fus and tufA, and the 5' half of the tufA gene. Further blotting experiments to define the insert location more precisely are underway. These results will be published in detail later.

Missense errors in vitro

We are studying the phenotypes of wild-type and error-prone EF-Tu in an in vitro translation elongation system. This system operates at 37°C, has an elongation rate and accuracy close to those measured in vivo, and is composed of individually purified components (Wagner et al, 1982; Ehrenberg et al, 1986). The results presented below on the in vitro phenotypes of EF-Tu mutants will be published in more detail later. We have purified EF-Tu from our various wild-type and error-prone mutant strains. We first asked whether the accuracy phenotypes observed in vivo for frameshift and nonsense errors are reproduced in vitro for missense errors. The errors measured are for different leucine tRNA species competing with an equal amount of tRNA Phe for polyU message. Wild-type S.typhimurium EF-Tu (TuA,TuB or pure TuB) has an in vitro missense error indistinguishable from that of E.coli EF-Tu (approximately 5 x 10^{-4} tRNA leu4, 1 x 10^{-4} tRNA Leu2). Error-prone EF-Tu (TuA8,TuB103 or pure TuB103) has approximately a four-fold increase in missense errors with each of the tRNA Leu species. Mixtures of mutant and wild-type EF-Tu (TuA8,TuB or TuA,TuB103) isolated from strains carrying one mutant and one wild-type tuf gene cause missense error levels for each tRNA Leu species intermediate between the wild-type and the pure error-prone level. The increase in errors caused by TuA8 and TuB103 is similar. When the wild-type value is subtracted as a background it appears that TuA8 and TuB103 contribute additively to the total error of the double mutant. The observation that pure TuB103 causes a four-fold error increase characteristic of the TuA8,TuB103 mixture, while pure wild-type TuB shows the normal TuA,TuB wild-type error, argues strongly that MudJ has indeed inactivated tufA as argued in the previous section. Thus our results on in vitro missense errors caused by TuA8 and TuB103 appear to mimic our earlier in vivo error results. This suggests that it could be productive to use this in vitro translation system to search for the cause of the increase in errors.

in vitro translation analysis

Missense errors in the polyU system are errors of aminoacyl-tRNA selection. The accuracy (A) of this selection is the inverse of the error which we measure. The accuracy of aminoacyl-tRNA selection on the ribosome can be formally divided into two components, an initial selection (I) and an accuracy enhancement of this initial selection by kinetic proofreading (F), as proposed by Hopfield (1974), Ninio (1975) and experimentally demonstrated by Thompson and Stone (1977), Ruusala et al (1982). Thus A = IxF. The present experimental evidence is that with wild-type components the contributions of I and F to the overall accuracy are approximately equal. The proofreading factor F is composed of fc (proofreading of cognate ternary complex) and fw (proofreading of non-cognate ternary complex) where F = fw/fc. With wild-type components the magnitude of fw is much greater than fc. Thus after the initial selection and GTP hydrolysis, incorrect ternary complexes are rejected by the ribosome more frequently than correct ternary complexes, giving a corresponding enhancement of accuracy over that achieved in the initial selection. From this description it is clear that the number of GTP's hydrolysed after initial selection , is related to the number of peptide bonds successfully made, by the magnitude of the intervening proofreading factor. The measurement of proofreading in translation is discussed in detail by Ehrenberg et al (1986). Previous measurements made in this laboratory on a set of ribosomal mutants with accuracy phenotypes ranging from error-prone to error-restrictive (summarized in Andersson et al, 1986) show that for these mutants the change in accuracy is accounted for by a change in the intensity of proofreading. In addition the magnitude of fc is directly correlated with the kinetic efficiency of translation and with growth rate.

We have begun measuring proofreading and kinetic efficiency for our mutant and wild-type EF-Tu species and present here some preliminary results. We calculate fc, fw and F independently, by measuring the number of GTP's hydrolyzed on EF-Tu during translation per cognate or non-cognate amino acid incorperated into protein. We find that wild-type EF-Tu (mixed or pure TuB) from S.typhimurium has proofreading characteristics similar to EF-Tu from E.coli. Our error-prone EF-Tu (TuA8,TuB103 or TuB103) differs from the wild-type in several ways. The proofreading of cognate ternary complex (fc) appears to be significantly increased while the proofreading of non-cognate ternary complex (fw) shows a relatively small decrease. Our initial measurement of the overall proofreading factor (F) suggest that it too may be decreased. However the decrease in F does not appear to be large enough to account for the decrease in accuracy displayed by these mutants. Thus it appears tentatively that these error-prone EF-Tu mutants decrease the accuracy of the initial selection of ternary complex on the ribosome but also influence proofreading. Further measurements are required to establish the precise magnitude of these changes. We have also measured an increased intrinsic GTPase activity apparently associated with these error-prone EF-Tu species. Because of this GTPase activity we are cautious in our interpretation of proofreading results based on the ratio of GTP's hydrolyzed per peptide bond, until we are sure that we understand how it influences our results. The k_{cat} of elongation and the K_m of the ternary complex-ribosome interaction are both decreased by the error-prone EF-Tu species. In vivo we have measured only a marginally decreased elongation rate for B-galactosidase but a significant (28%) decrease in growth rate associated with the error-prone tuf genes.

GROWTH-IMPAIRED MUTANTS OF EF-Tu

We have isolated a novel class of kirromycin resistant tuf mutations that display a temperature sensitive phenotype. They were originally identified as cold sensitive mutants on the basis that they fail to grow on solid media below 20°C. The cs mutations occur in both the tufA and tufB genes. The cs phenotype is at least partially recessive as strains carrying one wild-type and one tuf^{cs} allele grow at low temperature. This phenotype is not simply one of cold sensitivity as these mutants are significantly impaired in their growth at other temperatures. Depending on the particular allele or combination of alleles, the growth rate is reduced 30-50% at 37°C and 50-90% at 26°C. These mutants provide tentative evidence that the tufB gene may not be essential in S.typhimurium. Strains with a $tufB^{cs}$ and a MudJ inactivated tufA, or a $tufA^{cs}$, have a cold sensitive phenotype, whereas equivalent strains with a wild-type tufA and a $tufB^{cs}$ do grow at low temperature.

We have begun in vitro analysis on EF-TuB401, isolated from a cold sensitive strain carrying a MudJ inactivated tufA gene. Our initial results show that the k_{cat} for elongation supported by the mutant TuB401 is reduced (but only to the same level as that of the error-prone TuB103) while the K_m of the ternary complex ribosome interaction shows little change. This suggests the possibility that TuB401 is rate-limiting some reaction in translation, however other explanations are possible. At 20°C there is a four-fold reduction in k_{cat} for elongation but the same relative reduction is seen with error-prone and wild-type EF-Tu. A striking feature of this mutant is that it has a high K_d for the release of GDP from Tu.GDP. In our wild-type (E.coli and S.typhimurium) and error-prone EF-Tu's the K_d is 0.011 s^{-1} while in EF-TuB401 this is increased seven to eight-fold. This characteristic of the mutant may not neccessarily be related to the in vivo growth impairment. In vivo, in the presence of EF-Ts, the K_d is expected to be about 3000-fold higher than 0.011 s^{-1}. More intriguingly we have preliminary evidence that the amount of active EF-Tu in some cold sensitive strains may be low.

In conclusion, we have isolated and partially characterized some novel mutants of EF-Tu influencing translational accuracy and growth rate. We are hopeful that the analysis of mutant forms of EF-Tu defective in specific partial reactions in translation may help in understanding the role of ternary complex in translation.

ACKNOWLEDGEMENTS

We thank Måns Ehrenberg for helpful discussions on in vitro experiments. This work was supported by the Swedish Natural Science Research Council and by the Swedish Cancer Society.

REFERENCES

Andersson, D.I., van Verseveld, H.W., Stouthamer, A.H. and Kurland, C.G., 1986, Suboptimal growth with hyperaccurate ribosomes, Arch. Microbiol., 144:96.

Ehrenberg, M., Kurland, C.G. and Ruusala, T., 1986, Counting cycles of EF-Tu to measure proofreading in translation, Biochemie, 68:261.

Hopfield, J.J., 1974, Kinetic proofreading: a new mechanism for reducing errors in biosynthetic processes requiring high specificity, Proc. Nat. Acad. Sci. USA, 71:4135.

Hughes, D., 1984, External suppression of +1 and -1 frameshift mutations: a genetic analysis in bacteria, Ph.D. Thesis, Dublin University.

Hughes, D., 1986, The isolation and mapping of EF-Tu mutations in Salmonella typhimurium, Mol. Gen. Genet., 202:108.

Hughes, D., 1987, Mutant forms of tufA and tufB independently suppress nonsense mutations, J. Mol. Biol., 197:611.

Hughes, D., Atkins, J.F. and Thompson, S., 1987, Mutants of elongation factor Tu promote ribosomal frameshifting and nonsense readthrough, EMBO J., 6:4235.

Kurland, C.G., 1979, Reading frame errors on ribosomes, in: 'Nonsense mutations and tRNA suppressors,' J.E. Celis and J.D. Smith, eds., Academic press, New York.

Ninio, J., 1975, Kinetic amplification of enzyme discrimination, Biochemie, 57:587.

Ruusala, T., Ehrenberg, M., and Kurland, C.G., 1982, Is there proofreading during polypeptide synthesis?, EMBO J., 1:741.

Thompson, R.C. and Stone, P.J., 1977, Proofreading of the codon-anticodon interaction on ribosomes, Proc. Nat. Acad. Sci. USA, 74:198.

Vijgenboom, E. and Bosch, L., 1987, Transfer of plasmid-borne tuf mutations to the chromosome as a genetic tool for studying the functioning of EF-TuA and EF-TuB in the E.coli cell, Biochemie, 69:1021.

Wagner, E.G.H., Jelenc, P.C., Ehrenberg, M., and Kurland, C.G., 1982, Rate of elongation of polyphenylalanine in vitro, Eur. J. Biochem.,122:193

Wolf, H., Chinali, G. and Parmeggiani, A., 1977, Mechanism of inhibition of of protein biosynthesis bi kirromycin, Eur. J. Biochem., 75:67.

THE ELONGATION FACTOR EF-Tu FROM *E. COLI* ACTIVATES THE tRNA-*TUFB* OPERON IN *TRANS* BY BINDING TO A *CIS*-ACTING REGION UPSTREAM OF THE PROMOTER

E. Vijgenboom, L. Nilsson, A. Talens and L. Bosch
Department of Biochemistry
Leiden University
Wassenaarseweg 64, 2333 AL Leiden
The Netherlands

A highly representative member of the group of guanine nucleotide binding proteins is the polypeptide chain elongation factor Tu of *E. coli* (EF-Tu). This translational factor is a multifunctional protein able to bind, beside GDP and GTP, a relatively large number of ligands, such as tRNA, ribosomes, the elongation factor EF-Ts and antibiotics like kirromycin and pulvomycin. EF-Tu is also involved in the replication of RNA phages as one of the host donated subunits of the viral RNA replicase (Miller et al., 1977; Bosch et al., 1983; Bosch et al., 1986). EF-Tu therefore is an attractive object for studies of the relationship between structure and function. Considerable progress has been made with the elucidation of the three-dimensional structure of EF-Tu as is reported elsewhere in this volume (Jurnak et al., 1989; Nyborg et al., 1989). Genetic studies revealed that EF-Tu is encoded by two genes: *tufA* and *tufB*, located some 660 kbp apart on the *E. coli* chromosome (Jaskunas et al., 1975). The two genes have been cloned and sequenced (An and Friesen, 1980; Yokota et al., 1980; Hudson et al., 1981) so that a firm experimental basis has been laid for structure/function studies. They are part of two operons quite different in character (compare Fig. 1).
The coding sequences of the two *tuf* genes are almost identical and the gene products, designated EF-TuA and EF-TuB, differ in their C-terminal amino acid only (residue 393). That of EF-TuA is Gly, that of EF-TuB is Ser (Arai et al., 1980; An and Friesen, 1980; Yokota et al., 1980). This complicates the above mentioned studies considerably. Are both gene products functionally equivalent and are both *tuf* genes essential for bacterial growth or is one of them dispensable? Here we address these questions and show that *tufA* is essential but *tufB* is dispensable. Furthermore we demonstrate that the expression of *tufB* is dependent on that of *tufA* and that EF-Tu activates the *tufB* operon *in trans* by binding to a *cis*-acting region upstream of the promoter.

Selective inactivation of the *tuf* genes

For the selective inactivation of either *tufA* or *tufB* advantage is taken from the availability of *tuf* mutations conferring resistance to the antibiotic kirromycin (Van de Klundert et al., 1978; Fisher et al., 1977). This antibiotic binds EF-Tu with high specificity. In doing so it alters the conformation of EF-Tu.GDP to such an extent that this protein is unable to leave the ribosome during polypeptide synthesis. As a result the

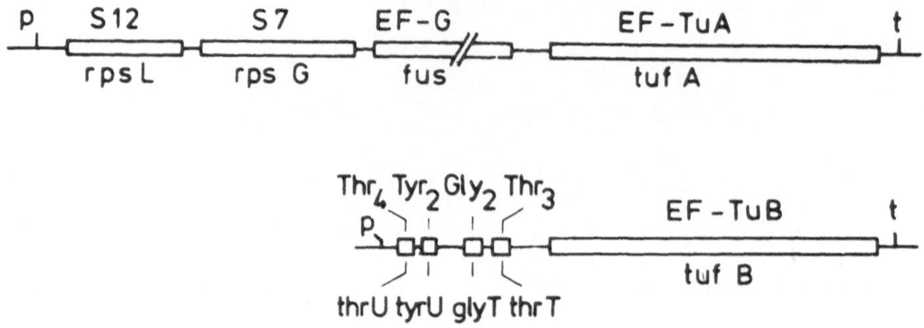

Fig. 1. The *str* operon and the tRNA-*tufB* operon.

ribosome is immobilized on the messenger RNA (Wolf et al., 1974; Parmeggiani and Sander, 1980). This makes the antibiotic a very effective inhibitor of protein synthesis, since the stuck ribosome blocks the movement of all trailing ribosomes. A consequence of this mechanism is that phenotypic expression of kirromycin resistance requires alteration of both *tuf* genes and that sensitivity to the antibiotic dominates kirromycin resistance (Van de Klundert et al., 1978; Fisher et al., 1977).

Selective inactivation of one of the *tuf* genes can thus be achieved by infecting an *E. coli* strain harbouring a kirromycin resistant and a wild type *tuf* gene with the bacteriophage Mu. Insertion of Mu into the wild type *tuf* gene alters the phenotype and can thus be selected for in the presence of kirromycin. Cells resistant to kirromycin are screened for Mu immunity (Van de Klundert et al., 1978; Vijgenboom and Bosch, 1987; Van der Meide et al., 1982). Table I shows some growth characteristics of two *E. coli* strains that have undergone such a Mu insertion directed into *tufB* (strain PM816, line 1) or into *tufA* (strain EV102, line 2). Strain PM816 harbours a kirromycin resistant *tufA* and a wild type *tufB*. The kirromycin resistant allele carries a mutation that substitutes threonine for alanine-375 (Duisterwinkel et al., 1981).

Approximately 18 hours after Mu infection, kirromycin resistant colonies with a diameter of 1 mm appeared on plates and after screening for Mu immunity the frequency of Mu insertion appeared to be 10^{-5} to 10^{-6}. It can be concluded that inactivation of *tufB* is compatible with growth, as was also found after inactivation with amber mutations (Van de Klundert et al., 1978). Strain EV102 (Vijgenboom et al., 1987) harbours a wild type *tufA* and a kirromycin resistant *tufB*. Upon Mu infection no kirromycin resistant colonies, displaying immunity to Mu, could be detected. This result indicates that inactivation of *tufA* is lethal.

Complementation of the defective tufA by plasmid-borne tuf genes

The loss of chromosomal *tufA* can be complemented with plasmid-borne *tufA*. To show this, EV102 cells were transformed with the plasmid $pGp82^R$ (Fig. 2a), harbouring a kirromycin resistant *tufA*. As can be concluded from Table I, line 3, kirromycin resistant colonies appeared at about the same time after Mu infection as with the control strain PM816 and the frequency of Mu insertion was virtually identical. This complementation was not affected by deleting sequences downstream of *tufA* on the plasmid. By putting the plasmid-borne *tufA* under the control of the *lac* promoter it could be demonstrated that complementation was due to transcription of the coding sequence of *tufA*. This proves that the loss of cell viability upon Mu insertion into *tufA* was due to inactivation of *tufA per se* rather than to inactivation of another non-identified essential gene.

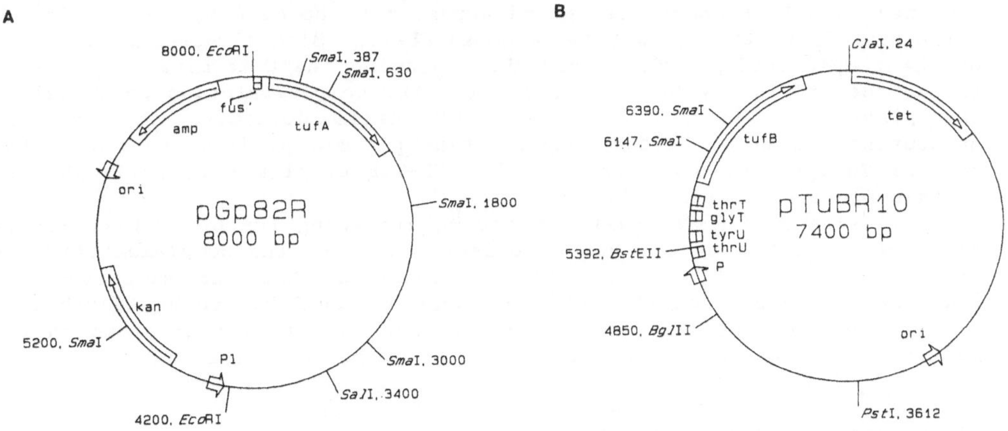

Fig. 2. Genetic maps of the plasmids pGp82^R (A) and pTuB$^R_{10}$ (B). Restriction sites are indicated at the outer rings, with distances in bp from EcoRI site in *fus* (A) or in ClaI (B). Arrows indicate direction of transcription; amp, ampicillin resistance; kan, kanamycin resistance; ori, origin of replication; tet, tetracycline resistance; PL, P_L promoter of bacteriophage lambda.

Table I

Formation of kirromycin colonies, immune to Mu, following Mu infection

	Strain	Plasmid	Colony appearance (hr) *)	Frequency Mu insertion
1	PM816		18	10^{-5} to 10^{-6}
2	EV102			none
3	EV102	pGp82^R	18	10^{-5} to 10^{-6}
4	EV102	pTuB$^R_{10}$	25	10^{-7}
5	EV102	pGp82^R (Gly-393--->Ser)	18	10^{-5} to 10^{-6}
6	EV102	pTuB$^R_{10}$ (Ser-393--->Gly)	25	10^{-7}

Complementation with pTuB$^R_{10}$ (Fig. 2b), harbouring a kirromycin resistant *tufB* was much less effective. Using the complementation assay described above, colony appearance was found to be retarded by seven hours and the frequency of Mu insertion to be lowered by at least one order of magnitude (Table I, line 4).

How is this difference in complementation behaviour of plasmid-borne *tufA* and *tufB* to be explained? It cannot be ascribed to a functional difference between the *tuf* gene products EF-TuA and EF-TuB. EF-TuB was isolated in a homogeneous state and appeared to be as competent during *in vitro* protein synthesis as (homogeneous) EF-TuA. Site-directed mutagenesis of the complementing *tufA* on pGp82^R, replacing the C-terminal Gly-393 of the encoded EF-TuA by Ser, did not alter the complementation potential of the plasmid (Table I, line 5). EV102 cells defective in *tufA* and harbouring the thus treated complementing plasmid produces EF-TuB as the sole EF-Tu species, demonstrating that EF-TuB by itself is fully able to sustain protein synthesis *in vivo*.

Site-directed mutagenesis of pTuB$^R_{10}$ replacing the C-terminal Ser of the encoded EF-TuB by Gly had no effect either on the complementation by this plasmid (Table I, line 6). We conclude that the difference in complementation between plasmid-borne *tufA* and *tufB* has to be ascribed to a difference in *tuf* gene expression, rather than to a difference in the ability of EF-TuA and EF-TuB to sustain polypeptide synthesis.

EF-Tu acts as a *trans*-activator of *tufB*

The results described so far are best explained by assuming that *tufA* defective cells are unable to express *tufB* effectively, if at all. Accordingly, complementation by plasmid-borne *tufA* is superior to that by plasmid-borne *tufB*. The implication of this interpretation is that the expression of *tufB* is dependent on that of *tufA* and thus on EF-Tu, acting as a *trans*-activator of *tufB*. Bacterial growth of *tufA* defective cells complemented by pTuB$^R_{10}$, although very slow (Table I, line 4) is not completely abolished as is that of the non-complemented cells (Table I, line 2). Multiple copies of plasmid-borne *tufB* are responsible for the production of a limited amount of *trans*-activator, just sufficient for strongly retarded growth.

Wild type cells from the strain LBE 1001, used for most of our studies (Van der Meide et al., 1983a; Van der Meide et al., 1983b), produce approximately equal amounts of EF-TuA and EF-TuB. Reduction of the total cellular amount of EF-Tu to 50% by inactivation of *tufB* is compatible with growth (Van der Meide et al., 1982; Van der Meide et al., 1983a). Upon inactivation of *tufA*, however, the cellular EF-Tu content drops below 50%, resulting in abolishment of growth. This could be demonstrated by modulating the poor complementation by pTuB$^R_{10}$ further down (Van Delft et al., 1987a). A correlation was found between the complementation and the transcription of the tRNA-*tufB* operon on the plasmid. These experiments led to the conclusion that the contribution of the chromosomal *tufB* of *tufA* defective cells must be much lower than 50% and that *tufB* expression is severely depressed when no actively functioning *tufA* is present.

The target of the *trans*-activator lies upstream of the tRNA-*tufB* operon

In a search for the target of the *trans*-activating EF-Tu we deleted various segments of the plasmid-borne tRNA-*tufB* operon and found a promoter upstream region that enhances transcription five to ten fold. This is illustrated in Figure 3, showing that transcription drops when deletions, started at position -500 upstream of the transcription initiation site, are extended from position -150 to -60.
The residual transcription activity is then about 15% of the original value (Van Delft et al., 1987a). Similar *cis*-acting sequences, called activator elements, have been found by others (Lamond and Travers, 1983; Gourse et al., 1986; Bauer et al., 1988) upstream of the P1 promoter of the rRNA operon *rrnB* and of the promoters of the *tyrT* and the *leuV* operons.

Interestingly, an EF-Tu.GDP preparation obtained by affinity chromotography and about 95% pure binds specifically to the activator

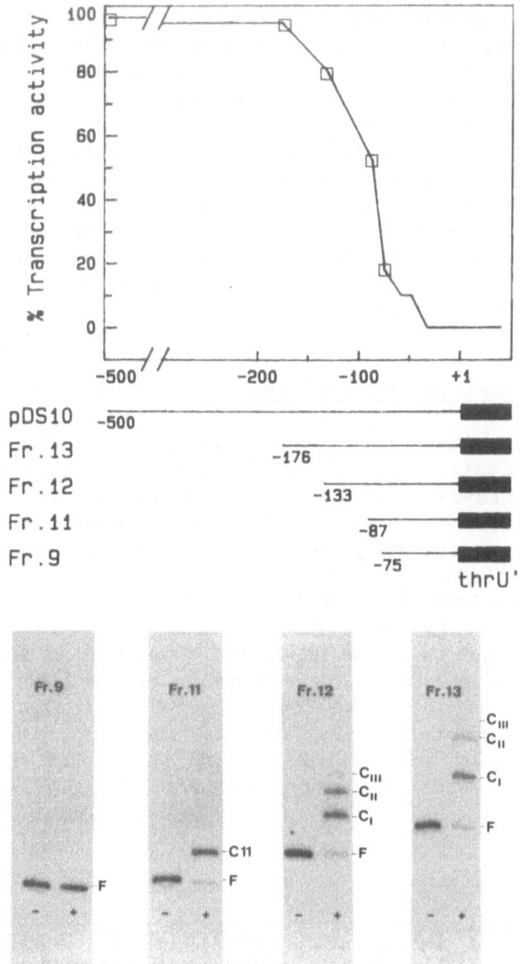

Fig. 3. Transcription activities of cells transformed with plasmid derivatives harbouring deletions in the upstream region of the tRNA-*tufB* operon (upper part) and electrophoresis of complexes of EF-Tu.GDP and DNA fragments (lower parts). Transcription activities arfe those reported by van Delft et al. (1987a). Nucleotide positions upstream of the transcription start site are plotted on the abscissa. The DNA fragments, used for EF-Tu.GDP binding, illustrated in the middle part, are derived from the plasmid deletion derivatives used for *in vivo* transcription. The endpoints of the deletions are indicated by a quadrangle (upper part) and by a number in the middle part. The lower part demonstrates EF-Tu.GDP/DNA binding with the electrophoretic retardation technique. (F: free DNA; C_I, C_{II} and C_{III}: EF-Tu.GDP/DNA complexes).

region upstream of the tRNA-*tufB* operon *in vitro*. End-labeled DNA fragments derived from the DNA region in question (Figure 3, middle part) were incubated with the EF-Tu.GDP preparation at 37°C for 10 min, whereafter complex formation was studied with the electrophoretic retardation technique. As is illustrated in Figure 3, lower part, three protein/DNA complexes can thus be detected with the DNA fragments 12 and 13. Only one complex is observed with fragment 11 and no complex with fragment 9. The entire binding sequence lies downstream of position -134 and footprinting delimits this sequence further to position -45. In fact the binding region and the activator region coincide. When DNA fragment 12 is incubated with increasing amounts of EF-Tu.GDP, complexes I, II and III appear successively, indicating that at least three protein molecules can bind to the activator element.

Upon purification to homogeneity the EF-Tu.GDP preparation loses its binding capability. Evidence exists, however, that EF-Tu itself is involved in the binding. First kirromycin abolishes complex formation entirely and second a protein can be recovered from the complexes with DNA that migrates with the same rate as EF-Tu during SDS gel electrophoresis (our unpublished results). Most likely EF-Tu binding to DNA requires an additional component to do so.

Other targets upstream of the tyrT and the rrnB operon

As pointed out above, similar *cis*-acting activator regions are found upstream of the promoters of at least three other stable RNA operons: the *tyrT* (Lamond and Travers, 1983), the *leuV* (Bauer et al., 1988) and the *rrnB* (Gourse et al., 1986) operon. The three regions do not show striking sequence similarities but they all are AT-rich and adopt unusual physical conformations, most likely involving bending of the DNA helix. This is concluded from anomalous electrophoretic mobilities of the DNA fragments in non-denaturing gels (compare for instance Fig. 4).

Recently we found that the EF-Tu.GDP preparation described above also binds to the activator containing fragments derived from the *tyrT* and the *rrnB* operons. Conceivably the three activator elements share common features of helix geometry that enable them to bind a common *trans*-activator. Coordinated expression of the three, if not all stable RNA operons, by such a common regulator may thus appear a serious possibility.

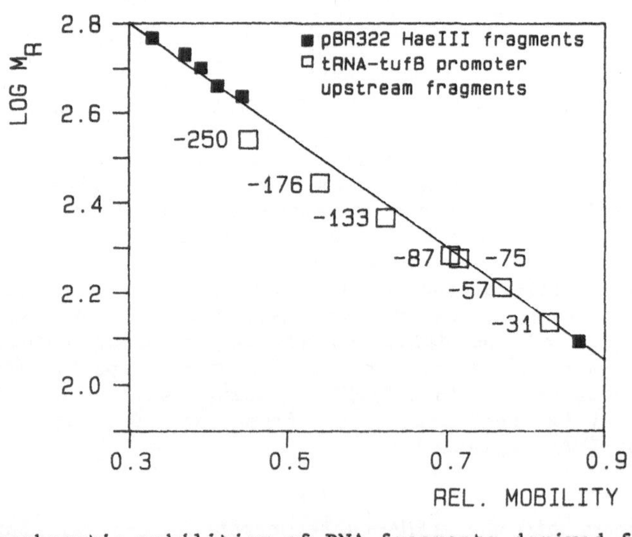

Fig. 4. Electrophoretic mobilities of DNA fragments derived from the tRNA-*tufB* operon. The fragments derived from the operon all have their termini in the activator region as indicated. Their other terminus is position + 109. The logarithm of their relative molecular mass (log M_R) is plotted against their relative electrophoretic mobility in a 5% polyacrylamide gel (30 acrylamide:1 bisacrylamide) at approximately 14 V/cm in Tris-acetate buffer (□). pBR322 HaeIII DNA fragments are run as markers (■).

Transactivation of *tufB* is an autocatalytic process

Expression of *tufB* leads to the synthesis of the *trans*-acting EF-Tu. *TufB* expression therefore is an autocatalytic process. *TufA* defective cells, poorly complemented by plasmid-borne *tufB*, gradually overcome the *trans*-activator deficiency and when reaching the logarithmic phase grow as fast as cells complemented by plasmid-borne *tufA*. This means that exponentially growing *tufA* defective cells are able to meet their EF-Tu demands by solely expressing *tufB*. Reversal to retarded growth occurs when the cells are stored at 4°C for twenty days. The autocatalytic character of *tufB* control is an essential feature of this autogenous type of regulation. Its experimental demonstration strongly supports the role of EF-Tu as a *trans*-activator of *tufB*.

CONCLUSIONS

Studies of the relationship between structure and function of EF-Tu depend to a high extent on information concerning the regulation of the two EF-Tu encoding genes: *tufA* and *tufB*. Until recently such information was virtually not available. In the last few years we therefore have performed various studies (Bosch et al., 1983; Bosch et al., 1986; Vijgenboom and Bosch, 1987; Van der Meide et al., 1982; Duisterwinkel et al., 1981; Van der Meide, et al., 1983a; Van der Meide et al., 1983b; Van Delft et al., 1987a; Van Delft et al., 1987b; Van Delft et al., 1988a; Van Delft et al., 1988b; Van Delft and Bosch, 1988) to shed some light on these questions. So far the emphasis has been mainly on the tRNA-*tufB* operon. Here we have dealt with one single aspect of the regulation of this operon only: *Trans*-activation by the elongation factor itself. As such it thus is a form of autogenous control.

The results presently obtained have yielded firm evidence for such an autogenous control of *tufB* through *trans*-activation by EF-Tu. Regarding the mode of action of the *trans*-activator a number of questions remain. Although the data available so far indicate that EF-Tu.GDP, with a purity of about 95%, binds to the activator element upstream of the promoter of the tRNA-*tufB* operon, it is clear that further purification to homogeneity eliminates all binding activity. Apparently, an additional component is required for the binding. So far the nature of this component has remained obscure. Further deepening of our insight into the mode of action may also be expected from *in vitro* studies of *trans*-activation. The finding that the DNA sequence involved in EF-Tu binding and the activator region coincide, strongly suggests that we are dealing with regulation of the entire tRNA-*tufB* operon, rather than with that of the *tufB* gene only.

Our data seem to add an additional function to the many functions of this remarkable member of the guanine nucleotide binding proteins. The role of *trans*-activator may have interesting implications. Various effector molecules have been reported to exert conformational and functional effects on EF-Tu. How do they affect *trans*-activation?
The finding that the EF-Tu.GDP preparation studied here also binds specifically to the activator elements of the *tyrT* and *rrnB* operons has come as a surprise. It indicates that EF-Tu may act as a common signal transducer for the regulation of three, if not all stable RNA operons. If so, this may endow EF-Tu with a key function in controlling the synthesis of the translational machinery.

ACKNOWLEDGEMENTS

The research of the authors is supported in part by the Commission of the European Communities, Biotechnology Action Programme (BAP) Directorate-General "Science, Research and Development". One of them (L.N.) is the recipient of a long term EMBO Fellowship.

REFERENCES

An, B., and Friesen, J. D., 1980, The nucleotide sequence of *tufB* and four nearby tRNA structural genes of *Escherichia coli*, Gene, 12:33.

Arai, K., Clark, B. F. C., Duffy, L., Jones, M. D., Kaziro, Y., Laursen, R. A., L'Italien, J., Miller, D. L., Nagarkatti, S., Nakamura, S., Nielsen, K. M., Petersen, T. E., Takahashi, K., and Wade, M., 1980, Primary structure of elongation factor Tu from *Escherichia coli*, Proc. Natl. Acad. Sci. USA, 77:1326.

Bauer, B. F., Kar, E. G., Elford, R. M., and Holmes, W. M., 1988, Sequence determinants for promoter strength in the *leuV* operon of *Escherichia coli*, Gene, 63:123.

Bosch, L., Kraal, B., Van Noort, J. M., Vijgenboom, E., Van Delft, J. H. M., and Talens, A., 1986, Novel functions of EF-Tu during polypeptide synthesis and *tuf* gene expression, in: Hardesty, B. and Kramer, G. (Eds.), Structure, Function and Genetics of Ribosomes, Springer-Verlag, p. 658.

Bosch, L. Kraal, B., Van der Meide, P. H., Duisterwinkel, F. J., and Van Noort, J. M., 1983, The elongation factor EF-Tu and its two encoding genes, in: Cohn, W.E. and Moldave, K. (Ed.), Progress in Nucleic Acids Research and Molecular Biology, Acad. Press, Vol. 30, p. 91.

Duisterwinkel, F. J., de Graaf, J. M., Kraal, B., and Bosch, L., 1981, A kirromycin resistant elongation factor EF-Tu from *Escherichia coli* contains a threonine instead of an alanine residue in position 375, FEBS Lett., 3:130.

Fisher, E., Wolf, H., Hantke, K., and Parmeggiani, A., 1977, Elongation factor Tu resistant to kirromycin in an *Escherichia coli* mutant altered in both *tuf* genes, Proc. Natl. Acad. Sci. USA, 74:4341.

Gourse, R. L., de Boer, H. A., and Nomura, M., 1986, DNA determinants of rRNA synthesis in *E. coli*: Growth rate dependent regulation feedback inhibition, upstream activation, antitermination, Cell, 44:197.

Hudson, L., Rossi, J., and Landy, A., 1981, Dual function transcripts specifying tRNA and mRNA, Nature, 294:422.

Jaskunas, S. R., Lindahl, L., Nomura, M., and Burgess, R. R., 1975, Identification of two copies of the gene for the elongation factor EF-Tu in *E. coli*, Nature, 257:458.

Jurnak, F., Nelson, M., Yoder, M., Heffron, S., and Miu, S., 1989, Progress on the three-dimensional structural determination of trypsin-modified EF-Tu-GDP, this book.

Lamond, A. J., and Travers, A. A., 1983, Requirement for an upstream element for optimal transcription of a bacterial tRNA gene, Nature, 305:248.

Miller, D. L., and Weissbach, H., 1977, Aminoacyl-tRNA transfer factors, in: Weissbach, H. and Pestka, S., Molecular Mechanisms of Protein Biosynthesis, Acad. Press., p. 324.

Nyborg, J., La Cour, T., Kjelgaard, M., Thirup, S., Jensen, M., and Clark, B. F. C., 1989, New structural data on elongation factor-Tu: GDP based on X-ray crystallography, this book.

Parmeggiani, A., and Sander, G., 1980, in: Topics of Antibiotic Chemistry, 5:165.

Van Delft, J. H. M., Marinon, B., Schmidt, D. S., and Bosch, L., 1987a, Transcription of the tRNA-*tufB* operon of *E. coli*: activation, termination and antitermination, Nucl. Acids Res., 15:9515.

Van Delft, J. H. M., Schmidt, D. S., and Bosch, L., 1987b, The tRNA-*tufB* operon. Transcription termination and processing upstream from *tufB*, J. Mol. Biol., 197:647.

Van Delft, J. H. M., Verbeek, H. M., de Jong, P. J., Schmidt, D. S., Talens, A., and Bosch, L., 1988a, Control of the tRNA-*tufB* operon in *Escherichia coli*. 1. rRNA gene dosage effects and growth-rate dependent regulation, Eur. J. Biochem., 175:355.

Van Delft, J. H. M., Talens, A., de Jong, P. J., Schmidt, D. S. and,

Bosch, L., 1988b, Control of the tRNA-*tufB* operon in *Escherichia coli*. 2. Mechanisms of the feedback inhibition of *tufB* expression studied *in vivo* and *in vitro*, Eur. J. Biochem., 175:363.
Van Delft, J. H. M., and Bosch, L., 1988, Control of the tRNA-*tufB* operon in *Escherichia coli*. 3. Feedback inhibition of *tufB* expression by an EF-Tu with a deletion in the guanine-nucleotide-binding domain, Eur. J. Biochem., 175:375.
Van de Klundert, J. A. M., Van der Meide, P. H., Van de Putte, P., and Bosch, L., 1978, Mutants of *Escherichia coli* altered in both genes coding for the elongation factor Tu, Proc. Natl. Acad. Sci. USA, 75:4470.
Van der Meide, P. H., Vijgenboom, E., Dicke, M., and Bosch, L., 1982, Regulation of the expression of *tufA* and *tufB*, the two genes coding for the elongation factor EF-Tu in *E. coli*, FEBS Lett., 139:325.
Van der Meide, P. H., Vijgenboom, E., Talens, A., and Bosch, L., 1983a, The role of EF-Tu in the expression of *tufA* and *tufB* genes, Eur. J.Biochem., 130:397.
Van der Meide, P. H., Kastelein, R. A., Vijgenboom, E., and Bosch, L., 1983b, *Tuf* gene dosage effects on the intracellular concentration of EF-TuB, Eur. J. Biochem., 130:409.
Vijgenboom, E., and Bosch, L., 1987, Transfer of plasmid-borne *tuf* mutations to the chromosome as a genetic tool for studying the function of EF-TuA and EF-TuB in the *E. coli* cell, Biochimie, 69:1021.
Vijgenboom, E., Nilsson, L. and Bosch, L., 1988, The elongation factor EF-Tu from *E. coli* binds to the upstream activator region of the tRNA-*tufB* operon, Nucl. Acids Res., 16, in press.
Wolf, H., Chinali, G., and Parmeggiani, A., 1974, Kirromycin, an inhibitor of protein biosynthesis that acts on elongation factor Tu, Proc. Natl. Acad. Sci. USA, 71:4910.
Yokota, T., Sugisaki, H., Takanami, M., and Kaziro, Y., 1980, The nucleotide sequence of the cloned *tufA* gene of *Escherichia coli*, Gene, 12:25.

STRUCTURE-FUNCTION RELATIONSHIPS OF THE GTP-BINDING DOMAIN OF ELONGATION FACTOR Tu

Pieter H. Anborgh, Robbert H. Cool, Eric Jacquet, Michael Jensen*, Giuseppe Parlato¶ and Andrea Parmeggiani

Laboratoire de Biochimie, Ecole Polytechnique, Laboratoire Associé du C.N.R.S. n° 240 F-91128 Palaiseau Cedex, France; *Division of Biostructural Chemistry, Aarhus University, D-8000 Aarhus, Denmark; ¶Istituto di Biochimica Fisica e Patologia Molecolare e Cellulare, Facoltà di Medicina e Chirugia, I-88100 Catanzaro, Italy

INTRODUCTION

Elongation factor Tu (EF-Tu) is a multifunctional enzyme, essential for protein synthesis in which it acts as the carrier of aa-tRNA to the ribosome·mRNA complex (Miller & Weissbach, 1977; Bosch et al., 1983; Parmeggiani & Swart, 1985). EF-Tu belongs to the class of guanine nucleotide-binding proteins which in recent years have been found to play a crucial role in controlling the transmission of information in fundamental processes of the eucaryotic and procaryotic cell, such as growth, hormone response, neurotransmission, membrane transport and protein synthesis (for refs : Masters *et al.*, 1986; Dever *et al.*, 1987; Gilman, 1987; Barbacid, 1987). In pathological processes guanine nucleotide binding proteins are involved in the oncogenic transformation of the human cell (Scolnick *et al.*, 1979) and are encoded by the HIV retrovirus (Guy *et al.*, 1987). In all of these proteins, the active form needed for the interaction with the various ligands, is the complex with GTP; their intrinsic GTPase activity cleaving the γ-phosphate is therefore determinant for controlling their activity. The primary structure of this class of proteins shows a pronounced homology which in most cases affects the first 150-180 N-terminal amino acid residues. Functional, immunological and structural studies, and comparison with nucleotide binding proteins have led to the identification within these homologies of a consensus sequence involved in the binding of the substrate GTP/GDP (McCormick *et al.*, 1985; Dever *et al.*, 1987).

EF-Tu has been taken as a reference model for this class of proteins, because it is A) the best studied guanine nucleotide binding protein and B) until recently it has been the only one, of which the threedimensional (3-D) structure of the GTP binding domain had been elucidated at high resolution by X ray diffraction analysis (la Cour *et al.*, 1985; Jurnak, 1985). The polypeptide chain of EF-Tu (393 amino acid residues, M.W. = 43 kDa) is folded in three distinct domains (Figure 1A), of which the N-terminal domain has primary and secondary structure common to other guanine nucleotide binding proteins. The 3-D model of this domain as part of a nicked EF-Tu·GDP complex displays an α/β type structure, a common situation for nucleotide binding proteins. The GDP binding pocket is constituted by four loops connecting the C-terminus of β-strands with the N-terminus of α-helices, including the consensus motifs responsible for binding the guanine nucleotide. The first two (<u>Gly18</u>-His-Val-Asp-His-<u>Gly-Lys24</u> and <u>Asp80</u>-Cys-Pro-<u>Gly</u>-His84) are involved in the interaction with the phosphoryl groups of GDP/GTP, the third (<u>Asn-135-Lys</u>-Cys-<u>Asp138</u>) and the fourth ones (<u>Ser173-Ala174</u>, especially concerning EF-Tu and the *ras* products) interact with the base. The 3-D model of the GTP binding domain of EF-Tu has been taken as a reference for tracing the 3-D structure of other GTP-binding proteins such as the *ras* product protein p21 (McCormick *et al.*, 1985), the bacterial initiation factor 2 (Cenatiempo *et al.*, 1987) and transducin (Hingorani & Ho, 1987). Recently the 3-D model of a truncated human p21 has also been elucidated at high resolution. Although it is still too early to draw a true comparison between the two structures, the authors (deVos *et al.*, 1988) report that part of the p21 model, such as the region of the first 25 residues and the spanning residues from 78 to 142, shares similarities with the homologous structures of the EF-Tu model.

The genetic, functional and structural situation of EF-Tu led us to start a research program, directed to investigate the structure-function relationships of EF-Tu and developed in collaboration with Prof. L. Bosch and Prof. B.F.C. Clark. The first priority was the isolation of the guanine nucleotide binding domain of EF-Tu, via genetic manipulations. Further goals were the isolation of EF-Tu factors deleted either of the Middle domain (EF-TuΔM) or of the C-terminal domain (EF-TuΔC), as well as the introduction of substitutions in the guanine nucleotide binding pocket, following the indications of the X-ray diffraction analysis. In this article we give a survey of the actual state of our research, paying particular attention to the isolation and characterization of the GTP binding domain of EF-Tu and to the replacement of critical residues in the guanine nucleotide binding site such as Val20→Gly. For details refer to Parmeggiani *et al.*, 1987; Swart *et al.*, 1987; Jacquet and Parmeggiani, 1988).

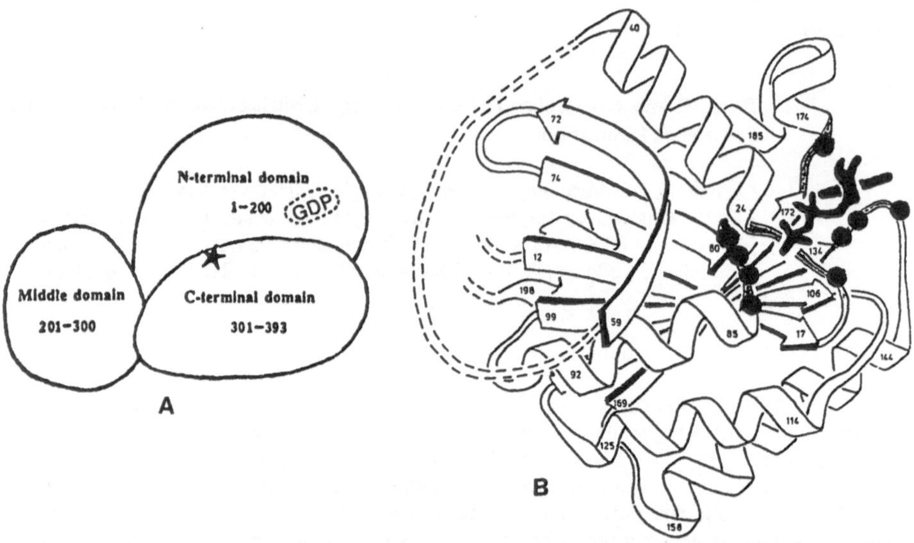

Figure 1A.- Scheme of the three domains of EF-Tu with the GDP binding site in the N-terminal domain (named G domain when separated). The star indicates position 375, responsible for kirromycin resistance. **B.**- Threedimensional structure of the N-terminal domain of nicked EF-Tu·GDP complex derived from X-ray diffraction analysis. The solid circles indicate the point substitutions in the four loops constituting the guanine nucleotide binding pocket. This figure is derived from Duisterwinkel *et al.*, 1984 and la Cour *et al.*, 1985.

EXPERIMENTAL AND DISCUSSION

Construction and Overproduction of the EF-Tu Mutants

All introduced mutations concerned the *tufA*, one of the two almost identical genes coding for EF-Tu in *E. coli*.. A 2kb DNA fragment containing the structural *tufA* gene, with or without the vicinal secondary promotor, was cloned in the pEMBL, a family of plasmids susceptible to be secreted from the cell as single stranded DNA (Dente *et al.*, 1983). Deletions and point substitutions of *tufA* were obtained via synthetic oligodeoxynucleotides hybridized to a single stranded *tufA* exposed in a gapped duplex (Kramer *et al.*, 1982). Selection of the clones containing the mutated plasmids was performed by cell colony hybridisation, followed by DNA sequencing. Recently the method of Kunkel (1985), adapted to a gapped duplex, has been used systematically in order to digest the wild-type single-strand DNA of the gapped duplex. This increases the percentage of mutant clones from 0.1-5 % to 10-20%. When the Kunkel variation was applied, DNA sequencing was in most cases sufficient for screening the transformants, making cell colony hybridisation superfluous. The mutated *tufA* gene was transferred to the expression runaway vector pCP40 (Remaut *et al.*, 1983) under control of a thermoinducible λP_L or to the vector pTTQ18 (Stark, 1987) under control of the *tac* promotor. The host cell for overproduction was usually *E. coli* strain 71/18, in the case of deleted EF-Tu factors, or in the case of point substitution in the intact EF-Tu *E. coli* PM455, a strain carrying only one active *tuf* gene (*tufA*), coding for a kirromycin resistant product (see below). We also used strain PM505, containing only one active *tuf* gene (*tufA*) coding for a kirromycin-sensitive EF-Tu, as well as the corresponding rec⁻ variants. These strains were constructed by the group of Prof. Bosch (van der Meide *et al.*, 1982 and 1983).

Problems Associated with the Overproduction and Isolation of Mutated EF-Tu factors

There are a great deal of difficulties in constructing and overproducing soluble EF-Tu mutants in the *E. coli* cell. The most frequent ones are: the progressive appearance of revertants, inefficient overproduction and occurrance of unsoluble overproduced mutant factors. These problems are in a way not surprising, if we consider that EF-Tu, the most abundant protein in *E. coli* (5-10% of the total cell proteins), is a crucial component for cell growth. Competition phenomena between the mutated product with altered functions and the chromosomic product is the most probable explanation for the toxicity of the overproduced EF-Tu mutant. In line with this is the general observation that cells expressing plasmid-borne mutated factors usually display a slower growth than strains carrying plasmid-borne wild-type EF-Tu. Procedures for isolating and producing mutant EF-Tu factors may therefore turn out to be complicated, time-consuming and not seldomly disappointing. *Rec⁻* strains have occasionally been used to reduce the occurrance of revertants. As a general rule, in the cell preparations presence of the mutation was again carefully controlled by DNA sequencing prior to the start of the purification procedure.

To improve overproduction and solubility of mutant factors, critical parameters such as induction kinetics and temperature conditions must be investigated in details. A product is considered to be soluble, if still present in the supernatant after centrifugation at 100,000g for at least 90 min.

Since chromatographic methods based on charge or molecular weight differences are unable to distinguish between point substitutions, for purifications of EF-Tu mutants differing from the wild-type factor by only one point substitution we routinely used the method set up in our laboratory to separate from each other the two mutant species produced by the chromosomic *tufA* and *tufB* (Swart *et al.*, 1987). This method is based on the different affinities for kirromycin of the two products and on the competition between the antibiotic and EF-Ts for binding to EF-Tu. It allows separation of two proteins regardless of charge or molecular weight differences, only requiring the use of cell strains with a chromosomic EF-Tu having a different kirromycin-sensitivity from the plasmid-borne product (Swart et al., 1987). This condition was fulfilled by utilizing strains PM455 or PM505 (see above).

Isolation of the GTP Binding Domain of EF-Tu and of EF-Tu Factors lacking the Middle or C-terminal Domain

The GTP binding domain of EF-Tu (G domain = EF-TuΔ(C,M) = 22 kDa) was isolated as a protein of 203 amino acid residues and consists of the first 202 N-terminal residues and the C-terminal Gly of EF-TuA (Parmeggiani *et al.*, 1987). The overproduced G domain (30-40 % of the total cell proteins) was in large part soluble (60-80 %). At -30 °C in 50% glycerol, 50 mM imidazole-acetate buffer, pH 7.8, 10 mM $MgCl_2$, 50 mM NH_4Cl, 5 mM dithiothreitol and, most important, 5 μM GDP was stable for at least several months. The basic properties of EF-Tu, such as binding of GDP/GTP and GTP hydrolysis are conserved. Striking differences become apparent with respect to the allosteric mechanisms regulating the activity of the intact molecule. The affinity for GDP has decreased more than thousand times whereas that of GTP only 10 times; as a result, one of the most typical characteristicts of EF-Tu, the ability to distinguish between GDP and GTP by a difference in affinity of almost three orders of magnitude, is practically abolished. These results are resumed in Table I which for comparison also report corresponding values reported for protein p21. Like EF-Tu, the G domain has intrinsic GTPase activity. This activity displays a multiple-round turnover with a linear course for several hours. The turnover rate of the G domain is higher than the intrinsic activity of EF-Tu, as a consequence of the new equilibrium situation which no longer favors the G domain complex with GDP over that with GTP.

We tested the action of several ligands of EF-Tu, such as aa-tRNA, ribosomes, elongation factor Ts (EF-Ts) and kirromycin, and found that only the ribosome was still able to exert on the G domain the same kind of effect (a stimulation) as on the intact molecule, though to a smaller extent. The influence on the GTPase activity and dissociation rates of the complexes with GTP and GDP exerted by the other ligands was small, as compared to the intact molecule, and of different kind. These results show that the Middle and C-terminal domain of EF-Tu are essential for the allosteric regulation of the guanine nucleotide binding and for the effects induced by aa-tRNA, EF-Ts or kirromycin. By contrast, the ribosome, whose action is mainly directed on the catalytic center of EF-Tu appears to have conserved in part its physiological action.

Table I

Apparent dissociation constants (K'_d) of the GTP and GDP complexes of the G domain relative to EF-Tu and p21

complex	K'_d (μM)				
	G domain[1]		**EF-Tu**[1]		**p21**[2]
temperature (°C)	0	30	0	30	30
GTP complex	6	4	0.5	0.6	0.012
GDP complex	2	3	0.001	0.006	0.013

1 Parmeggiani *et al.*, 1987 and unpublished results
2 Trahey *et al.*, 1987

We have also constructed the two combinations **EF-TuΔC** (33 kDa) and **EF-TuΔM** (33.5 kDa). So far only the former mutant, which is about 50 % soluble, has been overproduced and purified. Preliminary experiments indicate that the affinity for GDP and GTP of **EF-TuΔC** is lower than that shown by the G domain, by one order of magnitude. These results together with the results obtained with the G domain suggest that the elimination of the C terminal domain considerably influences the conformation of the guanine nucleotide binding site and that the presence of only the Middle domain cannot induce the physiological conformation of this site. The characterization of the other properties of **EF-TuΔC**, particularly the action of the different ligands of EF-Tu is in progress.

Point Substitutions Introduced in EF-Tu or its G domain

In this section we resume the EF-Tu mutants constructed up to now, mentioning the actual state of our research.

-EF-TuVal20→Gly and G domainVal20→Gly. See next section.

-EF-TuThr25→Ala. This mutant has been overproduced, but only partially purified. Preliminary results seems to indicate that it is not active in protein synthesis.

-EF-TuCys81→Gly. Considerable difficulties have been met in the overproduction of this mutant, due to frequent occurrance of revertants which emphasize the importance of this residue.

-G domainPro82→Thr. Overproduced and purified in preparative amounts. A very strong decrease of the GTPase activity (<10%) is associated with an increased affinity for the substrate GDP/GTP.

-G domainHis84→Gly. Overproduced and purified in preparative amounts. This residue plays an important role in the catalytic activity of EF-Tu. The affinity for GDP/GTP of the mutant is essentially unchanged, whereas the GTPase activity is strongly decreased (<10%) but is nevertheless enhanced to 15-20 % of the control with increasing the concentration of monovalent cations to 2 M.

-EF-TuAsn135→Asp and EF-TuAsn135/Lys136→Asp/Ile. Obtention of these two mutants is very difficult, due to the occurrance of revertants and to the insolubility of the overproduced proteins. However, it still remains an important aim of our work.

-EF-TuAsp138→Asn and G domainAsp138→Asn. This mutated EF-Tu has been overproduced and partially purified. Its complex with GDP shows a much faster dissociation rate than that of wild-type EF-Tu and can interact with XTP/XDP, in line with the observations of Hwang and Miller (10). The overproduced mutated G domain is mostly insoluble.

-G domainSer173→Gly. The overproduced, but not yet purified protein is partially soluble.

-G domainAsp80→Gly, G domainGln114→Glu, G domainGlu117→Gln, EF-TuHis66→Gly and EF-TuHis118→Gly, EF-TuAsp138→Trp. These mutants have yet to be overproduced.

Mutation of Val20, the Residue Homologous to Position 12 in p21

The first consensus element of the GTP binding pocket at the N-terminal end of EF-Tu includes residues 18-24 (Gly18-His-Val-Asp-His-Gly-Lys24) and is involved in the interaction with the phosphoryl groups of the substrate. According to primary sequence alignment, position 20 (Val) of EF-Tu is homologous to position 12 (Gly) of protein p21 (McCormick *et al.*, 1985). With respect to this position, therefore, wild-type EF-Tu (EF-TuVal20) corresponds to the oncogenic variant of p21 having Gly12 replaced by Val. In p21, replacement of Gly with Val induces oncogenic transformation of the mammalian cell (Tabin et al., 1982; Taparowsky et al., 1982). *In vitro* experiments show that this mutation strongly decreases the intrinsic GTPase activity of p21 and, as very recently shown, reduces the dissociation rates of the EF-Tu·GDP complex (Satoh *et al.*, 1988). This situation suggested that Val20 of EF-Tu could be an interesting position for investigating the function-structure relationships not only of the factor but also of other guanine nucleotide binding proteins. To this purpose we have substituted Val20→Gly and purified to homogeneity the overproduced EF-TuGly20 (Jacquet & Parmeggiani, 1988).

Table II

GTPase activity of mutant EF-TuGly20 and wild-type EF-TuVal20. Effect of kirromycin and ribosomes[1]

system	mmol GTP hydrolyzed·min^{-1}·mol^{-1} EF-Tu
EF-TuVal20	1.5
EF-TuVal20 plus ribosomes	7
EF-TuVal20 plus kirromycin	44
EF-TuVal20 plus kirromycin and ribosomes	150
EFTuGly20	0.4
EF-TuGly 20 plus ribosomes	0.3
EF-TuGly20 plus kirromycin	5.0
EF-TuGly20 plus ribosomes	5.5

[1] Derived from Jacquet and Parmeggiani (1988). The reaction mixture (100 µl) containing 50 mM imidazole acetate, pH 7.5, 5 mM $MgCl_2$, 50 mM KCl, 0.5 mM dithiothreitol, 4.2 µM GTP (specific activity, 1640 cpm·pmol^{-1}), 0.56 µM EF-TuGly20 or EF-TuVal20, and when indicated 50 µM kirromycin and /or 0.4 µM ribosomes was incubated at 37 °C. Samples (20 µl) were withdrawn at 15, 30, 45 and 60 min and the $^{32}P_i$ liberated was measured by the charcoal method.

Our results show that also in EF-Tu mutation Val20→Gly affects the interaction with GTP/GDP. However, differently from the p21, the GTPase activity is strongly inhibited (Table II). Since this replacement introduces a residue homologous to that found in the protooncogenic p21, this effect is opposite to that observed in p21. On the other hand as in EF-Tu, in p21 the presence of a Gly in the homologous position increases the dissociation rates of the EF-Tu·GDP and p21·GDP complex; in EF-Tu 25 times and in p21 2-3 times (Table III). In EF-Tu only, this phenomenon is associated with a remarkable increase of the association

Table III

Dissociation constants, association and dissociation rate constants for the complexes of EF-TuGly20 and wild-type EF-TuVal20 with GTP or GDP[1]

EF-Tu complexes	Association rate constants $10^{-4} \cdot k'_{+1}$ ($M^{-1} \cdot s^{-1}$)	Dissociation rate constants $10^{4} \cdot k'_{-1}$ (s^{-1})	Dissociation constant K'_d (nM)
EF-TuGly20·GTP (EF-TuVal20·GTP)	0.7 (1)	11.8 (59)	170 (590)
EF-TuGly20·GDP (EF-TuVal20·GDP)	260 (26)	53 (2.3)	2 (0.9)

[1] The values, determined at 0 °C, were derived from Jacquet and Parmeggiani (1988)

rate of EF-Tu·GDP, resulting in a situation reminiscent of the effect induced by EF-Ts on the EF-Tu·GDP complex (Fasano *et al.*, 1978). As compared to EF-Tu·GDP, the modifications in EF-Tu·GTP are different and of modest extent: the dissociation rate is decreased by four times and the association rate only slightly reduced. The equilibrium of the complexes with GDP and GTP, measured as the K'_d, show some alteration (between 2-4 times) which in part corresponds to that reported for the p21. It should be emphasized that in EF-TuGly20 the difference between the affinity for GTP and GDP, though reduced by about 7 times still remains of two orders of magnitude.

The observed differences between EF-Tu and p21, suggest a different situation in the primary and tertiary structure of the guanine nucleotide binding sites. Comparison of the first consensus element, involved

Table IV

Activity of EF-TuGly20 in poly(Phe) synthesis. Effect of elongation factor Ts[1]

system	mol Phe incorporated·$min^{-1} \cdot mol^{-1}$ EF-Tu
EF-TuVal20 plus EF-Ts	2.5
EF-TuVal20 minus EF-Ts	0.06
EF-TuGly20 plus EF-Ts	0.8
EF-TuGly20 minus EFTs	0.55

[1] Derived from Jacquet and Parmeggiani (1988). The reaction mixture (300 µl) contained 50 mM imidazole acetate, pH 7.5, 40 mM NH_4Cl, 50 mM KCl, 10 mM $MgCl_2$, 0.011 µM EF-TuGly20 or 0.011 µM wild-type EF-TuVal20, plus or minus 0.17 µM EF-Ts, 18 µg poly(U), 0.22 µM ribosomes, 0.15 µM EF-G, 0.5 mM dithiothreitol, 2 mM phosphoenolpyruvate, 9 µg pyruvate kinase, 1 mM GTP, 1 mM ATP, 2 µM $tRNA^{Phe}$, 5 µM [^{14}C]Phe (specific activity, 468 cpm·$pmol^{-1}$) and purified Phe-tRNA synthetase. Incubation was at 37 °C. At given times, a sample (50 µl) was pipetted onto a filter and the poly(Phe) synthesized measured.

in the interaction with the phosphate groups of the guanine nucleotide, show the presence of several variable residues (EF-Tu: G18-H-V-D-H-G-K24; p21: G10-A-G-G-V -G-K16), which together with the other structures of the two molecules appear to be determinant for inducing the specific differences observed in the common functions of the two proteins. We like here to mention that in both proteins, the residue(s) directly involved in the catalysis have yet to be identified. The 3-D studies seem to exclude for the p21 that the second consensus motif, the loop 57-61 is directly involved in the catalytic process (deVos *et al.*, 1988), a situation that may occur for the homologous element (loop 80-84) in EF-Tu.

Of the EF-Tu ligands, aa-tRNA and kirromycin show essentially the same action as on the wild-type, factor, whereas the ribosome is unable to stimulate the GTPase activity of EF-Tu. However, the observation that EF-TuGly20 is still active in poly(Phe) synthesis (Table IV), though at much lower rate than the wild-type factor, indicates that the physiological GTP hydrolysis induced by the interaction between the ternary complex EF-Tu·GTP·aa-tRNA and the ribosome·mRNA complex can still take place. Apparently the contraints imposed by the codon-anticodon interaction are able to compensate in part for the deficient interaction with the ribosome. This situation is reminescent of that reported for the EF-TuBo (an EF-TuBGly222→Asp), whose intrinsic GTPase activity can be enhanced by the ribosome only in the presence of mRNA, that is when the codon-anticodon interaction takes place (Swart *et al.*, 1987). Another interesting finding is that the activity of EF-TuGly20 in poly(Phe) synthesis is little affected by the presence of EF-Ts. Its absence decreases the rate of poly(Phe) synhesis by only 30 %, whereas with wild-type EF-TuVal20 poly(Phe) is essentially abolished (Table IV). This confirms the conclusion drawn from the dynamics of the EF-Tu·GDP complex that substitution Val20→Gly induces an EF-Ts-like conformation of the EF-Tu molecule (see above).

CONCLUSIONS

A great deal of work is still needed to enable us to draw a model comprehensive of the basic structure-function relationships in EF-Tu with respect to the interaction with the substrate, and the different ligands. Nevertheless, in spite of the considerable difficulties we start to have some essential data concerning the functional role of the three domains as well as the implications of the single loops and residues participating in the formation of the substrate binding site and in the catalytic activity. Difficulties were in a way expected, since we are dealing with a system which expresses and overproduces in *E. coli* an enzyme which is the most abundant protein and one of the essential enzymes of *E. coli* itself.

As main targets for future work, we will continue the characterization of the structure-function relationships of the domains as well as of the substrate binding pocket, paying particular attention to the structural and functional homologies with the other guanine nucleotide binding proteins. Moreover, mutations will be introduced such to affect A) the thermostability of EF-Tu and B) its sensitivity to proteolytic phenomena. Concerning the former aspect, we plan to substitute Val88 with Asp and Leu121 with Lys with the purpose to introduce a salt bridge which may affect the thermostability of the EF-Tu molecule. With respect to sensitivity to proteolysis, we will substitute Arg44 and Arg55 in order to construct a factor resistant to proteolytic phenomena, to which the wild-type molecule is exposed. This may allow us to isolate crystals of intact EF-Tu, avoiding the usually occurring proteolytic nicks. Other methods to facilitate the isolation of mutated EF-Tu, such as gene displacement or overproduction of the mutated EF-Tu in organisms other than *E. coli* will be explored.

ACKNOWLEDGEMENTS

In addition to the authors, a number of other scientists essentially contributed to the work described in this article. Drs. G. W. M. Swart and K. K. Mortensen carried out the engineering of the G domain and Dr. M. Merola participated in setting up the mutagenesis procedure, obtaining G domainAsp80→Gly. EF-TuHis66→Gly and EF-TuHis118→Gly were obtained by Dr. J. Jonak, G domainGlu114→Gln and G domainGln117→Glu by Dr. K. Harmark, EF-TuSer173→Gly by F. Gümüsel and EF-TuAsp138→Trp by A. Weijland. We are deeply indebted to Prof. L. Bosch and his collaborators for the gift of biological materials. We like to express to them as well as to Prof. B. F. C. Clark and Prof. R. Cortese, Drs. T. F. M. la Cour, J. Nyborg, M. Kjeldgaard and L. Dente our gratitude for the close collaboration during the course of this work. This research project has been carried out in the framework of the grant BAP-0066-F (CD) from the Biotechnology Action Program of the Commission of the European Community. One of us (E.J) is recipient of a long-term Fellowship from the "Association pour la Recherche sur le Cancer".

REFERENCES

Barbacid, M., 1987, *ras* Genes, *Ann. Rev. Biochem.* 56:779-827

Bosch, L., Kraal, B., van der Meide, P., Duisterwinkel, F.J., and van Noort, J. M., 1983, The elongation factor EF-Tu and its two encoding genes, *Prog. Nucleic Acid Res. Mol. Biol.*, 21:91-126

Cenatiempo, Y., Deville, F., Dondon, J. Grunberg-Manago, M., Sacerdot, C., Hershey, J. W. B., Hansen, H. F., Petersen, H. J. U., Clark, B. F. C., Kjeldgaard, M., la Cour, T. F. M., Mortensen, K. K., and Nyborg, J., 1987, The protein synthesis initiation factor 2 G domain. Study of a functionally active C-terminal 65-kilodalton fragment of IF-2 from *E. coli*, *Biochemistry*, 26:5070-5076

Dente, L., Cesareni, G., and Cortese, R., 1983, pEMBL: a new family of single stranded plasmids, *Nucleic Acids Res.*, 11:1645-165

Dever, T. E., Glynias, M. J., and Merrick, W. C., 1987, GTP-binding domain: Three consensus sequence elements with distinct spacing, *Proc. Natl. Acad. Sci. USA*, 84:1814-1818

Duisterwinkel, F. J., Kraal, B., De Graaf, J. M. Talens, A., Bosch, L., Swart, G. W. M., Parmeggiani, A., la Cour, T. F. M., Nyborg J., and Clark, B. F. C., 1984, Specific alterations of the EF-Tu polypeptide chain considered in the light of its three-dimensional structure, *EMBO J.*, 3:113-120

de Vos, A. M., Tong, L., Milburn, M. V., Matias, P. M., Jancarik, J., Noguchi, S., Nishimura, S., Miura, K., Ohtsuka, E., and Kim, S.-H., 1988, Three-dimensional structure of an oncogene protein: catalytic domain of human c-H-*ras* p21, *Science*, 239:888-893

Fasano, O., Bruns, W., Crechet, J. B., Sander, G., and Parmeggiani, A., 1978, Modification of elongation factor Tu-Guanine nucleotide interaction by kirromycin: a comparison with the effect of aminolacyl-tRNA and elongation factor Ts, *Eur. J. Biochem.*, 89:557-565

Gilman, A. G., 1987, G proteins: Transducers of receptor-generated signals, *Ann. Rev. Biochem.*, 56:615-649

Kramer, W., Shughart, K., and Fritz, H.-J., 1982, Directed mutagenesis of DNA cloned in filamentous phage: influence of hemimethylated GATC sites on marker recovery from restriction fragments, *Nucleic Acid Res.*, 10:6475-6485

Kunkel, T. A., 1985, Rapid and efficient site-specific-mutagenesis without phenotypic selection, *Proc. Natl. Acad. Sci. USA*, 82:488-492

Guy, B., Kieny, M. P., Riviere, Y., le Peuch, C., Dott, K., Girard, M., Montagnier, L., and Lecocq, J.-P., 1987, HIV F/3' *orf* encodes a phosphorylated GTP-binding protein resembling an oncogene product, *Nature* 330:266-269

Hingorani, V.N., and Ho, Y.K., 1987, A structural model for the α–subunit of transducin, *FEBS Lett.*, 220:15-21

Jacquet, E., and Parmeggiani, A., 1988, Structure-function relationships in the GTP-binding domain of EF-Tu: mutation of Val20, the residue homologous to position 12 in p21, *EMBO J.*, in print

Ivell, R., Sander, G. ,and Parmeggiani, A., 1981, Modulation by monovalent and divalent cations of the guanosine-5'-triphosphatase activity dependent on elongation factor Tu, *Biochemistry*, 20:6852-6859

Jurnak, F., 1985, Structure of the GDP domain of EF-Tu and location of the amino acids homologous to *ras* oncogene proteins, *Science*, 230:32-36

la Cour, T. F. M., Nyborg, J., Thirup, S., and Clark, B. F. C., 1985, Structural details of the binding of guanosine diphosphate to elongation factor Tu from *E. coli* as studied by X-ray crystallography, *EMBO J.*, 4:2385-2388

Masters, S., Stroud, R. M., and Bourne, H. R., 1986, Family of G protein a chains: amphipathic analysis and predicted structure of functional domains, *Protein Engineering* 1:47-54

McCormick, F., Clark, B. F. C., la Cour, T. F. M., Kjeldgaard, M., Norskov-Lauritsen, L., and Nyborg, J., 1985, A model for the tertiary structure of p21, the product of the ras oncogene, *Science* 230:70-82

Miller, D. L., and Weissbach, H., 1977, Factors involved in the transfer of aminoacyl-tRNA to the ribosomes, in : "Molecular mechanisms in protein biosynthesis", Weissbach, H. and Pestka, S. (eds), Academic Press, New York

Parmeggiani, A. and Swart, G. W. M., 1985, Mechanism of action of kirromycin-like antibiotics, *Ann. Rev. Microbiol.*, 39:557-577

Parmeggiani, A., Swart, G. W. M., Mortensen, K. K., Jensen, M., Clark, B. F. C., Dente, L., and Cortese, R., 1987, Properties of a genetically engineered G-domain of elongation factor Tu, *Proc. Natl. Acad. Sci. USA*, 84:3141-3145

Remaut, E., Tsao, H., and Fiers, W., 1983, Improved plasmid vectors with a thermoinducible expression and temperature-regulated runaway replication, *Gene*, 22:103-112

Satoh, T., Nakamura, S., Nakafuku, M., and Kaziro, Y., 1988, Studies on *ras* proteins. Catalytic properties of normal and activated *ras* proteins purified in the absence of protein denaturants, *Biochim. Biophys. Acta*, 949:97-109

Scolnick, E. M., Papageorge, A. G., and Shih, T. Y., 1979, Guanine nucleotide binding activity as an assay for *src* protein of rat-derived murine sarcome virus, *Proc. Natl. Acad. Sci. USA* , 76:5335-5339

Seeburg, P. H., Colby, W. W., Capon, D. J., Goeddel, D. V., and Levinson, A. D., 1984, Biological properties of human c-Ha-*ras*1 genes mutated at codon 12, *Nature (London)*, 312:71-75

Stark, M.J.R., 1987, Multicopy expression vectors carrying the *lac* repressor gene for regulated high-level expression of genes in *E. coli*, *Gene*, 51:255-267

Swart, G.,W.,M., Parmeggiani, A., Kraal, B., and Bosch, L., 1987, Effects of the mutation Gly-222→Asp on the functions of elongation factor Tu, *Biochemistry* , 26:2047-2054

Tabin, C. J., Bradley, S. M., Bargmann, C. I., Weinberg, R. A., Papageorge, A. G., Scolnick, E. M., Dhar, R., Lowy, D. R., and Chang, E.H., 1982, Mecanism of activation of a human oncogene, *Nature (London)*, 300:143-149

Trahey, M., Milley, R. J., Cole, G. E., Innis M., Paterson, H., Marshall, C. J., Hall, A., and McCormick, F.,1987, Biochemical and biological properties of the human N-*ras* p21 protein, *Mol. Cell. Biol.*, 7: 541-544

Taparowsky, E., Suard, Y., Fasano, O., Shimizu, K., Goldfarb, M., and Wigler, M., 1982, Activation of the T24 bladder carcinoma transforming gene is linked to a single amino acid change, *Nature (London)*, 300:762-765

van der Meide, P. H., Vijgenboom, E., Dicke, M., and Bosch, L., 1982, Regulation of the expression of *tufA* and *tufB*, the two genes coding for the elongation factor EF-Tu in *E. coli*, *FEBS Lett.*, 139:325-330

van der Meide, P. H., Kastelein, R. A., Vijgenboom, E., and Bosch, L., 1983, *tuf* gene dosage effects on the intracellular concentration of EF-TuB, *Eur. J. Biochem.*, 130:409-417

A MUTATION THAT HINDERS THE GTP INDUCED AMINOACYL-tRNA BINDING OF ELONGATION FACTOR TU

Yu-Wen Hwang, Frances Jurnak* and David L. Miller

Molecular Biology Department, NYS Institute for Basic Research, 1050 Forest Hill Rd., Staten Island, New York 10314, USA. *Department of Biochemistry, University of California, Riverside, California, 92521, USA

INTRODUCTION

Elongation factor Tu (EF-Tu) is a member of the guanine-nucleotide binding protein family. This group of proteins include translational factors[1], the ras gene family[2], and signal transducing proteins[3]. Guanine-nucleotide binding proteins exhibit diverse functions but possess common structural elements. Comparing the peptide sequences of these proteins reveals several regions with remarkable amino acid sequence similarity[4-6]. These regions have been ascribed to the guanine-nucleotide binding domains[4-6].

GTP induces the active conformations of these proteins. A subsequent reaction of the active conformation with the target factor stimulates the hydrolysis of GTP and produces the inactive GDP-bound form. The peptide sequence similarity together with the generality of the recycling mechanism argue that the molecular details of the GTP-induced conformational change might be common to all guanine-nucleotide binding proteins.

EF-Tu promotes the binding of aa-tRNA to the ribosomes when activated by GTP. Interaction with programmed ribosomes stimulates GTP hydrolysis, which produces the inactive GDP-bound state of EF-Tu. EF-Ts catalyzes the guanine-nucleotide exchange that recycles the EF-Tu[7]. During the elongation process, EF-Tu interacts sequentially with GTP, aa-tRNA, the ribosome, GDP and EF-Ts. Because of its multiple functions, EF-Tu provides a good model not only for studying guanine-nucleotide binding but also for studying protein-protein and protein-nucleic acid interactions. Models of EF-Tu's guanine-nucleotide binding domains have been derived by X-ray crystallography[8,9]. According to these models, conserved residues 18-26 are located adjacent to the pyrophosphate group, conserved residues 79-85 positions near the beta-phosphate and residues 135-138 interact directly with substituents on the guanine ring.

The X-ray models implicate the involvement of specific amino acid residues in interacting with guanine-nucleotides. The contribution of these residues in guanine-nucleotide binding can be evaluated by amino acid substituion using site-directed mutagenesis. Previously, we have verified the specific interaction between Asp-138 of EF-Tu and the 2-amino group of guanine-nucleotide by demonstrating that changing Asp-138 to Asn leads to the switch of base specificity from guanine to xanthine[10]. In this study, we described the properties of a mutant altered at residue Gly-83, a conserved glycine residue in guanine-nucleotide binding proteins.

MATERIALS AND METHODS

Mutant Construction. Site specific mutant Ala-83 was constructed by a modification of the procedure of Zoller and Smith as described previously[10]. A 15-mer oligonucleotide (GTGCGCCGGGCAGTC) with a single base mismatch (underlined) with the *tuf*A gene was used as the mutagenic primer. The mutant was screened by hybridizing it to the mutagenic primer as outlined below. Single stranded phages from appropriately plated plaques were transferred to nitrocellulose membrane and processed according to Maniatis, et. al.[11]. Prehybridization was carried out in a solution containing 6X SSC, 5X Denhardt's solution and 0.1% SDS at 48 ^{o}C for 2 hr. Hybridization was performed under the same conditions as prehybridization except with added ^{32}P-labeled probe ($5x10^6$ CPM/ml of hybridization solution). After hybridization, the membrane was washed in 6X SSC at room temperature for 3 min and then washed in the same solution at 58 ^{o}C for 15 min. The final washing temperature was empirically determined as it was substantially higher than the predicated Tm. The reason for this deviation might be due to high GC content in this region. Positive clones were identified by autoradiography. Several positive clones were selected and sequenced by the dideoxynucleotide chain termination method[12] to confirm the mutation site. A 16-mer (GGATGGGCTCACGAGT) complementary to a region about 100 bps downstream from the mutation site was used as the primer for sequencing. Subsequently, the EcoRI-HindIII fragment that contained the EF-Tu Ala-83 mutation was cleaved from M13 phage and subcloned into plasmid vector pTZ18U[14] as described[10].

Maxicell Expression. The procedures used for maxicell expression were as described[10] except that baterial strain MV1190 (Δ(*lac-pro*)Δ(*srl-rec*A)::Tn10 *thi* *sup*E/F' *tra*D36 *pro* *lac*I^qZΔM15) was used instead of strain SR58 (F^- *uvr*B5 *rec*A56 *rps*L *arg* *his* *leu* *pro* *thi*).

EF-Ts EF-Tu Complex Dissociation and Phe-tRNA Binding Assays. These assays were performed exactly as previously described[10]. GTP was treated with phosphoenolpyruvate (PEP) and pyruvate kinase to remove contaminating GDP before use in the complex dissociation assays. The amount of GDP in GTP was measured as described[14]. The residual GDP in the GTP that used in this report was less then $4X10^{-4}$ mole fraction.

Kirromycin Sensitivity Assay. *E. coli* strain HW123, a strain containing kirromycin resistant *tuf*A gene, a deleted *tuf*B gene and *lac*I^q repressor, was used as the host for testing kirromycin sensitivity instead of the previously described strain LBE2012[10,15]. The construction of strain HW123 will be described in detail elsewhere (Hwang et. al., in preparation). The assays were carried out by transforming strain HW123 to ampicillin resistance with the testing plasmid and then screening the transformants for kirromycin sensitivity on YT plates containing the indicated amounts of kirromycin. EDTA at a concentration of 2 mM was included in the plates to prevent resistant cells from growing due to transport failure.

Trypsin Cleavage. Tryptic cleavages of EF-Tu GTP complex were performed by digesting 60 ug of crude maxicell lysate proteins in 40 ul of 50 mM Tris-HCl, pH 7.4, 10 mM $MgCl_2$, 50 mM NH_4Cl, 1mM DTT, 2mM GTP, 5mM PEP and 10 ug pyruvate kinase with 0.075 ug of TPCK-treated trypsin at 32 ^oC for various time periods. The sample mixtures were pre-incubated for 2 min at 32 ^oC without maxicell lysate then with lysate for additional 5 min at 32 ^oC before the addition of trypsin. Digestions were stopped by quickly mixing an aliquot of the reaction mixture with the same amount of gel loading buffer (62.5 mM Tris-HCl, pH 6.8, 2% SDS, 5% beta-mercaptoethanol, 10% glycerol and 0.005% bromophenol blue) and boiling at 100 ^oC for 3 min. The digested samples were subjected to eletrophoresis on 10% polyacryamide gel and then visualized by autoradiography. For tryptic cleavage of EF-Tu GDP complex, the reaction mixtures were similarily prepared except that GTP was substituted by GDP and PEP was omitted. Pyruvate kinase was initially excluded in the pre-incubation mixture; however, the same amount of kinase as in the EF-Tu GTP experiments was added just prior to trypsin addition. RNase A treatments of crude lysate were performed by adding 5 ug of heat-treated RNase A per 60 ug of crude maxicell lysate (total proteins) and incubating for 10 min at 32 ^oC.

RESULTS

The single amino acid substitution (Gly to Ala) at position 83 of EF-Tu from E. coli was constructed by oligonucleotide site-directed mutagenesis. The mutant EF-Tu gene was then subcloned into the plasmid vector pTZ18U to facilitate expression of the mutant protein.

The mutant protein was labeled with [^{35}S]-methionine by the maxicell technique which specifically labeled the plasmid harboring genes; therefore, the biochemical properties of the mutant EF-Tu can be assessed by following the radioactivities in the crude lysate without the need for extensive purification. Labeled proteins were released by lysozyme and osmotic shock. Polyacrylamide gel electrophoresis of the lysate revealed the 43 KDa EF-Tu (Figure 2 Lanes 5 and 6) and 31 KDa beta-lactamase (not shown), encoded by the ampicillin resistant gene as the two major labeled bands.

The ability of mutant EF-Tu to bind EF-Ts was examined by gel filtration chromatography. Under the conditions employed[10], EF-Tu by itself elutes at fraction 28 and EF-Ts·EF-Tu complex appears at fraction 26. As evident in Figure 1, both the wild type and mutant Ala-83 were able to respond to EF-Ts and emerge as the EF-Ts·EF-Tu complex. Apparently, Ala-83 mutation does not affect the EF-Ts binding of EF-Tu. The concentrations of guanine nucleotides required to dissociate the EF-Ts·EF-Tu complex were measured. The assays were performed by first isolating

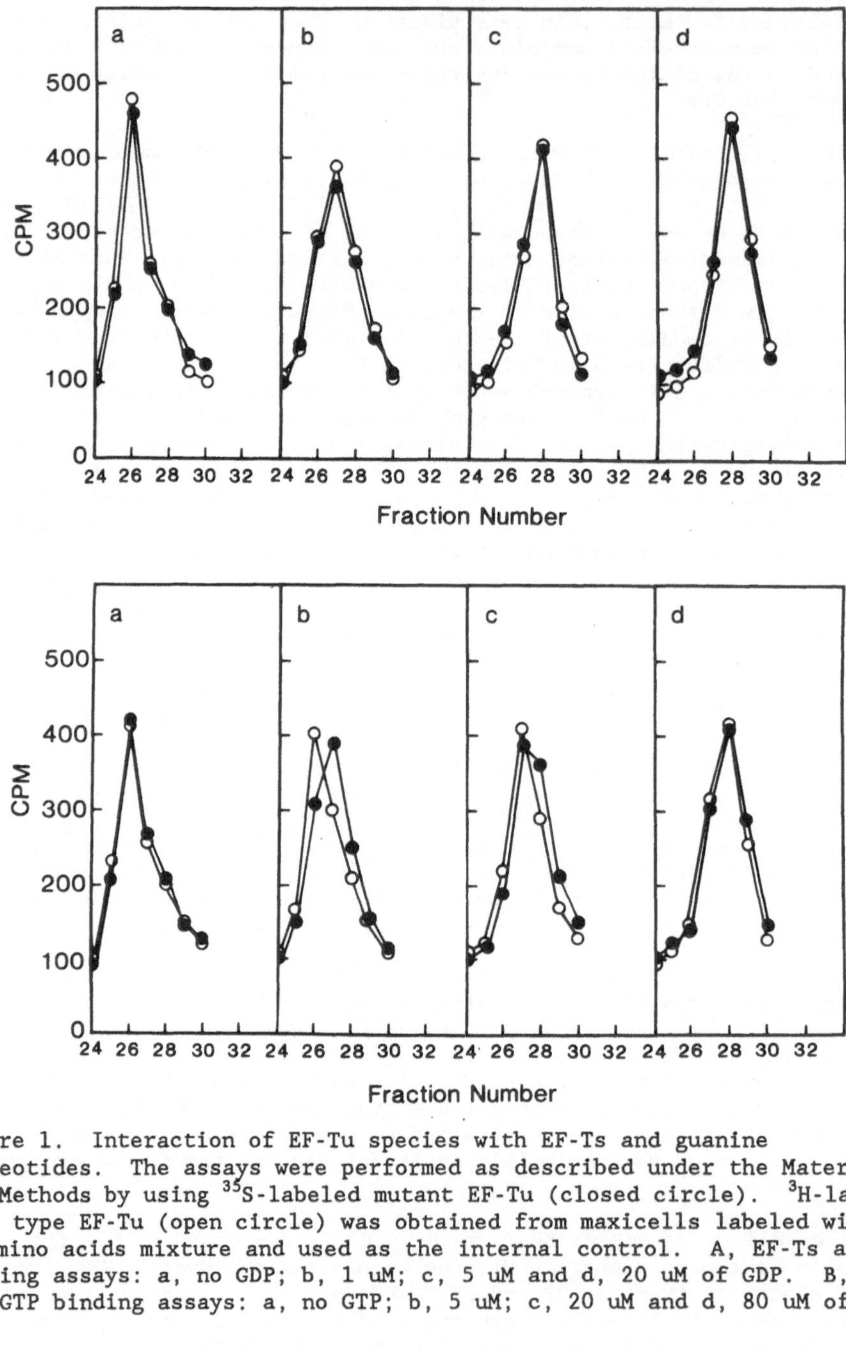

Figure 1. Interaction of EF-Tu species with EF-Ts and guanine nucleotides. The assays were performed as described under the Materials and Methods by using ^{35}S-labeled mutant EF-Tu (closed circle). ^{3}H-labeled wild type EF-Tu (open circle) was obtained from maxicells labeled with a ^{3}H-amino acids mixture and used as the internal control. A, EF-Ts and GDP binding assays: a, no GDP; b, 1 uM; c, 5 uM and d, 20 uM of GDP. B, EF-Ts and GTP binding assays: a, no GTP; b, 5 uM; c, 20 uM and d, 80 uM of GTP.

the EF-Ts·EF-Tu complex and then reeluting the complex with indicated amounts of guanine nucleotide to determine the profile of dissociation. The profile of complex dissociation induced by GDP appeared to be similar in the wild type and mutant (Figure 1A). As can be estimated from Figure 1A, 50% complex dissociation (peaked at fraction 27 and with equal amounts of radioactivities in fracton 26 and 28) occurred at a GTP concentration of about 1 uM for both the wild type and mutant. We also measured the ability of GTP to dissociate EF-Ts·EF-Tu complex (Figure 1B). EF-Tu has a lower affinity for GTP than GDP[16]; therefore, a higher concentration of GTP will be needed to dissociate the complex. The estimated GTP concentration for 50% complex dissociation is about 18 uM for the wild type and about 10 uM for the mutant Ala-83. It appeared that the mutant Ala-83 might have a higher affinity for GTP than the wild type; but, the differences toward GTP diminished as the GTP concentration increased.

We then examined the interaction of the mutant with Phe-tRNA. At 10 uM GTP, a concentration normally employed for tRNA binding assay, the wild type EF-Tu reacts extensively with Phe-tRNA to form ternary complex (Table 1). On the other hand, mutant Ala-83 was devoid of this activity under the same conditions even through the affinity for GTP is similar to the wild type. Furthermore, we performed the Phe-tRNA binding with higher GTP concentrations. The mutant Ala-83 was found to be unable to interact with Phe-tRNA <u>in vitro</u> with GTP concentrations of up to 1 mM (Table 1).

The <u>in vivo</u> function of the mutant was subsequently assessed by testing its ability to confer kirromycin sensitivity onto the kirromycin resistant host. The mode of action of kirromycin requires EF-Tu to bind guanine nucleotide, aa-tRNA and antibiotic[17]. This quaternary complex then binds irreversibly to ribosomes and inactivates translation[17]. The sensitive EF-Tu is dominant over the resistant trait[17]; thus, introducing an active EF-Tu into the kirromycin resistant host should convert the host to a sensitive phenotype. Suprisingly, the wild type as well as mutant Ala-83 were able to convert the phenotype of strain HW123 (Table 2). In fact, as it is evident in Table 2, the mutant Ala-83 is as effective as the wild type in this activity.

Limited digestion of EF-Tu by trypsin produces a 38 KDa polypeptide which resulted from single tryptic cleavage at Arg-58 of EF-Tu. The rate of cleavage at this position is more rapid on EF-Tu·GTP than on EF-Tu·GDP[18]. Thus, we investigated the effect of GTP on the tryptic cleavage rate of the mutant Ala-83. Samples containing crude maxicell lysate from the wild type and mutant were prepared as described in the Materials and Methods. As shown in Figure 2A, the rate of tryptic cleavage on mutant Ala-83 was increased by GTP (Lanes 6-10); whereas, cleavage of the wild type was not affected by GTP (Lanes 1-5). Since the assays were performed by using crude cell extract, we reasoned that the wild type EF-Tu might be protected against trypsin by the aa-tRNA in the lysate. Therefore, the crude lysates were pre-treated with RNase A before tryptic digestion. The tryptic cleavage patterns of RNase A pre-treated lysates were shown in Figure 2B. The results clearly indicated that the wild type was protected by aa-tRNA in the reaction mixture, since the removal of aa-tRNA by ribonuclease concomitantly removed the resistance of the wild type EF-Tu·GTP complex against digestion by trypsin. The results also indicated that mutant Ala-83 was not protected against trypsin by aa-tRNA (Figure 2B). This observation is consistent with the <u>in vitro</u> Phe-tRNA binding

Table 1. Ternary Complex Formation Activities of WT and Mutant Ala-83.

[GTP](uM)	Ternary Complex Formed (%) WT	Ala-83
10	68	2
100	nd	1
500	nd	0
1000	nd	2

nd: not determined

Table 2. Kirromycin Sensitivity of WT and Mutant Ala-83.

[kirromycin](uM)	Control	+pWT	+pAla-83
100	R	S	S
50	R	S	S
25	R	S	S

R: resistant ; S: sensitive.
Control: Strain HW123 alone.
+pWT and +pAla-83: Strain HW123 transformed with plasmids harboring WT and mutant _tuf_A genes, respectively.

activities of mutant Ala-83 (Table 1). Nevertheless, we failed to detect difference in the rates of tryptic cleavage on the wild type and mutant Ala-83 EF-Tu regardless of the bound nucleotide.

DISCUSSION

We have constructed a single amino acid substitution (Gly to Ala) mutant at position 83 of EF-Tu from _E. coli_. The mutant protein was specifically labeled by maxicells and analyzed _in vitro_. This substitution does not appear to impair the ability of EF-Tu to interact with EF-Ts, GDP and GTP as demonstrated in the EF-Ts EF-Tu complex dissociation assays (Figure 1). Although mutant Ala-83 possesses normal GTP affinity, this mutant is unable to form a stable ternary complex with Phe-tRNA and GTP (Table 1).

Gly-83 is located at the C-region consensus sequence[4]. This region is composed of sequence Asp-X-X-Gly and lies near the beta-phosphate of the bound guanine nucleotide[8,9]. The Asp residue in this region has been suggested to interact with the beta-phosphate via a Mg^{+2} ion, but the role of the conserved glycine is not clear. The corresponding glycine residue is conserved in all known guanine nucleotide binding proteins, a fact which suggests that Gly-83 of EF-Tu might be involved in a common mechanism shared by the diverse proteins rather than in a direct interaction with aa-tRNA. Because glycine exhibits a broader range of phi

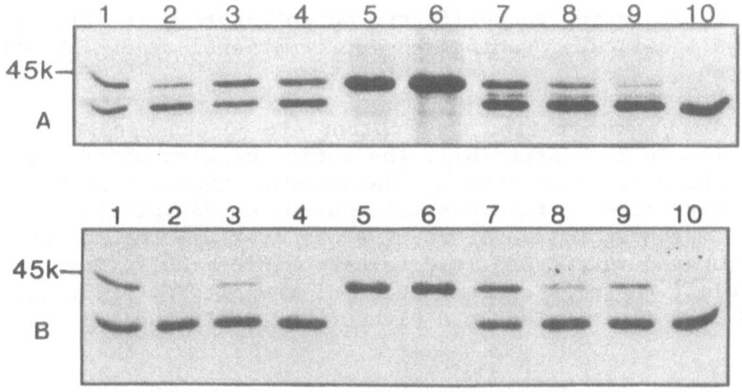

Figure 2. Tryptic cleavage of wild type and mutant EF-Tu. The experiments were performed as described under the Materials and Methods. A, Without RNAase A treatment; B, with RNAase A treatment. Lane 1, wild type EF-Tu GDP with trypsin digestion for 5 min; Lane 2, same as the Lane 1 but with 10 min digestion; Lane 3, wild type EF-Tu GTP with trypsin digestion for 5 min; Lane 4, same as the Lane 3 but with 10 min digestion; Lane 5, wild type without trypsin; Lane 6, mutant without trypsin; Lane 7, mutant EF-Tu GDP with trypsin digestion for 5 min; Lane 8, same as the Lane 7 but with 10 min digestion; Lane 9, mutant EF-Tu GTP with trypsin digestion for 5 min; Lane 10, same as the lane 9 but with 10 min digestion.

and psi backbone dihedral angles than other amino acids, the role of an invariant glycine in guanine nucleotide binding proteins might be to provide a flexible pivot point for a GTP-induced conformational change.

The conformational difference between EF-Tu·GTP and EF-Tu·GDP has been demonstrated by tritium exchange[19], fluorescent dye-binding[20] and differential protease sensitivity[18]. Since only the last technique is feasible with maxicell crude lysate, we performed the trypsin cleavage experiments. Trypsin has been shown to cleave EF-Tu·GTP faster than EF-Tu·GDP[18]. As evident in Figure 2, the rate of tryptic cleavage on mutant Ala-83 was not only increased by GTP but the rate was indistinguishable from the wild type. These results suggested that mutant EF-Tu undergoes, at least, a partial conformational change upon GTP complexation. However, because the GTP form of the mutant EF-Tu does not bind to aa-tRNA, the molecular basis for the GTP-induced tRNA binding of EF-Tu is probably not related to the GTP-induced tryptic cleavage rate enhancement. Not only Phe-tRNA failed to bind EF-Tu; aa-tRNA mixtures from E. coli crude lysate also failed to interact with mutant EF-Tu and protect it from trypsin digestion. Alternatively, the lack of activities of mutant Ala-83 toward aa-tRNA might be due to the steric hindrance imposed by the extra methyl group at position 83 of EF-Tu. This argument is unlikely if one considers the physical size of aa-tRNA and possible multiple-site interactions between EF-Tu and aa-tRNA; but, we were unable to rule it out at present.

We are currently trying to purify the mutant protein in large quantities, which should facilitate probing the conformational change by other physical means.

It was surprising to find that mutant Ala-83 was sensitive to kirromycin (Table 2). After all, the action of kirromycin seems to require the binding of aa-tRNA[17]. The binding constant of mutant Ala-83 to aa-tRNA has not been rigorously measured, a possibility exist that high intracellular concentration of aa-tRNA may overcome the aa-tRNA binding deficiency of the mutant and lead to kirromycin sensitivity. Alternatively, the high concentration of the altered EF-Tu GTP may allow it to bind directly to ribosomes without the assistance of aa-tRNA.

ACKNOWLEDGMENTS

This project was supported by United States Public Health Service Grant GM30800.

REFERENCES

1. Y. Kaziro, The Role of Guanosine 5'-Triphosphate in Polypeptide Elongation, Biochim. Biophy. Acta., 505:95 (1978).
2. M. Barbacid, ras Genes, Ann. Rev. Biochem., 56:779 (1987).
3. A. G. Gilman, G Proteins: Transducers of Receptor Generated Signals, Ann. Rev. Biochem., 56:615 (1987).
4. K. R. Holliday, Regional Homology in GTP-Binding Proto-Oncogene Products and Elongation Factors, J. Cyclic Nucleotide Protein Phosphorylation Res., 9:435 (1984).
5. T. E. Dever, M. J. Glynias, and W. C. Merrick, GTP-Binding Domain: Three Consensus Sequence Elements with Distinct Spacing, Proc. Natl. Acad. Sci. USA, 84:1814 (1987).
6. F. McCormick, B. F. C. Clark, T. F. M. la Cour, M. Kjeldgaard, L. Norskov-Lauritsen, and J. Nyborg, A Model for the Tertiary Structure of p21, the Product of the ras Oncogene, Science, 230:78 (1985).
7. D. L. Miller, and H. Weissbach, Factors Involved in the Transfer of Aminoacyl-tRNA to the Ribosome, in:"Molecular Mechanism of Protein Biosynthesis" H. Weissbach and S. Peska, ed., Academic Press, New York (1977).
8. T. F. M. la Cour, J. Nyborg, S. Thirup, and B. F. C. Clark, Structural Details of the Binding of Guanosine Diphosphate to Elongation Factor Tu from E. coli as Studied by X-ray Crystallography, EMBO J., 4:2385 (1985).
9. F. Jurnak, Structure of the GDP Domain of EF-Tu and Location of the Amino Acids Homologous to ras Oncogenic Proteins, Science, 230:2385 (1985).
10. Y. W. Hwang, and D. L. Miller, A Mutantion that Alters the Nucleotide Specificity of Elongation Factor Tu, a GTP Regulatory Protein, J. Biol. Chem., 262:13081 (1987).
11. T. Maniatis, E. F. Fritsch, and J. Sambrook, Molecular Cloning, Cold Spring Harbor Laboratory, Cold Spring Harbor, NY (1982).
12. F. Sanger, S. Nicklen, and A. R. Coulson, DNA Sequencing with Chain-Terminating Inhibitors, Proc. Natl. Acad. Sci. USA, 74:5463 (1977).

13. D. A. Mead, E. Szczesna-Skorupa, and B. Kemper, Single-Stranded DNA "Blue" T7 Promoter Plasmids: a Versatile Tandem Promoter System for Cloning and Protein Engineering, Protein Engineering, 1:67 (1986).
14. Y. W. Hwang, and D. L. Miller, A Study of Kinetic Mechanism of Elongation Factor Ts, J. Biol. Chem., 260:11498 (1985).
15. J. A. M. Van De Klundert, P. H. Van der Meide, P. Van De Putte, and L. Bosch, Mutants of E. coli Altered in Both Genes Coding for the Elongation Factor Tu, Proc. Natl. Acad. Sci. USA, 75:4470 (1978).
16. D. L. Miller, and H. Weissbach, Studies on the Purification and Properties of Factor Tu from E. coli, Arch. Biochem. Biophys., 141:26 (1970).
17. A. Parmeggiani, and G. W. M. Swart, Mechanism of Action of Kirromycin-Like Antibiotics, Ann. Rev. Microbiol., 39:557 (1985).
18. J. Douglass, and T. Blumenthal, Conformational Transition of Protein Synthesis Elongation Factor Tu Induced by Guanine Nucleotides, J. Biol. Chem., 254:5383 (1979).
19. M. Printz, and D. L. Miller, Evidence for Conformational Changes in Elongation Factor Tu Induced by GTP and GDP, Biochem. Biophys. Res. Commun., 53:149 (1973).
20. L. J. Crane, and D. L. Miller, Guanosine Triphosphate and Guanosine Diphospahte as Conformation-Determining Molecules. Differential Interaction of a Fluorescent Probe with the Guanosine Nucleotide Complexes of Bacterial Elongation Factor Tu, Biochem., 13:933 (1974).

THE APPLICATION OF FLUORESCENT AND PHOTOSENSITIVE ANALOGUES OF GUANINE NUCLEOTIDES TO THE FUNCTION AND STRUCTURE OF G-BINDING PROTEINS

J.F. Eccleston, T.F. Kanagasabai, D.P. Molloy, S.E. Neal and M.R. Webb

National Institute for Medical Research, Mill Hill
London NW7 1AA, U.K.

In order to understand the mechanism of the interaction of a guanine nucleotide-binding protein at the molecular level it is necessary to determine the number of key intermediates of the process, to measure their rates of interconversion and ideally to gain structural information about the changes occurring during transitions between the intermediates. Our work has concentrated on defining the kinetic mechanism of elongation factor Tu (EF-Tu) on interaction with mRNA programmed ribosomes (Eccleston et al., 1985) and in the GTPase of $p21^{N\text{-}\underline{ras}}$ (S.E. Neal, J.F. Eccleston, A. Hall, & M.R. Webb, unpublished results). These studies depended on measurements of GTP cleavage and, in the first case, peptide bond formation. However, these measurements at most give limited information about binding steps, protein isomerisations and dissociation steps which are essential features in the processes mediated by these GTPases. This information can be obtained by the use of spectroscopic probes which can be introduced into many different positions in the system.

We have chosen to modify the guanine nucleotide since this gives stoichiometric labelling and the probe is close to the GTPase site. Ideally, the probe needs to be able to report on events without participating in them, probably an unattainable ideal. However, even if modification of the nucleotide affects the elementary rate constants of a process, it is likely to proceed via the same mechanism if the analogue mediates the biological process of interest. For example, the 5'-triphosphate of the chromophoric nucleoside 2-amino-6-mercaptopurine riboside (thioGTP) forms a functional ternary complex with EF-Tu and aminoacyl-tRNA and the release of thioGDP from EF-Tu is catalysed by EF-Ts, even though these analogues bind to EF-Tu weaker than GDP (Eccleston, 1981). We have used the absorption changes of thioGDP on binding to EF-Tu to investigate the mechanism of the EF-Ts mediated process and shown that it proceeds via an isomerisation of an EF-Tu.thioGDP.EF-Ts ternary complex (Eccleston et al., 1988).

The inherent sensitivity of fluorescence prompted us to investigate the use of fluorescent analogues of guanine nucleotides, particularly in order to investigate the events occurring after the EF-Tu.GTP.aminoacyl-tRNA ternary complex has bound to the ribosome. The interaction with these analogues with p21 is also being investigated. Fluorophores which are sensitive to their environment are most useful for this work but fluorescent analogues which do not have this property can nevertheless be

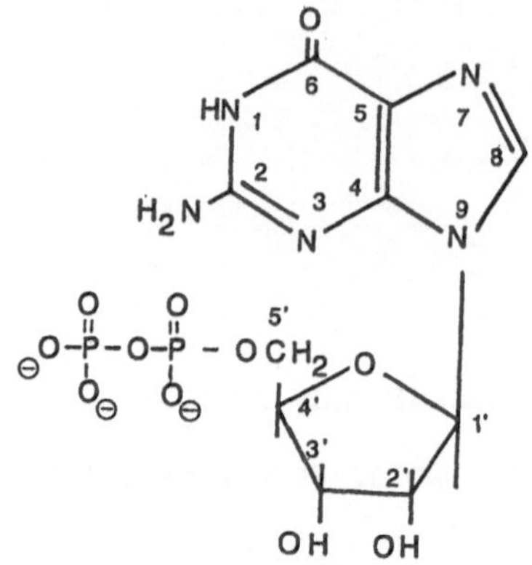

FIG 1. Structure of GDP

used to yield important information about biological systems. For example, they can be used for hydrodynamic measurements of the G-protein.nucleotide complex and complexes with other components of the system (Eccleston et al., 1987), for investigations of solvent accessibility of the nucleotide binding site (Hiratsuka, 1984), for distance measurements based on energy transfer (Stryer, 1978), and potentially for visualising the location of G-proteins in organised systems by the highly sensitive digitized fluorescence microscopy systems now available.

Examination of the structure of GDP (fig. 1) shows that modifications of guanine nucleotides can be divided into two classes, those of the purine ring and those of the ribose moiety. These are each discussed together with their application to EF-Tu and $p21^{N\text{-}ras}$.

Modifications of purine ring

The possibilities of modifying the purine ring are limited and the fluorescence properties of the nucleotide are fixed by the chemical modification possible. Also, the crystal structure of EF-Tu (Jurnak, 1985, LaCour et al., 1985) suggests specific interactions between certain positions of the purine ring and the protein which can be crucial to binding.

The 2-amino group of GDP can be replaced by a 2-azido group to give a fluorescent (and potentially photo-reactive) analogue (Weigand & Kaleja, 1976) which binds to EF-Tu with a 13% decrease in fluorescence although it binds 100-fold weaker than GDP (Eccleston, 1981). Studies with this analogue are complicated by its existence as an equilibrium mixture of the 2-azido nucleotide and the tetrazole. The 6-oxy group of GDP can be replaced with a 6-thio group again with a reduced affinity for EF-Tu and also for $p21^{N\text{-}ras}$. This analogue is not fluorescent but shows measureable perturbations of its absorption spectrum on binding to EF-Tu (Eccleston, 1981, 1984) and has been used for the studies described above. Since its absorption spectrum overlaps tryptophan fluorescence, it has the potential for quenching tryptophan fluorescence by energy transfer, and although this has been useful in other systems (Eccleston & Trentham, 1977), the effect is very small in EF-Tu.

2-aminopurine nucleoside analogues are fluorescent (Ward et al., 1969) but show no detectable binding to EF-Tu or to $p21^{N\text{-}ras}$, indicating the importance of the 6-oxy group in the protein-nucleotide interaction in both cases. Methylation of N-7 of guanine nucleotides results in quaternisation

Table 1. Relative dissociation constants (K') of guanosine nucleotide analogues to EF-Tu and p21$^{N\text{-}\underline{ras}}$ compared to GDP

Analogue	EF-Tu	p21$^{N\text{-}\underline{ras}}$
6thioGDP	95[1]	57
2-aminopurine riboside-5'diphosphate	>1000	>1000
8Br-GDP	26[2]	10
IDP	130[3]	
7-methyl GDP	>1000	>1000
2'-dGDP	3[3]	
Bz_2-GDP	1.1	2.7
Meant-GDP	1.4	1.1
Fluram-GDP	15[4]	

[1]Eccleston (1981) [2]Wittinghofer et al. (1977) [3]Miller & Weissbach (1977) [4]Eccleston et al. (1987). The original papers should be consulted for exact conditions of the determinations. Those reported here for EF-Tu are as in Eccleston (1981) at 37°C and for p21$^{\underline{ras}}$ in 50 mM tris HCl, 5 mM $MgCl_2$, 50 mM NaCl, 5 mM DTT, pH 7.5 at 37°C.

of this nitrogen with a formal positive charge and has strong fluorescence (Kulikowska, 1986) but no binding of the 5' diphosphate analogue of this nucleotide could be detected with either EF-Tu or p21$^{N\text{-}\underline{ras}}$.

The potential for modifying the 8-position of nucleotides is vast, and a wide variety of such analogues have been described. The key intermediate in the synthesis of these analogues is the 8-Br derivative since it can easily be synthesised and the bromine can then be replaced by nucleophiles (Kapuler & Reich, 1971). For example, reaction with thiourea produces the 6-SH derivative (Caretta et al., 1985) which can be reacted with thiol reagents such as iodoacetamide derivatives or epoxides. These analogues have been used extensively in studies of cGMP dependent systems where they often show tighter binding to the protein (or at least mimic the biological process at lower concentrations) than cGMP itself. However, 8Br-GDP binds to EF-Tu 26-fold weaker than GDP (Wittinghofer et al., 1977) and to p21$^{N\text{-}\underline{ras}}$ 10-fold weaker. A facile explanation for this difference is that in EF-Tu the nucleotide is bound in the anti conformation whereas substitution at the 8-position of a nucleotide favours the syn conformation, the form of cGMP postulated to bind to protein kinase (Weber et al., 1987). However the situation is likely to be more complex than this.

The etheno derivatives of nucleosides formed by reaction with haloacetaldehydes are generally highly fluorescent and have been extensively used, in particular the adenine nucleotide derivatives (Leonard, 1984). However, the N^2, 1-ethenoguanosine derivatives produced by this reaction are non-fluorescent. A recent report has described the synthesis of N^2, 3-ethenoguanosine derivatives which are highly fluorescent (Kusmierek et al., 1987) but information on their interactions with G-proteins is not available. Derivatives based on reaction of the thiol group with fluorescent thiol reagents have been described (Leonard & Tolman, 1975) but in view of the considerations described above are unlikely to bind to G-proteins.

A summary of the relative binding affinities of the analogues to EF-Tu and p21$^{N\text{-}\underline{ras}}$ is shown in Table 1.

<u>General approaches to modification of the ribose moiety</u>

The X-ray structure of EF-Tu shows that the 2',3' cis diol of GDP projects away from the protein into the solvent suggesting that this group can be modified without losing tight binding. This has been confirmed by attaching large groups to these positions with little or no change in the binding constant to either EF-Tu or p21$^{N\text{-}\underline{ras}}$ (Table 1).

A wide variety of modifications to the ribose moiety can be made by two major synthetic routes; by reaction of the 2'(3')-amino-2'(3')-deoxyguanosine nucleotide or by direct acylation of the 2' or 3' hydroxyl groups.

Syntheses have been described for the chemical synthesis of 2'-amino-2'-deoxyguanosine (Hobbs & Eckstein, 1977) and 3'-amino-3'-deoxyguanosine (Faulhammer et al., 1986). The former nucleoside is also produced by the bacterium Enterobacter cloacae (Nakanishi et al., 1974). The amine can then be reacted with a wide variety of fluorescent amine reactive compounds to produce the required derivative. The fluorescamine derivative of 2'-amino-2'-deoxyguanosine-5'-diphosphate (Eccleston et al., 1987) and a spin labelled derivative of 3'-amino-3'-deoxyguanosine-5'-diphosphate have been described (Faulhammer et al., 1986).

Direct acylation can be achieved by reaction of nucleotides with isatoic or methyl isatoic anhydride to produce the 2'(3')-anthraniloyl or 2'(3')-methyl anthraniloyl derivatives respectively (Hiratsuka, 1983). The interaction of these derivatives of GDP with EF-Tu and p21$^{N\text{-}\underline{ras}}$ is described below.

A more general approach to acylation of the 2'(3')-positions is based on the work of Jeng and Guillory (1975) who showed that the imidazolidate of carboxylic acids is a good acylating agent of the 2'(3')-positions of nucleosides. This method was originally used by them for the synthesis of potentially photoaffinity labels of ATP-dependent systems and has since been used for the synthesis of fluorescent derivatives of ATP (see below).

It should be noted that nucleotides acylated on the ribose moiety consist of an equilibrium mixture of the 2' and 3' acyl derivatives as shown in fig. 2 (Griffin et al., 1966, Cremo & Yount, 1987) which may complicate distance and polarisation measurements.

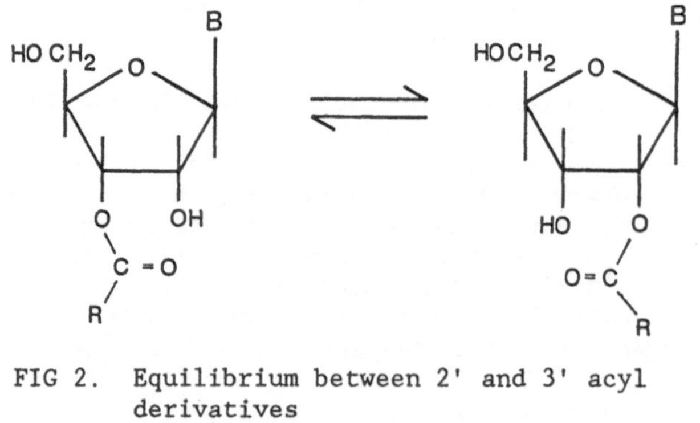

FIG 2. Equilibrium between 2' and 3' acyl derivatives

A synthesis has recently been described by Braxton and Yount (1988) in which the 2'3' carbonate derivative of the nucleotide is formed which is reacted with an amine to produce a carbamate derivative. Reaction with a diamine results in the formation of a derivative with a reactive amine group separated from the ribose moiety by a spacer arm, again allowing the introduction of an amine specific fluorophore.

Choice of fluorophores

The methods described above show that in general the modification of the purine ring results in fluorophores with defined properties and weaker binding to the protein than guanine nucleotides. However, modification of the ribose moiety by the methods described gives a general approach to the synthesis of fluorescent analogues in which the properties of the fluorophore can be chosen for the particular application required, and binding constants similar to the physiological nucleotide are retained.

We are primarily concerned with the synthesis of environmentally sensitive fluorescent analogues of guanine nucleotides so that small perturbations cause large changes in fluorescence. It is difficult to predict the properties of a fluorophore since they are governed by many factors and an empirical approach is probably best followed although theoretical considerations have been discussed by Weber and Farris (1979).

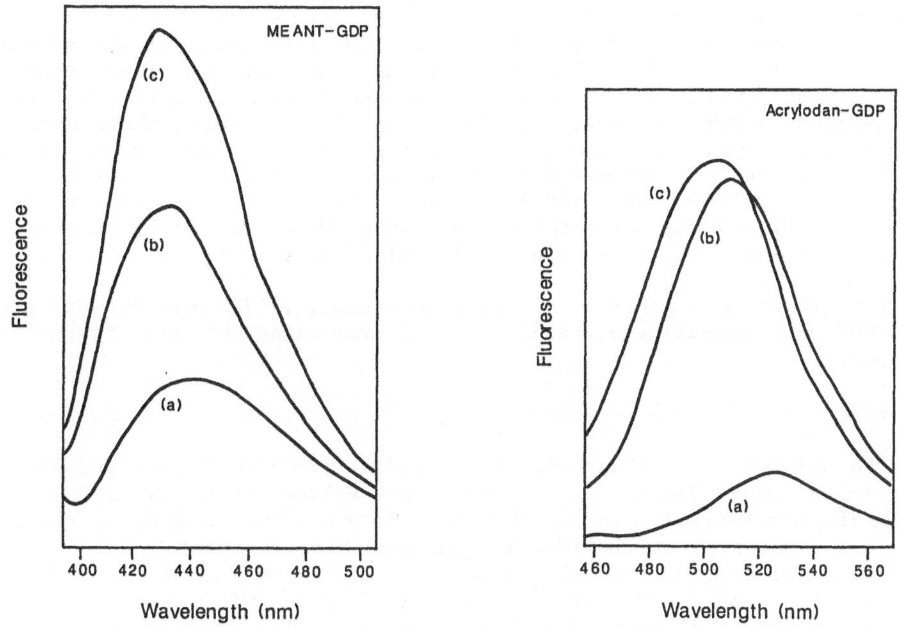

FIG 3. Effect of ethanol on the emission spectra of meant-GDP and acrylodan-GDP. The solvent conditions were (a) water, (b) 50% ethanol, (c) 90% ethanol. All solutions also contained 50 mM tris HCl, pH 7.5. Spectra were recorded at 20°C. Excitation wavelengths were 350 nm for meant-GDP and 400 nm for acrylodan-GDP.

Based on this work, Prendergast et al., (1983) synthesised 6-acryloyl-2-dimethylaminonaphthalene (Acrylodan) which reacts with thiol groups to produce fluorophores highly sensitive to environment.

We have synthesised a carboxylic acid derivative of acrylodan by reaction with thioglycollic acid which could then be introduced into guanine nucleotides by following activation with carbonyl diimidazole as described above (acrylodan-GDP). The fluorescence of the acrylodan-GDP and 2'(3')O methylanthraniloyl-GDP (meant-GDP) solvent conditions was first investigated to give a measure of their sensitivity to environment. Fig. 3 shows that the fluorescence of these nucleotides is perturbed on moving from water to ethanol with acrylodan-GDP showing the greatest effect.

Interaction of fluorescent analogues of guanine nucleotides with EF-Tu and p21

The fluorescence emission spectrum of EF-Tu.meant-GDP was recorded before and after displacement of meant-GDP by excess GDP. Fig. 4A shows that the intensity of meant-GDP when bound to EF-Tu is approximately twice that of the meant-GDP free in solution and is shifted by 7 nm to the blue.

The excitation polarisation spectrum of EF-Tu.meant-GDP at 20°C gave a value for polarisation at 380 nm of 0.310 compared to a value of 0.475 for meant-GDP in glycerol at 0°C. These steady-state polarisation values are related to the rotational parameters of the protein-nucleotide complex by the Perrin equation.

$$1/P - 1/3 = (1/Po - 1/3)(1 + 3\tau/\rho_h)$$

where P is the observed polarisation of the complex fluorescence, Po the limiting or intrinsic polarisation, τ the fluorescence lifetime and ρ_h the harmonic mean of the Debye rotational relaxation times of the principal axes of rotation. The fluorescence lifetimes of the EF-Tu.meant-GDP complex of meant-GDP in solution are not described by simple exponentials (T.L. Hazlett & D.M. Jameson, unpublished work). Fitting their data from multi-frequency phase and modulation fluorometry techniques to two lifetime models, they obtained values of 5.9 ns (F = 0.83) and 2.7 ns (F = 0.17) for the EF-Tu.meant-GDP complex and 3.8 ns (F = 0.92) and 1.5 ns (F = 0.08) for free meant-GDP. Using only the major component value for EF-Tu.meant-GDP in the above equation, we obtain a value of 28 ns for ρ_h

The effect of acrylamide on the fluorescence of EF-Tu.meant-GDP and of meant-GDP was investigated and analysed according to the Stern-Volmer equation.

$$Fo/F = 1 + K_{SV} [Q]$$

where Fo and F are the fluorescence intensities in the absence and presence of quencher respectively, K_{SV} is the Stern-Volmer quenching constant and [Q] is the concentration of acrylamide. In the case of dynamic quenching $K_{SV} = k_q\tau$ where k_q is the bimolecular quenching constant. Values of K_{SV} for the EF-Tu.meant-GDP complex and meant-GDP were 1.5 M^{-1} and 2.7 M^{-1} respectively. Using the fluorescence lifetimes of the major component of EF-Tu.meant-GDP and meant-GDP given above we obtain values of 2.5×10^8 $M^{-1}s^{-1}$ and 7.1×10^8 $M^{-1}s^{-1}$ respectively.

These fluorescent measurements show that although the intensity of the fluorophore is perturbed on binding of meant-GDP to EF-Tu, presumably by interacting with specific amino acid residues, it is able to undergo considerable local motion since the value of ρ_h is lower than expected for

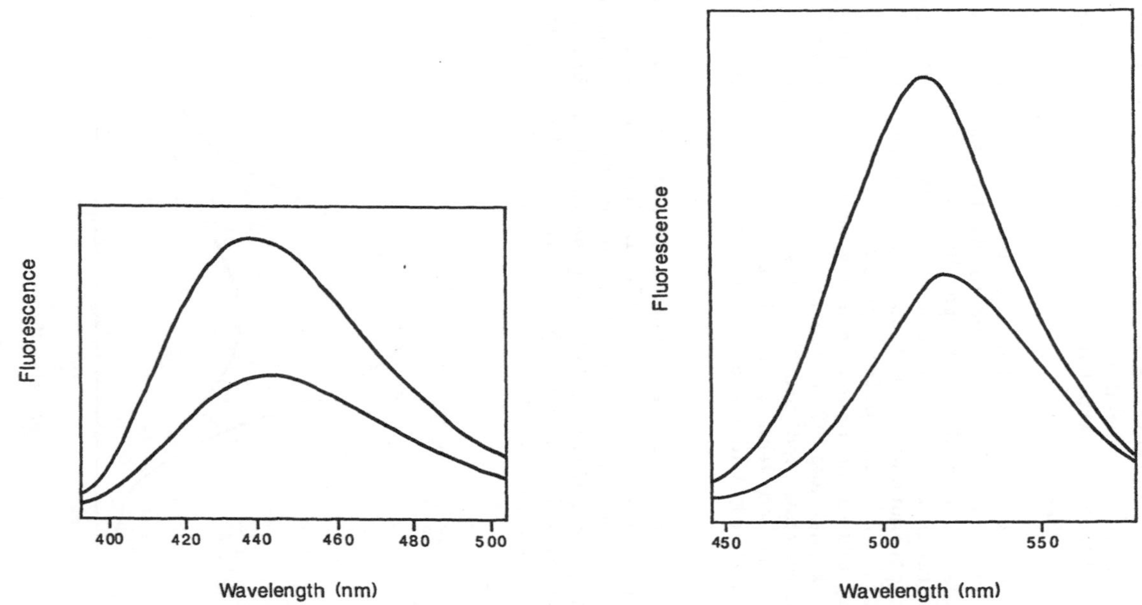

FIG 4A. Changes in the emission spectrum of meant-GDP after displacement from EF-Tu by GDP. The upper trace is the emission spectrum of 10 μM EF-Tu.meant-GDP (excitation 350 nm). The lower trace was recorded after the addition of 300 μM GDP. The buffer was 50 mM tris HCl, 10 mM $MgCl_2$, pH 7.6 at 20°C.

4B. As 4A except that initial solution contained 10 μM acrylodan-GDP. Excitation was at 380 nm.

a protein of this size and than measured previously by studies of the single tryptophan of EF-Tu and of EF-Tu.fluram-GDP (Jameson et al., 1987, Eccleston et al., 1987). The difference between the results with fluram-GDP and meant-GDP is probably due to the longer distance between the fluorophore and ribose moiety in the latter case and the increased possibilities of bond rotation although a more detailed analysis of the relative contributions of local and global motion of EF-Tu.meant-GDP will be achieved by time resolved anisotropy measurements. The quenching data shows that the meant-fluorophore in EF-Tu.meant-GDP is accessible to solvent molecules. Initial measurements on the EF-Tu.meant-GTP and EF-Tu.meant-GMP.PNP complexes show reduced fluorescence enhancements and lower polarisations probably reflecting structural differences in the nucleotide binding site of the three complexes.

Preliminary investigations of the interaction of meant-GDP and meant-GTP with $p21^{N\text{-}\underline{ras}}$ have been made. Meant-GDP binds to the normal (Gly 12) protein 3.2-fold more weakly than GDP although the corresponding values for the Asp 12 and Val 12 mutants are 1.7 and 6.6 respectively. On binding to the normal protein and both mutants, the fluorescence intensity is enhanced by a factor of 2.3 (fig. 5). The effect of acrylamide on the $p21^{N\text{-}\underline{ras}}$, meant-GDP fluorescence was also measured and gave a value for K_{SV} of 2.1 M^{-1} for both the Gly 12 and Asp 12 proteins. This is higher than the values for the EF-Tu complexes and probably results from the fluorophore being highly accessible to solvent although a more definite conclusion awaits the determination of the lifetime of the fluorophore in the p21.meant-GDP complex.

Initial experiments show that acrylodan-GDP undergoes a 7 nm blue shift and an increase in intensity of 2.2 fold in binding to EF-Tu (fig. 4B). This is of similar magnitude to that of meant-GDP although the environmental sensitivity of acrylodan-GDP is greater than that of meant-GDP based on the effects of ethanol (fig. 3B). This probably results from the greater distance between the acrylodan fluorophore and ribose moiety.

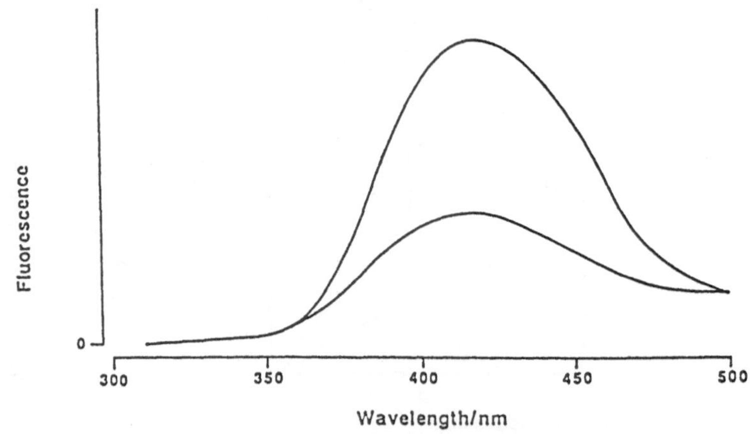

FIG 5. Change in emission spectrum of meant-GDP after displacement from $p21^{N\text{-}\underline{ras}}$ by GDP. The upper trace is the emission spectrum of 5 μM $p21^{N\text{-}\underline{ras}}$.meant-GDP in 50 mM tris HCl, 50 mM NaCl, 5 μM $MgCl_2$, 10 mM DTT at 22°C. The lower spectrum was recorded 35 min after the addition of 300 μM GDP.

Photosensitive analogues of guanine nucleotides

The considerations described above for the design of fluorescent analogues of guanine nucleotides also apply to photosensitive analogues.

8-azidoguanine nucleotides have been used as photoaffinity labelling agents on some G-binding proteins (Potter & Haley, 1983) but have the disadvantage of requiring photolysis at short wavelengths, and also may bind weaker than the normal nucleotide. The chemical affinity label fluorosulphonyl benzoylguanosine is an analogue of guanosine nucleotides (Colman, 1983) but we have not been able to detect any interaction between it and EF-Tu. The introduction of the (4-benzoyl)benzoyl (Bz_2) moiety onto the ribose of nucleotides offer several advantages over these approaches because photoactivation of the Bz_2 is achieved at wavelengths of greater than 300nm so that photo-damage to the protein is avoided. Also, the excited state of the benzophenone carbonyl group reacts very slowly with water and so can be repeatedly formed to give, in theory, higher levels of incorporation than photoaffinity labels which generate nitrenes or carbenes.

We have synthesised Bz_2-GDP and Bz_2-GTP by procedures similar to those described for the adenine analogues by Cremo and Yount (1987). Proof of the structures was obtained by hydrolysis in 50 mM NaOH at 20°C for 15 mins when stoichiometric amounts of GTP or GDP and (4 benzoyl)benzoic acid were obtained.

Although we have demonstrated tight binding of Bz_2-GDP to EF-Tu and p21^N-ras^ (Table 1), we have not yet obtained covalent incorporation into these proteins.

Comparison of the nucleotide binding sites of EF-Tu and p21^N-ras^ and other G proteins

The known sequences of EF-Tu and p21 and their preliminary structure determinations have led to discussions of the similarities of the nucleotide-binding site in the two proteins (reviewed by Jurnak, (1988). The interactions of a variety of modified guanine nucleotide analogues reported here with EF-Tu and p21 are significant in this respect. Inspection of the modifications of the 2, 6, 7 and 8-positions (Table 1) shows qualitatively the same effect in both proteins. Changing the 6-oxy group to a 6-thio group results in a decrease in the binding whereas complete removal of a substituent group at this position gives an analogue which shows no detectable binding. Methylation of the N-7 also results in loss of detectable binding and substitution of Br at C-8 reduces the binding affinity of the nucleoside 5'-diphosphate.

Introduction of relatively large groups at the 2'(3')-position of GDP has very little effect on the binding of the nucleotide to either EF-Tu or p21^N-ras^. The crystal structure shows the 2' 3' cis diol of GDP bound to EF-Tu to be projecting out into the solvent and these results are consistent with a similar situation in p21^N-ras^. A detailed study of the fluorescent properties of a fluorophore attached to the ribose moiety of the nucleotide in both protein.nucleotide complexes is not yet available.

Conclusions

The results presented here show that the introduction of fluorophores into the ribose moiety of guanine nucleotides has little effect on the binding of the nucleotide to EF-Tu or p21^ras^ but, given a suitable fluorophore, provides a sensitive reporter group to give kinetic and structural information about the nucleotide binding site of these proteins. It is likely that this approach is a general one for studying the interactions of the wide range of G-proteins which are involved in a variety of biological control processes.

REFERENCES

Braxton, S., and Yount, R.G., 1988, The synthesis of a novel class of ribose-modified nucleotide analogues. I. Affinity purification of skeletal myosin subfragment-1, Biophys. J., 53:178a.

Caretta, A., Cavaggioni, A., and Sorbi, R.T., 1985, Binding stoichiometry of a fluorescent cGMP analogue to membranes of retinal rod outer segments, Eur. J. Biochem., 153:49-53.

Colman, R.F., 1983, Affinity labelling of purine nucleotide sites in proteins, Ann. Rev. Biochem., 52:67-91.

Cremo, C.R., and Yount, R.G., 1987, 2'-Deoxy-3'-O-(4-benzoylbenzoyl)- and 3'(2')-O-(4-benzoylbenzoyl)-1,N^6-ethanoadenosine 5'-diphosphate, fluorescent photoaffinity analogues of adenosine 5'-diphosphate. Synthesis, characterization, and interaction with myosin subfragment 1, Biochemistry, 26:7524-7534.

Eccleston, J.F., 1981, Spectroscopic studies of the nucleotide binding site of elongation factor Tu from Esherichia coli. An approach to characterizing the elementary steps of the elongation cycle of protein biosynthesis, Biochemistry, 20:6265-6272.

Eccleston, J.F., 1984, A kinetic analysis of the interaction of elongation factor Tu with the guanosine nucleotides and elongation factor Ts, J. Biol. Chem., 259:12997-13003.

Eccleston, J.F., and Trentham, D.R., 1977, The interaction of chromophoric nucleotides with subfragment 1 of myosin, Biochem. J., 163:15-29.

Eccleston, J.F., Dix, D.B., and Thompson, R.C., 1985, The rate of cleavage of GTP on the binding of Phe-tRNA elongation factor Tu.GTP to poly(U)-programmed ribosomes of Escherichia coli, Biol. Chem., 260:16237-16241.

Eccleston, J.F., Gratton, E., and Jameson, D.M., 1987, Interaction of a fluorescent analogue of GDP with elongation factor Tu: steady state and time-resolved fluorescence studies, Biochemistry, 26:3902-3907.

Eccleston, J.F., Kanagasabai, T.F., and Geeves, M.A., 1988, The kinetic mechanism of the release of the nucleotide from elongation factor Tu promoted by elongation factor Ts determined by pressure relaxation studies, J. Biol. Chem., 263:4668-4672.

Faulhammer, H.G., Denninger, G., Hartl, P.J., Azhayer, A.V., Schwoerer, M., and Sprinzl, M., 1986, Spin-labelled analogues of GDP and GTP as site-specific reporter groups for guanosine nucleotide-binding proteins, Biochim. Biophys. Acta, 884:182-190.

Griffin, B.E., Jarman, M., Reese, C.B., Sulston, J.E., and Trentham, D.R., 1966, Some observations relating to acyl mobility in aminoacyl soluble ribonucleic acid, Biochemistry, 5:3638-3649.

Hiratsuka, T., 1983, New ribose-modified fluorescent analogues of adenine and guanine nucleotides available as substrates for various enzymes, Biochim. Biophys. Acta, 742:496-508.

Hiratsuka, T., 1984, Distinct structures of ATP and GTP complexes in the myosin ATPase, J. Biochem. Tokyo, 96:155-162.

Hobbs, J.B., and Eckstein, F., 1977, A general method for the synthesis of 2'-azido-2'-deoxy- and 2'-amino-2'-deoxyribofluranosyl purines, J. Org. Chem., 42:714-719.

Jameson, D.M., Gratton, E., and Eccleston, J.F., 1987, Intrinsic fluorescence of elongation factor Tu in its complexes with GDP and elongation factor Ts, Biochemistry, 26:3894-3901.

Jeng, S.J., and Guillory, R.J., 1975, The use of aryl azido ATP analogues as photoaffinity labels for myosin ATPase, J. Supramol. Struct., 3:448-468.

Jurnak, F., 1985, Structure of the GDP domain of EF-Tu and location of the amino acids homologous to ras oncogene proteins, Science, 230:32-36.

Jurnak, F., 1988, The three-dimensional structure of c-H-ras p21: implications for oncogene and G protein studies, Trends in Biochem. Sci., 13:195-198.

Kapuler, A.M., and Reich, E., 1971, Some stereochemical requirements of Esherichia coli ribonucleic acid polymerase. Interaction with conformationally restricted ribonucleoside 5'-triphosphates: 8-bromoguanosine, 8-ketoguanosine, and 6-methylcytidine triphosphates, Biochemistry, 10:4050-4061.

Kulikowska, E., Bzouska, A., Wierzchowski, J., and Shugar, D., 1986, Properties of two unusual, and fluorescent, substrates of purine-nucleoside phosphorylase: 7-methylguanosine and 7-methylinosine, Biochim. Biophys. Acta, 874:355-363.

Kusmierek, J.T., Jensen, D.E., Spengler, S.J., Stolarski, R., and Singer, B., 1987, Synthesis and properties of N^2, 3-ethenoguanosine and N^2, 3-ethenoguanosine 5'-diphosphate, J. Org. Chem., 52:2374-2378.

LaCour, T.F.M., Nyborg, J., Thirup, S., and Clark, B.F.C., 1985, Structural details of the binding of guanosine diphosphate to elongation factor Tu from Esherichia coli as studied by X-ray crystallography, EMBO J., 4:2385-2388.

Leonard, N.J., and Tolman, G.L., 1975, Fluorescent nucleosides and nucleotides, Ann. NY Acad. Sci., 255:43-58.

Leonard, N.J., 1984, Etheno-substituted nucleotides and coenzymes: fluorescence and biological activity, CRC Crit. Rev. Biochem., 15:125-199.

Miller, D., and Weissbach, H., 1977, Factors involved in the transfer of aminoacyl-tRNA to the ribosome, in "Molecular mechanisms of protein biosynthesis," H. Weissbach and S. Pestka, eds., Academic Press, London.

Nakanishi, T., Tomita, F., and Suzuki, T., 1974, Production of a new aminonucleoside "9-(2'-amino-2'-deoxypentofuranosyl) guanine" by Aerobacter-sp, Agri. Biol. Chem., 38: 2465-2469.

Potter, R.L., and Haley, B.E., 1983, Photoaffinity labeling of nucleotide binding sites with 8-azidopurine analogues: techniques and applications, Methods Enzymol., 91:613-633.

Prendergast, F.G., Meyer, M., Carlson, G.C., Iida, S., and Potter, J.D., 1983, Synthesis, spectral properties, and use of 6-acryloyl-2-dimethylaminonapthalene (Acrylodan). A thiol-selective, polarity-sensitive fluorescent probe, J. Biol. Chem., 258:7541-7544.

Stryer, L., 1978, Fluorescence energy transfer as a spectroscopic ruler, Ann. Rev. Biochem., 47:819-846.

Ward, P.C., Reich, E., and Stryer, L., 1969, Fluorescence studies of nucleotides and polynucleotides, J. Biol. Chem., 244:1228-1237.

Weber, G., and Farris, F.J., 1979, Synthesis and spectral properties of a hydrophobic fluorescent probe: 6-propionyl-2-(dimethylamino) mapthalene, Biochemistry, 18:3075-3078

Weber, I.T., Steitz, T.A., Bubis, J., and Taylor, S.S., 1987, Predicted structures of cAMP binding domains of Type I and II regulatory subunits of cAMP-dependent protein kinase, Biochemistry, 26:343-351.

Weigand, G., and Kaleja, R., 1976, Fluorescent guanosine-nucleotide analogues suitable for photoaffinity-labelling experiments, Eur. J. Biochem., 65:473-479.

Wittinghofer, A., Warren, W.F., and Leberman, R., 1977, Structural requirements of the GDP binding site of elongation factor Tu, FEBS Lett., 75:241-243.

AFFINITY LABELING OF THE GDP/GTP BINDING SITE IN *THERMUS THERMOPHILUS* ELONGATION FACTOR TU*

Marcus E. Peter and Mathias Sprinzl

Laboratorium für Biochemie und Bayreuther Institut für makromolekulare Forschung, Universität Bayreuth, Postfach 10 12 51, D-8580 Bayreuth, FRG

SUMMARY

Affinity labeling of *T. thermophilus* EF-Tu•GDP(GTP) complexes by *in situ* oxidation with periodate results in specific modification of lysine residues 52, 137 and 325. Residue 52 of the native EF-Tu is modified by both GDP_{oxi} and GTP_{oxi}, but it cannot be modified by GDP_{oxi} if the protein is cleaved at arginine 59. Residue 137 is preferentially modified by GTP_{oxi} and is essentially unreactive toward GDP_{oxi}. Cleavage of EF-Tu at position 59 renders an amino acid residue be very sensitive to reaction with GDP_{oxi} which is essentially unreactive in the native protein. Residue 325 is a minor reaction site accessible from both GDP_{oxi} and GTP_{oxi} in the native protein. EF-Tu nicked at position 59 does not show any reaction with GDP_{oxi} at lysine 325. Photoirradiation leads to crosslinking of the guanine moiety with region 181-190 of *T. thermophilus* EF-Tu. When the above affinity labeling results are fitted into the available three dimensional models of *E. coli* EF-Tu•GDP and H-ras protooncogen product p21•GDP complexes the following points emerge:

1) There is a high homology in the folding of the structural domains in both proteins.
2) The loop region connecting helix A with the ß-sheet b (La Cour et al., 1985) of elongation factor Tu is placed in the vicinity of the bound GDP/GTP and in analogy to p21 it probably forms a part of a binding pocket for the nucleotide. This loop corresponds to the "effector loop" of G-proteins.
3) Cleavage at position 59 of EF-Tu leads to a conformational change resulting in altered reactivity of GDP_{oxi} towards lysine residues adjacent to the nucleotide binding pocket.

INTRODUCTION

All GDP/GTP binding proteins have a high degree of sequence homology in their nucleotide binding regions (Gilman, 1987; Stryer and Bourne, 1986). In two instances, elongation

---
* Dedicated to Professor Fritz Cramer on his 65th birthday.

factor Tu from E. coli (Jurnak 1985; la Cour et al., 1985) and H-ras protooncogen product p21 (de Vos et al., 1988) the three dimensional structure of the conserved nucleotide binding region around the GDP binding site was determined by X-ray structural analysis. The GDP binding pocket seems to be different in the two structures (Jurnak, 1988). The reason for this difference may be due to the different function of the proteins or alternatively may be a consequence of their structural modification in the course of purification and crystallization. A well known structural modification of the crystallized EF-Tu•GDP from E. coli is a cleavage at the conserved arginine residue 59. This position is highly susceptible to proteolytic cleavage in all elongation factors (Möller et al., 1987) leading to nicked proteins which are active in binding of GDP, GTP and EF-Ts (Wittinghofer et al., 1980) but have strongly reduced affinity to aminoacyl-tRNA (Masuda et al., 1985; Wittinghofer et al., 1980). We attempted to resolve this problem by studying the nucleotide binding site of EF-Tu by affinity labeling. Temperature-stable EF-Tu from the bacterium Thermus thermophilus was used in these studies. Its sequence has been recently determined in two laboratories revealing 70 % identity to EF-Tu from E. coli (Seidler et al., 1987; Kushiro et al., 1987). A feature of this protein is the presence of 10 additional amino acid residues in its "tight" GDP binding domain. These ten residues are not found in the primary structures of other bacterial EF-Tu and up to now are found only in EF-Tu isolated from chloroplasts.

We address three main questions in this communication:

1) Are the structures of the native and nicked EF-Tu identical with regard to the GDP binding site?
2) Which structural changes occur as a consequence of GTP hydrolysis on EF-Tu?
3) Where are the additional 10 amino acid residues of T. thermophilus EF-Tu located with respect to the nucleotide binding site?

Affinity labeling of GDP/GTP binding proteins

For obvious reasons we attempted to use "zero distance" affinity label probes, obtained by in situ modification of EF-Tu-bound GDP and/or GTP. This was achieved by two methods

a) by photoactivation of the nucleotide at 257 nm and
b) by in situ periodate oxidation of the 2′,3′cis diol group in GDP/GTP.

In both cases first the radioactively-labeled nucleotides were bound to the apoprotein. ^{32}P-labeled nucleotides were generally used. For the sequencing work we used uniformely labeled ^{14}C nucleotides (U^{14}C-nucleotides), since in oxidized GTP or GDP the label localized on the nucleobase or on the phosphate atom can be easily lost by β-elimination (Steinschneider, 1971). A in situ modification strategy was important to achieve selectivity of the reaction. Chemical or photochemical alteration of the nucleotide lowers its binding affinity to EF-Tu and increases its dissociation rate from the complex (Wittinghofer et al., 1977). Consequently unspecific reactions on the surface of the protein can occur. For these reasons complexes of EF-Tu•GDP or EF-Tu•GTP were prepared in 2:1 (EF-Tu : nucleotide) stoichiometry, oxidized with periodate in situ and successively reduced by sodium cyanoborohydride. Alternatively photocrosslinking of the complexes by irradiation at 257 nm was performed (scheme 1):

$$\text{1)}\quad \text{EFTu}\cdot\text{GDP (GTP)} \xrightarrow[\text{1 min}]{NaIO_4} \text{EFTu}-\text{GDP}_{oxi}\ (\text{GTP}_{oxi})$$

$$\text{EFTu}-\text{GDP}_{oxi}\ (\text{GTP}_{oxi}) \xrightarrow[\text{1 min}]{NaBH_3CN} \text{EFTu}-\text{GDP}_{red}\ (\text{GTP}_{red})$$

$$\text{2)}\quad \text{EFTu}\cdot\text{GDP (GTP)} \xrightarrow[\text{257 nm, 5 min}]{h\cdot\nu} \text{EFTu}-\text{GDP (GTP)}$$

RESULTS

Photoaffinity labeling

Radioactive nucleotides became covalently bound to the proteins after irradiation. The yield and specificity of the reaction can be followed by polyacrylamide gel electrophoresis of the cyanogen bromide (CB) fragments obtained from irradiated EF-Tu (Fig. 1). There are 6 CB-fragments (out of 12) visible on the gel after Coomassie blue staining. Several small fragments cannot be stained. Autoradiography of the gel shows radioactivity at the band CB4. This belongs to peptide 183-223 of T. thermophilus EF-Tu. Fig. 1B shows the influence of the radiation intensity and 1C the influence of GDP excess on the relative yield of labeling at particular cyanogen bromide fragments. With limited light-dosage and without

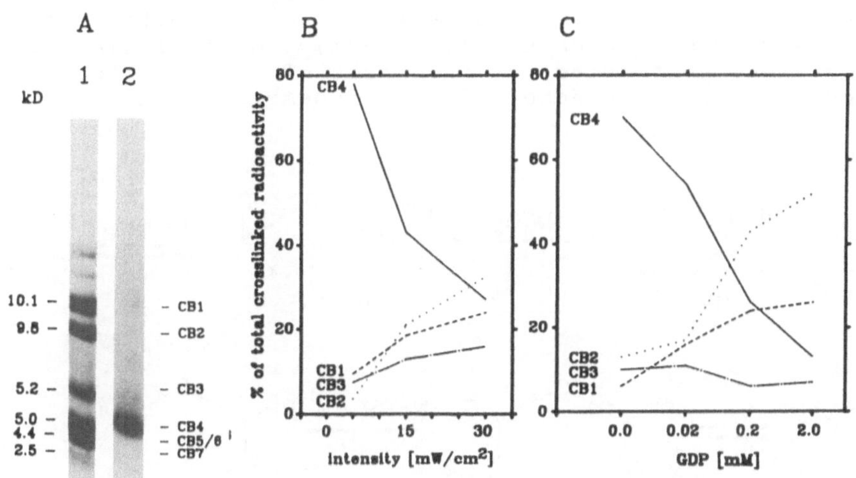

Fig. 1. Photocrosslinking of GDP to T. thermophilus EF-Tu. Reaction was performed as described (Peter et al., 1988), A is PAGE of the cyanogen bromide (CB) fragments obtained from crosslinked EF-Tu.
A: 1= Coomassie blue staining; 2= autoradiography. B: decrease of specificity as a function of light intensity. C: decrease of specificity as a function of GDP excess.

excess of GDP only CB4 is labeled significantly (Fig. 1A). Increasing irradiation intensity or using excess of GDP leads to increased unspecific reaction on other CB fragments. Control experiments which were performed with radioactive ATP resulted in a statistical distribution of radioactivity in CB fragments even without an excess of nucleotide and/or short irradiation times. CB4 was not labeled by ATP. Oligopeptide CB4 (183-223) corresponds to residues which form helix F in the tertiary structure of EF-Tu from E. coli (see below). A loop region on the N-side of this helix is extended by 10 additional amino acids in T. thermophilus EF-Tu and theoretically may be located in the vicinity of the bound guanosine residue. An analogous region in the three dimensional model of p21 was indeed identified as one of the guanine-interacting loops of the protein (de Vos et al., 1988), and very recently it was shown that a mutation in this loop of p21 results in a strong increase of the GDP dissociation rate from the p21•GDP complex (Feig and Cooper, 1988).

Due to both chemical and photochemical instability of the photochemically induced covalent EF-Tu•GDP complexes we were not able to identify the amino acid residue linked to guanine. A likely interpretation of our results would place the photoreaction site in the region 181-190 of T. thermophilus EF-Tu with threonine 188 as the most likely candidate for crosslinking (Shetlar, 1981). Study of the yield of photocrosslinking as a function of the GDP or GTP form of the protein or as a function of nicked versus unnicked Tu are in progress.

Chemical labeling with oxidized nucleotides

Periodate-oxidized ribonucleotides react preferentially with lysine residues of proteins. The initially formed labile Schiffs´base can be stabilized by reaction with cyanoborohydride which reduces imminium systems and is quite unreactive toward aldehydes (Borch et al., 1971). The following reaction scheme was postulated on the basis of model studies with monomers (Lowe and Beechey, 1982) (Scheme 2):

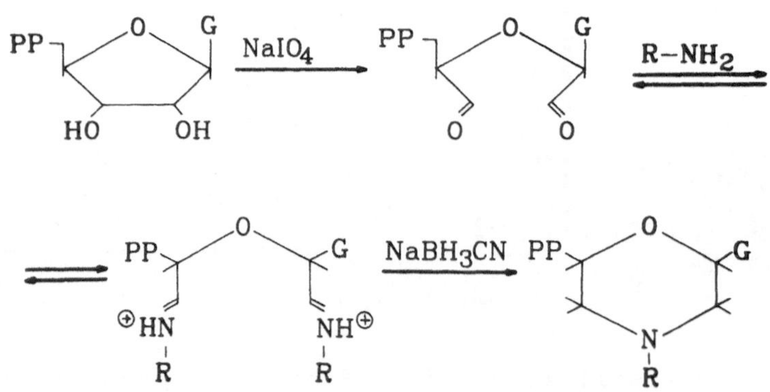

The precise mechanism of the reaction with proteins is, however, not known. Usually it is difficult to isolate and chemically characterize the covalently linked product due to its lability. After many unsuccessfull experiments we opti-

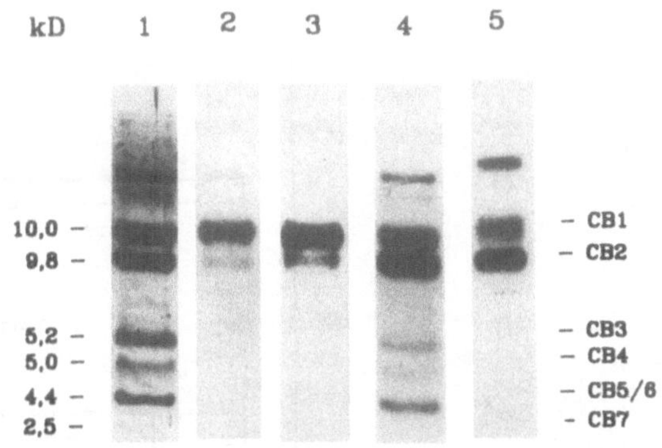

Fig. 2. Cyanogen bromide fragments of EF-Tu from T. thermophilus. 1: Native EF-Tu (stained with Coomassie blue), 2: [U-^{14}C]GTP$_{oxi}$; 1 min-labeled EF-Tu, 3: [U-^{14}C]GDP$_{oxi}$; 1 min-labeled EF-Tu, 4: [U-^{14}C]GDP$_{oxi}$; 60 min-labeled EF-Tu and 5: [^{3}H]CDP$_{oxi}$; 1 min-labeled EF-Tu.

mized the conditions for the isolation of the peptides crosslinked to GDP$_{oxi}$ by gel permeation and high performance chromatography.

In Fig. 2 the cyanogen bromide fragments of GDP$_{oxi}$ and GTP$_{oxi}$ labeled EF-Tu were separated by PAGE and visualized by staining and autoradiography, respectively. Different conditions for labeling were used resulting in different labeling yields. From the autoradiography of different CB fragment preparations the effect of GTP hydrolysis, time, and unspecific binding of CDP$_{oxi}$ to EF-Tu on the yield and site of the reaction was demonstrated (Fig. 2). Detailed chemical analysis of the modified lysine residues was performed on the CB-peptides from experiments in lane 2 and 3 (Fig. 2). Lysine residues 52, 137 and 325 were identified as major reaction sites by Edman degradation with detection by radioactiviy and HPLC analysis (Peter et al., 1988).

After chemical assignment of the crosslinking sites a fast method was developed to study the yield of crosslinking under different reaction conditions. For this purpose a limited trypsin hydrolysis of EF-Tu was suitable. Limited hydrolysis of EF-Tu from T. thermophilus with trypsin provides, as in the case of E. coli EF-Tu (Wittinghofer et al., 1980) a pattern of characteristic fragments which were identified by PAGE and sequence analysis (Fig. 3).

Native unmodified T. thermophilus EF-Tu has 405 amino acid residues. The first proteolytic cleavage occurs at arginine 59 providing the 1-59 and 60-405 fragments (Fig. 3). The small fragment 1-59 is further degraded at arginine 57, lysine 52 and lysine 45. The main fragment 60-405 is further cleaved at residue 275. Autoradiography of the limited digest derived from affinity labeled EF-Tu•[^{32}P]GDP$_{oxi}$ revealed the labeling pattern of the intact protein. There was very weak labeling of the 60-405 fragment containing residues 137 and 325, which was then hydrolyzed to the two smaller fragments 60-275 containing Lys 137, and 276-405, containing Lys 325. The small fragment 1-59 was most strongly labeled. Fragments 1-45 and 1-52 which are visible in the stained gel are not visible by autoradio-

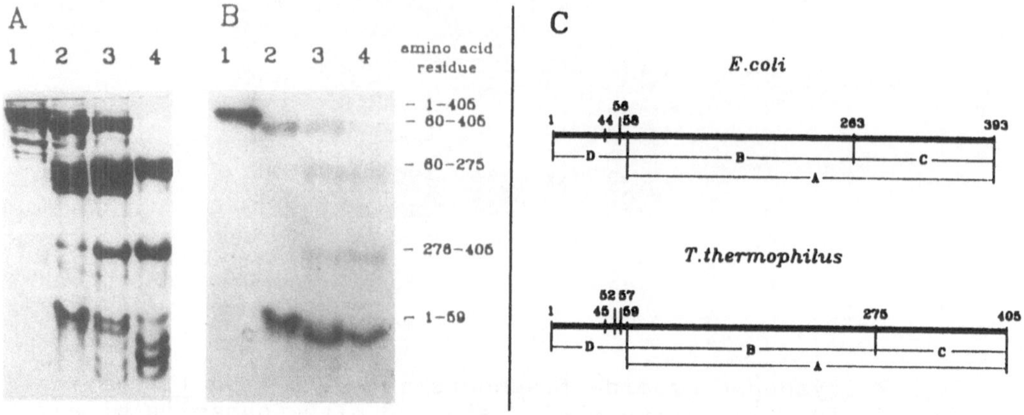

Fig. 3. Limited tryptic digest of [β-^{32}P]GDP_{oxi}-labeled EF-Tu. A: Coomassie blue-staining. B: Autoradiography. Lane 1: Electrophoresis of the labeled polypeptide and lanes 2,3,4: of the labeled peptide after treatment with trypsin for 10 sec, 1 min and 2 hrs as described previously (Peter et al., 1988). C: Schematic representation of the initial tryptic cleavage sites on the polypeptide chain of EF-Tu from E.coli and T.thermophilus.

graphy, clearly indicating the presence of crosslinking at lysine residue 52. Thus limited digestion with trypsin, and autoradiography allows us to screen for crosslinking at positions 52, 137 and 325.

Effect of GTP hydrolysis on affinity labeling

For investigation of a conformational change of EF-Tu upon binding of GTP or GDP, respectively, affinity labeling was performed according to the following scheme (scheme 3):

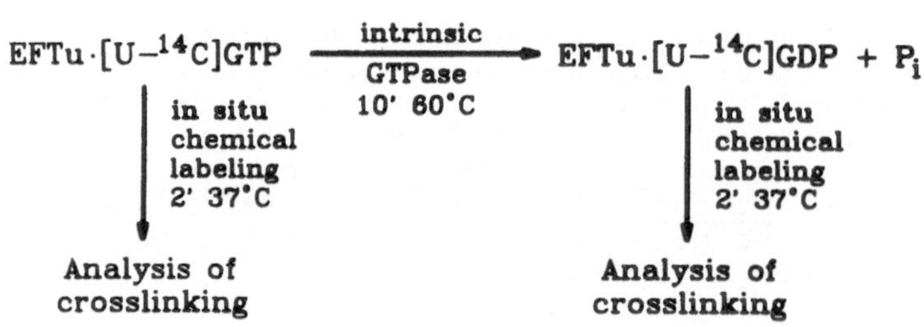

Affinity labeled EF-Tu was cleaved with cyanogen bromide and the fragments separated by gel permeation chromatography. Results depicted in Fig. 4 demonstrate the difference in labeling by GTP_{oxi} and GDP_{oxi}. Whereas with GTP_{oxi} the Lys 137 residue which is located in a short CB polypeptide (Fig. 4A, pool b) was heavily labeled this was not the case with EF-Tu•GDP_{oxi}. A similar interpretation can be derived from analysis of the limited trypsin digestion of GDP_{oxi}-labeled EF-Tu (Fig. 3 and data not shown). The relative yields for labeling of different EF-Tu complexes are summarized in Table 1.

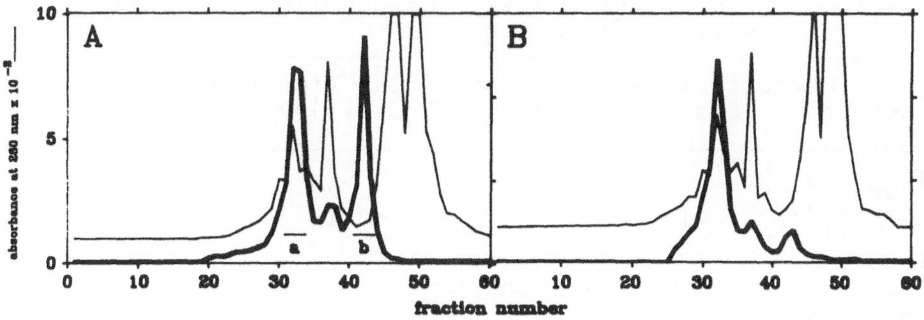

Fig. 4. Gel permeation chromatography of the cyanogen bromide fragments obtained by cleavage of EF-Tu labeled with [U-^{14}C]GTP$_{oxi}$ (A) and [U-^{14}C]GDP$_{oxi}$ (B). Radioactivity (▬) and A_{280} (—) profiles are shown.

Effect of proteolytic cleavage at arg 59 on the affinity labeling of particular lysine residues

Crystallization of E. coli EF-Tu was performed with a protein containing a nick on the invariant arginine 59 (La Cour et al., 1985). Having a rapid assay available to analyze the modification of the nucleotide binding site by oxidized nucleotides, we addressed the question, whether the proteolytic cleavage at Arg 59 changes the accessibility of lysine residues 52, 137 and 325, the following reaction scheme was developed for this purpose (scheme 4):

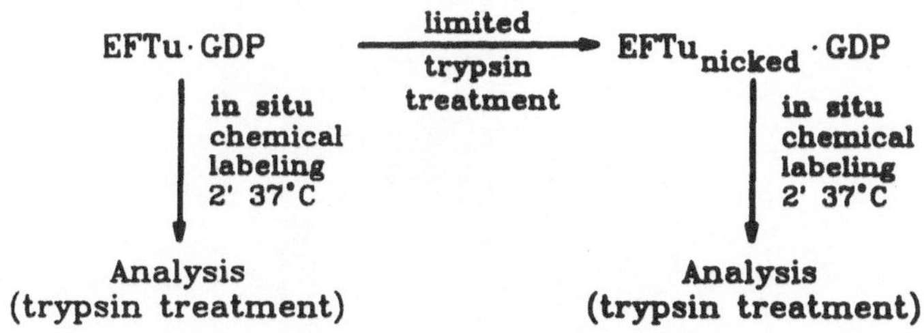

The results of these experiments are shown in Fig. 5. In the case of the native EF-Tu GDP complex the in situ reaction leads to heavy labeling of fragment 1-59. As expected the small remainder of the radioactivity is distributed between fragments 60-275 and 276-405 which may correspond to modified lysines 137 and 325, respectively.

Nicked EF-Tu reacts with GDP$_{oxi}$ almost exclusively via a lysine residue in the tryptic fragment 60-275. In this respect the EF-Tu(nicked)·GDP complex resembles the situation with the native EF-Tu·GTP complex, where Lys 137 was reactive. In this case, however, reaction occured also on lysine 52 (Fig. 3 and Table 1). The proteolytic cleavage at Arg 59 probably disturbs

Table 1. Modification of _T. thermophilus_ EF-Tu by affinity labeling

Method	Percent of total radioactivity attached to EF-Tu[a] Lys 52	Lys 137	Lys 325	CB4
GDP_{oxi}[b]	56	11	22	2
GTP_{oxi}[a,b]	41	38	10	2
$GDP_{oxi}(h\cdot\nu)$[c]	-	-	-	79

[a] Yields and locations of the reactive sites were determined as described by Peter et al., 1988.
[b] Labeling with chemical method see scheme 1 (1)
[c] Labeling with photochemical method see scheme 1 (2)

the location of lysine 52 in the vicinity of the nucleotide binding site. Remarkable is also the reactivity change of Lys 325 in the native and nicked protein. Although this residue is much less reactive as compared to Lys 137 and Lys 52, still a significant increase of reactivity is observed by comparing the native protein with its nicked variant (fragment 276-405 in lanes 2B and 4B, respectively).

CONCLUSIONS

Location of the crosslinking sites in the 3-structure of EF-Tu is shown in Fig. 6. Lysine 137 is part of a highly conserved sequence Asn-Lys-X-Asp present in all EF-Tu molecules sequenced up to now. This residue is located in the nucleotide binding site as revealed by X-ray structural analysis and

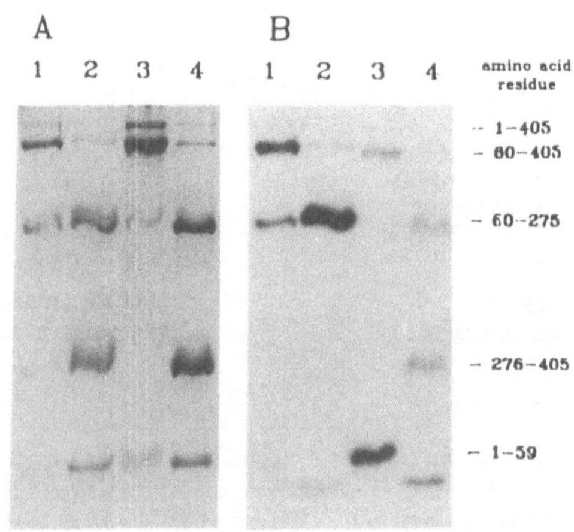

Fig. 5. Polyacrylamide (PAGE) electrophoretic analysis of $[^{32}P]GDP_{oxi}$-labeled EF-Tu. Lanes 1 and 2= EF-Tu containing a nick at position 59, lanes 3 and 4 native EF-Tu. The method of limited trypsin digest (Fig. 3) was used to analyze the positions of labeling. Lanes 1 and 3= 30 sec hydrolysis with trypsin (after affinity labeling), lanes 2 and 4= 30 min hydrolysis with trypsin. A: Coomassie blue-staining of the gel, B: autoradiography of A.

is reactive in the GTP form of the protein or after the cleavage at Arg 59 in its GDP form.

Residue 52 is always a hydrophilic amino acid and is surrounded by a sequence which is common to all elongation factors in this region: Ile-Asp-Lys(52)-Ala-Pro-Glu-Glu. This sequence is part of a loop which could not be localized in the 3D-structure of E. coli EF-Tu probably due to its intrinsic mobility as a result of the nick at position 59 (La Cour et al., 1985). Our results demonstrate that this loop and especially the hydrophilic residue 52 is in the vicinity of the nucleotide binding site. The cleavage of this loop at position 59 changes the geometry of the nucleotide binding region and results in a different distribution in the affinity labeling yield among the reactive lysine residues.

The loop which is partly composed of the 10 additional amino acids present in T. thermophilus EF-Tu residues 181-190 is probably in the vicinity of the guanine base of GDP(GTP). This model receives further support from the fact that this loop contains the sequence Lys-Thr-Arg (187-189) which is also present in the sequence p21 (Dhar et al., 1982) being a part of the second guanine binding loop of its G-binding site (de Vos et al., 1988).

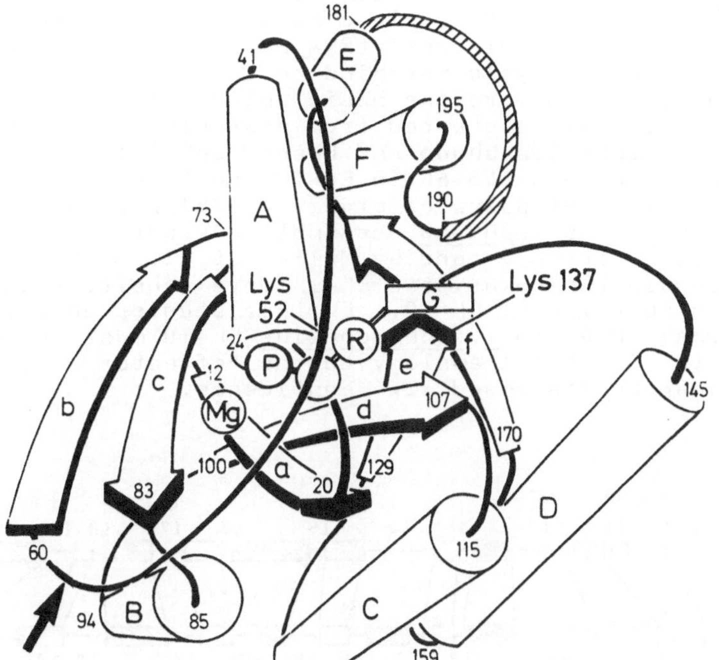

Fig. 6. The G-domain model of T. thermophilus EF-Tu constructed according to the three dimensional models of E. coli EF-Tu, and incorporating the additional information from this work. The lysine residues which crosslink to the oxidized guanosine-nucleotides 52, 137 and the region 181-190 are indicated in model. The minor reaction site-lysine residue 325 is located outside the G-domain. The arrow indicates the nick at Arg 59.

Lysine 325 is located in domain III of E. coli EF-Tu (Jurnak, 1985) and is not part of the GTP binding domain. This residue was identified as an additional reaction site of GDP_{oxi} (GTP_{oxi}). Since the reactivity of this residue is sensitive to the nick at position 59 it is possible that it also reflects a specific site with proximity to the nucleotide binding center of the protein.

The available three-dimensional structures of EF-Tu•GDP from E. coli and p21•GDP, together with the results of this work, allows a comparison of the G-domains of both proteins. By doing so in addition to some sequence homology, a homology in the secondary and tertiary structures of both proteins emerges (Fig. 7). In p21 four loops are essential for GDP binding. Affinity labeling has now identified three of these four loops as a GDP binding site in T. thermophilus EF-Tu. The fourth loop which is located in the proximity of the nucleotide phosphate groups would not be able to react either with photoactivated guanine ring or with the oxidized ribose of GDP/GTP for steric reasons.

The loop consisting of amino acid residues 45-60 in T. thermophilus EF-Tu is structurally homologous to the so called "effector loop" in G-proteins and p21 (de Vos et al., 1988). This loop was implicated in the binding of effector molecules like adenylate cyclase in G-proteins (Bourne, 1986). Indeed binding of aminoacyl-tRNA which is the "effector" in the case of EF-Tu cycle may bind to this region (Laursen et al., 1981). Furthermore the basic amino acid residues located in this region are protected from proteolysis by the binding of aminoacyl-tRNA (Jacobson and Rosenbusch, 1977). The important role of residues 45-60 in EF-Tu function is reinforced by the observations of Gulewicz et al., 1981 who showed that EF-Tu•GDP from T. thermophilus does not interact with EF-Ts when cleaved at position 59 and by Möller et al. (1987) who found this region in the eukaryotic EF-1 more sensitive to tryptic cleavage by binding of EF-1ß. Further studies on the function of residues 40-60 in T. thermophilus EF-Tu must be performed before its functional analogy to the "effector region" of G-proteins can be discussed with confidence.

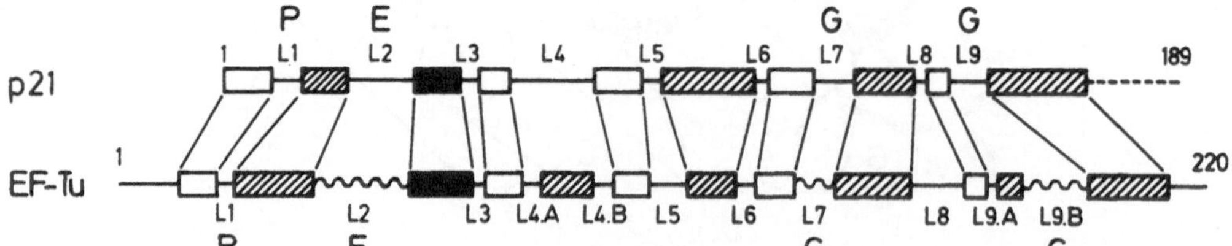

Fig. 7. Comparison of the secondary structure-elements of the G-domain from T. thermophilus EF-Tu with p21. Hatched boxes indicate the α-helices, open boxes ß-strands and the full boxes correspond to the only antiparallel ß-strand in each protein. The loops are numbered from L1-L9 according to de Vos et al., 1988. P: phosphate binding loop, E: effector loop, G: guanine binding regions. Affinity labeled positions are marked by arrows.

Acknowledgements: This work was supported by the Deutsche Forschungsgemeinschaft, SFB 213, D5. We thank S. Bachmann for assistance.

REFERENCES

Borch, R.F., Bernstein, M.D. and Dupont Durst, H., 1971, The cyanohydridoborate anion as a selective reducing agent, J. Am. Chem. Soc., 93:2897.

Bourne, H.R., 1986, One molecular machine can transduce diverse signals, Nature, 321:814.

De Vos, A.M., Tong, L., Milburn, M.V., Matias, P.M., Jancarik, J., Noguchi, S., Nishimura, S., Miura, K., Ohtsuka, E. and Kim, S.-H., 1988, Three-dimensional structure of an oncogene protein: catalytic domain of human c-H-ras p21, Science, 239:888.

Dhar, R., Ellis, R.W., Shih, T.Y., Oroszlan, S., Shapiro, B., Maizel, J., Lowy, D. and Scolnick, E., 1982, Nucleotide sequence of the p21 transforming protein of Harvey Murine Sarcoma Virus, Science, 217:934.

Feig, L.A. and Cooper, G.M., 1988, Relationship among guanine nucleotide exchange, GTP hydrolysis and transforming potential of mutated ras proteins, Mol. Cell Biol., 8:2472.

Gilman, A.G., 1987, G proteins: transducers of receptor-generated signals, Ann. Rev. Biochem., 56:615.

Gulewicz, K., Faulhammer, H.G. and Sprinzl, M., 1981, Properties of native and nicked elongation factor Tu from Thermus thermophilus HB8, Eur. J. Biochem., 121:155.

Halliday, K.R., 1984, Regional homology in GTP-binding protooncogene products and elongation factors, J. Cycl. Nucl. Prot. Phosph. Res., 9:435.

Jacobson, R.G. and Rosenbusch, J.P., 1977, Limited protelysis of elongation factor Tu from Escherichia coli. Multiple intermediates, Eur. J. Biochem., 77:409.

Jurnak, F., 1985, Structure of the GDP domain of EF-Tu and location of the amino acids homologous to ras oncogene proteins, Science, 230:32.

Jurnak, F., 1988, The three-dimensional structure of c-H-ras p21: implications for oncogene and G protein studies, Trends in biochemical Science, 13:195.

Kushiro, A., Shimizu, M. and Tomita, K.-I., 1987, Molecular cloning and sequence determination of the tuf gene coding for the elongation factor Tu of Thermus thermophilus HB8, Eur. J. Biochem., 170:93.

La Cour, T.F.M., Nyborg, J., Thirup, S. and Clark, B.F.C., 1985, Structural details of the binding of guanosine diphosphate to elongation factor Tu from E. coli as studied by X-ray crystallography, EMBO J., 4:191.

Laursen, R.A., L'Italien, J.J., Nagarkatti, S. and Miller, D.L., 1981, The amino acid sequence of elongation factor Tu of Escherichia coli, J. Biol. Chem., 256:8102.

Masuda, E., Louie, A. and Jurnak, F., 1985, Effect of trypsin modification of the Escherichia coli elongation factor, Tu, on the ternary complex with aminoacyl-tRNA, J. Biol. Chem., 260:8702.

Möller, W., Schipper, A., and Amons, R., 1987, A conserved amino acid sequence around Arg 68 of Artemia elongation factor 1 is involved in the binding of guanine nucleotide and aminoacyl transfer RNAs, Biochimie, 69:983.

Peter, M.E., Wittmann-Liebold, B. and Sprinzl, M., 1988, Affinity labeling of the GDP/GTP binding site in Thermus thermophilus elongation factor Tu, Biochemistry, 27:9132.

Seidler, L., Peter, M., Meißner, F. and Sprinzl, M., 1987, Sequence and identification of the nucleotide binding site for the elongation factor Tu from Thermus thermophilus HB8 , Nucl. Acids Res., 15:9263.

Shetlar, M.D., 1981, Cross-linking of proteins to nucleic acids by ultraviolet light, Photochem. Photobiol. Rev., 5:105.

Steinschneider, A., 1971, Effect of methylamine on periodate-oxidized adenosine 5′-phosphate, Biochem., 10:173.

Stryer, L. and Bourne, H.R., 1986, G proteins: a family of signal transducers, Ann. Rev. Cell Biol., 2:391.

Wittinghofer, A. and Warren, W.F., 1977, Structural requirements of the GDP binding site of elongation factor Tu, FEBS Lett., 75:241.

Wittinghofer, A., Frank, R. and Leberman, R., 1980, Composition and properties of trypsin-cleaved elongation factor Tu, Eur. J. Biochem., 108:423.

CHARACTERIZATION OF ELONGATION FACTOR Tu FROM BACILLUS SUBTILIS MODIFIED BY AFFINITY LABELLING

Jiří Jonák, Karel Karas, and Ivan Rychlík

Institute of Molecular Genetics, Czechoslovak Acad. Sci.
Flemingovo nám. 2
166 37 Prague 6, Czechoslovakia

INTRODUCTION

The ability of aminoacyl-tRNA to complex with elongation factor Tu is essential since the ternary complex EF-Tu.GTP.aminoacyl-tRNA is the way aminoacyl-tRNA is carried to the A site of mRNA-programmed ribosomes. During the reaction, GTP in the complex is hydrolysed to GDP, and a stable form of EF-Tu.GDP, which has a low affinity for aminoacyl-tRNA, is released from the ribosome (Kaziro, 1978). The role of GTP might be described as that of an effector which controls the function of the protein binding site for aminoacyl-tRNA.

A model of the tertiary structure of the GDP-binding domain of EF-Tu from E. coli as revealed by high-resolution crystal X-ray diffraction studies has been described (La Cour et al., 1985, Jurnak, 1985). The protein elements that take part in the nucleotide binding are located in four loops connecting parallel β-strands with α-helices. The GDP ligand is linked to the protein via a Mg^{2+} ion which forms a salt bridge between the β-phosphate of GDP and the side chain of aspartic acid 80. Data from modification, photooxidation, and cross-linking studies (Miller et al., 1971, Sedláček et al. 1971, Kaziro, 1978, Duffy et al., 1981, Jonák et al., 1984) indicate that the other principal ligand of EF-Tu, aminoacyl-tRNA, binds at least in part, in this area too, its binding site spanning approximately from His-66 to His-118. In particular, it has been found that modification of cysteine-81, the residue located next to the functionally important Asp-80, diminishes significantly the affinity of the factor for the polynucleotide and especially the affinity for its 3′ aminoacylated terminus (for review see e.g. Jonák et al., 1980, Jonák et al., 1984). These results suggest that the binding sites for GDP (GTP) and aminoacyl-tRNA in EF-Tu from E. coli are in close vicinity or possibly overlap. Furthermore, La Cour and coworkers (1985) have predicted that the loop consisting of residues Asp-80 to His-84 is likely to move to accomodate an additional γ-phosphate residue when GTP, rather than GDP, is bound. Thus, structural properties of this region of EF-Tu appear to have key importance for the function of the protein.

To elucidate the question which molecular groups and interactions participate in the recognition of GDP, GTP and aminoacyl-tRNA by other elongation factors a study with the affinity labelled elongation factor Tu from a taxonomically distant procaryote, B. subtilis, was performed.

RESULTS AND DISCUSSION

Characterization of EF-Tu from B. subtilis

Elongation factor Tu from B. subtilis was purified from an extract of frozen cells by modified classical procedures and obtained in virtually homogeneous form. In an electrophoretic experiment on SDS-polyacrylamide slab gel, the mobility of the final product indicated an apparent relative molecular mass of about 50,000 (Fig. 1). EF-Tu from B. subtilis thus appears to be by about 7,000 daltons larger than E. coli EF-Tu. The yield of B. subtilis EF-Tu ranged between 8 - 12 mg of protein of indicated purity per 20 g of wet bacterial cells.

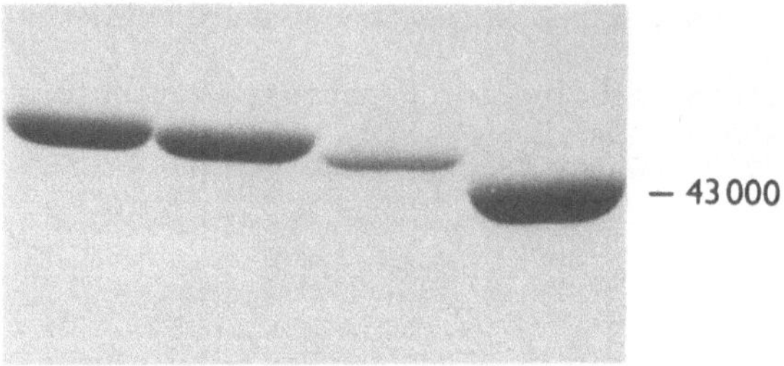

Fig. 1. SDS-polyacrylamide gel electrophoresis of elongation factor Tu from different bacteria. Left to right: B. subtilis, B. stearothermophilus, S. aureofaciens, E. coli

Screening of restriction digests of chromosomal DNA from B. subtilis with the /^{32}P/ labelled tufA probe from E. coli indicated that in contrast to E. coli there is only one gene for elongation factor Tu in B. subtilis genome. (Fig. 2)

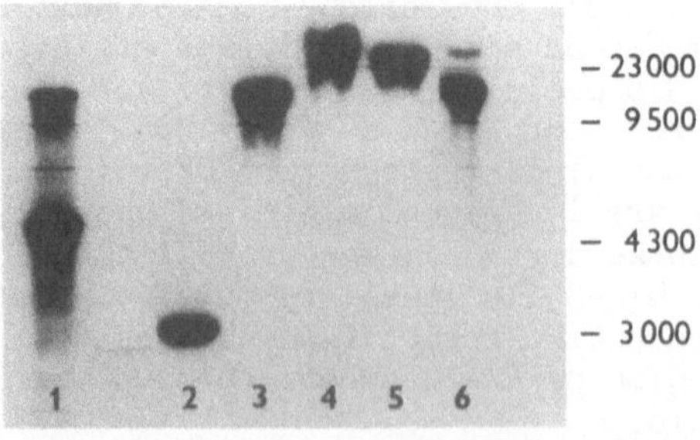

Fig. 2. Hybridization of bacterial chromosomal DNA against a /^{32}P/-labelled E. coli tuf A probe. Lane 1: E. coli DNA, EcoRI-cleaved; lanes 2 - 6: B. subtilis DNA, cleaved by EcoRI, BamHI, SalI, SmaI and BglII, respectively

Table 1. Amino acid composition of EF-Tu from *B. stearothermophilus* (Jonák et al., 1986), *B. subtilis* and *E. coli* (Jones et al., 1980). B.ST., *B. stearothermophilus*, B. SUB, *B. subtilis*, n.d., not determined.

Amino acid	B.ST.	B.SUB.	E. coli
	(residues per molecule)		
Lys	25-26	26	23
His	12	14	11
Arg	25	23	23
Cys	3	3	3
Asx	36-37	42	31
Thr	34-35	37-38	30
Ser	15-16	21-22	11/10
Glx	60-63	61-64	45
Pro	25-27	17-18	20
Gly	41-43	43-44	40/41
Ala	37	32-33	27
Val	42-43	43	37
Met	13	13	10
Ile	26-27	26	29
Leu	27-29	28	28
Tyr	11	12-13	10
Phe	13-14	14-15	14
Trp	1-2	n.d.	1

The amino acid composition of *B. subtilis* EF-Tu obtained after 20 h hydrolysis and based on M_r of 50,000, is shown in Table 1. For comparison, the amino acid composition of another EF-Tu of bacillary origin, EF-Tu from *B. stearothermophilus* together with that of the classical EF-Tu from *E. coli* are also given here. There is a great deal of similarity between the *E. coli* and the bacillary EF-Tu´s, but some significant differences exist. The relative content of Ser residues in *B. subtilis* EF-Tu is much higher, whereas the content of Ile and Pro residues is lower, than in the *E. coli* factor. A lower relative content of Ile was also found in *B. stearothermophilus* EF-Tu. All three elongation factors contain identical amounts of cysteine residues, viz. 3 cysteines per molecule of either factor.

The N-terminus of the protein is apparently blocked because no amino acid residue could be removed by the normal Edman degradation procedure. The protein has a peptide bond sensitive to cleavage, which may sometimes occur during the preparation procedure. This is probably due to the action of one of the many proteolytic enzymes present in *B. subtilis* cells. This phenomenon of a limited cleavage was analysed by electrophoresis in the presence of SDS (Fig. 3h). It results in the loss of material of about 9,000 daltons and the formation of a large fragment of EF-Tu of relative molecular mass of approximately 40,500. This fragment firmly sticks to the intact EF-Tu and copurifies with it until the final stage. The fragment can be degraded by the Edman procedure, this giving the N-terminal sequence Gly-Leu/Ile-Thr which indicates that the limited digestion took place in the N-terminal half of the molecule. In *E. coli* EF-Tu, the same sequence Gly-Ile-Thr can be found at position 59 to 61 preceded by Arg-58, and it is one of the two peptide bonds sensitive to limited cleavage of the factor by trypsin. Our results thus suggest that an analogous topography exists in *B. subtilis* EF-Tu, but the fragment removed from the N-terminus

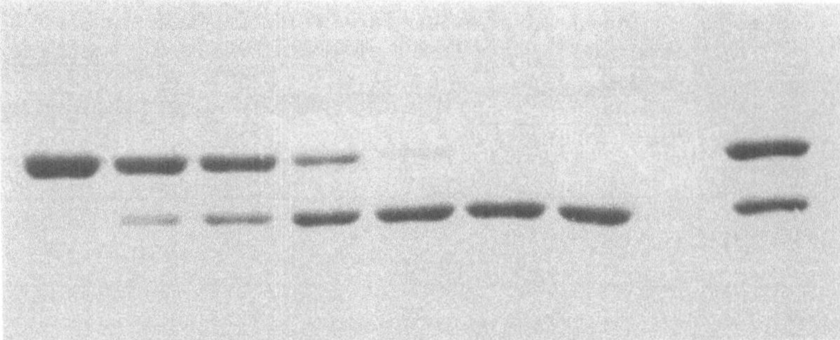

a b c d e f g h

Fig. 3. Limited digestion of B. subtilis EF-Tu by trypsin as revealed by SDS polyacrylamide gel electrophoresis. Lanes a - g illustrate the results of the digestion for 0, 1, 3, 10, 20 and 40 min, respectively. Lane h: a preparation of EF-Tu containing a partially cleaved factor

is by about 20 amino acid residues longer than that in E. coli. This analogy was further confirmed in vitro. Digestion of the B. subtilis factor by trypsin at 4° C resulted in rapid formation of a fragment which had the same mobility as that generated during the purification process (Fig. 3a-g). The GDP binding and exchange activity of EF-Tu.GDP was not affected by the digestion (data not shown). Thus, the integrity of the susceptible peptide bond in EF-Tu from B. subtilis appears to be not essential for the interaction with GDP.

Labelling of EF-Tu from B. subtilis by N-Tosyl-L-phenylalanyl chloromethane

In our previous studies with E. coli and B. stearothermophilus elongation factors Tu (Jonák et al., 1982, Jonák et al., 1986), we had found, that the region containing the functionally important Asp-80 can be specifically labelled by N-Tosyl-L-phenylalanyl chloromethane (TPCK) and then isolated and identified. We applied the same strategy of labelling by TPCK to B. subtilis EF-Tu to probe structural-functional relationships in its molecule. The elongation factor was treated with radioactive TPCK and the effect of the reagent on EF-Tu is summarized in Table 2.

Table 2. Effect of TPCK on EF-Tu from B. subtilis. Cysteine was determined as cysteic acid, TPCK incorporation was measured as in Jonák et al., (1980), phenylalanine polymerization as in Jonák et. al., (1986).

Elongation factor Tu	Cysteine (residues/mole EF-Tu)	TPCK incorporated (residues/mole EF-Tu)	Phenylalanine polymerized (pmoles)
CONTROL	2.99	-	5.1
TPCK - treated	2.15	0.99	0.2

The treatment destroyed the ability of EF-Tu to promote protein synthesis, the loss of about one cysteine residue per molecule of the protein was detected by amino acid analysis, and radioactivity measurements revealed that approximately one molecule of the reagent was incorporated per molecule of the elongation factor. To identify this functionally important site of the protein, the /^{14}C/ TPCK-inactivated EF-Tu was digested with trypsin and the resulting peptides were separated by HPLC on a CGC Separon Six C_{18} reverse-phase column with a gradient of methanol. Fig. 4 shows that a new distinct peak (indicated by arrow), not detected in the hydrolysate of the untreated factor, appeared in the chromatographic profile and it was the only radioactive fraction among all those collected. The amino acid composition of the TPCK-labelled peptide was identical with the amino acid composition of the corresponding peptide isolated from E. coli TPCK-labelled EF-Tu and the complete homology of both peptides was confirmed by sequencing. Thus, it was found that in B. subtilis EF-Tu the TPCK-labelled peptide had the sequence: His-Tyr-Ala-His-Val-Asp-Cys-Pro-Gly-His-Ala-Asp-Tyr-Val-Lys and the target site of the reagent was its cysteine residue. As had been pointed out before, in E. coli EF-Tu this sequence forms one of the four principal binding loops of its GDP-binding domain.

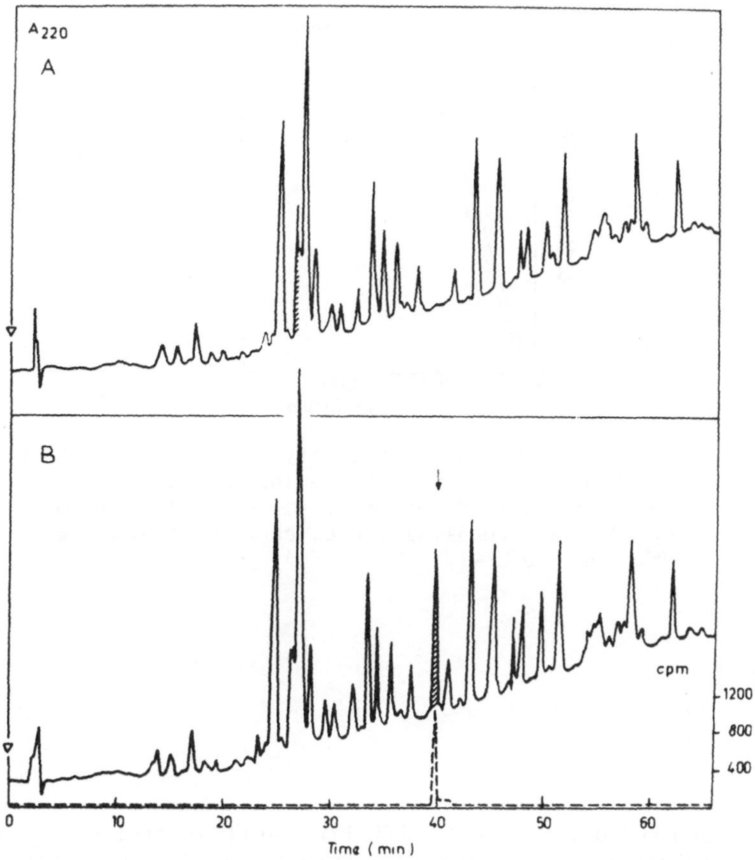

Fig. 4. HPLC of tryptic digest of EF-Tu from B. subtilis. Absorbance at 220 nm was recorded, and radioactivity (---) in a 100 µl aliquot of each fraction was measured: (A) control EF-Tu, (B) TPCK-labelled EF-Tu

According to our results, the integrity of this sequence appears to be essential also for the function of the bacillary elongation factor because alkylation of the cysteine residue results in complete loss of the polymerization activity of the factor. The complete homology of the sequence in this region can be further extended to EF-Tu from B. stearothermophilus (Jonák et al., 1986) and chloroplast EF-Tu (Montandon and Stutz, 1983) and the corresponding sequences of other known EF-Tu´s are very similar in this region too. The sequence Asp-Cys/Ala-Pro-Gly-His is always conserved. The above data illustrate again that even though the bacillary as well as the other known elongation factors Tu differ from E. coli EF-Tu in their amino acid composition and molecular mass, evolutionary constraints on function have maintained this structure from organism to organism.

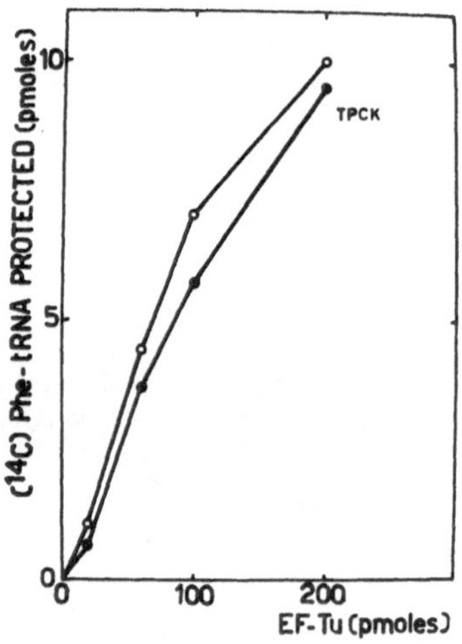

Fig. 5 Effect of TPCK on the activity of EF-Tu to protect the ester bond between the amino acid residue and the 3´terminal adenosine of aminoacyl-tRNA from an alkaline hydrolysis. o-o, control EF-Tu; •-•, TPCK-treated EF-Tu

As had been shown in Table 2, TPCK-treated EF-Tu from B. subtilis is inactive in protein synthesis. Protein synthesis is a complex process in which EF-Tu is involved in several partial reactions such as binding of GDP, GTP, aminoacyl-tRNA, splitting of GTP etc. We tried to determine which partial reaction is the most susceptible to the TPCK modification.

In the E. coli or B. stearothermophilus system, which had been studied before, it had been clearly established that TPCK-modified EF-Tu had lost the ability to catalyse polymerization because its activity to interact with aminoacyl-tRNA had been abolished (Richman and Bodley, 1972, Sedláček et al., 1974, Jonák et al., 1986). The aminoacyl-tRNA binding activity has been routinely measured by the hydrolysis protection assay (Pingoud and Urbanke, 1980). When the TPCK-treated B. subtilis EF-Tu was tested in this assay, to our surprise, the factor, fully inactive in polymerization, was almost not at all inhibited in its hydrolysis protection ability (Fig. 5). The results of this assay thus indicated that, in contrast to E. coli or B. stearothermophilus EF-Tu's, its ability to interact with aminoacyl-tRNA was not affected. TPCK-EF-Tu was then screened for other functions and of

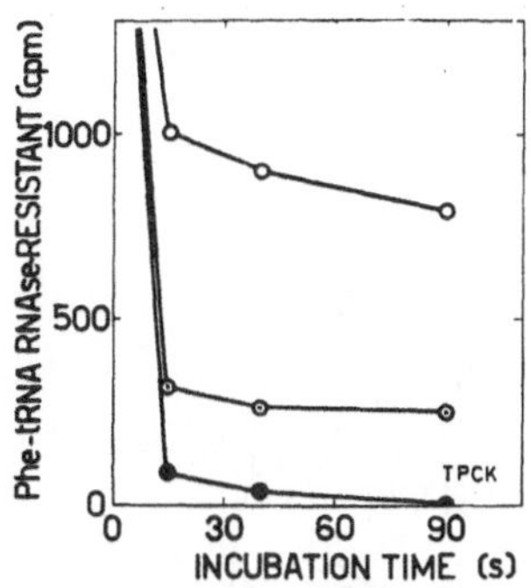

Fig. 6. Effect of TPCK on the activity of EF-Tu to protect aminoacyl-tRNA from the digestion by pancreatic RNase, o-o, control EF-Tu; ⊙-⊙, control EF-Tu, a 5 fold lower concentration; ●-●, TPCK-treated EF-Tu

all the individual functions tested only the activity of the factor to catalyze the transfer of aminoacyl-tRNA onto the ribosome was inhibited to a similar extent as its polymerization activity (data not shown). Therefore, the ability of TPCK-EF-Tu to interact with aminoacyl-tRNA was tested again, but this time using the RNase-resistance assay (Knowlton and Yarus, 1980). The results shown in Fig. 6 clearly indicate that the modified factor lost completely the protecting activity against the RNase. $/^{14}C/$ labelled phenylalanyl-tRNA was digested in its presence almost to the same extent as in the complete absence of the protein. Thus this assay supplied the unequivocal answer: modification of B. subtilis EF-Tu by TPCK results in the loss of its aminoacyl-tRNA binding activity.

How can this finding be reconciled with the negative results of the hydrolysis protection assay? The available data offer the following explanation. The treatment with TPCK made it possible for the first time to identify and to distinguish between two distinct binding sites for aminoacyl-tRNA on the molecule of EF-Tu (Fig. 7). One for the 3´ terminal CC polynucleotide structure (target site for the RNase) and the other for the ester bond of the 3´ terminal aminoacyladenosine of tRNA. Our results indicate that in B. subtilis EF-Tu the former binding site can be selectively disabled by modification with TPCK, while the latter binding site is only slightly affected. They further suggest that the 3´ end polynucleotide structure of aminoacyl-tRNA represents functionally more important binding site for EF-Tu from B. subtilis than the ester bond area, because the lack of the interaction at the former site completely abolished the proteosynthetic activity of EF-Tu even though the interaction at the latter binding site was retained. Thus, despite the complete sequence homology in the functionally essential region around Asp-80 the involvement of its individual amino acid residues in the interaction with aminoacyl-tRNA might vary between E. coli EF-Tu, B. stearothermophilus and EF-Tu from B. subtilis.

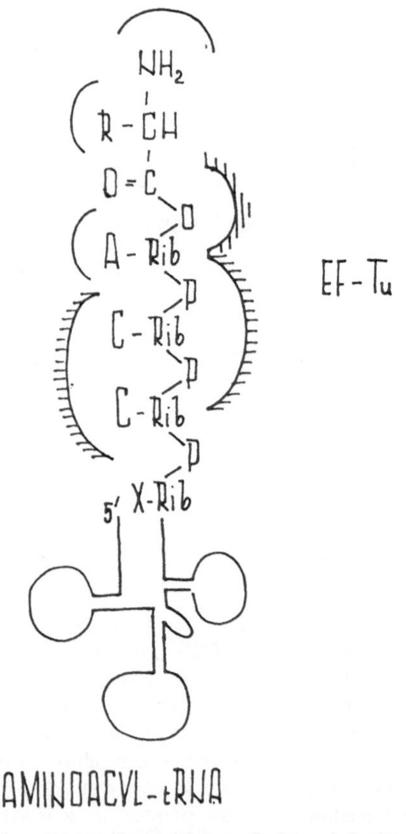

Fig. 7. A schematic representation of two binding sites for aminoacyl-tRNA on the molecule of EF-Tu from B. subtilis as revealed in the study with the TPCK-labelled factor. See text for details. Three other regions of the 3´terminus of aminoadyl-tRNA known to be recognized by EF-Tu are also depicted

REFERENCES

Duffy, L. K., Gerber, L., Johnson, A. E., and Miller, D. L., 1981, Biochemistry, 20:4663.

Jonák, J., Smrt, J.,A., Holý, A., and Rychlík, I., 1980, Eur. J. Biochem., 105:315.

Jonák, J., Petersen, T. E., Clark, B. F. C., and Rychlík, I., 1982, FEBS Lett., 150:485.

Jonák, J., Petersen, T. E., Meloun, B., and Rychlík, I., 1984, Eur. J. Biochem., 144:295.

Jonák, J., Pokorná, K., Meloun, B., and Karas, K., 1986, Eur. J. Biochem., 154:355.

Jones, M. D., Petersen, T. E., Nielsen, K. M., Magnusson, S., Sottrup-Jensen, L., Gausing, K., Clark, B. F. C., 1980, Eur. J. Biochem., 108:507.

Jurnak, F., 1985, Science, 230:32.

Kaziro, Y., 1978, Biochim. Biophys. Acta, 505:95.

Knowlton, R. G., and Yarus, M., 1980, J. Mol. Biol., 139:721.

La Cour, T. F. M., Nyborg, J., Thirup, S., and Clark, B. F. C., 1985, EMBO J., 4:2385.

Miller, D. L., Hachmann, J., and Weissbach, H., 1971, Arch. Biochem. Biophys., 144:115.

Pingoud, A., and Urbanke, C., 1980, Biochemistry, 19:2108.

Richman, N., and Bodley, J. W., 1973, J. Biol. Chem., 248:381.

Sedláček, J., Jonák, J., and Rychlík, I., 1971, Biochim. Biophys. Acta, 254:478.

Sedláček, J., Rychlík, I., and Jonák, J., 1974, Biochim. Biophys. Acta, 349:78.

EFFECTS OF KIRROMYCIN ON THE ELONGATION FACTOR EF-Tu AND ITS INTERACTIONS WITH GDP OR GTP AND tRNA. THE APPLICATION OF ZONE-INTERFERENCE GEL ELECTROPHORESIS, A NEW METHOD FOR THE ANALYSIS OF WEAK COMPLEXES

Barend Kraal, Jan Pieter Abrahams and Leendert Bosch

Department of Biochemistry
Leiden University
Wassenaarseweg 64, 2333 AL Leiden
The Netherlands

INTRODUCTION

All the members of the GTP-binding protein family behave like molecular switches. They can have two conformations: in the presence of bound GTP the complex is in the "on" conformation and, after GTP hydrolysis, the GDP containing complex is in the "off" position. The switching process can become blocked by natural ways, such as phosphorylation or ADP-ribosylation, as well as by artificial tricks, such as mutagenesis at strategic positions of the protein chain or by the use of non-hydrolyzable GTP-analogues.
In the case of the bacterial peptide chain elongation factor EF-Tu it was discovered by Parmeggiani's group, that the antibiotic kirromycin is such an effector (for a review see Parmeggiani and Swart, 1985). Upon binding it strongly enhances the intrinsic GTPase activity of the protein and forces the latter into an aberrant conformation. As a result, the EF-Tu.kirromycin complex remains firmly stuck on the ribosome and thus blocks the peptide chain elongation cycle. Whereas normally EF-Tu.GTP but not EF-Tu.GDP forms ternary complexes with aminoacyl-tRNA, the binding of kirromycin enhances the formation of the latter.

A part of the elongation cycle is shown in Fig. 1. For an efficient binding of the complex EF-Tu.GTP.aminoacyl-tRNA to the ribosomal A-site it is obligatory, that the ribosome.mRNA complex has a filled P-site (panel A). After the initial recognition step and codon-anticodon interaction

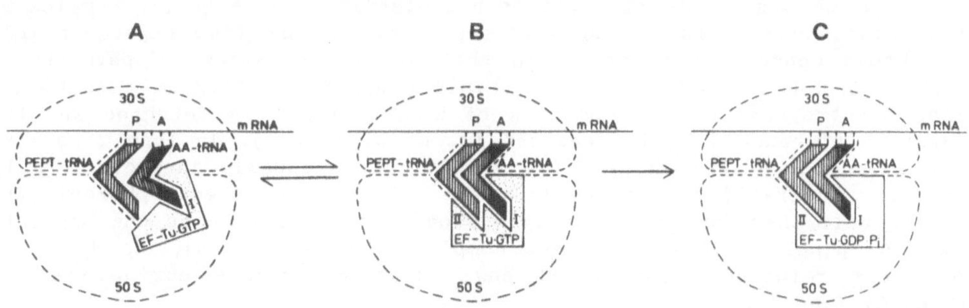

Fig. 1. Model for the recognition of an EF-Tu.GTP.aminoacyl-tRNA complexed to an mRNA programmed ribosome. For further details see text.

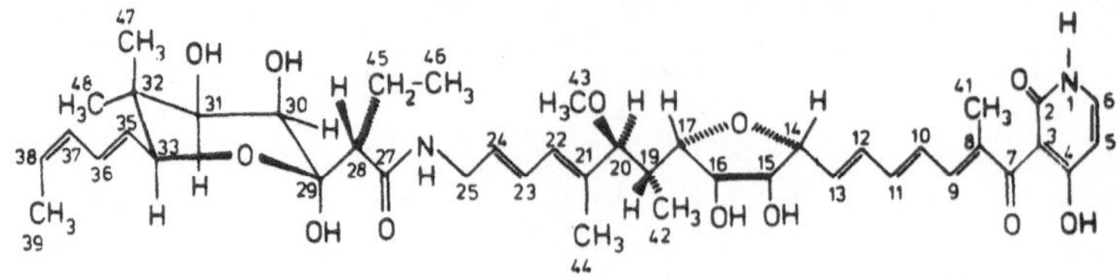

Fig. 2. Structure of the antibiotic kirromycin. The N^1-methyl derivative is called aurodox.

(panel B), the GTPase centre on EF-Tu is activated (panel C). Normally, the resulting EF-Tu.GDP falls off the ribosomal complex. In the presence of kirromycin, however, this does not occur.

The drawings of Fig. 1 suggest a direct interaction between EF-Tu and both tRNA molecules. Indeed, from previous cross-linking experiments with 3' oxidized tRNA (a zero-length cross-linker) we know, that the 3' end of A-site bound tRNA is close to Lys-237 of EF-Tu (site I in Fig. 1). On the other hand, the 3' end of P-site bound tRNA was exclusively cross-linked to Lys-208 in site II (Van Noort et al., 1985). Both lysine residues are located in domain II of the three-dimensional EF-Tu structure (see Chapter 1, Fig. 1 and Table 1). In the absence of ribosomes no such cross-linking occurs, unless kirromycin is added. Again, the same two lysine residues became cross-linked, whereas the effector molecule kirromycin itself could be cross-linked to Lys-357 in domain III (Van Noort et al., 1984). Experiments with EF-Tu.GDP/GTP.kirromycin complexes and increasing amounts of aminoacyl-tRNA strongly suggested that such complexes can simultaneously bind two tRNA molecules as well (Van Noort et al., 1986). In Fig. 2 the kirromycin structure is illustrated. The N^1-methyl derivative is called aurodox and has been mostly used in the present study; no significant functional differences between the two have been observed.

ZONE-INTERFERENCE GEL ELECTROPHORESIS: A NEW METHOD TO ANALYZE WEAK COMPLEXES

In order to better characterize the effect of kirromycin on the interactions between EF-Tu and tRNA molecules, we have developed a new gel electrophoretic method for the analysis of weak macromolecular complexes under equilibrium conditions (Abrahams et al., 1988). The method was inspired by the classical Hummel-Dreyer gel permeation chromatography (Hummel and Dreyer, 1962) and can in fact be considered as a hybrid method between retardation gel electrophoresis (Garner and Revzin, 1981) and equilibrium dialysis. The general principle is shown in Fig. 3, panels A-D, lane 1. During electrophoresis in a cooled 2% agarose gel, a complex ML of macromolecule M and ligand L is migrating all the time through a zone with known concentration of L. In this way dissociation is permanently counteracted by association under rapid dynamic equilibrium conditions. After electrophoresis the position of M in the gel is detected and the migration distance d_{exp} is measured (panel D, lane 1). In lane 2 no zone of L is added. Weak complexes ($K_d > 10^{-6}$ M) are usually characterized by very short half-life times and from the onset of the electrophoresis the complex will therefore immediately dissociate and migrate as the separate M and L (panel C, lane 2). Tight complexes ($K_d < 10^{-9}$ M), on the other hand, are relatively long-lived and hardly dissociate during the run (panel C, lane 3).

From the measured distances and the concentrations of L in a series of zones, the dissociation constant K_d of the complex ML in lane 1 can be determined as follows.

Depending on the value of [L], during a certain fraction of time the component M is migrating as a complex, while the rest of the time it is migrating freely. Therefore, in a dynamic equilibrium we can define d_{exp} by equation (1), in which d_{ML} is the migration distance of the complex ML and d_M that of the free component M.

$$d_{exp} = \frac{[ML]}{[ML] + [M]} d_{ML} + \frac{[M]}{[ML] + [M]} d_M \qquad (1)$$

Please note that [L] in the band of migrating molecules M is the same as in the rest of the zone. This means that [L] does not have to be corrected for complexed L in the equation $K_d=[M][L]/[ML]$. This simplifies substitution of the latter into (1), resulting in equation (2).

$$\frac{d_{exp} - d_M}{[L]} = - \frac{d_{exp} - d_{ML}}{K_d} \qquad (2)$$

This is a Scatchard-like formula: when $(d_{exp}-d_M)/[L]$ is plotted against d_{exp}, a straight line results with a slope of $-1/K_d$ and an intercept of d_{ML} at the horizontal axis (compare Fig. 5).

Interestingly, the concentration [M] does not play a role in (2). In practice, this is a great advantage: in the case of labile macromolecules unknown partial inactivation does not influence the measurement! Other general advantages of the present method for K_d determination are: its high sensitivity (dependent on the autoradiography, immunoblotting or staining technique used) and its speed (electrophoresis time 20 min).

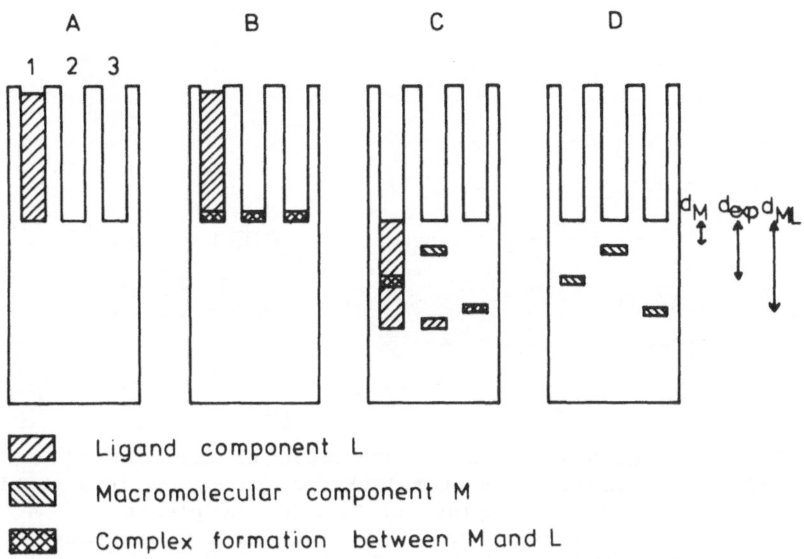

Fig. 3. Schematic comparison of zone-interference gel electrophoresis (lane 1, sample with weak complex) with gel retardation (lane 3, sample with tight complex). In lane 2 the same complex is applied as in lane 1, without interfering zone of ligands.
(*A*) Application of a zone with known concentration (one of a series) of ligand L in lane 1. (*B*) Application of samples ML. (*C*) End of electrophoresis. (*D*) Detection of components M.

CHARACTERIZATION OF VARIOUS EF-Tu COMPLEXES BY THEIR EQUILIBRIUM KINETICS

The zone-interference patterns of combinations of two different EF-Tu complexes and tRNA preparations are shown in Fig. 4. In the panels A and C no significant effect on the EF-Tu migration is observed with increasing concentrations of deacylated tRNA in the concentration range studied. For EF-Tu.aurodox.GDP.aminoacyl-tRNA (panel B) one can estimate the K_d at a first glance already to be around 10 μM. When d_{exp} is in the middle of d_M and d_{ML}, equation (2) shows that the [L] of the particular zone equals the K_d value. For the EF-Tu.GTP complex <u>two</u> bands appear, which will turn out below to correspond to complexes with GTP and GDP, respectively.

In Fig. 5, the migration data from Fig. 4 are plotted. For the EF-Tu.aurodox.aminoacyl-tRNA complex with GDP a K_d of 11 μM is found and for the complex with GTP a value of 3 μM, both at 9°C. The emerging of the additional band in the experiment with the GTP complex can be easily explained. The intrinsic GTPase activity of the EF-Tu molecule is stimulated by aurodox and the interaction of tRNA (see Parmeggiani and Swart, 1985). When the analogue guanosine 5'-O-(3-thiotriphosphate) is used, only the faster moving band appears (result not shown). Fig. 4D thus illustrates an additional important advantage of the zone-interference method: even two complexes may be analyzed together in the same sample.

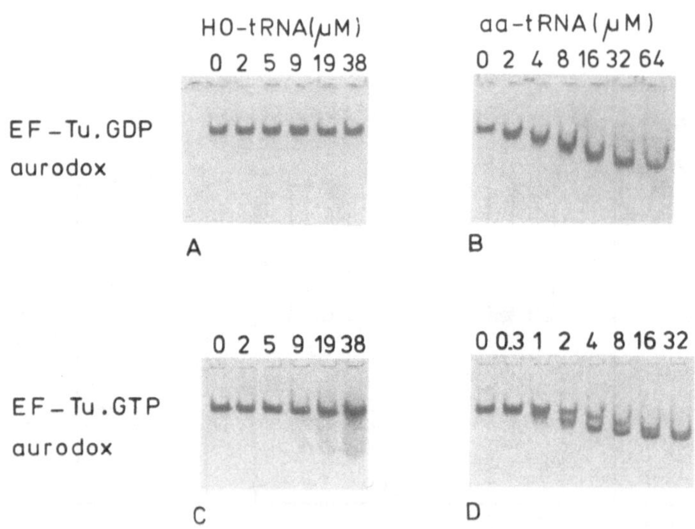

Fig. 4. Zone-interference gel electrophoresis of mixtures of various *E. coli* EF-Tu complexes and total tRNA preparations in a vertical 2% agarose gel system. Samples containing complexes of 4 μM EF-Tu.GDP.aurodox mixed with 4 μM deacylated tRNA (*A*) or aminoacyl-tRNA (*B*), and complexes of 4 μM EF-Tu.GTP.aurodox mixed with 4 μM deacylated tRNA (*C*) or aminoacyl-tRNA (*D*) were applied at the bottom of zones containing either deacylated tRNA or aminoacyl-tRNA in the concentrations indicated. The electrophoresis buffer contained 20 mM Tris acetate pH 7.6/3.5 mM Mg acetate/1 μM aurodox and either 10 μM GDP (in *A*,*B*) or 10 μM GTP (in *C*,*D*). During electrophoresis the gel temperature was about 9°C. Protein bands were stained with Coomassie Brilliant Blue. For further details see Abrahams et al., 1988.

The present value of about 10 µM for the GDP complex is more or less comparable to that of that of 3 µM (25°C) with Phe-$tRNA_{yeast}$ as found in a hydrolysis protection assay (Pingoud et al., 1982), and of 0.9 µM (6°C) with Phe-$tRNA_{E. coli}$ as found by fluorescence titration (Johnson et al., 1986).

Literature values for the corresponding GTP complex are not so easily available. An earlier estimate by means of the hydrolysis protection assay (Pingoud et al., 1978) suggested a two-fold weakening by kirromycin of the binding of aminoacyl-tRNA to EF-Tu.GTP, *i.e.* a much higher affinity than the one found here. On the other hand, a study of the effect of aminoacyl-tRNA (sub)structures on the GTPase stimulation with kirromycin (Parlato et al., 1981) pointed to a K_d value of around 1 µM. In experiments to be published elsewhere, we confirmed our value of about 3 µM, using on one hand the GTPase stimulation assay just mentioned and, on the other hand, a fluorescence titration assay (in collaboration with Sprinzl and coworkers).

No indications were found in Fig. 4 that pointed to the formation of EF-Tu.aurodox complexes with two simultaneously bound tRNA molecules (Van Noort et al., 1986). We cannot exclude the possibility, however, that EF-Tu complexes with one and two tRNA molecules may not differ much in electrophoretic mobility, the effect of the higher negative charge of the latter complex being compensated by a larger Stokes' radius.

In Fig. 6 the equilibrium between EF-Tu.GDP and aurodox is analyzed by application of agarose gel electrophoresis in a horizontal system. Several groups have reported a K_d value of about 1 µM for the weak complex (Parmeggiani and Swart, 1985, and references therein). In our case, complex formation can be analyzed, because the aurodox containing complex moves faster than EF-Tu.GDP alone. Apparently, the complex has got a higher negative charge, since its molecular size is still about the same

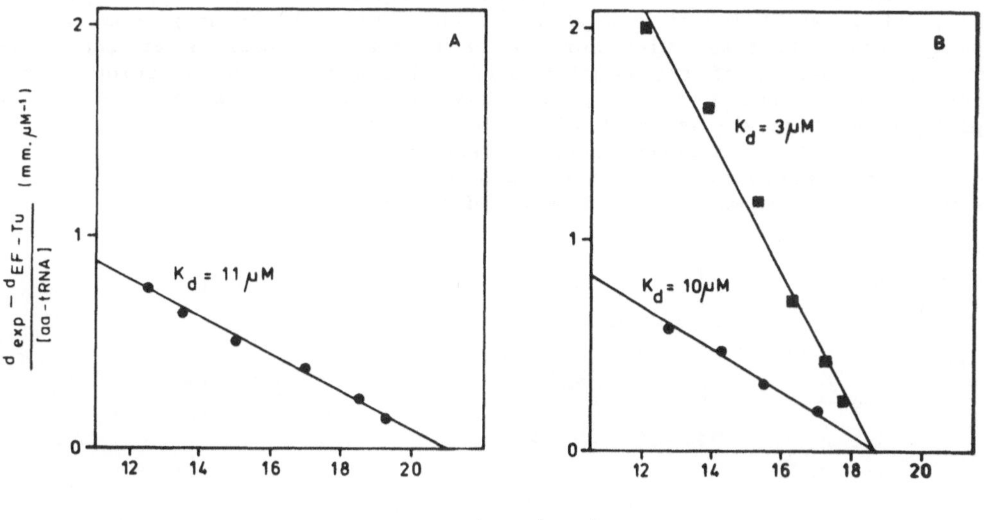

Fig. 5. Graphical calculation of K_d values from the experimental data in Fig. 4 by use of equation (2). In (*A*) the data from Fig. 4B for the equilibrium between EF-Tu.GDP.aurodox and aminoacyl-tRNA are plotted. In (*B*), the data from Fig. 4D are plotted, corresponding to EF-Tu.GTP.aurodox.aminoacyl-tRNA ([]) together with EF-Tu.GDP.aurodox.aminoacyl-tRNA (O).

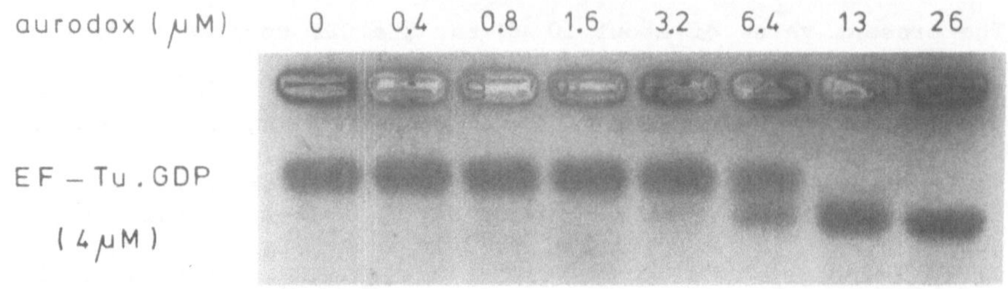

Fig. 6. Horizontal agarose gel electrophoresis of EF-Tu.GDP and aurodox mixtures at concentrations indicated. Apart from aurodox, gel and buffer compositions were the same as used in Fig. 4. For technical details see Abrahams et al., 1988.

as judged by gel permeation chromatography (not shown). The samples were incubated with aurodox in the concentrations indicated; zones with aurodox could not be applied here, because the latter does not migrate as it lacks electric charge. Now the surprise of the gel pattern is, that the band of complexed EF-Tu does not appear to be smeared out but shows a clear jump. In fact, it is a normal retardation pattern (compare Fig. 3, lane 3). The explanation here is, that the complex has extremely slow association as well as slow dissociation rates, as previously reported by Eccleston (1981) using stop flow measurements to be $4.8 \times 10^3\ M^{-1}s^{-1}$ and $1.2 \times 10^{-3}\ s^{-1}$, respectively.

In a similar way, very slow kinetics seem to operate in the case of EF-Tu.GDP.aminoacyl-tRNA, as can be seen in Fig. 7. Whereas no significant affinity is seen for deacylated tRNA, there is an immediate jump of the EF-Tu band to the same migration position as reached by EF-Tu.aurodox.GDP.aminoacyl-tRNA only under saturating concentrations of tRNA in the zone (compare Fig. 4B). For EF-Tu.GDP.aminoacyl-tRNA, K_d values in the range of 10 μM have been reported (Pingoud et al., 1982; Johnson et al., 1986), which is not much higher than reported by those groups for the same complex together with aurodox or kirromycin. Looking at the half-maximum intensity of the complex band and the tRNA concentration in the corresponding zone in Fig. 7, we arrive at a similar value; a Scatchard plot calculation yields about 20 μM (not shown). The implication of such a value is, that the apparently slow dissociation rate (as deduced from the migration behaviour) is counteracted by a very slow association rate, several orders of magnitude below a diffusion controlled one.

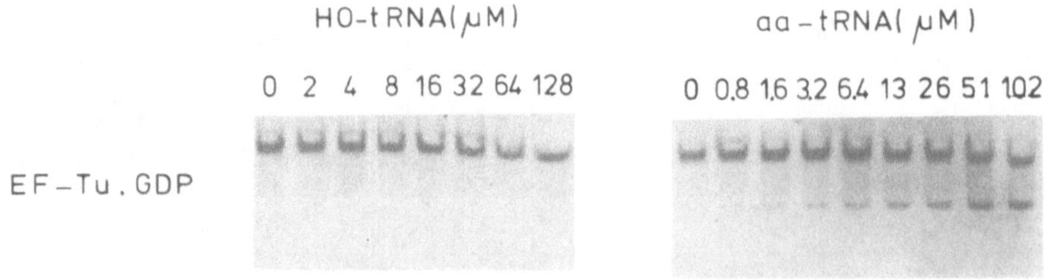

Fig. 7. Zone-interference gel electrophoresis of samples with either 4 μM EF-Tu.GDP plus deacylated-tRNA (*A*) or 4 μM EF-Tu.GDP plus aminoacyl-tRNA (*B*). The samples were applied at the bottom of zones containing the corresponding tRNA in the concentrations indicated. For further details see Fig. 4.

Another surprise in the gel pattern is, that about half of the EF-Tu molecules are non-reactive in tRNA complex formation and remain in the top band position. These molecules are not denatured, however, and fully active in nucleotide binding, while they all become complexed with aminoacyl-tRNA in the presence of aurodox (Fig. 4B). The interpretation of the phenomenon is not so easy. It might be explained by the assumption of two EF-Tu conformations, an "open" and a "closed" one, being in slow equilibrium with each other.

CONCLUDING REMARKS

In order to get a coherent picture of the effects of kirromycin/aurodox on the elongation factor EF-Tu and its interactions with GDP or GTP and aminoacyl-tRNA, in Fig. 8 a cyclic scheme is shown of all the equilibria involved. In fact it is a free energy diagram with all the non-bound components in the top left corner and with the all-embracing complex in the bottom right corner. The difference between upper and lower value for each pair of K_d values reflects the free energy contribution of the γ-phosphate. The difference in K_d between the left and right column of complexes in Fig. 8 is due to the effect of kirromycin/aurodox. The binding of the latter levels the differences between the complexes with GTP and GDP and strongly accelerates their equilibrium kinetics. If one wishes to look upon EF-Tu.kirromycin as a molecular switch, one can see from Fig. 8 that it is frozen in the "off" position with regard to its tRNA-binding conformation. The addition of the antibiotic levels down the aminoacyl-tRNA affinity of EF-Tu.GTP for at least three orders of magnitude.

Having all these thermodynamic values in mind, can one now understand why kirromycin does immobilize EF-Tu so strongly on the ribosomal complex?

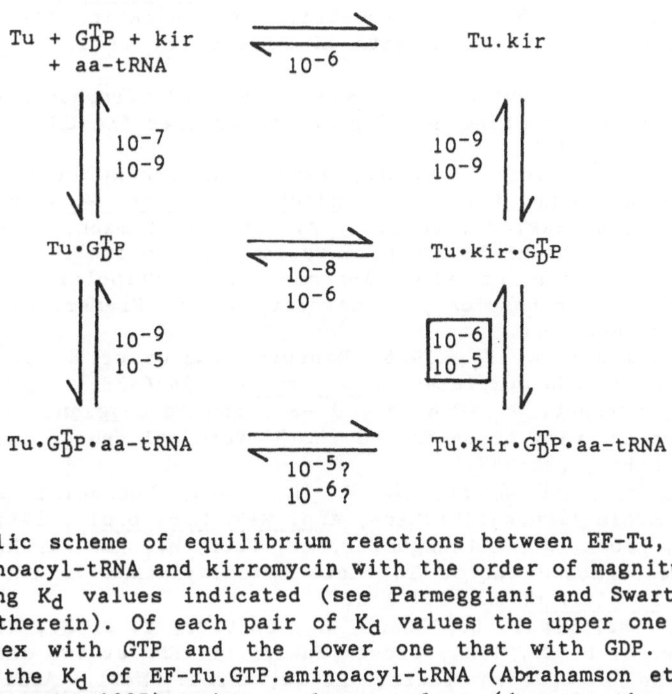

Fig. 8. Cyclic scheme of equilibrium reactions between EF-Tu, GTP or GDP, aminoacyl-tRNA and kirromycin with the order of magnitude of the corresponding K_d values indicated (see Parmeggiani and Swart, 1985, and references therein). Of each pair of K_d values the upper one regards the EF-Tu complex with GTP and the lower one that with GDP. More recent studies on the K_d of EF-Tu.GTP.aminoacyl-tRNA (Abrahamson et al., 1985; Louie and Jurnak, 1985) point to lower values (down to the subnanomolar range) than the reports before 1985. The boxed values originate from the experiments just described. The question mark refers to calculated values without experimental confirmation.

For a long time it has been thought that, after GTP hydrolysis, release of EF-Tu.kirromycin.GDP does not occur because of its relatively high affinity for the aminoacyl-tRNA at the ribosomal A-site (compare Fig. 1). In the meantime, this has appeared not to be true outside the ribosome and now it is becoming clear, that also EF-Tu.kirromycin.GTP has a dramatically reduced affinity for aminoacyl-tRNA. Both complexes display much faster dissociation rates (Fig. 4) than the corresponding ones without kirromycin (Fig. 7), and could therefore leave the ribosome much quicker.
Why does kirromycin immobilize EF-Tu so strongly....? The answer is yet difficult to give. Other interactions with components of the ribosomal complex are undoubtedly involved. Our attention should be focused on experiments dealing with that aspect.

ACKNOWLEDGEMENTS

We would like to thank Dr. A. Parmeggiani for his stimulating discussions. J.P.A. is supported by a grant from the Netherlands Foundation of Scientific Research (N.W.O.).

REFERENCES

Abrahams, J. P., Kraal, B., and Bosch, L. 1988, Zone-interference gel electrophoresis: a new method for studying weak protein-nucleic acid complexes under native equilibrium conditions, Nucl. Ac. Res., in press.

Abrahamson, J. K., Laue, Th. M., Miller, D. L., and Johnson, A. E., 1985, Direct determination of the association constant between EF-Tu.GTP and aminoacyl-tRNA using fluorescence, Biochemistry, 24:692.

Eccleston, J. F., Spectrophotometric and kinetic studies on the interaction of aurodox with EF-Tu from E. coli, J. Biol. Chem., 256:3175.

Garner, M. M., and Revzin, A., 1981, A gel electrophoresis method for quantifying the binding of proteins to specific DNA regions, Nucl. Ac. Res., 9:3047.

Hummel, J. P., and Dreyer, W. J., 1962, Measurement of protein-binding phenomena by gel filtration, Biochim. Biophys. Acta, 63:530.

Johnson, A. E., Janiak, F., Dell, V. A., and Abrahamson, J. K., 1986, The aminoacyl-tRNA.EF-Tu.GTP ternary complex and its role in aminoacyl-tRNA selection at the ribosome, in: "Structure, function and genetics of ribosomes", B. Hardesty and G. Kramer, eds., Springer-Verlag, New York.

Louie, A., and Jurnak, F., 1985, Kinetic studies of E. coli EF-Tu.GTP. aminoacyl-tRNA complexes, Biochemistry, 24:6433.

Parlato, G., Guesnet, J., Crechet, J.-B., and Parmeggiani, A., 1981, The GTPase activity of EF-Tu and the 3' terminal end of aminoacyl-tRNA, FEBS Lett., 125:257.

Parmeggiani, A., and Swart, G. W. M., 1985, Mechanism of action of kirromycin-like antibiotics, Ann. Rev. Microbiol., 39:557.

Pingoud, A., Block, W., Wittinghofer, A., Wolf, H., and Fischer, E., 1982, The elongation factor Tu binds aminoacyl-tRNA in the presence of GDP, J. Biol. Chem., 257:11261.

Van Noort, J. M., Kraal, B., Bosch, L., La Cour, T. F. M., Nyborg, J. and Clark, B. F. C., 1984, Cross-linking of tRNA at two different sites of the elongation factor EF-Tu, Proc. Natl. Acad. Sci. USA, 81:3969.

Van Noort, J. M., Kraal, B., and Bosch, L., 1985, A second tRNA binding site on EF-Tu is induced while the factor is bound to the ribosome, Proc. Natl. Acad. Sci. USA, 82:3212.

Van Noort, J. M., Kraal, B., and Bosch, L., 1986, The GTPase center of elongation factor Tu is activated by occupation of the second tRNA binding site, Proc. Natl. Acad. Sci. USA, 83:4617.

FACTORS AND RIBOSOMES: THEIR COUPLING AND MODE OF SIGNAL PROCESSING

Wim Möller and Reinout Amons

Department of Medical Biochemistry
University of Leiden
P.O.Box 9503, 2300 RA Leiden
The Netherlands

INTRODUCTION

This chapter is concerned with the function of elongation factors as revealed by studies on factor-ribosome interactions. To try presenting a comprehensive picture is impossible given the formidable complexity of the ribosome and the many allosteric interactions between the components. Rather we have chosen a few critical issues, which are amenable to reasonable interpretation. One is that factors possess a GTPase in association with the ribosome. We know or pretend to know that mainly a small distinct area of the 50S ribosome induces this activity and that as a result factor interactions with the ribosome occur in an orderly fashion. We are confident that bacterial 50S ribosomal A protein (L12) is involved in translocation events. Although we are aware of the danger of trying to reduce complex problems to a small facet, the structure and mobile properties of the 50S ribosome stalk guarantee an active role for their constituents in GTPase-related functions.

In the second part of this chapter we try to unravel the thread running through the workings of guanine nucleotide binding enzymes and exchanging enzymes in fields as divergent as protein biosynthesis and visual excitation. This endeavour shows the extraordinary tenacity by which the cell adhers to γ-phosphate hydrolysis of mononucleotides as a means to drive organellar reactions via repetitive association and dissociation of nucleotide binding proteins. The great variability of this principle represents much of the beauty of cellular biochemistry in protein synthesis, microtubule formation and muscle contraction.

1. Ribosome linked GTPase of factors

Elongation factors are multifacetted proteins which act on the ribosome and supply a rich source of structural information. Elongation factor EF-Tu is a guanine nucleotide binding protein which hydrolyzes GTP in a complex system composed of the factor, ribosomes, transfer RNA and messenger RNA. It also can hydrolyze GTP in the absence of ribosomes. However, the biological significance of this non-ribosome-induced GTP hydrolysis of factors is unclear. In the cell substantial GTP hydrolysis by EF-Tu alone would seem disastrous for the energy housekeeping. The great value of kirromycin studies, however, is that they prove that this protein can hydrolyze GTP in the presence of this antibiotic (Wolf et al., 1974) and that as a result at least part of the

GTPase center is likely to be situated there. The picture is somewhat analogous to the ribosome-uncoupled GTPase activity of elongation factor G. This EF-G dependent hydrolysis in the presence of ribosomes but without template and transfer RNA is also non-physiological but does show an clear dependency on L12 in triggering the hydrolyzing activity (Kischa et al., 1971; Hamel and Nakamoto, 1972). In isolation, no single component of the ribosome has been shown to be able to trigger this GTPase. Decisive for insight in the actual function of GTP hydrolysis in protein synthesis has been that EF-Tu.GTP reacts with aminoacyl tRNA but EF-Tu.GDP does not. Equally, factor.GTP complexes have a great affinity to ribosomes but factor.GDP complexes do not. Therefore, a structural transition in EF-Tu upon substitution of GDP for GTP was indicated. The first direct evidence for such a conformational change in EF-Tu was given independently by Miller and by Kaziro (Printz and Miller 1973 and Arai et al., 1974). Since the configuration of inorganic phosphate as produced from GTP in a EF-G.ribosome system becomes inverted with respect to the same phosphorus atom in intact GTP, direct transfer of water without the formation of an intermediate phosphoenzyme was proposed (Webb and Eccleston, 1981). It was considered unlikely that GTP hydrolysis is intrinsic to translocation and support was given to hydrolysis as a means to dissociate factors from the ribosome. In other words, after factors settle on the ribosome, stable ribosome.factor.GTP complexes are converted into instable ribosome.factor.GDP complexes (Kaziro, 1973, Möller, 1974).

2. The GTPase trigger on the ribosome

For the sake of a fast and efficient synthesis of proteins, apparently nature has selected GTP hydrolysis as an essential step in the control of factor-ribosome interactions. In this way alternate binding of the two elongation factors EF-Tu and EF-G to a small 50S ribosomal region comprising proteins L10, L11 and L12 was proposed. Reconstitution and cross-linking experiments by many groups identified protein L12 as the first protein involved in GTPase-related functions on the ribosome (Möller, 1974). Immunoelectronmicroscopy studies by the group in Poustchino (Girshovich et al., 1981) have localized the contact area of bacterial elongation factors at the base of the stalk of the 50S subunit. Our proposed localization of the pentameric $(L12)_4$.L10 complex

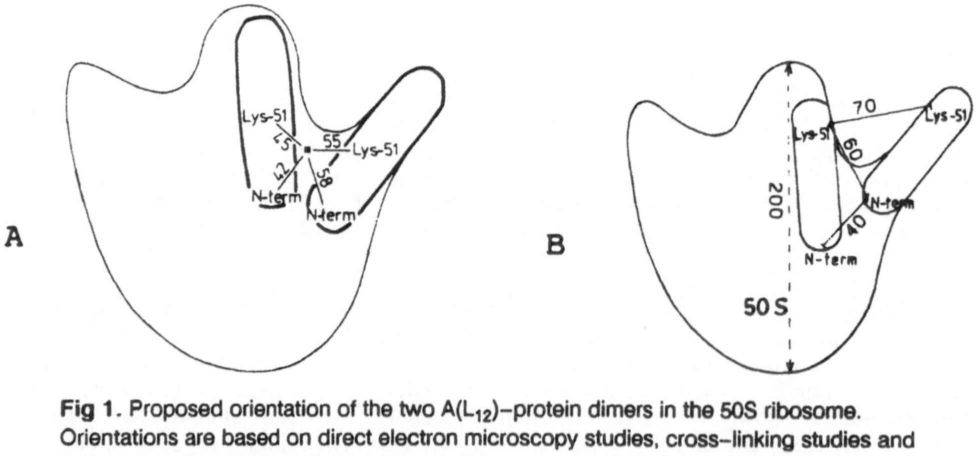

Fig 1. Proposed orientation of the two A(L_{12})-protein dimers in the 50S ribosome. Orientations are based on direct electron microscopy studies, cross-linking studies and energy transfer studies.
A) orientation based on measured distances from a fluorescent label on Cys-70 of L_{10} (▪)
B) orientation via direct distance measurements between the two A-protein dimers.
Distances are expressed in A. Elongation Factors (EF-Tu and EF-G) bind to an overlapping area, lying near the base (N-terminal part) of the tetrameric A-protein structure.

with respect to this stalk is given in Fig.1; for more details and references see Möller and Maasen 1985. We have found that when in the ternary EF-G.GTP.ribosome complex, either GTP was replaced by azido-salicyl-GTP, or EF-G by formylazidophenoxybutyrimidate-labeled EF-G, subsequent photochemical cross-linking mainly labeled the 50S ribosomal protein L11 (Maassen and Möller, 1974, 1978, 1981). Since protein L11 lies also at the base of the stalk (Traut et.al., 1983) the results of cross-linking and electronmicroscopy are consistent. Furthermore, the proteins L12 and L11 bind adjacent to one another in a short region around position 1067 of 23S ribosomal RNA (Schmidt et al., 1981, Beauclerk et al., 1984) and it is precisely this region which is protected by elongation factor EF-G against chemical attack by dimethyl sulphate and kethoxal (Moazed et al., 1988).

The combined evidence supports that factors bind to a protein-23S RNA region located at the base of the 50S stalk and facing the 30S subunit. To draw more farreaching conclusions would be dangerous at this moment. Site-directed mutagenesis studies on ribosomal RNA and ribosomal proteins may shed more light on this problem.

3. <u>L12 as a modulator of tRNA movement in the ribosome</u>

Ribosome research covering almost 20 years, indicates that the function of protein L12 correlates with tRNA displacements on the ribosome rather than with GTP hydrolysis as such. It has been proposed that the two rod-like acidic protein dimers, each carrying at their caboxyterminal part a globular region, could fill up cavities created by tRNAs leaving or occupying A or P site (Möller and Maassen, 1985). Speculation about a CSAM model of the type depicted in Fig.2 finds its origin in the exceptional freedom of rotation of L12 in the ribosome (Gudkov et al., 1982, Cowgill et al., 1984) and in the proficiency of this model to incorporate a great number of experimental facts.

The model includes a variable stalk region and would explain that an aminoterminal rigid fragment of L12 binds to ribosomes but fails to activate protein synthesis (Van Agthoven et al., 1975). The globular carboxyterminal part of bacterial L12 contains many acidic groups which could mimic phosphate groups in tRNA and thus lower electrostatic barriers for tRNAs. This grouping of negative charges in the carboxy-terminal part is even more apparent in eukaryotic L12. It is remarkable that in both prokaryotes and eukaryotes protein L12 possesses an alanine-rich hinge in the middle of the molecule (Liljas, 1982). In support of our model we can add that point mutations at position 74 (Gly → Asp) or at position 82 (Glu → Lys) of bacterial L12 results in decreased growth and increased non-sense codon read-through <u>in vivo</u> (Kirsebom et al., 1986). Mutant analysis of activating regions in transcription factors

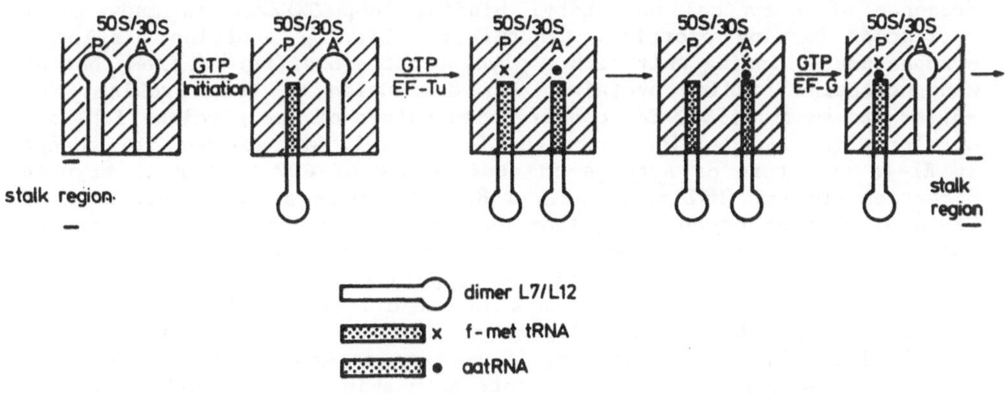

Fig 2. CSAM MODEL (complementary surface arm model)

Table 1. Phosphoryl Binding Loop of Guanine Nucleotide Binding Proteins.

Elongation Factor EF-1α (EF-Tu)	
Artemia	GHVDS G KST
Yeast	GHVDS G KST
Human	GHVDS G KST
Mucor racemosus	GHVDS G KST
Mouse	GHVDS G KST
Drosophila	GHVDS G KST
Xenopus	GHVDS G KST
E. Coli	GHVDH G KTT
T. thermophilus	GHVDH G KTT
Euglena chloroplast	GHVDH G KTT
Yeast mitochondrial	GHVDH G KTT
E. coli IF_2	GHVDH G KTT
Hormone Receptor	
G_o - protein	GAGEG G KTS
Transducin	GAGEG G KTS
Ras protein	
Oncogen P_{21}	GAVGV G KSA
Proto-oncogen P_{21}	GAGGV G KSA
H/K viral oncogen P_{21}	GARGV G KSA
Adenylate kinase	GGPGS G KGT
Concensus sequence	GXXXX G KS/T

also suggests that an excess of negative charge is important for their function (Kakidani and Ptashne, 1988). Just like in eukaryotic L12, the activator for gene expression in GAL4 is characterized by a *large* excess of acidic residues and the overall negative charge rather than the precise sequence dictates the effect of activation. Acidic blobs may function in both transcription and translation as subtle modulators of DNA/RNA movements to their centra of action.

4. Conformational switch in EF-1α

Comparative sequence studies have strengthened the idea of the presence of a typical phosphoryl binding loop GXXXXGK in many mono-nucleotide binding proteins. Dinucleotide binding proteins display a pyrophosphate binding unit with a glycine pattern different from that of mononucleotide binding proteins. For discussion of this problem, see Möller and Amons, 1985. The conservative nature of the glycine-rich loop of elongation factors is documented in Table 1. We searched for change in El-1α structure of Artemia on replacement of GDP by GTP in trypsin digestion studies (Möller et al., 1987). There is a local unfolding of EF-1α at residue 68 on going from GDP to GTP; aminoacyl tRNA increases this local exposure; the response is specific for GTP and not for ATP, CTP or UTP. The sequence around the trypsin sensitive region concerned is compared for different organisms in Table 2; for further references, Möller et al., 1987. Residue Arg-68 of EF-1α of Artemia corresponds to Arg-58 of E.coli EF-Tu; this Arg-58 residue is selectively attacked by trypsin in bacterial EF-Tu also. Since both aminoacyl and uncharged tRNA enhance trypsin cleavage at Arg-68 of EF-1α, tRNA as such apparently exposes residue Arg-68 in EF-1α.

Table 2. Amino acid sequence around common tryptic cleavage site in EF-1α, EF-Tu, EF-2 and EF-G proteins.

Protein	Residue number	sequence
Human EF-1α	(64-75)*	KAERERGITIDI
Artemia EF-1α	(63-74)	KAERERGITIDI
Xenopus EF-1α [a]		KAERERGITIDI
Drosophila EF-1α [a] [b]		KAERERGITIDI
Yeast EF-1α	(64-75)*	KAERERGITIDI
Mucor racemosus EF-1α	(64-75)*	KAERERGITIDI
Yeast mitochondrial EF-Tu	(92-103)	PEERARGITIST
Euglena gracilis chloroplast EF-Tu	(54-65)	PEERARGITINT
E. coli EF-Tu	(53-64)	PEERARGITINT
Hamster EF-2	(61-72)	KDEQERCITIKS
E. coli EF-G	(53-64)	EQEQERGITITS

* Numbering includes initiator methionine;

[a] Krieg P.A., Kentner C. & Melton D., unpublished;

[b] Hovemann B., personal communication.

5. The form of elongation factor I changes during development of Artemia

In the last ten years the two parts of Artemia elongation factor 1, EF-1α and EF-1βγ have been thoroughly purified, characterized, cloned and sequenced (see Van Hemert et al., 1984, and Maessen et al. 1986, 1987). EF-1α binds aminoacyl tRNA to the ribosome. EF-1β catalyses the actual exchange of GDP for GTP on EF-1α while isolated EF-1γ, a strongly aggregating protein, has no clear-cut function but see paragraph 7. Intriguing is that changes in the form of EF-1 occur during a period in which protein synthesis in the Artemia cysts starts again. In the cyst

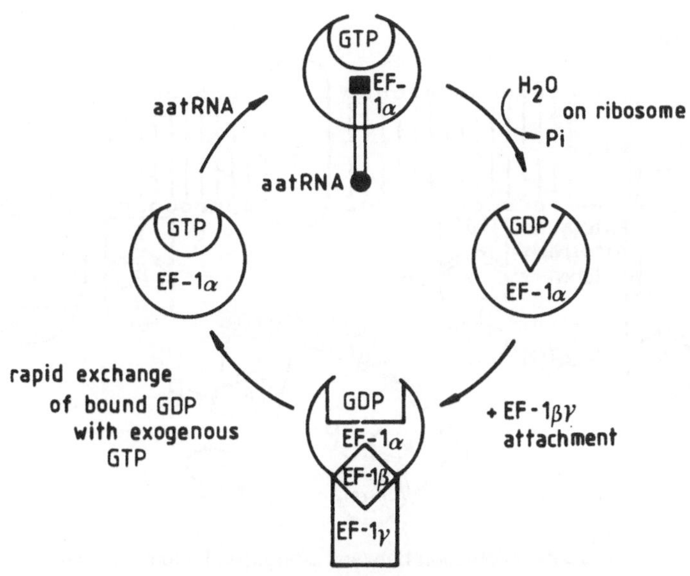

Fig 3. Recycling Mechanism of EF-1α.

most of Artemia's EF-1 is present as a high molecular weight form EF-1H which is an oligomeric 1:1:1 complex of EF-1α, EF-1β and EF-1γ. In the free-swimming nauplii the light form of EF-1L (or EF-1α) has become dominant. It has been proposed that the change in the observed size distribution reflects the break-down of a storage form of EF-1α (Slobin and Möller, 1975). A good correlation between the level of protein synthesis and the degree of EF-1H to EF-1L conversion has been reported in the development of wheat seeds (Sacchi et al., 1984). Specific changes in the relative amounts of the two forms of EF-1 during the recovery from a dormant state into an active state of protein synthesis are therefore indicated in two different cryptobiotic systems. Present research on Artemia is directed towards elucidating the fine structure of EF-1αβγ and its expression during development. The advantage which Artemia offers to those of us who are interested in protein synthesis, development and gene regulation should be viewed against the paucity of data on its genetics and the impermeability of the cyst. For a short review of EF-1 from Artemia see Möller et al., 1987b.

6. EF-βγ seems essential for high elongation rates

The function of EF-1βγ in protein synthesis is to replace GDP for GTP after EF1-α.GDP has left the ribosome (Fig.3 and Kaziro, 1978). The enzymatic interaction of EF-1βγ with EF-1α.GDP has been investigated kinetically in Artemia according to a procedure of Chau et al., 1981 (Janssen and Möller 1988b). To our best knowledge, our investigation presents the first detailed kinetic analysis of an eukaryotic exchange factor. It appears that EF-1βγ is needed to bring the conversion of EF-1α.GDP into EF-1α.GTP at a level which is high enough to be compatible with the known elongation rates in vivo; moreover the data give additional weight to EF-1βγ being a regulatory protein in translation (Janssen and Möller, 1988b). In this connection it is important to note that nucleoside diphosphate phosphotransferase (NDP), an ubiquitous enzyme contaminant in many enzyme preparations, according to us cannot convert EF-1α.GDP into EF-1α.GTP via direct transfer of a γ-phosphate moiety from ATP or GTP. The possibility that the enzyme NDP associates specifically with EF-1βγ cannot be excluded (Janssen and Möller, 1988a). The occurrence of a nucleotide exchange enzyme in ribosomal tRNA binding rather than tRNA translocation reactions may be related to an advantage of a high accuracy of proof reading of aminoacyl tRNAs at the entry site of the ribosome.

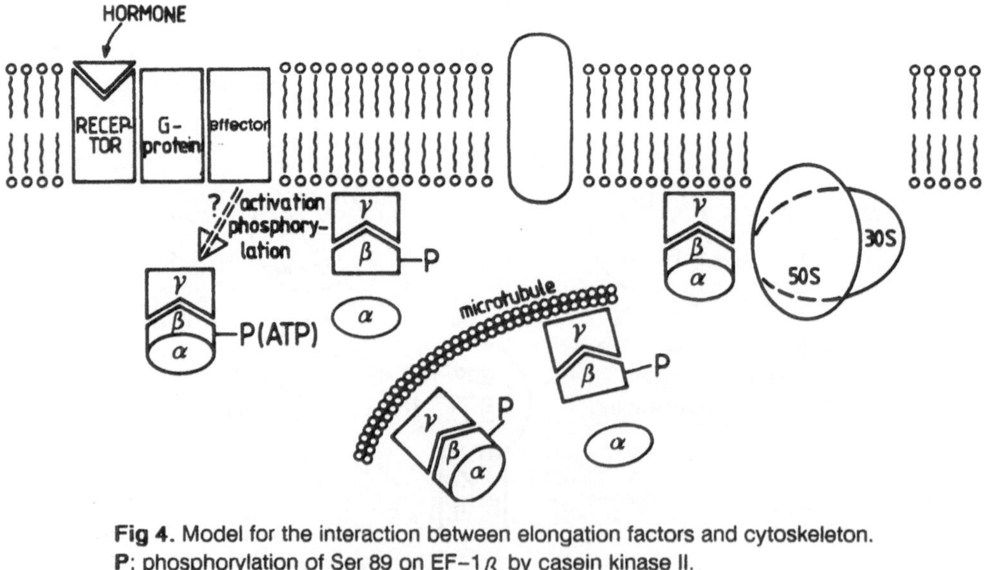

Fig 4. Model for the interaction between elongation factors and cytoskeleton. **P**; phosphorylation of Ser 89 on EF-1β by casein kinase II.

7. EF-1γ behaves as if it anchors EF-1βγ to the cytoskeleton-membrane system

Elongation factor 1 is notorious for its high aggregation behaviour which has complicated its isolation and structure determination in the past. *Artemia* EF-1α normally does not aggregate in aqueous solutions while EF-1βγ does. Even in 1% cholate solutions, EF-1βγ elutes from Sephacryl S300 columns as an aggregate having an apparent molecular weight of about 250,000. Isolated EF-1γ aggregates even more strongly; EF-1β, however, is very soluble in aqueous solvents. *In vitro* experiments demonstrate that EF-1γ interacts specifically with tubulin and membranes, thereby supplying a possible anchor for part of the protein synthetic machinery to the cytoskeleton of the cell (Janssen and Möller, 1988a and Fig.4).

8. Phosphorylation of Artemia EF-1β by casein kinase II

Ejiri and Honda first reported the occurrence of an ATP- and GTP-dependent autophosphorylation of the EF-1β chain in EF-1H (EF-1αββ'γ) from wheat embryos (Ejiri and Honda, 1985). We also found a selective autophosphorylation of EF-1β in EF-1βγ preparations of *Artemia* cysts. Our studies show that the site of phosphorylation is an unique serine residue at (89)Ser-Asp-Glu-Glu-Asp-Glu-Glu(95) phosphorylated by endogenous casein kinase of type II. Most of Ser-89 of EF-1β is phosphorylated as such. Contrary to Ejiri and Honda we did not find an effect of cyclic AMP or cyclic GMP on the degree of phosphorylation. The kinase activity obeys the rules of substrate recognition by casein kinase II and phosphorylation is prevented by heparin and 2,3-diphosphoglycerate. The autophosphorylation effect is only seen with EF-1βγ and not with EF-1β alone, making EF-1γ a candidate for the kinase, as suggested by Ejiri and Honda. However, on using purified casein kinase from *Artemia* (a gift from Dr. H. Slegers from Antwerpen) no correspondence in Mr was found between this kinase and EF-1γ; moreover, no clear ATP-binding site

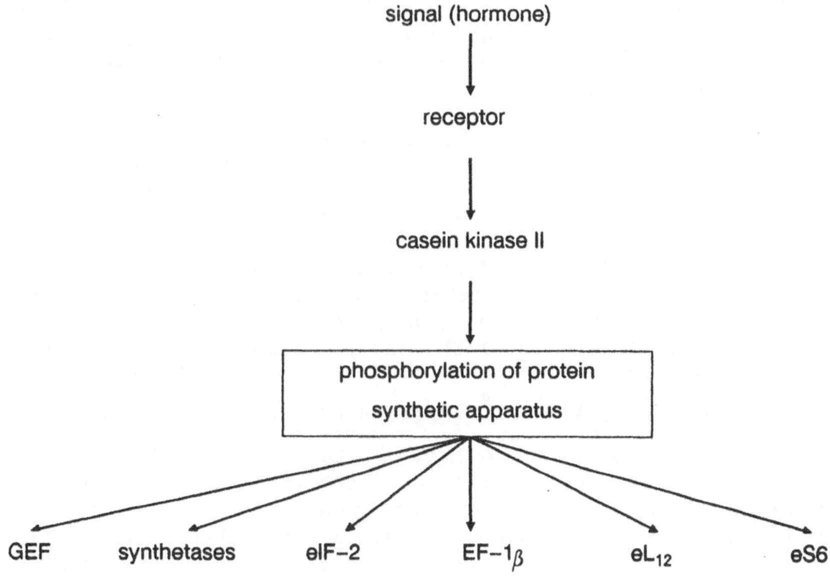

Fig 5. Model for the regulation of protein biosynthesis.

GEF guanine nucleotide exchange factor in initiation step of protein synthesis

eIF-2 eukaryotic initiation factor 2.

eL_{12} eukaryotic 60S ribosomal protein (acidic P_2 protein).

eS6 eukaryotic 40S ribosomal protein.

was found in the published protein sequence of EF-1γ (Maessen et al., 1987) or detected by affinity labeling (unpublished results). Although one cannot exclude that EF-1γ is a hitherto unknown kinase it seems more likely that casein kinase II is a contaminant of EF-1βγ preparations.

Be it as it may, we set out to explore the functional effect of the phosphorylation of EF-1β and found that dephosphorylation of Ser-89 results in an one hundred percent increase of the catalytic nucleotide exchange stimulatory activity of EF-1β (Janssen et al., 1988c). The phosphorylation effect was reversible that is to say rephosphorylation of the dephosphorylated form of EF-1β gave rise to a reduction of 50% of its activity.

Casein kinase II phosphorylates, amongst others, several proteins which belong to the protein synthetic apparatus (see Fig.5). Presently the function of the phosphorylation of EF-1β in protein synthesis is not clear. The same kinase has been reported to stimulate the activity of a GEF-protein which functions in guanine nucleotide exchange during the initiation stage of protein synthesis (Dholakia and Wahba, 1988). In vivo studies of casein kinase may throw light on this problem. With the finding that EF-Tu and EF-Ts form part of the RNA polymerase which replicates bacteriophage Qβ (Blumenthal et al., 1972), one could consider EF-1β to be also involved in transcriptional regulation of eukaryotes.

9. Probing guanine nucleotide binding and exchange sites

Recently, a plethora of sequence similarities between different guanine nucleotide binding proteins has appeared in the literature. Out of the cloning and sequencing of many proteins, universal principles about the structure of guanine nucleotide binding domains arose. Less attention has been paid to the fact that protein synthesis, visual excitation, G protein-dependent hormone action and perhaps even the p21 ras function, all obey reaction pathways containing a guanine nucleotide exchange protein, guanine nucleotide binding protein and a guanine nucleotide hydrolysis regulating protein. The nucleotide exchange protein is often a receptor which acts via an external signal; in eukaryotic protein synthesis this exchange occurs via a protein EF-1β, which gets phosphorylated by casein kinase II at a specific serine residue (Janssen et al., 1988c).

Schemes indicating these common features of guanine nucleotide binding proteins are given in Fig.6. The presentation follows that of the signal transduction route of transducin, as given by Fung et al. 1981. On comparing the different routes one may wonder whether polypeptide chain elongation could be construed as a blue print for the manner in which GTP was selected in cellular evolution.

With respect to visual excitation, cf. Ho and Hingorani, this book, the laboratory of Chabre (Deterre et al., 1988) has discovered that the T_α subunit of transducin reacts with a small inhibitory protein I (Mr 11kD) of phosphodiesterase (PDE) thereby releasing the phosphodiesterase activity according to:

$$2T_\alpha + I_2.PDE \rightarrow 2T_\alpha I + PDE^* \text{ (*activated state).}$$

The point we wish to make here is that in protein biosynthesis, activation of the effector, in this case the ribosome, also proceeds via the intervention of a small effector modulating protein namely the protein L12 (Möller and Maassen, 1985). In bacterial protein synthesis however, L12 protein (Mr 12kD) remains anchored to the ribosome during the course of the factor binding process; also no homologies between L12 and I-protein are apparent. In eukaryotic protein synthesis a pool of unphosphorylated, non-ribosomal bound eL12 (P2-protein) has been found in the cytosol; phosphorylation of eL12 enhances the protein synthetic activity of the ribosomal cores after reconstitution (MacConnell and Kaplan, 1982). The associative state of cytosolic eL12 is still unknown.

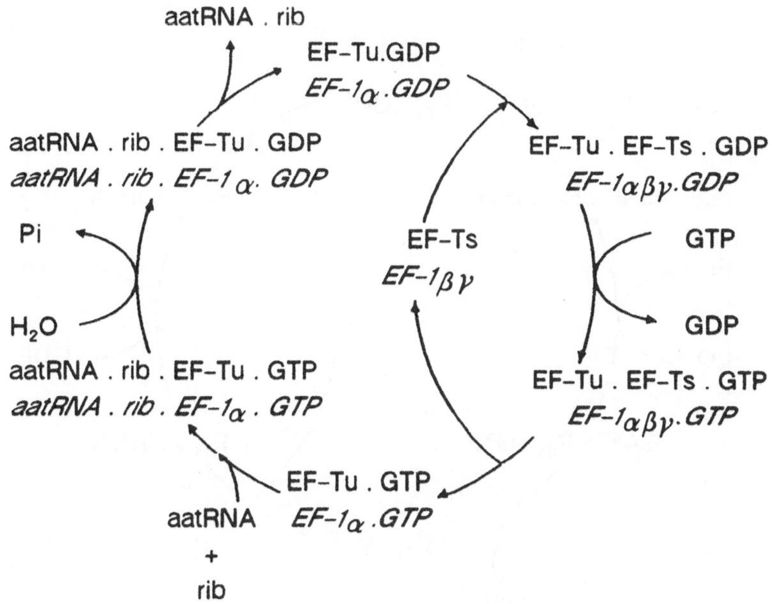

Fig 6.A Peptide chain elongation

EF-Tu = prokaryotic Elongation Factor Tu
EF-Ts = prokaryotic Elongation Factor Ts
EF-1α = eukaryotic Elongation Factor 1_α
EF-1βγ = eukaryotic Elongation Factor $1_{\beta\gamma}$
rib = ribosome

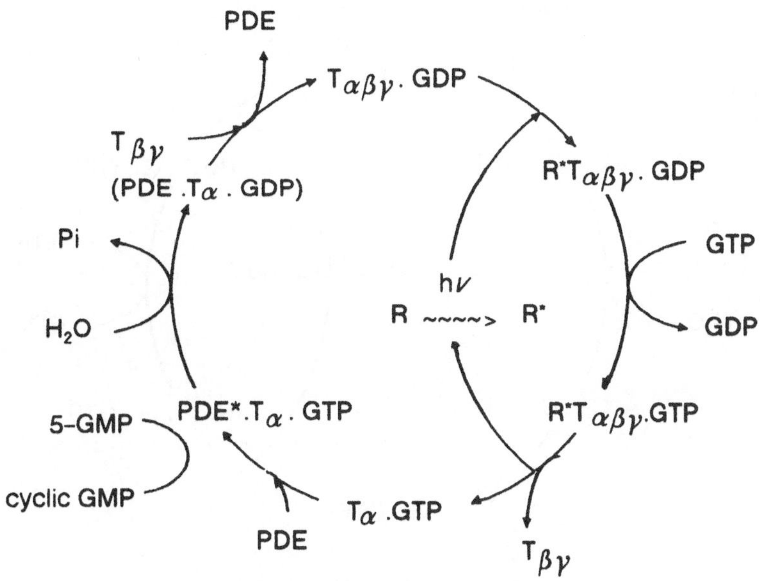

Fig 6.B Visual exitation

R* = photolyzed rhodopsin
T = transducin
PDE* = phosphodiesterase (activated)
adapted from Fung et al (1981)

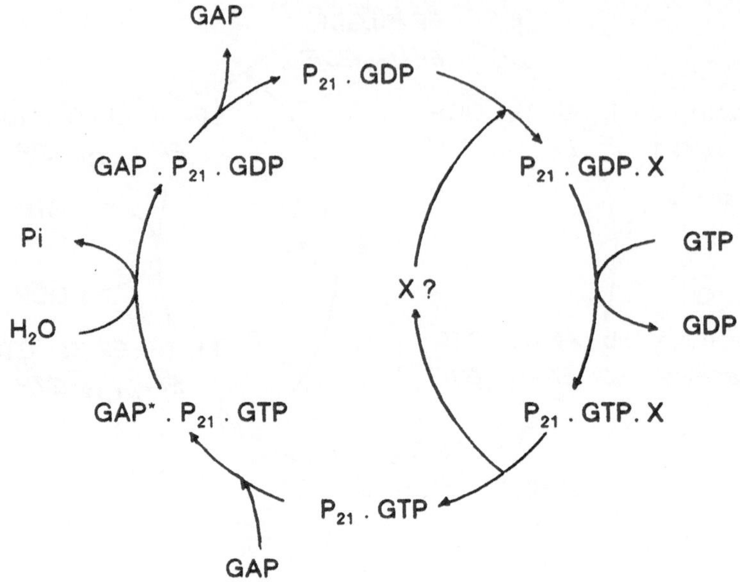

Fig 6.C Ras P_{21} action
P_{21} = RAS protein, 21 kD
X = hypothetical exchange factor
GAP * = activated GAP protein

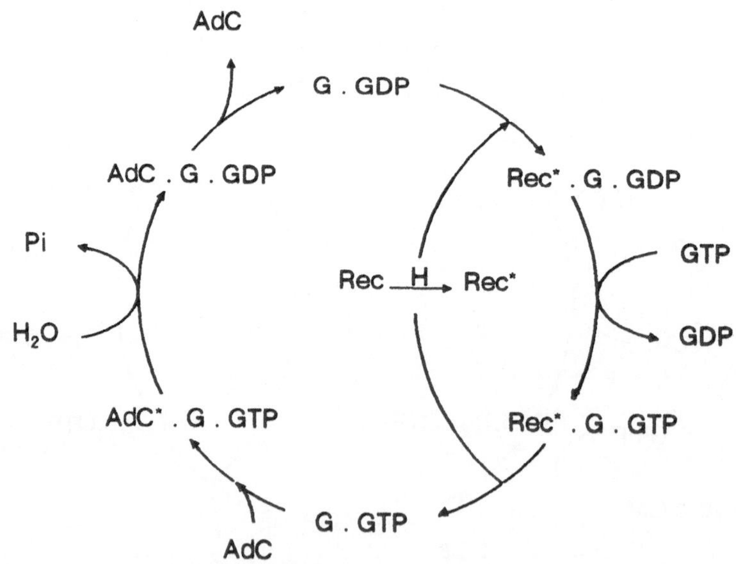

Fig 6.D Hormone action
G = G–protein
H = hormone
Rec* = activated receptor (with hormone)
AdC* = activated adenylate cyclase

10. p21-GAP protein interactions and GTP hydrolysis

The presence of ras genes in non-cancerous cells and the capacity of their protein product p21 to split GTP at a low rate have drawn wide attention. Even more remarkable has been the discovery of the GAP-protein (Trahey and McCormick, 1987) which enhances the GTPase activity of N-ras protein but does not influence the activity of oncogenic ras-mutants. Recent results of the same group indicate that GAP may be the ras effector protein implying that GAP acts downstream from ras-p21 in signal transduction (Adari et al., 1988). So far no protein, including GAP, seems able to catalyze the exchange of GDP for GTP on p21 protein. The property of GAP to act catalytically with respect to stimulation of the GTPase activity of normal ras without affecting the GTPase of certain oncogenic mutants (Gibbs et al., 1988) is reminiscent of the manner in which ribosomes activate the GTPase of elongation factors (compare Fig.6A and 6D). It is possible that p21 participates in a cyclic system where biologically significant GTP hydrolysis is coupled to the repeated association of other components besides GAP. For instance in the presence of messenger RNA and aminoacyl tRNA, one ribosome can induce as many rounds of EF-Tu dependent GTP hydrolysis as there are amino acids in the growing chain (see reaction scheme in Fig.6A). Lack of aminoacyl tRNAs slows down this process (Miller, 1972). However, the rate of GTP hydrolysis in a complete system consisting of ribosomes EF-G, mRNA and aminoacyl tRNA is often at least as high when mRNA and aminoacyl tRNA are omitted (Modolell and Vazquez, 1975 and Parmeggiani and Sander, 1981). The detection of unknown components which change the GTP hydrolysis rate on the GAP-p21 complex may help to find a clue to the reason of loss of growth control in p21 mutants.

REFERENCES

Adari, H., Lowy, D.R., Willumsen, B.M., Der, C.J., McCormick, F., 1988, Science 240:518-521.

Arai, K-I., Kawakita, M., and Kaziro, Y., 1974, J.Biol.Chem., 249:3311-3313.

Beauclerck, A.A.D., Cundliffe, E., and Dijk, J., 1984, J.Biol.Chem., 259:6559-6563.

Blumenthal, T., Landers, T.A., and Weber, K., 1972, Proc.Natl.Acad.Sci. USA 69:1313-1317.

Chau, V., Romero, G., and Biltonen, R.L., 1981, J.Biol.Chem., 256:5591-5596.

Cowgill, C.A., Nichols, B.G., Kenny, J.W., Butler, P., Bradbury, E.M., and Traut, R.R., 1984, J.Biol.Chem., 259:15257-15263.

Deterre, P., Bigay, J., Forquet, E., Robert, M. Chabre, M., 1988, Proc.Natl.Acad.Sci.USA 85:2424-2428.

Dholakia, J.N., and Wahba, A.J., 1988, Proc.Natl.Acad.Sci.USA 85:51-54.

Ejiri, S., and Honda, H., 1985, Biochem.Biophys.Res.Commun. 128:53-60.

Fung, B.K-K., Hurley, J.B., and Stryer, L., 1981, Proc.Natl.Acad.Sci.USA 78:152-156.

Gibbs, J.B., this volume.

Girschovich, A.S., Kurtskhalia, T.V., Ovchinnikov, Y.A., Vasiliev, V.D., 1981, FEBS Lett. 130:54-59.

Gudkov, A.T., Gongadze, G.M., Bushuev, V.N., and Okon, M.S. 1982, FEBS Lett. 138:229-232.

Hamel, E., and Nakamoto, T., 1972, J.Biol.Chem. 247:6810-6817.

Ho, Y.K., and Hingorani, V.N., this volume.

Janssen, G.M.C., and Möller, W., 1988a, Eur.J.Biochem. 171:119-129.

Janssen, G.M.C., and Möller, W., 1988b, J.Biol.Chem. 263:1773-1778.

Janssen, G.M.C., Maessen, G.D.F., Amons, R., Möller, W., 1988c, J. Biol.Chem. in press.

Kakidani, H., and Ptashne, M., 1988, Cell 52:161-167.

Kaziro, Y., 1973, In: Nakao, M., and Packer, L. (Eds), Organization of energy-transducing membranes, pp.187-200, University Park Press, Tokyo.

Kaziro, Y., 1978, Biochim.Biophys.Acta 505:95-127.

Kirsebom, L.A., Amons, R., and Isaksson, L.A., 1986, Eur.J.Biochem. 156:669-675.

Kischa, K., Möller, W., and Stöffler, G., 1971, Nature New Biol. 233:62-63.

Liljas, A., 1982, Prog.Biophys.Mol.Biol. 40:161-228.

Maassen, J.A., and Möller, W., 1974, Proc.Natl.Acad.Sci.USA 71:1277-1280

Maassen, J.A., and Möller, W., 1978, J.Biol.Chem. 253:2777-2783.

Maassen, J.A., and Möller, W., 1981, Eur.J.Biochem. 115:279-285.

MacConnell, W.P., Kaplan, N.O., 1982, J.Biol.Chem. 257:5359-5366.

Maessen, G.D.F., Amons, R., Maassen, J.A., and Möller, W., 1986, FEBS Lett. 208:77-83.

Maessen, G.D.F., Amons, R., Zeelen, J.P., and Möller, W., 1987, FEBS Lett. 223:181-186.

Miller, D.L., 1972, Proc.Natl.Acad.Sci.USA 69:752-755.

Moazed, D., Robertson, J.M., and Noller, H.F., 1988, Nature 334:362-364.

Modolell, J., and Vazquez, D., 1975, In: Arnstein, H.R.V., (Ed), MTP International Reviews on Science, vol.7, Synthesis of Amino Acids and Proteins, pp. 137-178, Butterworths, London.

Möller. W., 1974, In: Nomura, M. Tissières, A., and Lengyel, P. (Eds), Ribosomes, pp. 711-731, Cold Spring Harbor Laboratory.

Möller, W., and Amons, R., 1985, FEBS Lett. 186:1-7.

Möller, W., and Maassen, J.A., 1985, In: Hardesty, B., and Kramer, G. (Eds), Structure, function and genetics of ribosomes, pp. 309-325, Springer Verlag New York, Berlin, Heidelberg, London, Paris, Tokyo.

Möller, W., Amons, R., Janssen, G., Lenstra, J.A., Maassen, J.A., 1986, In: Decleir, W., Moens, L., Slegers, H., Jaspers, E., and Sorgeloos, P. (Eds), Artemia Research and its applications. Vol.2, Physiology, Biochemistry, Molecular Biology pp. 451-469. Universa Press, Wetteren, Belgium.

Möller, W., Schipper, A., and Amons, R., 1987, Biochimie 69:983-989.

Parmeggiani, A., and Sander, G., 1981, Mol.Cel.Bioch. 35:129-158.

Printz, M.P., and Miller, D.L., 1973, Biochem.Biophys.Res.Commun. 53:752-755.

Sacchi, G.A., Zocchi, G., and Cocucci, S., 1984, Eur.J.Biochem. 139:1-4.

Schmidt, F.J., Thomson, J., Lee, K., Dijk, J., Cundliffe, E., 1981, J.Biol.Chem. 256:12301-12305.

Slobin, L.I., and Möller, W., 1975, Nature 258:452-454.

Trahey, M., and McCormick, F., 1987, Science 238:542-545.

Traut, R.R., Lambert, J.M., and Kenny, J.W., 1983, J.Biol.Chem. 258: 14592-14598.

Van Agthoven, A.J., Maassen, J.A., Schrier, P.I., and Möller, W., 1975, Biochem.Biophys.Res.Commun. 64:1184-1191.

Van Hemert, F.J., Amons, R., Pluijms, W.J.M., Van Ormondt, H. and Möller, W., 1984, EMBO J. 5:1109-1113.

Webb, M.R. and Eccleston, J.F., 1981, J.Biol.Chem. 256:7734-7737.

Wolf, H., Chinali, G., and Parneggiani, A., 1974, Proc.Natl.Acad.Sci.USA 71:4910-4914.

THE STRUCTURE AND REGULATION OF MAMMALIAN INITIATION FACTOR eIF2

John W. B. Hershey, Vinay K. Pathak, Heidemarie Ernst, Markus Hümbelin and Randal J. Kaufman*

Department of Biological Chemistry
School of Medicine
University of California
Davis, CA 95616, USA

* Genetics Institute
Cambridge, MA 02140, USA

INTRODUCTION

The process of protein synthesis on ribosomes is promoted by soluble protein factors that act during the initiation, elongation and termination phases of translation (for reviews, see Moldave, 1985; Pain, 1986). A striking characteristic of these factors is that most of them bind GTP or GDP. Since the structure and mechanism of action of these proteins have been studied extensively for the past 20 years, considerable knowledge of their mechanisms of action has accummulated which may be relevant to understanding how G-proteins function in general. We are concerned here with one of the mammalian factors, initiation factor eIF2.

THE INITIATION PATHWAY

eIF2 is composed of three non-identical subunits: α (36.1 kDa; pI = 5.5); β (38.3 kDa; pI = 5.7); and γ (52 kDa; pI = 8.8). Its major function is to bind the initiator methionyl-tRNA (Met-$tRNA_i$) to the ribosome. The factor first forms a binary complex with GTP, and then forms a ternary complex with Met-$tRNA_i$. The ternary complex is an obligate intermediate in the initiation pathway, and binds to 40S ribosomal subunits. The 40S·Met-tRNA complex, called a 43S preinitiation complex, is more stable when some other initiation factors (eIF3, eIF4C) are present, but its formation does not require the presence of mRNA. The binding of eIF2 and Met-$tRNA_i$ is mutually dependent on each other; neither forms a stable 40S complex in the absence of the other. Next, mRNA in association with a number of other initiation factors (eIF4A, eIF4B, eIF4F) interacts at its 5'-terminus with the 43S preinitiation complex. The 40S subunit scans the mRNA, and a stable 40S initiation complex forms at the initiation codon, AUG. The complementary interaction of the anticodon of the Met-$tRNA_i$ and the AUG establishes the reading frame on the mRNA. The 40S initiation complex is short-lived, interacts rapidly with another initiation factor (eIF5) and the 60S ribosomal subunit, and is converted to an 80S initiation complex. This junction reaction requires the hydrolysis of the eIF2-bound GTP to GDP + P_i, and bound initiation factors are ejected from the ribosome. eIF2 leaves as a binary complex with GDP. In order for eIF2 to function catalytically and form another ternary complex, GDP must be exchanged for GTP. The guanosine nucleotide exchange reaction is quite slow under physiological conditions and

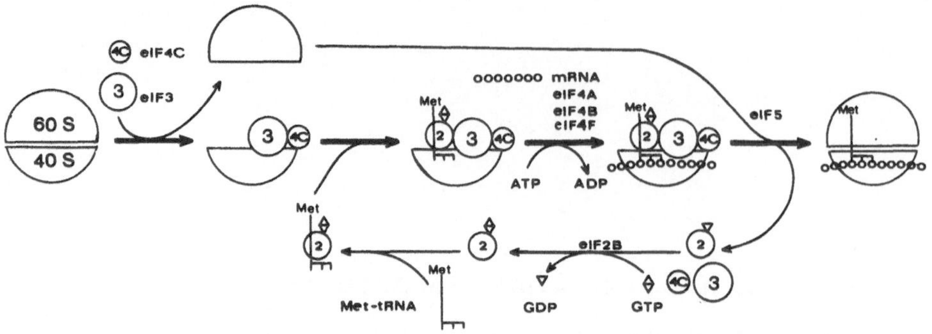

Fig. 1. Scheme for the pathway of initiation of protein synthesis.

requires catalysis by another factor, eIF2B (aka. GEF). The pathway is depicted in Figure 1.

STRUCTURE/FUNCTION OF eIF2

It is apparent from the above description that eIF2 interacts with a large number of translational components, namely: GTP; Met-tRNA$_i$; 40S ribosomal subunits; eIF2B; and possibly mRNA. The role of the individual subunits in these interactions is not yet clear. In order to dissociate the trimeric eIF2 complex, harsh denaturing conditions such as 6 M urea are required, and the resulting subunits are not able to reassociate to form active complexes. To shed light on the structure/function of eIF2, we have cloned and sequenced human cDNAs encoding the α- and β-subunits (Ernst et al., 1987; Pathak et al., 1988a) and we are in the process of trying to clone the γ-subunit cDNA. Northern blot analyses of HeLa mRNAs show that a single size-class (1.6 kb) of mRNAs exists for both the α- and β-subunits. The human gene for eIF2α has been cloned and its promoter structure is being studied in a collaborative arrangement with Dr. Brian Safer's laboratory at the NIH. There is a single locus for eIF2α in the human genome, whereas it appears that up to three gene loci exist for eIF2β.

eIF2α is the smallest of the three subunits and is an acidic protein. It contains 315 amino acid residues and is highly conserved within mammalian species, exhibiting 99% sequence identity between humans and rats. The protein exhibits a high degree of secondary structure as determined by the Garnier algorithm in the Microgenie™ program (Figure 2), but no especially noteworthy structures are discernable. It contains no consensus sequences for guanine nucleotide binding sites. The precise functional role of eIF2α, other than its regulation by phosphorylation (see below), is yet to be elucidated.

The eIF2β subunit also is an acidic protein that is known to be phosphorylated *in vivo*. The subunit is readily cleaved by proteases, resulting in preparations of eIF2 that

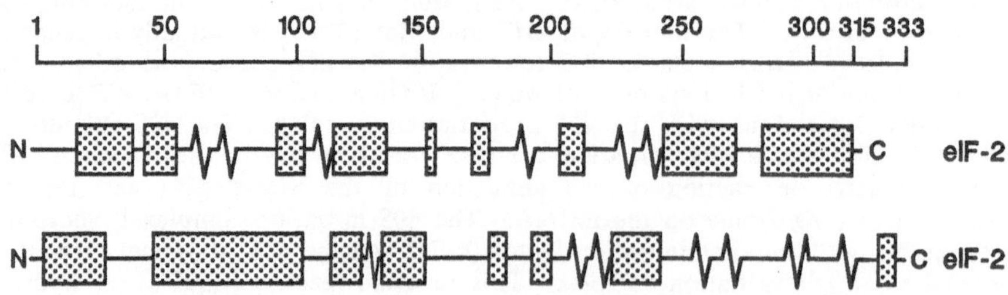

Fig. 2. Secondary structure predictions for eIF2α and eIF2β. The amino acid sequences derived from the cloned cDNAs were evaluated by the algorithm of Garnier by using the Microgenie™ program. Zig-zag lines indicate β-sheets; coils indicate α-helices; horizontal lines represent unstructured regions of the proteins.

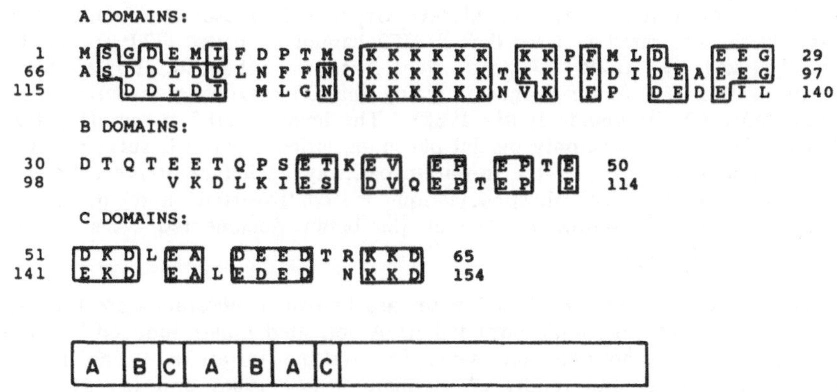

Fig. 3. Sequence analysis of the N-terminal region of eIF2β. The repeating domains of the N-terminal region of eIF2β are aligned to maximize matches of identical residues. Identical residues and conserved substitutions are boxed. The long retangular box at the bottom of the figure indicates the eIF2β coding region in which the N-terminal 154 residues of the protein consist of the three domains in the order shown.

are deficient or lacking in eIF2β (Harbitz and Hauge, 1979; Stringer *et al.*, 1979). Its sequence is less rich in secondary structure (Figure 2), but nevertheless has a number of striking features. 1) The N-terminal half of the 333 amino acid residue protein is very hydrophilic, with 55% of the residues being charged. Three blocks of 6-8 Lys residues are separated by domains rich in acidic residues; the acidic domains themselves are related to one another, suggesting that the coding region was formed by partial gene duplication. The organization of the repeating basic and acidic domains is shown in Figure 3. 2) Two of the three consensus sequences thought to be responsible for guanosine nucleotide binding are found, namely DEEG and NKKD, although the sequence for binding the phosphoryl portion, GSSSSGK, is not present. That eIF2β participates in GTP binding is consistent with affinity crosslinking experiments which label both the β and γ subunits (but not α) (Kurzchalia *et al.*, 1984; Anthony *et al.*, 1986). It is possible that GTP spans both subunits. 3) A zinc finger motif is present near the C-terminus, although zinc is not detected in our preparations of eIF2. The zinc finger motif shares considerable sequence identity with the zinc finger domain of methionyl-tRNA synthetase from *E. coli* (Posorske *et al.*, 1979).

The zinc finger motif and poly-Lys blocks suggest that eIF2β may interact with RNA, e.g. Met-tRNA$_i$, rRNA and/or mRNA. However, assignment of a critical role for eIF2β is complicated by the finding that eIF2 apparently lacking the β-subunit nevertheless forms a ternary complex and promotes the binding of Met-tRNA$_i$ to 40S subunits (Harbitz and Hauge, 1979; Stringer *et al.*, 1979). In recent work from Tom Donahue's laboratory (Donahue *et al.*, 1988), yeast eIF2 was shown to influence initiator codon selection, since a mutant form of the β-subunit enables the cells to initiate translation at the Leu codon, UUG. It is also noteworthy that yeast eIF2β, which shares 42% amino acid sequence identity with the human cognate protein, also possesses the poly-Lys blocks and zinc finger motif in the same locations in the protein. Mutations affecting initiation codon selection lie in the zinc finger motif. These results reinforce the view that the conserved Lys blocks and zinc finger motif are important for eIF2 function.

REGULATION OF eIF2 FUNCTION BY PHOSPHORYLATION

The phosphorylation of the α-subunit of eIF2 causes a severe inhibition of the initiation phase of protein synthesis in rabbit reticulocytes incubated in the absence of heme (reviewed by Ochoa, 1983; Pain, 1986; Proud, 1986). Lysates contain an eIF2α kinase named the heme controlled repressor (HCR) which is negatively regulated by heme. In the absence of heme, HCR is activated and phosphorylates the α-subunit, resulting in a steady-state level of eIF2α phosphorylation of about 30%. P-eIF2 is capable of promoting the binding of Met-$tRNA_i$ to 40S ribosomal subunits, but it has been directly demonstrated *in vitro* that P-eIF2 cannot exchange GTP for GDP in the reaction catalyzed by eIF2B. Instead, P-eIF2 shows enhanced affinity for eIF2B, thereby preventing this factor from catalyzing the GTP exchange reaction with non-phosphorylated eIF2 (Rowlands *et al.*, 1988). The level of eIF2 is significantly higher than that of eIF2B; therefore only partial phosphorylation of eIF2 is sufficient to inhibit eIF2B activity and subsequently the initiation pathway. Whether or not other functions of eIF2 are affected by phosphorylation is controversial and unclear. The phosphorylation of eIF2 represents one of the better documented cases of metabolic control by protein kinases.

At least two different protein kinases are known to phosphorylate the α-subunit of eIF2: HCR and DAI, a double-stranded RNA activated kinase induced by interferon. The two kinases phosphorylate the same Ser residue in eIF2a. The residue was identified as Ser-48 in one report (Wettenhall *et al.*, 1986) and Ser-51 in another (Colthurst *et al.*, 1987). We mutated the eIF2a cDNA, converting Ser-48 and Ser-51 to Ala-48 or Ala-51, respectively The wild-type and mutant forms were expressed *in vitro* by SP6 polymerase and the resulting capped mRNAs were translated in a rabbit reticulocyte lysate. Exogenous HCR and DAI phosphorylated the wild-type and Ala-48 mutant forms but not the Ala-51 form, suggesting that Ser-51 is the correct site of phosphorylation for these highly specific protein kinases (Pathak *et al.*, 1988b).

An *in vivo* approach to identifying the physiologically significant site of phosphorylation is to transfect cells with an expression vector carrying the eIF2α cDNA altered so that Asp rather than Ser occurs at positions 48 or 51. The Asp residue, carrying a negative charge, might mimic phosphoserine and thereby affect the activity of eIF2. The wild-type or mutant eIF2α cDNA was inserted downstream from the Adenovirus major late promoter with most of its tripartite leader intact (Kaufman, 1985). The structure of the expression vector is shown in Figure 4. Co-transfection of COS-1 cells with the eIF2α vector and a similar vector expressing DHFR results in about 25% of the cells expressing the vector genes. The transfected cell population can be purified by fluorescent staining with fluorescein-methotrexate and sorting with a fluorescence activated cell sorter. Such cells express wild-type and the Asp-48 mutant 10-20 fold over endogenous eIF2α. Immunoblotting of lysate proteins resolve by 2-dimensional gels and *in vitro* phosphorylation studies indicate that both forms are phosphorylated to a high extent. However, transfection with the Asp-51 mutant results in no overproduction of eIF2α (even though high levels of its mRNA are produced), but rather causes an inhibition of global protein synthesis. The result supports the view that phosphorylation at Ser-51 but not at Ser-48 causes inhibition of initiation. Such inhibition appears to occur at quite low levels of the mutant protein. This is expected if the Asp-51 mutant, either as the free subunit or exchanged into endogenous eIF2, is able to tie up eIF2B in a stoichiometric interaction. The synthesis of only 0.2 to 0.3 copies

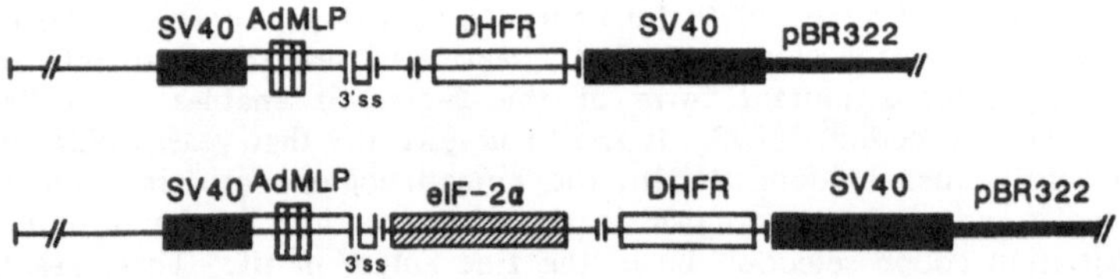

Fig. 4. DHFR and eIF2α expression vectors. The DHFR expression vector (pD61) has been described elsewhere (Kaufman, 1985). It is a pBR322-based plasmid which contains the SV-40 origin and enhancer element. Transcription is driven by the adenovirus major late promoter which is followed by most of the tripartite leader present on adenovirus late mRNAs. Following the DHFR gene is the SV-40 polyadenylation signal. The eIF2α expression vector, shown in the lower part of the figure, is basically the same as the DHFR vector, with the eIF2α cDNA (Ernst *et al.*, 1987) inserted between the tripartite leader and the DHFR gene.

of eIF2α (Asp-51) per endogenous eIF2 should be sufficient, yet would not readily be detected compared to the 10-20 copies for wild-type eIF2α.

Phosphorylation of eIF2α correlates with inhibition of the rate of initiation of protein synthesis in cells subjected to a variety of physiological stresses. We have applied Western immunoblotting to quantitate the extent of eIF2α phosphorylation in mammalian cells. The cells are lysed directly into high urea buffer to minimize alterations in phosphorylation states, and lysate proteins are fractionated by high resolution IEF/SDS-polyacrylamide gel electrophoresis. After transfer to nitrocellulose, eIF2α variants are detected with specific antibodies and a radiolabeled second antibody. Examples where eIF2α phosphorylation correlates with the inhibition of protein synthesis are: heat shock (Ernst *et al.*, 1982; Duncan and Hershey, 1984); serum deprivation (Duncan and Hershey, 1985); amino acid starvation (Clemens *et al.*, 1987); treatment with a variety of chemicals (Duncan and Hershey, 1987); and virus infections (Schneider and Shenk, 1987). The multiplicity of examples suggests that eIF2α phosphorylation may be a general pathway for regulating global rates of protein synthesis in mammalian cells. However, correlative information does not provide proof that this mechanism actually operates in intact cells, nor that eIF2α phosphorylation is the sole or primary cause of translational inhibition in the cases noted.

In order to study the role of phosphorylation in intact cells, we have used transfected COS-1 cells with plasmid vectors carrying cDNAs encoding forms of eIF2α with Ala substituted for Ser at positions 48 and 51. The rationale is that the Ala-51 mutant, when overexpressed, might interfere with eIF2α kinases in the cell, thereby preventing phosphorylation of the endogenous eIF2α. Two experimental systems were used to test the possible effects of eIF2α forms on protein synthesis: the synthesis of DHFR from plasmid vectors; and the synthesis of viral proteins in cells infected with mutant forms of adenovirus-2 lacking VA-RNA genes. High levels of VA-I RNA have been shown to stimulate the synthesis of DHFR and viral proteins by preventing the activation of DAI (Kaufman and Murtha, 1987; Schneider and Shenk, 1987). We asked whether or not overexpressed eIF2α could substitute for VA-I RNA in promoting DHFR or adenovirus protein synthesis.

COS-I cells were co-transfected with vectors designed to express DHFR and eIF2α, and transfected cells were enriched by staining with a fluorescein derivative of methotrexate and sorting with a fluorescence activated cell sorter. DHFR synthesis is low in the absence of VA-I RNA (Fig. 5, lane 2), but is stimulated in its presence (lane 1). In the absence of VA-I RNA, overexpression of the wild-type eIF2α had no effect (lane 3). However, overexpression of the Ala-48 or Ala-51 mutant forms (lanes 4 and 5) stimulated DHFR synthesis about 5 to 10 fold, to levels similar to those obtained with VA-I RNA. It is noteworthy that neither of the Ala mutant forms of eIF2α nor VA-I RNA affect global rates of protein synthesis. Apparently both interfere with DAI activation in the microenvironment of plasmid-expressed mRNAs, possibly due to regions of dsRNA generated by bidirectional transcription of the plasmids. A similar specific inhibition of mRNA translation was observed *in vitro* with mRNAs containing partially double-stranded structures (De Benedetti and Baglioni, 1984). Our results show clearly that both Ala mutants affect DHFR synthesis, and suggest that this may occur by interfering with the phosphorylation of endogenous eIF2. Since the Ala-48 mutant serves as a phosphorylation substrate for the eIF2α kinases both *in vitro* and *in vivo*, it is not clear how these mutant eIF2α proteins act to stimulate DHFR synthesis.

To test for the effect on adenovirus protein synthesis, long-term transfected cell lines were constructed which overexpress the wild-type and Ala-48 mutant forms of eIF2α. Infection of the Ala-48 cell line with adenovirus mutants lacking the VA-I and/or VA-II genes led to enhanced yields of adenovirus proteins and virions. This preliminary result provides further evidence that eIF2α phosphorylation contributes to control of protein synthesis in viral-infected cells.

Experiments are in progress to construct a long-term cell line expressing the Ala-51 mutant form of eIF2α. Such cell lines may be useful in determining whether or not eIF2α phosphorylation is necessary for inhibition of protein synthesis in cells subjected

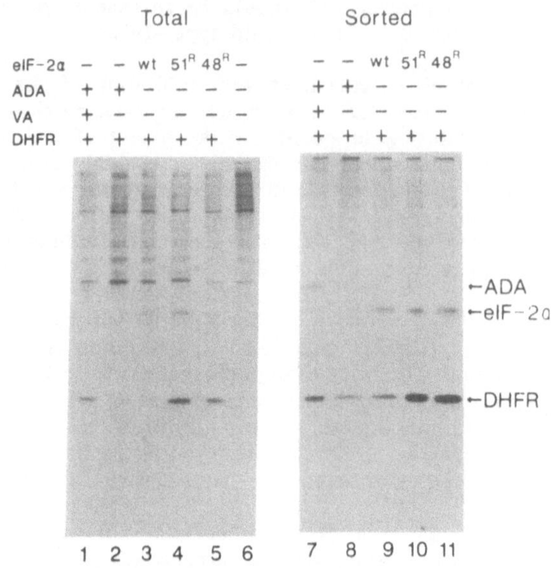

Fig. 5. The effect of eIF2α mutant forms on DHFR synthesis. COS-1 cells were cotransfected with pD61 expressing DHFR (see Fig. 4) and a similar plasmid expressing either murine adenosine deaminase (ADA) or eIF2α (see Fig. 4), as indicated at the top of the figure. Three forms of eIF2α were expressed: wild-type (wt); Ser-51 mutated to Ala-51 (51^R); and Ser-48 mutated to Ala-48 (48^R). Transfected cells were analyzed as a total population (lanes 1 - 6) or following sorting by a FACS (lanes 7 - 12). Unsorted and sorted cells were pulse-labeled with [^{35}S] methionine, and lysate proteins were fractionated by SDS-PAGE. The figure shows a photograph of the autoradiogram.

to inhibitory conditions. The use of mutagenesis coupled with transfection is a general approach for studying the role of phosphorylation in the regulation of metabolic pathways and may be particularly applicable to studies of G-proteins.

REFERENCES

Anthony, D.D., Dever, T.E., Abramson, R.D., Lobur, M., and Merrick, W.C., 1986, Affinity labeling of protein synthesis factors. Fed. Proc., 45:1768.

Clemens, M.J., Galpine, A., Austin, S.A., Panniers, R., Henshaw, E.C., Duncan, R., Hershey, J.W.B., and Pollard, J., 1987, Regulation of polypeptide chain initiation in CHO cells with a temperature-sensitive leucyl-tRNA synthetase: Changes in phosphorylation of initiation factor eIF-2 and in the activity of the guanine nucleotide exchange factor GEF. J. Biol. Chem., 262:767-771.

Colthurst, D.R., Campbell, D.G., and Proud, C.G., 1987, Structure and regulation of eukaryotic initiation factor eIF2. Sequence of the site in the α subunit phosphorylated by the haem-controlled repressor and by the double-stranded RNA-activated inhibitor. Eur. J. Biochem., 166:357-363.

De Benedetti, A., and Baglioni, C., 1984, Inhibition of mRNA binding to ribosomes by localized activation of dsRNA-dependent protein kinase. Nature, 311:79-81.

Donahue, T.F., Cigan, A.M., Pabich, E.K., and Valavicius, B.C., 1988, Mutations at a zinc(II) finger motif in the eIF-2β gene in yeast alter translation initiation codon selection during the scanning process. Cell, in press.

Duncan, R., and Hershey, J.W.B., 1984, Heat-shock-induced translational alteractions in HeLa cells. J. Biol.Chem., 259:11882-11889.

Duncan, R., and Hershey, J.W.B., 1985, Regulation of initiation factors during translational repression caused by serum deprivation. J. Biol. Chem., 260:5493-5497.

Duncan, R.F., and Hershey, J.W.B., 1987, Translational repression by chemical inducers of the stress response occurs by different pathways. Arch. Biochem. Biophys., 256:651-661.

Ernst, H., Duncan, R.F., and Hershey, J.W.B., 1987, Cloning and sequencing of complementary DNAs encoding the α-subunit of translational initiation factor eIF-2. J. Biol. Chem., 77:1286-1290.

Ernst, V., Baum, E.Z., and Reddy, P, 1982, in Heat Shock: from Bacteria to Man (Schlesinger, M.J., Ashburner, M., and Tissieres, A., eds.) pp. 215-225, Cold Spring Harbor Laboratory, NY.

Harbitz, I., and Hauge, J.G., 1979, Purification and properties of eIF-2 from pig liver. Methods Enzymol. 60:240-246.

Kaufman, R.J., 1985, Identification of the components necessary for adenovirus translational control and their utilization in cDNA expression vectors. Proc. Natl. Acad. Sci. USA, 82:689-693.

Kaufman, R.J., and Murtha, P., 1987, Translational control mediated by eukaryotic initiation factor-2 is restricted to specific mRNAs in transfected cells. Molec. Cell. Biol. 7:1568-1571.

Kurzchalia, T.V., Bommer, V.A., Babkina, G.T., and Karpova, G.G., 1984, GTP interacts with the γ-subunit of eukaryotic initiation factor eIF-2. FEBS Lett., 175:313-316.

Moldave, K., 1985, Eukaryotic protein synthesis. Annu. Rev. Biochem. 54:1109-1149.

Ochoa, S., 1983, Regulation of protein synthesis in eukaryotes. Arch. Biochem. Biophys., 223:325-349.

Pain, V.M., 1986, Initiation of protein synthesis in mammalian cells. Biochem. J., 235:625-637.

Pathak, V.K., Nielsen, P.J., Trachsel, H., and Hershey, J.W.B., 1988a, Structure of the β subunit of translational initiation factor eIF-2. Cell, in press.

Pathak, V.K., Schindler, D., and Hershey, J.W.B., 1988b, Generation of a mutant form of protein synthesis initiation factor eIF-2 lacking the site of phosphorylation by eIF-2 kinases. Molec. Cell. Biol., 8:993-995.

Posorske, L.H., Cohn, M., Yanagisawa, N., and Auld, D.S., 1979, Methionyl-tRNA synthetase of *Escherichia coli*. A zinc metalloprotein. Biochim. Biophys. Acta, 576:128-133.

Proud, C.G., 1986, Guanine nucleotides, protein phosphorylation and the control of translation. Trends Biochem., 11:73-77.

Rowlands, A.G., Panniers, R., and Henshaw, E.C., 1988, The catalytic mechanism of guanine nucleotide exchange factor action and competitive inhibition by phosphorylated eukaryotic initiation factor 2. J. Biol. Chem., 263:5526-5533.

Schneider, R.J., and Shenk, T., 1987, Impact of virus infection on host cell protein synthesis. Annu. Rev. Biochem., 56:317-332.

Stringer, E.A., Chaudhuri, A., and Maitra, U., 1979, Purified eukaryotic initiation factor 2 consists of two polypeptide chains of 48,000 and 38,000 daltons. J. Biol. Chem., 254:6845-6848.

Wettenhall, R.E.H., Kudlicki, W., Kramer, G., and Hardesty, B., 1986, The NH_2-terminal sequence of the α and γ subunits of eukaryotic initiation factor 2 and the phosphorylation site for the heme regulated eIF-2α kinase. J. Biol. Chem., 261:12444-12447.

STRUCTURE, FUNCTION AND GENETICS OF *ras* PROTEINS

STRUCTURE OF THE HUMAN *RAS* GENE FAMILY

P. Chardin, N. Touchot, A. Zahraoui, V. Pizon, I. Lerosey, B. Olofsson and A. Tavitian

INSERM U-248, 10, Avenue de Verdun 75010 Paris France

SUMMARY

To study the structure of the *ras* gene family, we devised an original oligonucleotide strategy and isolated cDNAs for several new *ras*-related proteins : the *ral*A protein (50% a.a. identity with *ras*) and the *rab* proteins, related to the yeast YPT and SEC4 proteins. These new isolates, as well as drosophila D-*ras*3 and *rho* probes were then used to precise the structure of the *ras* family, in human.

At present, the *ras* family includes three main branches represented by *ras*, *rho* and *rab* genes. 1) The first branch includes *ras*H, *ras*K and *ras*N/ R-*ras*/*ral*A and *ral*B/rap1A and rap1B/rap2. 2) So far, the second branch only includes the *rho*A, *rho*B and *rho*C proteins sharing 80-90% aa identity, it is likely that other proteins remain to be discovered in this branch. 3) The third branch includes *rab*1 (homologous to yeast YPT)/*rab*2/*rab*3A and *rab*3B/*rab*4/*rab*5/*rab*6 (homologous to the YPT2 protein of fission yeast). A large array of evidence coming from yeast studies implicate these proteins in secretion.

All *ras*-related proteins share four regions of high homology corresponding to the GTP binding site,in positions 10-17, 57-63, 113-120 and 143-149 ; however significant differences are found in these regions : *rho* proteins have Gly 12 but not Gly 13, *rab* proteins have Gly 13 but not Gly 12 (*rab*3 and *rab*5 have neither Gly 12 nor Gly 13) and rap proteins have a Thr instead of Gln 61, suggesting that these proteins might differ in their GTPase activities and/or GTP/GDP exchange rates. All *ras*-related proteins possess a Cystein near the C-terminus but closely surrounding sequences are specific, suggesting a role in the different intracellular locations of these proteins. Other external regions are differentially conserved in each branch such as region 32-42, known as the putative "effector" region, for the *ras* branch (stricIty identical in *ras* and rap 1 proteins) while the most conserved external loop in the *rab* proteins is around position 63-73.

We have expressed several of these proteins in E. Coli and confirmed their GTP binding ability ; we are now studying their biochemical and functional specificities.

INTRODUCTION

At least four distinct groups of GTP binding proteins share some structural homologies and biochemical analogies : 1°) the α subunits of the transducing G proteins (usually associated with βγ subunits); 2°) the elongation factors involved in protein synthesis; 3°) the ADP Ribosylation Factors: ARFs, one being needed for ADP ribosylation of the α subunits of G proteins by Cholera Toxin and 4°) the *ras* family proteins.

We have studied the genetic structure of this latter family, in human.

The H-*ras*, K-*ras* and N-*ras* genes code for 21kd GTP/GDP binding proteins possessing a weak GTPase activity and transiently anchored to the inner face of the plasma membrane by a palmitic acid covalently linked to their C-terminus.

Transforming alleles of the H-*ras*, K-*ras* or N-*ras* genes are frequently found in human tumours, where the acquisition of a transforming potential is usually due to a point mutation resulting in the substitution of amino acids (a.a.) 12, 13 or 61. *Ras* proteins are found in organisms as different as yeast and man, and their high phylogenic conservation indicates that they certainly play an essential role, however the precise biochemical function of the *ras* proteins is not understood (Barbacid, 1987).

The fortuitous discovery of two proteins : *rho* in the marine snail Aplysia (Madaule and Axel, 1985) and YPT in yeast (Gallwitz et al., 1983) sharing ≈ 30% homology with *ras* proteins, suggested the existence of a large family.

To study the structure of the *ras* family, we followed two different approaches :

1°/ A sequence of 6 a.a. : DTAGQE in position 57-62 of p21*ras*, was strictly conserved in the *ras* proteins from various organisms as well as in *rho* and YPT. Using an original oligonucleotide strategy, we took advantage of this conserved sequence to isolate several new members of the *ras* superfamily.

2°/ The cDNAs isolated by this first strategy were then used under low stringency conditions to search for related genes. We also searched for a human homolog of the *Drosophila Melanogaster* Dras3 gene.

The corresponding protein was of special interest to us because it was closely related to H, K and N-*ras*, even identical in the "effector" region but possessed a NTAGTE sequence around position 61, the only exception to the first criteria that we used to define a *ras*-related protein.

RESULTS

1°/ Using the oligonucleotide strategy we first isolated a simian cDNA for the *ral*A protein possessing ≈ 50% a.a. identity with H, K or N-*ras* (Chardin and Tavitian, 1986). We then used this probe to search for human *ral* cDNAs under low stringency conditions and isolated the human *ral*A cDNAs encoding a protein with only one amino acid difference from simian *ral*A, we also isolated cDNAs encoding the human *ral*B protein possessing ≈ 85% a.a. identity with human *ral*A, most of the differences being located in the C-terminal end. A situation very similar to the one already observed for the human H, K and N-*ras* proteins.

We also searched for human homologs of the drosophila D-*ras*3 gene. By use of the D-*ras*3 cDNAs we isolated human cDNAs encoding the *rap*1A protein sharing ≈ 70% a.a. identity with D-*ras*3 and ≈ 50% a.a. identity with H, K or N-*ras* and the *rap*2 protein sharing ≈ 50% a.a. identity with both D-*ras*3 and H, K or N-*ras* (Pizon et al., 1988a)

The rap1A cDNA, used under low stringency conditions, enabled us to isolated human cDNAs for the rap1B protein, very closely related to rap1A : ≈ 95% a.a. identity (Pizon et al. 1988b).

2°/ In the *rho* branch, we used the *rho*6, *rho*9 and *rho*12 cDNAs isolated by P. Madaule (1985) to isolate complete cDNAs encoding these three closely related proteins that we named *rho*A (*rho*12), *rho*B (*rho*6) and *rho*C (*rho*9) sharing ≈ 90% a.a. identity.

3°/ The extensive screening of a rat brain cDNA library by the oligonucleotide strategy enabled us to isolate ≈ 50 positive clones, 14 were studied in detail and sequenced in the oligonculeotide hybridizing region 1 was a K-*ras* cDNA, 3 were H-*ras* cDNAs, 6 did not possess the exact oligonucleotide sequence and were not *ras*-related, 4 encoded new *ras*-related proteins that we named *rab*1, 2, 3 and 4. *Rab*1 is a mammalian homolog of the yeast YPT protein (≈ 70% a.a. identity) while *rab*2, 3 and 4 possess ≈ 45% a.a. identity with *rab*1 and are clearly in the same branch of the *ras* family.

The rat cDNAs encoding *rab*1, 2, 3 and 4 were then used to search for human cDNAs under low stringency conditions. We isolated human *rab*1, 2, 3 and 4 cDNAs encoding proteins sharing more than 95% a.a. identity with their rat counterparts. A cDNA coding for a protein closely related to the first *rab*3A was also isolated, we named it *rab*3B (≈ 75% a.a. identity with *rab*3A).

Two other cDNAs encoding proteins with ≈ 45% a.a. identity with any of the other *rab* proteins were also isolated, and named *rab*5 and *rab*6.

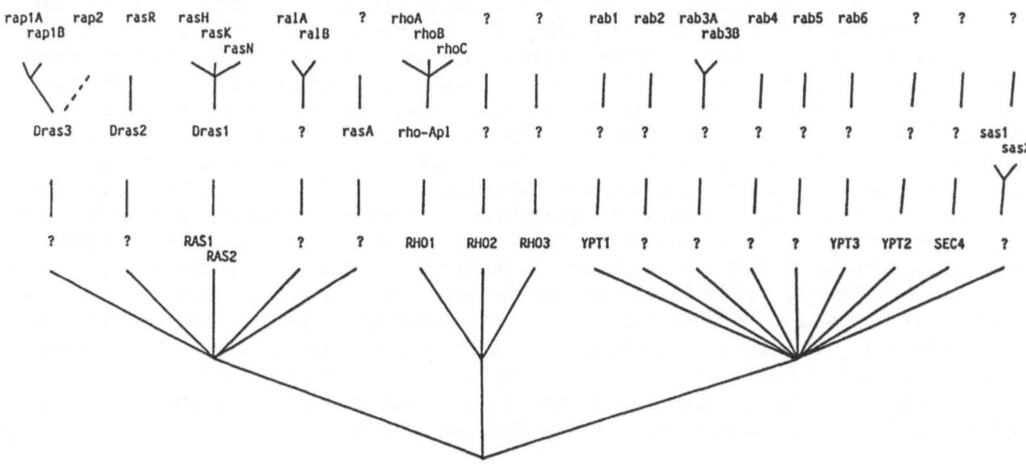

Structure of the *ras* superfamily

So far, the *ras* superfamily might be subdivided into three main branches represented by *ras*, *rho* and *rab*. On this figure the lower line represents yeast genes (Upper case letters), the middle line : Drosophila, Aplysia or Dictyostelium genes, and the upper line: mammalian genes.

1°/ In human, the *ras* branch includes the three classical *ras* proteins H-*ras*, K-*ras* and N-*ras* : ≈ 85% a.a. identity (reviewed by Barbacid, 1987) ; the *ral A and ral* B proteins : ≈ 85% a.a. identity and ≈ 50% identity with any of the *ras* proteins (Chardin and Tavitian, 1986 and unpublished results) ; the R-*ras* protein : ≈ 55 % identity with *ras* or *ral* and closely related to drosophila D-*ras*2 (65% a.a. identity) (Lowe et al, 1986), the *rap*1A and *rap*1B proteins 95% identical and sharing ≈ 55% identity with *ras*, the *rap*2 protein ≈ 50% a.a. identity with *ras* (Pizon et al., 1988 a and b) *Rap*1A and *rap*1B are clearly human homologs of the drosophila D-*ras*3 protein, *rap*2 has 60% a.a. identity with *rap*1A or *rap*1B but is not much more closely related to D-*ras*3 than to the other *ras* or *ras*-related proteins, however *rap*2 has a Threonine in position 61, an exception only found in *rap*1A, *rap*1B and D-*ras*3, at present the evolutionary relationships of this protein are not clear. Another gene : Apl-*ras* from Aplysia has no mammalian counterpart but it is likely that a mammalian homolog of Apl-*ras* might exist. This prediction and other recent discoveries suggest that the *ras* branch might include around a dozen of different proteins.

2°/ In human, the *rho* branch includes the three closely related *rho* proteins : *rho* A, (Yeramian et al., 1987) *rho* B and *rho* C (Chardin et al., 1988) : ≈ 85% a.a. identity and ≈ 30% when compared to *ras*.

The yeast counterpart of these human *rho* proteins is *RHO1*, but a second protein : *RHO*2 possessing ≈ 50% identity with *RHO*1 has also been found and might be considered as a yeast *rho*-like protein (Madaule et al., 1987). Another yeast gene previously known as CDC42 has recently been found to encode a *rho*-like protein (D. Johnson, personal communication) that could be named RHO3. It would be surprising that such *rho*-like proteins do not exist in mammals although they have not been found as yet.

3°/ In human, the third branch is represented by the *rab* proteins : *rab*1, 2, 3A, 3B, 4, 5 and 6 possessing ≈ 35-50% a.a. identity between each other except *rab*3A and *rab*3B that are much more closely related (≈ 78% a.a. identity). *Rab*1 is a mammalian homolog of the S. Cerevisiae YPT1 protein and *rab*6 is a human homolog of the S. Pombe YPT2 protein (D. Gallwitz, personal communication). Two other proteins from S. Cerevisiae : SEC4 (Salminen and Novick, 1987) and from S. Pombe : YPT3 (M. Yamamoto personal communication) as well as two proteins from Dictyostelium: SAS1 and 2 (A. Kimmel personal communication) clearly belong to this *rab* branch.

It is likely that some of the *rab* 2, *rab* 3, *rab* 4 and *rab*5 genes have yeast counterparts that remain to be discovered, while reciprocally, the yeast YPT3 and SEC4 genes and the Dictyostelium SAS genes probably have mammalian counterparts, suggesting that the *rab* branch might include also at least a dozen of different proteins.

Expression

Most of the *ras* superfamily genes are expressed in all tissues, although several of these genes are expressed at higher levels in one or a few organs (Leon et al., 1987, Olofsson et al., 1988). High levels of expression are frequently observed in brain ; moreover, one of the *rab* genes: *rab*3A appears to be expressed at high levels in brain while it is hardly detected in any other organ (Olofsson et al., 1988).

Each gene displays a specific pattern of expression in the different organs suggesting a complex regulation. A detailed expression study by in-situ hybridization would provide informations of high interest.

Structural properties of the *ras* superfamily proteins

A three dimensional model of the p21*ras* protein has recently been deduced from X-ray cristallography studies (de Vos et al., 1988) and it is of high interest to locate the major conserved regions on this model.

1./ Four regions : the Glycine rich region in position 10-20, the DTAGQE region in position 57-62 the LVGNK-DL region in position 113-120 the ETSAK region in position 143-147 and a few scattered amino-acids including F28 are involved in GTP binding and are highly conserved in all the proteins of the *ras* family. However significant differences are found at "critical" positions : the *rho* proteins have an Alanine in position corresponding to *ras* Gly 13, the *rab* proteins do not possess a Glycine in position corresponding to *ras* Gly 12 (*rab*3A, *rab*3B and *rab*5 have neither Gly 12 nor Gly 13).

R-*ras* has an Alanine instead of a Threonine in position 144 known to be of high importance since a yeast *RAS*2 protein might be activated by this kind of mutation (Camonis and Jacquet, 1988).

Even region 61 which had been considered as a hall mark of *ras* proteins is not strictly conserved : *rap* proteins possess a Threonine instead of the Glutamine found in *ras* position 61.

2./ Region 32-42 known as the "effector" region is strictly conserved between *ras* and *rap*1 proteins while there is one a.a. change in *rap*2 and R-*ras* and three differences in *ral*.

Nevertheless this region is not highly conserved between the different *rab* proteins, suggesting that they do not follow a similar functionnal scheme than the proteins of the *ras* branch.

Only one amino acid : Thr35 is conserved in all the proteins of the *ras* family, suggesting a special role for example as a Ser/Thr Kinase substrate.

3./ Regions 53-56, 63-82 and 97-104 are all located on the same part of the molecule close to loop 32-42 and are ≈ 70% homologous between the different proteins of the *ras* branch. We would predict that these regions are involved in the interaction with a large protein that remains to be characterized. It is noteworthy that small deletions might be introduced in each of these regions taken separately without impairing the transforming potential of a v-H-*ras* p21 (Willumsen et al., 1986) unfortunately the same kind of data is not available for a normal H-*ras* protein. In these regions Tyrosine 71 is conserved in all proteins of the *ras* family, and might be important as a substrate for regulation by Tyrosine Kinases.

4./ Two small sequences in position 47-49 and 161-164 are located close to the C-terminal tail of p21 *ras*, anchoring it to the plasma membrane, and are frequently conserved. These regions might play a role in the interaction with a putative "receptor" or any other kind of membrane bound protein.

5./ The 20-30 last C-terminal amino acids are highly variable and confer their specificity to the different proteins of a same sub-branch such as H, K and N-ras, most of the proteins of the *ras* branch have many basic amino acids such as Lysines and Arginines in their C-terminal end.

This characteristic is not found in the *rab* branch of the family.

6./ All the *ras* family proteins possess at least one Cysteine close to their C-terminal end, suggesting a fatty-acylation and membrane localization. However the conservation of specific motifs, from yeast to man, in each sub-branch strongly suggest a role in the targetting of these proteins to different sub-cellular membrane systems.

Biochemical properties of the *ras* superfamily proteins

The high conservation of the GTP binding region strongly suggested that all of these proteins were GTP binding proteins. In fact, the GTP binding ability, and low GTPase activity, have been demonstrated for *ras*, *rho*, YPT, R-*ras*, *ral* and *rab* proteins (Tucker et al., 1986 ; Anderson and Lacal, 1987 ; Wagner et al., 1987 ; Lowe and Goeddel, 1987 ; Chardin et al., Zahraoui et al., in preparation). However, it is noteworthy that many of these proteins have an amino acid difference in positions known to be "critical" in the case of *ras* proteins. The *rho* proteins have an Alanine in position corresponding to *ras* Glycine 13 ; *rab* proteins do not possess a Glycine in position corresponding to *ras* Glycine 12 (*rab* 3 and *rab* 5 have neither Gly 12 nor Gly 13). Even region 61, which had been considered as a hallmark of *ras* proteins is not strictly conserved : *rap* proteins possess a Threonine instead of the Glutamine found in *ras* position 61. As far as the conclusions drawn from the study of *ras* proteins might be transposed to the other proteins of this family we would predict that *rho*, *rab* and *rap* proteins are constitutively in a slightly more active state than *ras* proteins.

Preliminary results on *ral*, *rap*, *rho* and *rab*1, 2, 3, 4, 5 proteins expressed in E. Coli show that these proteins display very different GTP binding abilities after Western Blot transfer to nitrocellulose, furthermore careful studies on purified proteins show that several biochemical parameters such as the GTP/GDP exchange rate and the GTPase activity vary by at least one order of magnitude from one protein to the other, suggesting that although these proteins all bind GTP they possess significantly different biochemical properties

DISCUSSION

Transforming potential of *ras* related genes

While transforming alleles of the H, K or N-*ras* genes are found in 10 to 70% of human tumors, none of the *ras*-related genes has ever been isolated from the large number of tumors studied by the NIH/3T3 assay in many laboratories, indicating that *ras* related genes are not frequently implicated in tumor formation or can not be detected by this approach. In fact, it seems that neither R-*ras* (Lowe et al. 1987) nor *ral*A (Chardin et al., in preparation) are able to transform NIH/3T3 cells in the usual assay. Therefore the potential involvement of *ras* related genes in transformation remains an open question.

Function of the *ras* superfamily proteins

The first proteins characterized on the basis of their biological activity are H-*ras*, K-*ras* and N-*ras* : the activated proteins transform NIH/3T3 cells, and their most obvious effect, when microinjected are : an increase in membrane ruffling/fluid phase endocytosis/cell motility and the induction of mitosis, that might explain the loss of contact inhibition and the transformed phenotype.

In the fission Yeast Schizosaccharomyces Pombe, disruption of the *ras* 1 gene does not affect growth, but the cell shape is extensively deformed (shortened and swollen), futhermore *ras* 1 is essential for mating and required for efficient sporulation (Fukui et al., 1986).

The second protein isolated on a functional basis in yeast is SEC4 : this mutation results in accumulation of post-golgi vesicles (Salminen and Novick, 1987), and it has been shown that the SEC4 protein associates with secretory vesicles and the plasma membrane (Goud et al., 1988).

In S. Cervisiae, the YPT1 gene is located between the actin (ACT1) and ß-tubulin (TUB2) genes. The YPT protein is involved in the organization of the cytoskeleton during vegetative growth. YPT deficiency leads to larger cells, with two or several buds instead of one ; microtubules as well as the actin network are disorganized and nuclear integrity is lost (Schmitt et al., 1987 ; Segev and Botstein, 1987). YPT1 deficiency also causes a build-up of vesicles (Segev and Botstein, 1988). However YPT1 and SEC4 do not complement each other and it is unlikely that SEC4 and YPT1 fulfill similar roles. Therefore, it seems that the *ras* superfamily proteins play a critical role at the membrane/cytoskeleton interface and regulate some of the physical exchanges beween the cell and its environment. However, a direct involvement of the *ras* superfamily proteins in these events remains to be demonstrated.

Ras proteins possess some sequence homologies and some biochemical analogies with the α subunits of G proteins involved in many receptor/effector systems, as well as a common sub-membrane localization ; it was thus postulated that *ras* proteins might transduce signals from growth factor receptors to an effector such as phospholipase C or phospholipase A2. However, recent evidence indicate that the action of *ras* on these phospholipases is an indirect effect (Seuwen et al., 1988; Yu et al., 1988). It is likely that *ras* proteins are transiently activated by growth-factor receptors, but the biochemical basis of this activation is not understood. The main problem is that the careful study of any biochemical parameter in a *ras* transformed cell usually demonstrates a difference when compared to a normal cell, however it is well known that the biochemistry of a transformed cell is deeply altered (Warburg, 1920) and this kind of data provides little information, if any, on *ras* function. What we would like to know is the nature of the proteins directly interacting with p21 *ras*.

At present only one protein has been shown to interact directly with p21 *ras* : the GTPase Activating Protein (GAP). This protein is a 116 K cytoplasmic protein that can give rise to a 55K degradation product, still active and probably representing the "*ras* interacting" domain of the protein (Trahey et al., 1987, F. Mc Cormick personal communication).

An elegant genetic approach has localized an essential external loop of the p21*ras* protein that can not be deleted or mutated without loss of activity, even in a constitutively active protein : v-H *ras* p21 (Willumsen et al.,

1986; Sigal et al., 1986)) this region in position 32-42 is known as the "effector" binding site and recent evidence indicate that some mutations in this region impair the interaction with the GAP protein, and abolish the transforming potential, suggesting that GAP might be the effector (Adari et al., 1988; Calès et al., 1988) however for some mutants the correlation between diminished sensitivity to GAP and abolished transforming potential is no longer true (D. Lowy, personal communication) and other regions around position 60-70 also seem to be involved in GAP interaction (J.C. Lacal, personal communication).

It is noteworthy that at least two *ras*-related proteins *rap*1A and *rap*1B have exactly the same sequence than *ras* in the "effector" region while *rap*2 and R-*ras* display one difference and *ral* has three. It would of course be of major interest to know whether *rap*1A and *rap*1B really interact with the same "effector" than *ras*, and whether *rap*2, R-*ras* and *ral* also interact with identical or closely related effectors. As *rap* proteins possess in position 61 a Threonine, an a.a. leading to a slight activation of *ras*, it is conceivable that *rap* proteins are in a constitutively active state and interact with the same effector as *ras*, keeping it in a low, basal, activity level, while *ras* would be able to transiently boost this activity in response to an external signal, furthermore it is attractive to speculate that *rap* proteins interact with the same effector than *ras* but have an antagonist effect.

The availability of most of these proteins, expressed in E. Coli should enable us to test these hypothesis.

Acknowledgements

We wish to thank the following colleagues for their gifts of sequences prior to publication : D. Gallwitz and M. Yamamoto for YPT2 and YPT3 from S. Pombe, A. Kimmel for SAS1 and SAS2 from Dictyostelium and D. Johnson for CDC 42/*rho*3 from S. Cerevisiae. We also thank R. Kahn for helpful discussions on ARFs and their relationships to other GTP binding proteins.

References

Adari, H., Lowy, D.R., Willumsen, B.M., Der, C.J., and Mc Cormick, F. (198). Guanosine Triphosphatase Activating Protein (GAP) interacts with the p21 *ras* effector binding domain. Science, **240**, 518-521.

Anderson, P. and Lacal, J.C. (1987). Expression of the Aplysia californica *rho* gene in Escherichia coli : purification and characterization of its encoded p21 product. Mol. Cell. Biol. 7, 3620-3628.

Barbacid, M. (1987). *ras* Genes. Ann. Rev. Biochem. **56**, 779-827.

Bar-Sagi, D. and Feramisco, J.R. (1986). Induction of membrane ruffling and fluid-phase pinocytosis in quiescent fibroblasts by *ras* proteins. Science, 233, 1061-1068.

Calès, C., Hancock, J.F., Marshall, C.J. and Hall A. (1988). The cytoplasmic protein GAP is implicated as the target for regulation by the *ras* gene product. Nature, 332, 548-551.

Camonis, J. and Jacquet, M. (1988). A new *ras* mutation that supresses the CDC25 gene requirement for growth of Saccharomyces Cerevisiae. Mol. Cell. Biol., 8, 2980-2983.

Chardin, P. and Tavitian A. (1986). The *ral* gene : a new *ras* related gene isolated by the use of a synthetic probe. EMBO J. 5, 2203-2208.

Chardin, P., Madaule, P. and Tavitian A. (1988) Coding sequence of human *rho* cDNAs clone 6 and clone 9. Nucleic Acids Res. **16**, in the press.

de Vos, A.M., Tong, L., Milburn, M.V., Mathias, P.M., Jancarik, J., Noguchi, S., Nishimura, S., Miura, K., Ohtsuka, E. and Kim, S.-H. (1988) Science, **239**, 888-893.

Fukui, Y., Kozasa, T., Kaziro, Y., Takeda, T. and Yamamoto, M. (1986). Role of a *ras* homolog in the life cycle of schizo-saccharomyces pombe. Cell, **44**, 329-336.

Gallwitz, D., Donath, C. and Sander C. (1983). A yeast gene encoding a protein homologous to the human *c-has/bas* proto-oncogene product. Nature **306**, 704-707.

Goud, B., Salminen, A., Walworth, N.C., and Novick, P.J. (1988). A GTP-binding protein required for secretion rapidly associates with secretory vesicles and the plasma membrane in Yeast. Cell, **53**, 753-768.

Leon, J., Guerrero, I. and Pellicer, A. (1987). Differential expression of the *ras* gene family in mice. Mol. Cell. Biol. 7, 1535-1540.

Lowe, D., Capon, D., Delwart, E., Sakaguchi, A., Naylor, S. and Goeddel, D. (1987). Structure of the human and murine R-*ras* genes, novel genes closely related to *ras* proto-oncogenes. Cell **48**, 137-146.

Lowe, D. and Goeddel, D. (1987). Heterologous expression and characterization of the human R-*ras* gene product. Mol. Cell. Biol. 7, 2845-2856.

Madaule, P. and Axel, R. (1985). A novel *ras*-related gene family. Cell **41**, 31-40.

Madaule, P., Axel, R. and Myers, A. (1987). Characterization of two members of the *rho* gene family from the yeast Saccharomyces cerevisiae. Proc. Natl. Acad. Sci. USA **84**, 779-783.

Melançon, P., Glick, B.S., Malhotra, V., Weidman, P.J., Serafini, T., Gleason, M.L., Orci, L. and Rothman J.E. (1987).

Involvement of GTP-binding "G" proteins in Transport through the Golgi Stack. Cell, **51**, 1053-1062.

Mozer, B., Marlor, R., Parkhurst, S. and Corces, V. (1985). Characterization and Develomental Expression of a Drosophila *ras* oncogene. Mol. Cell. Biol. **5**, 885-889.

Olofsson, B., Chardin, P., Touchot, N., Zahraoui, A. and Tavitian A. (1988). Expression of the *ras*-related *ral*A, *rho*12 and *rab* genes in adult mouse tissues. Oncogene, **2**.

Pizon, V., Chardin, P., Lerosey, I., Olofsson B. and Tavitian A. (1988a). Human cDNAs *rap*1 and *rap*2 homologous to the *Drosophila* gene D*ras*3 encode proteins closely related to *ras* in the "effector" region. Oncogene, **2**,

Pizon, V., Lerosey, I., Chardin, P. and Tavitian A. (1988b). Nucleotide sequence of a human cDNA encoding a *ras*-related protein (*rap*1B). Nucleic. Acids. Res., **16**, 15.

Salminen, A. and Novick, P. (1987). A *ras*-like protein is required for a post-golgi event in yeast secretion. Cell **49**, 527-538.

Schejter, E. and Shilo, B.-Z. (1985). Characterization of functional domains of p21 *ras* by use of chimeric genes. EMBO J. **4**, 407-412.

Schmitt, A., Wagner, P., Pfaff, E. and Gallwitz, D. (1986). The *ras*-related YPT1 gene product in yeast : a GTP-binding protein that might be involved in microtubule organization. Cell **47**, 401-412.

Segev, N. and Botstein, D. (1987). The *ras*-like yeast YPT1 gene is itself essential for growth, sporulation, and starvation response. Mol. Cell. Biol. **7**, 2367-2377.

Segev, N., Mulholland, J. and Botstein, D. (1988) The Yeast GTP-binding YPT1 protein and a mammalian counterpart are associated with the secretion machinery. Cell, **52**, 915-924.

Seuwen, K., Lagarde, A. and Pouyssegur, J. (1988). Deregulation of hamster fibroblast proliferation by mutated *ras* oncogenes is not mediated by constitutive activation of phosphoinositide-specific phospholipase C. EMBO J. **7**, 161-168.

Sigal, I.S., Gibbs, J.B., D'Alonzo, J.S., and Scolnick, E.M. (1986), Proc. Natl. Acad. Sci. USA, **83**, 4725-4729.

Swanson, M., Elste, A., Greenberg, S., Schwartz, J., Aldrich, T. and Furth, M. (1986). Abundant expression of *ras* proteins in *Aplysia* neurons. J. Cell Biol. **103**, 485-492.

Touchot, N., Chardin P. and Tavitian, A. (1987). Four additional members of the *ras* gene superfamily isolated by an oligonucleotide strategy : molecular cloning of YPT-related cDNAs from a rat brain library. Proc. Natl. Acad. Sci. USA **84**, 8210-8214.

Trahey, M. and Mc Cormick, F. (1987). A cytoplasmic protein stimulates Normal N-*ras* p21 GTPase, but does not affect oncogenic mutants. Science, **238**, 542-545.

Tucker, J., Sczakiel, G., Feuerstein, J., John, J., Goody, R. and Wittinghofer, A. (1986). Expression of p21 proteins in Escherichia coli and stereochemistry of the nucleotide-binding site. EMBO J. **5**, 1351-1358.

Wagner, P., Molenaar, C., Rauh, A., Brökel, R., Schmitt, H. and Gallwitz, D. (1987). Biochemical properties of the *ras*-related YPT protein in yeast : a mutational analysis. EMBO J., **6**, 2373-2379.

Warren, G., Davoust, J. and Cockcroft, A. (1984). Recycling of transferrin receptors in A431 cells is inhibited during mitosis. EMBO J., **3**, 2217-2225.

Willumsen, B.M., Papageorge, A.G., Kung, H.F., Bekesi, E., Robins, T., Johnsen, M., Vass, W.C., and Lowy, D.R. (1986). Mutational analysis of a *ras* catalytic domain. Mo. Cell. Biol., **6**, 2646-2654.

Yeramian, P., Chardin P., Madaule, P. and Tavitian, A. (1987). Nucleotide sequence of human *rho* cDNA clone 12. Nucleic, Acids Res. **15**, 1869.

Yu, C.-L., Tsai, M.-H. and Stacey, D. (1988). Cellular *ras* activity and phospholipid metabolism. Cell **52**, 63-71.

Zahraoui, A., Touchot, N., Chardin, P. and Tavitian, A. (1988). Complete coding sequences of the *ras* related *rab*3 and 4 cDNAs. Nucleic. Acids. Res. **16**, 1204.

A MUTATIONAL ANALYSIS OF *RAS* FUNCTION

Berthe M. Willumsen[1], Hedy Adari[2], Ke Zhang[4], Alex G. Papageorge[4], James C. Stone[3], Frank McCormick[2] and Douglas R. Lowy[4]

[1]University Institute of Microbiology, Øster Farimagsgade 2A, DK1353, Copenhagen, Denmark; [2]Departmemt of Molecular Biology, Cetus Corporation, Emeryville, California, 94608 U.S.A.; [3]Jackson Laboratories, Bar Harbor, Maine 04609 U.S.A.; [4]Laboratory of Cellular Oncology, National Cancer Institute, Bethesda, Maryland 20892, U.S.A.

ABSTRACT

We have used linker insertion-deletion mutagenesis to study the Harvey murine sarcoma virus v-*ras*^{H} transforming protein. The mutants were characterized with respect to their ability to induce morphological transformation of NIH 3T3 cells and the capacity of their proteins to bind guanosine nucleotides, undergo post-translational processing, and localize to the plasma membrane. We have identified four non-overlapping segments that are dispensable for morphological transformation of NIH 3T3 cells, as well as several segments that are required for transformation and stability in mammalian cells and guanosine nucleotide binding. One essential segment that does not affect guanine nucleotide binding or stability, which appears to lie on the exterior of the protein and therefore may interact with the putative *ras* protein target, has been identified (the effector domain, Willumsen et al., 1986, Sigal et al., 1986). A selected group of these mutations, which leave the v-*ras*^{H} protein stable, processed and correctly localized, have been transferred to the c-*ras*^{H} allele; the proteins were expressed in *E. coli* and assayed for the susceptibility to acceleration of their intrinsic GTPase activities by the protein GAP (GTPase Activating Protein, Trahey and McCormick, 1987). The results show that only mutations in the effector domain destroy GAP susceptibility (Adari et al., 1988); however, not all mutations affect both activities coordinately. These results suggest that GAP as well as the effector mediating p21 transformation interact through the same region on p21. The identification of mutations that destroy transformation when present in the v-*ras*^{H} allele which do not destroy the GAP susceptibility of p21 protein from the c-*ras*^{H} allele raises the possibility that the two factors may not be the same.

INTRODUCTION

Ras genes are widely conserved in eukaryotes, from yeast to humans (Barbacid, 1987). Mammalian cells contain at least three *ras* genes: c-*ras*^{H}, c-*ras*^{K}, and c-*ras*^{N}; these genes encode similar 189 amino acid protein products which have been called p21 because they migrate in gels as molecules that are approximately 21 kd. The *ras* genes were first identified as the viral oncogenes of Harvey murine sarcoma virus (v-*ras*^{H}) and Kirsten (Ki) MuSV (v-*ras*^{K}), which are highly oncogenic versions of their normal cellular counterparts.

Members of this multigene family appear to serve essential, growth-related, physiological functions in eukaryotes (DeFeo-Jones et al., 1985, Kataoka et al., 1985, Mulcahy et al., 1985). They have also been implicated in the pathogenesis of a variety of human and animal tumors (Marshall et al., 1985). *Ras* genes can induce tumors in vivo and tumorigenic transformation of tissue culture cells in vitro. The mammalian *ras* proteins can induce morphological transformation of NIH 3T3 cells by overproduction of the normal *ras* protein product, by amino acid deletion, or by single amino acid substitution. In tumors, missense mutations appear to be the most common mechanism by which the genes become activated; the v-*ras*^{H} and v-*ras*^{K} genes of Ha-MuSV and Ki-MuSV, respectively, contain two missense mutations, either of which can independently activate the gene (Barbacid, 1987).

It has also been determined that the primary *ras* translation product (pro-p21) is synthesized in the cytosol and undergoes post-translational processing (to mature p21), leading to its translocation from the cytosol to the plasma membrane (Sefton et al., 1983). The mature p21 protein has a slightly faster mobility and contains palmitic acid linked near the C-terminus of the protein (Chen et al., 1985, Buss and Sefton, 1986). Genetic studies of v-*ras*^{H} have shown that both the processing and membrane association depend on a conserved cysteine residue (at amino acid 186) that in the primary translation product is located four amino acids from the C-terminus. Mutants that encode a protein lacking cysteine-186 cannot transform NIH 3T3 cells, and their proteins remain in the cytosol, do not change their migration rate, and fail to bind lipid (Willumsen et al., 1984a, 1984b). The precise biochemical nature of the processing event has not been elucidated; however, there is evidence in yeast and in mutant *ras* proteins in mammalian cells (Fujiyama et al., 1986, Lowy et al., 1988) that lipid attachment and the faster migration rate are separable.

It has not yet been determined how the *ras* proteins carry out their normal physiological functions, nor has the pathway by which they induce cellular transformation been elucidated. In the yeast *S. cerevisiae*, RAS apparently functions primarily by stimulating adenylate cyclase (Broek et al., 1985), but this does not seem to be the case for mammalian cells (Beckner et al., 1985, Levitzki et al., 1986) or for a different yeast (*S. pombe;* Fukui et al., 1986). Several presumably relevant features of the *ras* protein have been identified. They non-covalently bind guanosine nucleotides (GDP and GTP) and possess a GTPase activity that is analogous to that of the regulatory G proteins, with which they share some sequence homology (Barbacid, 1987). Activated versions of many *ras* proteins are associated with a significantly reduced GTPase activity, and genes activated by mutation of amino acid 59 from alanine to threonine possess an autophosphorylation activity, reflecting the transfer of the gamma phosphate of the bound GTP to the hydroxyl group of the threonine at position 59.

Recently a cytoplasmic activity which greatly stimulates the intrinsic GTPase activity of normal, but not oncogenic, p21 has been described (Trahey and McCormick, 1987). This protein is potentially a significant regulator of p21 activity, since all available data point to p21 stimulating when present in the GTP-bound form but being inactive in the GDP bound form (Field et al., 1987, Trahey and McCormick, 1987).

We are carrying out structure-function studies of p21 via a mutational analysis of the Ha-MuSV v-*ras*^{H} oncogene (Willumsen et al., 1984a, 1984b, 1985, 1986, Adari et al., 1988, Stone et al., 1988). The protein encoded by v-*ras*^{H} is identical to that encoded by the normal human or rodent c-*ras*^{H} gene except for two independently activating mutations, which are located at amino acid residues 12 and 59. These studies have enabled us to map segments in v-*ras*^{H} that are dispensable for morphological transformation as well as others that are essential for this function and for guanine nucleotide binding (Willumsen et al., 1986). The conclusions drawn from these earlier studies are in good agreement with the recently published crystal structure of p21 (de Vos et al., 1988). In addition, we have mapped a region involved in the interaction with GAP (Adari et al., 1988).

METHODS

v-*ras*^{H} mutants and expression

We have previously described the technique used to generate in frame v-*ras*^{H} mutants by deletion and linker insertion mutagenesis (Willumsen et al., 1984a). In summary, the mutants are constructed by the combination of two sequenced parts (N-terminal front ends and C-terminal tail ends) of v-*ras*^{H} through a BclI oligonucleotide linker. This linker results in the addition of three new amino acids within the protein at the site of the deletion. Two vectors, one for expression of the mutant genes in NIH 3T3 cells and one for expression in *E. coli*, have been used. The eukaryotic vector used for most of the mutants was developed by Jhappan et al. (1986). In this vector, the v-*ras*^{H} gene is located upstream from the SV40 sequences driving the gene giving rise to neomycin resistance; these two genes are flanked by a Moloney MuLV LTR. The use of the prokaryotic vector (pJCL-30, Lacal et al., 1984), which places the *ras* mutants under control of the lambda pL promoter and initiates *ras* protein synthesis from its authentic initiation AUG, has been described previously (Willumsen et al., 1986). The *E. coli* cells that harbor the expression vector contain a temperature-sensitive allele of the lambda repressor gene, cI857.

The NIH 3T3 cells and DNA transfection procedure have been previously described (Lowy et al., 1978). For immunoprecipitation, cultures transfected with Ha-MuSV mutants selected either for focus formation or for G418 resistance were metabolically labeled with ^{35}S-methionine (250 µCi/ml) in methionine-free medium. Extracts of whole cells were prepared and precipitated as previously described with a p21 monoclonal antibody - either Y13-238 or Y13-259 (Willumsen et al., 1984a). Mutants were expressed in *E. coli* and purified as described (Willumsen et al., 1986); GAP assays were carried out as described (Trahey and McCormick, 1987).

RESULTS AND DISCUSSION

Construction of insertion-deletion mutants

The *ras* proteins can be divided into at least three functional domains: the extreme C-terminus, which we call the membrane anchoring domain, is required for post-translational processing and membrane localization. A 20 amino acid segment that is called the major heterogenous region lies just upstream from the C-terminus (amino acid residues 160-185); it is highly divergent among different *ras* genes. The N-terminal 160 amino acids, which are highly conserved among mammalian *ras* proteins, represents the catalytic domain of the protein.

Because of the oligonucleotide linker used in constructing the insertion-deletion mutants, each mutated v-*ras*^{H} gene encoded three new amino acids at the side of the deletion. The designation of the proteins encoded by the various mutants as XNNNY specifies that the protein has the normal sequence from amino acid 1 through amino acid X, encodes 3 amino acids specified by the oligonucleotide linker (NNN) which is indicated with the one letter amino acid code and continues with the normal protein sequence from amino acid Y to 189 (the terminal amino acid). The eukaryotic vector placed the mutants under the regulatory control of a murine retroviral long terminal repeat (LTR). The vector also contained a linked selectable marker (giving neomycin-resistance) which in principle enabled us to study the *ras* proteins encoded by all mutants in NIH 3T3 cells, irrespective of their transforming capacity.

Representative mutants were also placed in a prokaryotic expression vector to analyze the GDP binding activities of purified mutant *ras* proteins produced in bacteria.

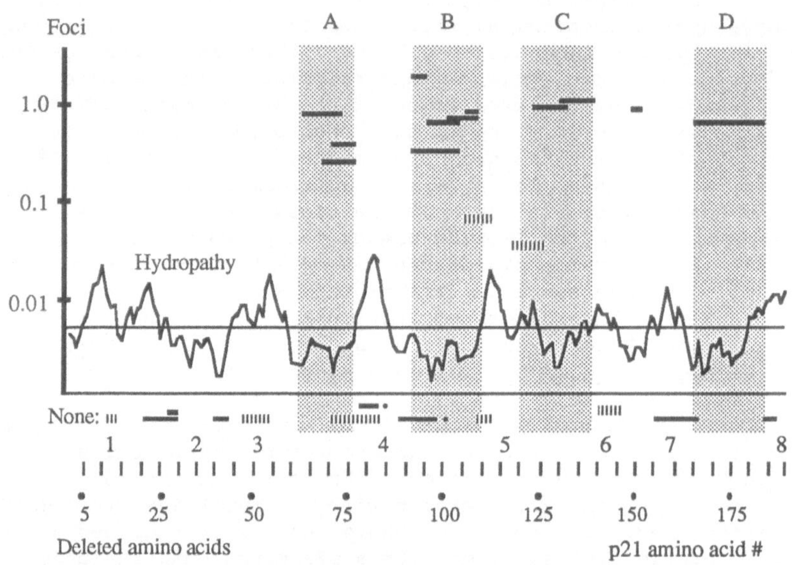

Figure 1. TRANSFORMATION BY DELETION MUTANTS. Transforming activity of representative *ras* mutants and hydropathic index of the *ras* protein (modified from Willumsen et al., 1986). The main horizontal axis represents the amino acid (1-189) in the *ras* protein; the vertical axis represents the relative NIH 3T3 focus forming activity of the mutants. Each mutant is represented by a horizontal line (solid for those mutant proteins that bind GDP, interrupted for those mutants whose proteins purified from *E. coli* do not bind detectable levels of GDP, bullet for untested); this line corresponds to the location and extent of the deletion. Segments designated A-D represent sequences that are not essential for transformation. Regions designated 1-8 contain sequences that are essential for transformation Data for mutants in A-C and 1-6 includes data from Willumsen et al., 1986; data for mutants in D, 7, and 8 are taken from from Willumsen et al., 1984a. The hydropathic index has been plotted according to Kyte and Doolittle (1982); hydrophobic regions are above the axis of the midpoint line, hydrophilic below.

Cell transformation

The capacity of the mutant v-*ras*^{H} genes in the eucaryotic vector to induce focal transformation of NIH 3T3 cells is shown in Fig. 1. Some mutants were transformation-competent and induced foci with an efficiency similar to that of the wild-type v-*ras*^{H} gene, the transforming capacity of other competent mutants was significantly reduced, and some mutants were transformation-defective (any mutant whose transforming activity was more than three orders of magnitude lower than that of the wild type gene will score as defective in this assay). The foci induced by mutants possessing a transforming efficiency less than 20% that of the wild type gene were generally detected later and remained smaller than those mutants that induced foci with an efficiency similar to that of the wild type gene.

Within the catalytic domain, we noted three different segments where deletions did not abolish the transforming activity of the gene (labeled A-C in Fig. 1). Each non-essential segment was quite large; A, B, and C were at least 13 (residues 64-76), 16 (93-108), and 19 (120-138) amino acids long, respectively. The heterogeneous region (segment D in Fig. 1) was also dispensable (Willumsen et al., 1985).

Given the evolutionary conservation of most v-*ras*^{H} amino acids, it might be expected that most deletions outside the heterogeneous region would abolish the biological activity of the gene. Indeed, six different regions within the catalytic domain were apparently essential for transformation (labeled 1-6 in Fig. 1), since lesions in each region rendered the genes defective. The precise boundaries of these essential regions are not defined. This is because the defective phenotype represents loss of a function, and we did not determine which of the variant amino acids in each mutant were responsible for the phenotype.

When an attempt was made to characterize the proteins expressed in NIH 3T3 cells by the transformation-defective mutants, it was found that only cells expressing mutants in essential region 2 and the C-terminus contained detectable levels of p21 protein. Since the epitope(s) recognized by the antibody used for detection of p21 is not deleted from mutants in the other essential regions (measured on the proteins as expressed in *E. coli*), we assume that the failure to detect such p21 proteins is due to an increased instability of the mutated proteins in mammalian cells. By contrast, those cells carrying mutants from essential region 2 and the C-terminus contained readily detectable mutant protein. Sub-cellular localization studies indicated further that proteins encoded by essential region 2 mutants migrated correctly to the membrane (Willumsen et al., 1986).

ras protein structure

Figure 1 also plots the hydropathic index of the wild-type protein, according to the method of Kyte and Doolittle (1982). For soluble globular proteins, interior portions generally map to the hydrophobic side of the midpoint line, while exterior portions are usually found on the hydrophilic side. The regions required for *ras* transforming function tend to fall on the hydrophobic side, with the notable exception of region 2, which is hydrophillic.

The crystal structure of the GDP bound form of a c-*ras*^{H} mutant protein (the region encoding amino acids 172-189 were deleted) has recently been determined (de Vos et al., 1988). The backbone structure of the protein consists of 6 β-strands, 4 α-helixes and 9 connecting loops. Our data assessing the biological activity of deletion mutants show that deletions involving the β-strands render the genes non-transforming and give rise to p21 proteins that do not accumulate in the transfected cells. The dispensable regions are located on the outside of the protein, as is the required region 2 (defined as the effector domain, see below). This is in full agreement with our earlier interpretation (Willumsen et al., 1986), using the p21 model structure based on the crystal structure of *E. coli* elongation factor Tu (McCormick at al., 1985, Jurnack, 1985).

Effector region and membrane anchor

Mutations in two essential regions produce stable protein when transfected into NIH 3T3 cells. One spans amino acid 186; mutation of this position leads to a non-processed, non-palmitoylated p21 protein which fails to localize to the membrane (Willumsen et al., 1984a,b). Mutations in the other region, maximally covering position 22 to 43, which is highly conserved among *ras* proteins, represents a required exterior portion of the protein. Mutations here give rise to stable, fully processed, membrane bound non-transforming p21. Since the v-*ras*^{H} derived p21 autophoshporylates by transferring the gamma phosphate of the bound GTP to residue 59 (a threonine), the presence of the phosphorylated form can be taken as an indication of the ability of the protein to bind GTP. p21 from mutants in this region are found phosphorylated in the cells (Willumsen et al., 1986). These mutants argue that this region participates in a different essential *ras* function, and we and others (Willumsen et al., 1986, Sigal et al., 1986) have speculated that the these mutant proteins are defective because they fail to interact with the putative target of the normal *ras* protein. The region (residues 32 to 42) has been called the effector region. These residues are conserved among all *ras* proteins, while previously published *ras*-like products are divergent in this region. Mutants defective in this region also fail, when expressed in yeast cells, to stimulate the yeast adenylate cyclase in a GTP-dependent manner (Sigal et al., 1986), and in yeast, a suppressor mutant to such an effector mutation has been mapped to the adenylate cyclase (Marshall et al., 1988).

Mutations Affecting Structural Integrety

None of the mutants that alter any of the β-strands, with the exception of those changing β-strand 2 which is part of the effector region, produce stable protein when transfected to NIH 3T3 cells. In addition, mutants resulting in changes in loop 3 (table 1) result in undetectable proteins. Thus, most of the required regions (which tend to be hydrophobic), are the residues involved in the formation of the β-sheets. This implies that disruption of interior portions alter the protein either by partially denaturing it, thereby reducing its half-life in the cells, or otherwise preventing the accumulation of p21 in the transfected cells.

Dispensable Regions

In contrast, several stretches of the external parts of the protein can be deleted without destroying the transforming capacity of the protein; this includes loop 4, α-helix 2, loop 7, α-helix 3 and part of α-helix 4 (table 1). There is, however, a requirement for a certain number of amino acids, since enlarging the deletions to span the entire non-essential region usually results in a non-transforming protein (for example 63SDQ73 and 68ADQ77 are both transforming, but 63SDQ77 is transformation-defective, table 1). Evolutionarily, region A is practically identical in all *ras* proteins, while regions B and C are less tightly conserved. For all or some of these segments, it remains possible that the specific sequences are necessary for interactions pertaining to the normal function of p21, but they are not essential in the activated, transforming protein studied here.

GTP-binding

The in vitro GDP binding activities of the mutants were determined from mutant protein synthesized in bacteria (Willumsen et al., 1986), in which all proteins were stable. Proteins from five of the six essential regions (1, 3, 4, 5 and 6) had impaired GDP-binding if p21 was isolated from inclusion bodies with the use of guanidine hydrochloride; however, when isolated with a non-denaturing extraction, mutants from region 6 showed low, but significant GDP binding. Two mutants from essential region 5 had some transforming activity, but their binding was markedly impaired (Fig. 1). When protein was isolated from cells from the few foci induced by the two mutant genes, there was a small amount of phosphorylated p21 present, indicating that these mutant proteins retained some GTP-binding activity in vivo. The few mutants from essential regions 3 and 4 had low (2% of wild-type) binding activity. Mutants from essential region 2 retained high GDP binding activity. Since mutant proteins may be able to

Table 1. Transforming activity of deletion mutants.

mutant number	amino acid structure[1]	focus formation[2]	GDP binding[3]	structural localization[4]	region (Figure 1)
pBW1423	wild type	+	+	NA	NA
pBW1200	21PDQ30	-†	-†	α1	2
pBW1506	48LIR49	-	nd	L3	3
pBW1303	63SDQ73	+	+	L4	A
pBW1418	68ADQ77	+	+	L4	A
pBW1304	63SDQ77	-	-	L4	A
pBW1399	78PDQ83	-	nd	β4	4
pBW1404	85TDQ87	+	nd	L5	4
pBW1267	92LIR96	+	+	α2	B
pBW1271	100LIR104	+	+	α2	B
pBW1248	101PDQ109	+	+	L6	B
pBW1244	106ADQ112	low	-	L6-β5	B&5
pBW1197	110LIR112	-	-	β5	5
pBW1234	112LIR113	-	-*	β5	5
pBW1220	119LIR126	low	-	L7	5&C
pBW1237	123LIR130	+	nd	L7-α3	C
pBW1238	123LIR132	+	+	L7-α3	C
pBW1239	130LIR139	+	low*	L8	C
pBW1306	140LIR146	-	-*	L9	6
pBW1313	146LIR151	-	-*	L9	6
pBW1316	148LIR151	+	low*	L9	6
pBW1236	165PDQ184	+	+*	NA	D

1 :The numbers indicate the v-*ras*H amino acids encoded by front- and tail- ends, respectively, the letters indicate the three amino acids specified by the oligonucleotide linker (indicated with the one letter amino acid code) which join the front- and tail-ends.

2 :A transforming activity equivalent to that of the wild type gene in the same vector is indicated by +: approximately 2000 focus forming units (ffu) per microgram of DNA. Low is 50 to 200 ffu/ug DNA. No foci obtained with 0.2 ug DNA is indicated by -

3 :GDP binding equal to that of wild type is indicated with +, low means 10-30% of that of wild type, unmeasurable is indicated with -. * means the measurement was done on protein which had not been denatured during isolation.

4 :Structure of location of deleted amino accids as determined by de Vos et al., 1988. αx: α-helix x; βy: β-strand y; Lz: loop z.

† :More than 90% of the protein is found phosphorylated in NIH3T3 cells

NA:Not Applicable.

fold and bind GDP in vivo, but be unable to renature in vitro, negative binding results obtained following extraction of bacterially synthesized protein with guanidine hydrocloride must be interpreted cautiously. However, these results are consistent with the hypothesis that GDP binding is necessary, but not sufficient, for efficient *ras*-mediated transformation.

The crystal structure of p21 shows that amino acids 28, 116, 119, 145 and 146 interact with the guanine moiety and that amino acids 11-16 are in close contact with the β-phosphate of the bound GDP (de Vos et al., 1988). We find that deletions of these codons are non-transforming and express unstable proteins in NIH 3T3 cells. An exception is phe 28; 21PDQ30 (table 1) produces a p21 where more than 90% of the protein is phosphorylated *in vivo*. Thus the destruction of the interaction with phe 28 is not sufficient to lead to a p21 negative for nucleotide-binding.

Sequences Defining Other Interactions

In order for p21 to mediate the growth signals that lead to transformation, it is assumed that the protein interacts with other cellular components, resulting in first the binding of the protein to GTP, and next to the transmission of the growth signal, via a GTP-induced conformational change.

Such interactive sites in proteins are often thought to be situated in the parts of the protein where the structural components (the α-helices and β-strands) join each other. Therefore, it may be of interest to analyze our mutants in the context of the joining regions as defined by the crystal structure. Such an examination suggests that the regions in the part of p21 included in the structure that might be involved in sequence-specific interactions with other cellular components are extremely limited. In the structure, the amino acid sequences believed to be taking part in the four β-strands are 1-10, 35-44, 51-57, 76-83, 109-116 and 139-144, and the α-helices are made from amino acids 14-26, 86-105, 126-135 and 150-171 (de Vos et al., 1988). Loops would thus include amino acids: 11-13, 27-34, 45-50, 58-75, 84-85, 106-108, 117-125, 136-138, and 145-150. Clearly, loops 10-14, and the beginning of loop 145-150 are involved in the GDP binding (de Vos et al., 1988). Residues 26-34 are important for transformation and include part of the effector region. A mutant, 48LIR49, is transformation defective, suggesting that these residues may be involved in sequence specific interactions. However, between positions 69 and 76, all amino acids can be deleted without a loss of transformation activity, and in region 83-94 some distortion is tolerated since mutant 85SDQ87 and 92LIR96 are transformation competent (table 1). In addition, mutants 101PDQ109, 123LIR132 130LIR139 and 148LIR151 are transformation competent, suggesting that a large part of the residues comprising the loops can be changed without loss of function.

Thus, with the exception of amino acids 34-63 there is little evidence that these joining regions are involved in functions that are essential for transformation.

GTPase Activating Protein

Recently, a protein which greatly stimulates the intrinsic GTPase activity of normal p21 (GTPase activating Protein, GAP) has been described (Trahey and McCormick, 1987). *In vivo*, the presumed growth-stimulating form of p21, GTP·p21, would be converted to GDP·p21 by this activity. Oncogenic p21 proteins, which generally have reduced intrinsic GTPase activities, are not stimulated by GAP (Trahey and McCormick, 1987), leading to a greatly increased half-life of oncogenic GTP·p21 species. One hypothesis for GAP function is that GAP is a down-regulator of p21, and that oncogenic versions of p21 avoid the extinguishing activity of GAP. Alternatively, GAP may get activated through the interaction with GTP·p21, and mediate the growth signal. Although the oncogenic forms of p21 do not display accelerated GTP-hydrolysis, they might still activate GAP.

Regions Dispensable for GAP function

Hoping that a determination of the regions of p21 which are important for the function of GAP might shed light on the question of the function of GAP, we undertook an analysis of mutant p21. Various well characterized mutations in v-*ras*^{H} were transferred to the c-*ras*^{H} allele, and the ability of GAP to stimulate the hydrolysis of GTP bound to the expressed p21 proteins was determined (Adari et al., 1988).

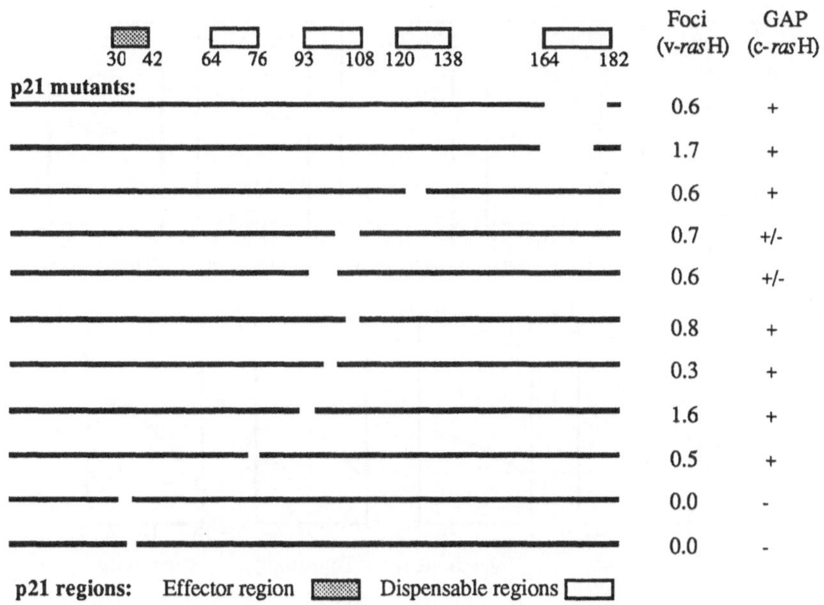

Figure 2. GAP SENSITIVITY OF P21 MUTANTS. Mapping the regions of p21 necessary for interaction with GAP. *E. coli* expressed p21 proteins were loaded with GTP and the ability of a GAP extract to accelerate the GTP hydrolysis was asessed.

The results (Figure 2) show that deletion of individual dispensable regions had no effect on the ability of p21 to interact with GAP; however, mutations in the effector region abolished the GTPase activation by GAP (Adari et al., 1988). Similar results on the effector domain have been obtained by Calès et al., 1988. As these findings suggest that the effector and GAP interact through the same region, it suggests that GAP may be the effector.

Separation of GAP and Effector activities

Since there as yet is no biochemical assay for effector function, the most direct way of asking whether GAP and effector function is the same is to make a series of mutations affecting one of the functions to different degrees and to determine if the other is affected with a similar hierarchy.

To this end, we made a series of mutations in the effector domain of p21 in the v-ras^H allele and measured their relative ability to transform NIH 3T3 cells (Stone et al., 1988). In addition, the mutants were transferred to the c-ras^H allele and assessed for the relative GAP activity. The results of the GAP assays are shown in figure 3, and the correlations between transformation activity and GAP stimulation in either allele in figure 4.

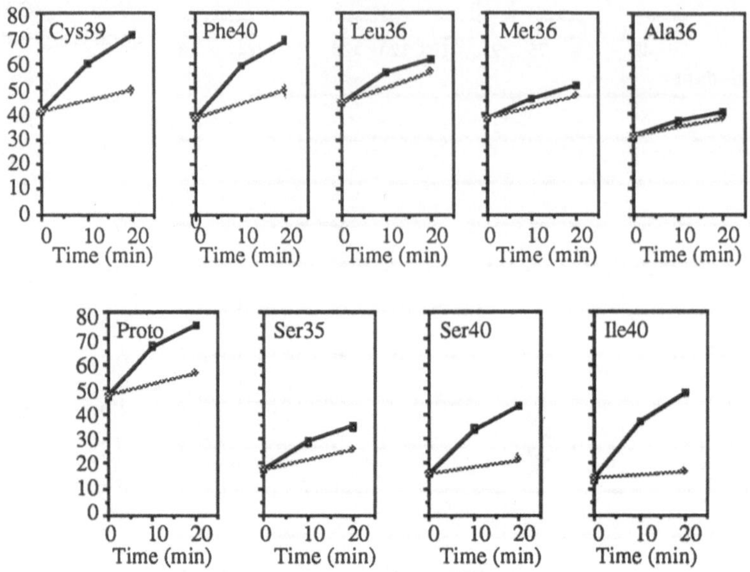

Figure 3. GTPase Accelleration for mutant p21 proteins. Acceleration of GTP hydrolysis by effector region point mutants. Mutants encoding proto-p21 with the indicated point mutations (*c-ras*H encodes thr35, ile36, ser39, and tyr40) were loaded with α-^{32}P-GTP and at time 0 a GAP extract added. Samples were withdrawn for precipitation with antibody 259, and the p21 bound nucleotide analysed by thin layer chromatography. The results are expressed as % conversion from GTP to GDP. Full lines: with GAP extract, shaded lines: without GAP extract.

As can be seen from figure 4, there are effector domain mutations that completely destroy the transforming activity of v-*ras*^{H}, whereas the same mutation, present in c-*ras*^{H}, leaves the GTPase activity of the encoded protein fully stimulated by GAP. The existence of several such mutations argues that GAP and the effector may be separate factors, interacting through the same domain. However, it is still possible that particular point mutations lead to a p21 which can interact and be stimlated to hydrolyze bound GTP by GAP, but which fail to transmit the signal to GAP.

CONCLUSIONS

All available evidence suggests that the p21 protein represents a molecular switch; a protein which can be turned ON and OFF by biochemical events and which, in the ON state, is able to activate other cellular components. Several other guanine-nucleotide binding proteins have such a function, including G-proteins of mammalian adenylate cyclase and transducin of the cGMP phosphodiesterase (Stryer and Bourne, 1986). The conserved amino acid sequence among these proteins, which centers around the GTP-binding pocket, suggests that once invented, the basic design was maintained: one conformation when GTP is bound, another when GDP is bound. Selection has then exploited this difference in conformation and coupled it to different pathways.

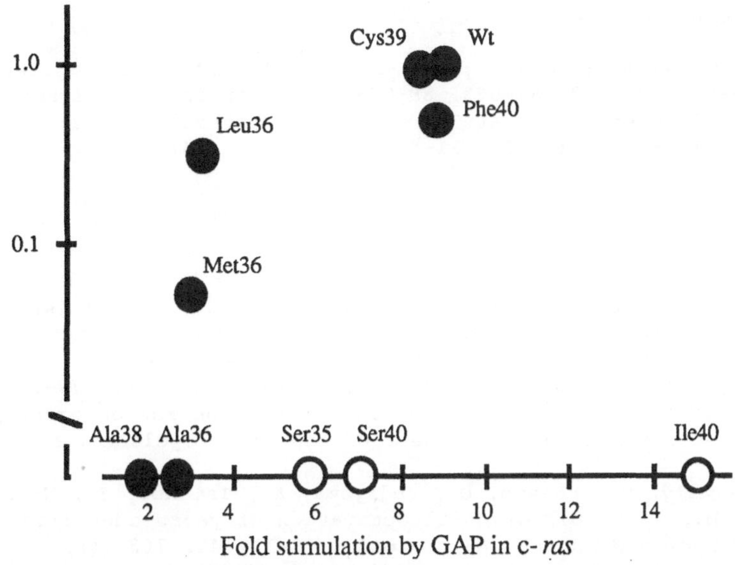

Figure 4. ACTIVITIES OF *RAS*^{H} EFFECTOR REGION POINT MUTANTS. Correlation between GAP sensitivity of the proto-oncogene version of the effector domain mutants with the transforming activity of the same mutations in the v-*ras*H gene. Fold stimulation is calculated as the initial rate of GTP hydrolysis in the presense of GAP divided with the rate of hydrolysis without GAP (Figure 3). Transformation activity was determined in the eucaryotic vector and standardized relative to v-*ras*^{H} via the efficiency of generation of G418 resistant colonies (Stone et al., 1988).

The mechanisms that regulate the availability of the active p21-species have been elucidated through the analysis of oncogenic mutants and, recently, of the GTPase Activating Protein. In theory, a mutation which captures the p21 in the active conformation, such that the face of the protein which interacts with the effector or target molecule is recognized by this, would be oncogenic, independent of the ability of p21 to bind GTP, GDP or other stimulating factors. Such a transforming, GTP-binding negative mutant has been described (Clanton et al., 1987). The existence of such mutants do not refute the importance of GTP-binding for biologic activity of the normal p21.

The interactions that result in the conversion of p21·GDP to p21·GTP are not understood. We speculate that the regions we have found to be dispensable for transformation may somehow be involved, since they are fairly well conserved among the *ras* genes and their location on the surface of the p21 protein makes them potentially accessible for interaction.

It is likely that other molecules in addition to GAP interact with p21. Indirect evidence in yeast suggests that the product of CDC25 interacts with the RAS gene product (Broek et al., 1987, Robinsion, et al., 1987). Analysis of *ras* mutants has provided a framework in which to understand the function of various regions of the protein and has led to the identification of proteins that interact with the *ras* protein. Further analysis of mutant *ras* should help to elucidate the significance of these other interactions.

ACKNOWLEDGEMENTS

Parts of this work have been supported by the Danish Cancer Society (84-068, 86-053, 86-149, 87-063, 88-078), the Danish Medical Research Council (12-5345 and 12-4663), the Danish Natural Science Research Council (11-5095, 11-5603, 11-6632) and NATO (84/165).

REFERENCES

Adari, H., Lowy, D. R., Willumsen, B. M., Der, C. J., McCormick, F., 1988, Guanosine triphosphatase activating protein (GAP) interacts with the p21 *ras* effector binding domain., Science, 240, 518-521.

Barbacid, M., 1987, *ras* genes, in Ann. Rev. Biochem., 56, 779-827.

Beckner, S. K., Hattori, S., Shih, T. Y., 1985., The *ras* oncogene product p21 is not a regulatory component of adenylate cyclase., Nature, 317, 71-73.

Broek, D., Samily, N., Fasano, O., Fujiyama, A., Tamanoi, F., Northup, J., Wigler, M., 1985, Differential activation of yeast adenylate cyclase by wild type and mutant RAS proteins., Cell, 41, 763-769.

Broek, D., Toda, T., Michaiel, T., Levin, L., Birchmeier, C., Zoller, M., Powers, S., Wigler, M., 1987, The S. cerevisiae CDC25 gene product regulates the RAS/adenylate cyclase pathway, Cell, 48, 789-799.

Buss, J. E., Sefton, B. M., 1986., Direct identification of palmitic acid as the lipid attached to p21-*ras*., Mol. Cell. Biol., 6, 116-122.

Calès, C., Hancock, J. F., Marshall, C., Hall, A., 1988, The cytoplasmic protein GAP is implicated as the target for regulation for the *ras* gene product, Nature, 332, 548-551.

Chen, Z-Q., Ulsh, L S., DuBois, G., Shih, T. Y., 1985, Posttranslational processing of p21 *ras* proteins involves palmitylation of the C-terminal tetrapeptide containing cysteine-186., J. Virol., 56, 607-612

Clanton, D. J., Lu, Y., Blair, D. G., Shih, T. Y., 1987, Structural significance of the GTP-binding domain of ras p21 studied by site-directed mutagenesis., Mol. Cell. Biol., 7, 3092-3097.

de Vos, A. M., Tong, L., Milburn, M. V., Matias, P. M., Jancarik, J., Noguchi, S., Nishimura, S., Miura, K., Ohtsuka, E., Kim, S-H., 1988, Three-dimentional structure of an oncogene protein: Catalytic domain of human c-Ha-*ras* p21, Science, 239, 888-893.

DeFeo-Jones, D., Tatchell, K., Robinson, L. C., Sigal, I. S., Vass, W. C., Lowy, D. R., Scolnick, E. M., 1985., Mammalian and yeast *ras* gene products: Biological function in their heterologous systems., Science, 228, 179-184.

Field, J., Broek, D., Kataoka, T., Wigler, M., 1987, Guanine nucleotide activation of, and competition between, RAS proteins from Saccharomyces cerevisiae., Mol. Cell. Biol., 7, 2128-2133.

Fujiyama, A., Tamanoi, F., 1986., Processing and fatty acid acylation of RAS1 and RAS2 proteins in Saccharomyces cerevisiae., Proc. Natl. Acad. Sci., 83, 1266-1270.

Fukui, Y., Kozasa, Kaziro, Y., Takeda, T., Yamamoto, M., 1986., Role of a *ras* homolog in the life cycle of Schizosaccharomyces pompe., Cell, 44, 329-336.

Jhappan, C., Vande Woude, G. P., Robins, T. S., 1986, Transduction of host cellular sequences by a retroviral shuttle vector., J. Virol., 60, 750-753.

Jurnack, F., 1985., Structure of the GDP domain of EF-Tu and location of the amino acids homologous to *ras* oncogene proteins., Science, 230, 32-36.

Kataoka, T., Powers, S., Camaron, S., Fasano, O., Goldfarb, M., Broach, J., Wigler, M., 1985., Functional homology of mammalian and yeast RAS genes., Cell, 40, 19-26.

Kyte J., Doolittle R. F., 1982, A simple method for displaying the hydropathic characted of a protein, J. Mol. Biol., 157, 105-132.
Lacal, J. C., Santos, E., Notario, V., Barbacid, M., Yamazaki, S., Kung, H-F., Seamans, C., McAndrew, S., Crowl, R., 1984, Expression of normal and transforming H-*ras* genes in Escherichia coli and purification of their encoded p21 proteins., Proc. Natl. Acad. Sci. USA. 81, 5305-5309.
Levitzki, A., Rudick, J., Pastan, I., Vass, W.C., Lowy, D.R., 1986, Adenylate cyclase activity of NIH 3T3 cell morphologically transformed by *ras* genes, FEBS Letters, 197, 134-138.
Lowy, D. R., Rands, E., Scolnick, E. M., 1978, Helper independent transformation by unintegrated Harvey Sarcoma virus DNA, J. Virol., 26, 291-298.
Lowy, D. R., Papageorge, A. G., Vass, W. C., Willumsen, B. M. (1988). Mutational analysis of *ras* processing and function. *In:* Cellular and Molecular Biology of Tumors and Potential Clinical Applications. Minna, J. D., Kuehl, M., eds., Alan R. Liss, Inc., New York, pp 203-212.
Marshall C:In Weiss R, et al (eds):"RNA Tumor Viruses. Molecular Biology of Tumor Viruses."New York:Supplement to 2nd edition, Cold Spring Harbor Laboratory, 1985, pp 487-558.
Marshall, M. S., Gibbs, J. B., Scolnick, E. M., Sigal, I. S., 1988, An adenylate cyclase from Sachharomyces cerevisiae that is stimulated by RAS proteins with effector mutations, Mol. Cell. Biol., 8, 52-61.
McCormick, F., Clark, B. F. C., La Cour, T. F. M., Kjeldgaard, M., Norskov-Lauritsen, L., Nyborg, J., 1985., A model for the tertiary structure of p21, the product of the *ras* oncogene., Science, 230, 78-82.
Mulcahy, L. S., Smith, M. R., Stacey, D. W., 1985., Requirements for *ras* proto-oncogene function during serum stimulated growth of NIH 3T3 cells., Nature, 313, 241-243.
Robinsion, L. C., Gibbs, J. B., Marshall, M. S., Sigal, I. S., Tatchell, K., 1987, CDC25: A component of the RAS-adenylate cyclase pathway in Saccharomyces cerevisiae., Science, 235, 1218-1221.
Sigal, I. S., Gibbs, J. B., D'Alonzo, J. S., Scolnick, E. M., 1986, Identification of effector residues and a neutralizing epitope of Ha-*ras*-encoded p21., Proc. Natl. Acad. Sci., 83, 4725-4729.
Stone, J. C., Vass, W. C., Willumsen, B. M., Lowy, D. L., 1988, p21-*ras* effector domain mutants constructed by "cassette" mutagenesis., Mol. Cell. Biol., 8, 3565-3569.
Stryer, L., Bourne, H., 1986, G proteins: A family of signal transducers, Ann. Rev. Cell Biol., 2, 389-417.
Trahey, M., McCormick, F., 1987, A cytoplasmic protein stimulates normal N-*ras* p21 GTPase, but does not affect oncogenic mutants., Science, 238, 542-545.
Willumsen, B.M., Papageorge, A.G., Kung, H.-F., Bekesi, E., Robins, T., Johnsen, M., Vass, W.C., Lowy, D.R.,1986, Mutational analysis of a *ras* catalytic domain, Mol. Cell. Biol., 6, 2646-2654.
Willumsen, B M., Christensen, A., Hubbert, N. L., Papageorge, A. G., Lowy, D. R., 1984a., The p21 *ras* terminus is required for transformation and membrane association., Nature, 310, 583-586.
Willumsen, B. M., Norris, K., Papageorge, A. G., Hubbert, N. L., Lowy, D. R., 1984b., Harvey murine sarcoma virus p21 *ras* protein: biological and biochemical significance of the cysteine nearest the carboxy terminus., EMBO J. 3, 2581-2585.
Willumsen, B. M., Papageorge, A. G., Hubbert, N., Bekesi, E., Kung, H-F, Lowy, D. R., 1985., Transforming p21 *ras* protein: flexibility in the major variable region linking the catalytic and membrane-anchoring domain., EMBO J., 4, 2893-2896.
Willumsen, B. M., Papageorge, A. G., Kung, H-F., Bekesi, E., Robins, T. S., Johnsen, M., Vass, W. C., Lowy, D. R., 1986, Mutational analysis of a *ras* catalytic domain, Mol. Cell. Biol., 6, 2646-2654.

ANALYSIS OF THE BIOCHEMICAL AND BIOLOGICAL ACTIVITIES OF DELETION MUTANTS OF THE H-RAS P21 PROTEIN SUGGEST THAT GAP IS AN ESSENTIAL COMPONENT OF ITS EFFECTOR FUNCTION

Armando Di Donato, Shiv K. Srivastava# and
Juan Carlos Lacal*

Laboratory of Cellular and Molecular Biology
National Cancer Institute. Bethesda, Maryland 20892
(#) Georgetown University Medical School
Washington D.C. 20007

SUMMARY

The anti-ras p21 monoclonal antibody Y13-259 has been shown to efficiently neutralize DNA synthesis induced by serum in quiescent cells (1), and phenotypically reverts cells transformed by a variety of oncogenes (2). Due to the putative biological relevance of the epitope recognized by the monoclonal antibody Y13-259, we have generated deletion mutants of the H-ras p21 protein which lack residues 58 to 63 or 64 to 68, and contain either glycine or arginine at position 12. In addition, those mutants carrying a deletion at position 64 to 68 also carried an activating substitution of Thr at position 59. None of the deleted proteins were recognized by monoclonal antibody Y13-259, and those mutants carrying activating mutations showed at least a 100-fold reduction in their transforming activities compared to their non-deleted counterparts. The mutant proteins carrying a normal 12 position also showed a decreased transforming activity when compared to the normal protein. Alterations observed in the in vitro GTPase or GTP-interchange properties in our deletion mutants were

(*) Present address: Instituto de Investigaciones Biomedicas, Facultad de Medicina, Universidad Autonoma, Madrid 28029, Spain

not consistent with their decreased transforming activities. Moreover, all the generated mutants showed normal palmitylation and membrane localization, known to be essential for biological activity of *ras* proteins. Recently, a protein, designated as GAP, has been described which is able to specifically increase the GTPase activity of normal *ras* p21 (3). We have found that GAP is unable to increase the GTPase activity of our deleted proteins. These observations strongly suggest that the recognition site for Y13-259 within the *ras* p21 molecule influences directly or indirectly the interaction of *ras* p21 with GAP, and suggest that this interaction is critical for the biological activity of *ras* proteins.

INTRODUCTION

The *ras* family of proto-oncogenes code for guanine nucleotide binding proteins, which are evolutionary conserved in mammals and lower eukaryotes (4,5). Point mutations affecting two major hot spots in the coding sequence of mammalian *ras* genes alter the biological properties of their encoded proteins (p21) in a manner that makes them potent transforming proteins (6-9). Due to the likely importance of *ras* proto-oncogenes in the neoplastic transformation, there have been intensive investigations of the structure and function of p21 proteins. *In vitro* mutagenized *ras* genes, as well as antibodies of defined specificities raised against p21, have been employed to identify functionally important regions of the p21 molecule (5,10). The residues in the p21 amino acid sequence which constitute the epitope for the monoclonal antibody Y13-259 have been of particular interest, since microinjection of this antibody into NIH/3T3 cells and certain other fibroblastic cell lines blocks serum-stimulated DNA synthesis (1). Antibody Y13-259 also inhibits the *ras*-stimulated adenylate cyclase activity in yeast (11) and causes reversion of the transformed phenotype induced by a number of oncogenes (2). Utilizing deletions or site directed mutagenesis of *ras* proteins, the epitope recognized by Y13-259 has been localized to a stretch of amino acid residues from position 63 to 73 in the p21 amino acid sequence (11,12).

RESULTS AND DISCUSSION

Earlier studies showed that antibody Y13-259 does not affect *in vitro* GTP-binding, autophosphorylation or GTPase activities of p21 proteins (12). However, GTP exchange by p21 proteins was found to be impaired by this antibody (13,14). In order to evaluate the functional significance of residues which form the epitope for antibody Y13-259 in p21 amino acid sequence, we have generated viral H-*ras* p21 deletion mutants lacking residues 58 to 63, or lacking residues 64 to 68 with Thr at position 59. One set of mutants contained the normal Gly at position 12, whereas the other set substituted arginine at position 12, a mutation which is sufficient to confer oncogenic activity to ras p21 protein (16).

The strategy utilized to construct ras p21 deletion mutants is described in Fig 1. The newly generated mutants harboring chimeric genes were placed under the transcriptional control of the Abelson murine leukemia virus (Ab-MuLV) LTR (19), and their transforming

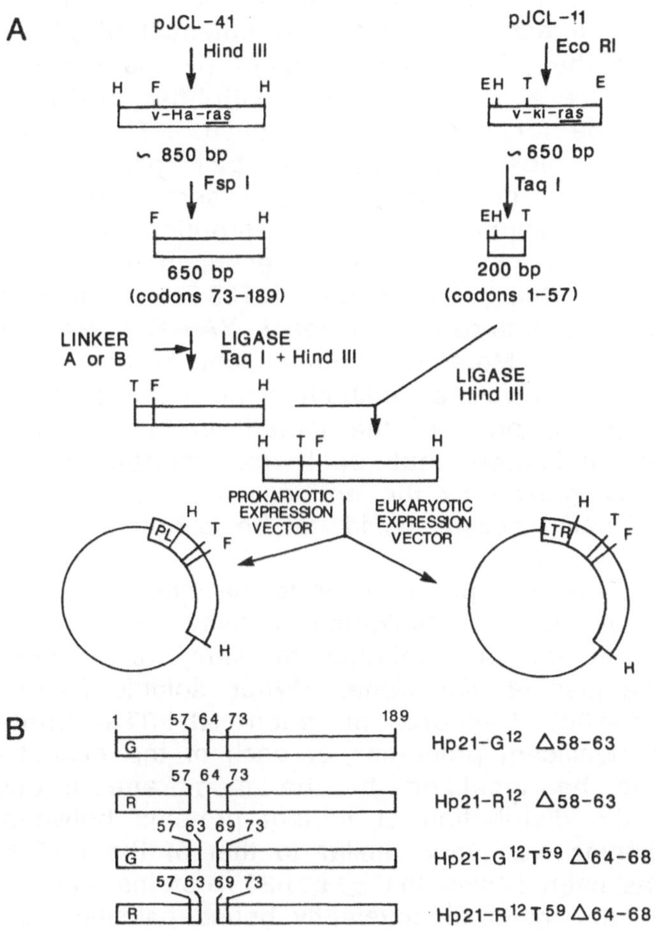

Fig 1. Construction of p21 deletion mutants

(A) Schematic representation of the construction of the different deletion mutants from DNA fragments of the viral H- or K-ras genes.

(B) Schematic representation of generated ras p21 proteins indicating the specific deleted regions in each product. (G) glycine, (R) arginine, (T) threonine.

activities were analyzed in the NIH/3T3 focus-forming assay. The deletion mutant p21 Arg^{12} (Δ58-63), with oncogenic substitution at position 12, or p21 Arg^{12} Thr^{59} (Δ64-68) mutant with oncogenic substitutions at both positions 12 and 59, showed at least 100-fold reduction in their transforming activities when compared to the v-H-ras oncogene encoding for p21 Arg^{12} Thr^{59} (Table 1). The transforming activity of these mutants was similar to the normal H-ras gene. Furthermore, the activity of the p21 Gly^{12} Thr^{59}(Δ64-68) mutant carrying the Thr^{59} activating mutation was reduced by 300 fold in comparison to the H-ras oncogene encoding p21 Gly^{12} Thr^{59}. In addition, transforming activity of deletion mutants with a normal glycine residue at position 12 was at least 10 fold reduced compared to the normal p21 protein. These results indicate that deletion of either residues 58 to 63 or residues 64 to 68 in p21, exerted a potent inhibitory effect on transforming potential of p21 protein.

To analyze the p21 protein encoded by the generated mutants, NIH/3T3 cells were co-transfected with the pSV2-neo plasmid carrying the gene for neomycin resistance. The resistant colonies were grown to mass culture, and p21 proteins analyzed by immunoprecipitation with various anti-p21 sera. When Y13-259 was used for immunoprecipitation, the p21 proteins encoded by deletion mutants were undetectable. However, endogenous p21 proteins were readily detected in the representative NIH/3T3 transfectants (Fig. 2A). In contrast, another anti-p21 monoclonal, YA6-172, recognized each of the mutant proteins. Moreover, both an anti H-ras p21 peptide (161-176) serum and a polyclonal antibody against bacterially expressed H-ras p21 recognized each of the mutant proteins (data not shown). These results indicated that each p21 mutant lacked antigenic determinants recognized by the antibody Y13-259, but their markedly reduced transforming activity could not be accounted for by low levels of expression.

The translocation of p21 proteins to the membrane has been shown to be essential for their biological activity (25,26). Therefore, we investigated whether our deletion mutants were able to properly localize to the plasma membrane. When soluble (S-100) and crude membrane (P-100) fractions of each NIH/3T3 tranfectant were analyzed , a significant proportion of each of the mutant p21 proteins was found to be localized to the membrane fraction (Fig 2B) Furthermore, the distribution of mutant proteins between S-100 and P-100 components was very similar to that of the wild type ras p21 product. It has been shown that ras p21 proteins become membrane associated as the result of covalently bound palmitic acid attached to Cys^{186} residue at the extreme carboxy terminus of the molecule (25,26). Fig. 2C shows that representative p21 mutants from each set were palmitylated. All of these results strongly suggested that p21 proteins lacking either residues 58 to 63 or residues 64 to 68 underwent the normal post-translational modifications of the wild type ras protein.

TABLE 1. Transforming activity of ras p21 genes

Protein	Foci/μg DNA*	Focus morphology
Arg^{12} Thr^{59}	100	Large
Gly^{12} Thr^{59}	30	Large
Gly^{12} Ala^{59}	1	Small
Gly^{12} Δ58-63	≤0.1	Small
Arg^{12} Δ58-63	1	Medium
Gly^{12} Thr^{59} Δ64-68	≤0.1	Small
Arg^{12} Thr^{59} Δ64-68	1	Small

*Data represent relative transforming activity of each ras gene, where 100% represents 1.0 x 10^4 ffu/μg DNA. Results are representatives of three independent experiments with similar results.

Since the mutant p21 proteins with reduced transforming activity displayed normal post-translational modification and subcellular localization, it was of interest to determine the effects of these deletions on the known in vitro biochemical properties of p21 protein. Towards this objective, the mutant genes were inserted into a prokaryotic expression vector and expressed in E. coli. The mutant proteins were then purified to apparent homogeneity as previously described (17,20), and analyzed for guanine nucleotide binding. As shown in Table 2, the mutant p21 proteins did not show any significant alteration of this activity arguing that residues 58 to 68 do not participate in the guanine nucleotide binding function of the protein, in good agreement with recent crystallographic studies of the ras p21 molecule (27).

Ras proteins have been shown to possess the ability to hydrolyze bound GTP at a very slow rate of 0.02 to 0.2 min-1 in vitro (28-31). Several ras p21 mutants with high transforming efficiency have been shown to exhibit decreased in vitro GTPase activity (28-31). As shown in Table 2, the newly generated p21 mutant ras proteins showed a rate of GTP hydrolysis that was reduced from 2.5 to 10-fold as compared to the normal p21 protein. The deletion mutants p21 Arg^{12} (Δ58-63) and p21 Arg^{12} Thr^{59}(Δ64-68) with reduced transforming activities, exhibited about 10-fold increase in their GTPase activities when compared to viral p21 Arg^{12} Thr^{59}. However, the p21 Gly^{12} Thr^{59} (Δ64-68) mutant with even lower transforming activity showed at least 4-fold reduction in its GTPase activity with respect to the normal protein, p21 Gly^{12} Ala^{59}. Furthermore, the p21 Gly^{12} (Δ58-63) mutant without any oncogenic amino acid substitution showed

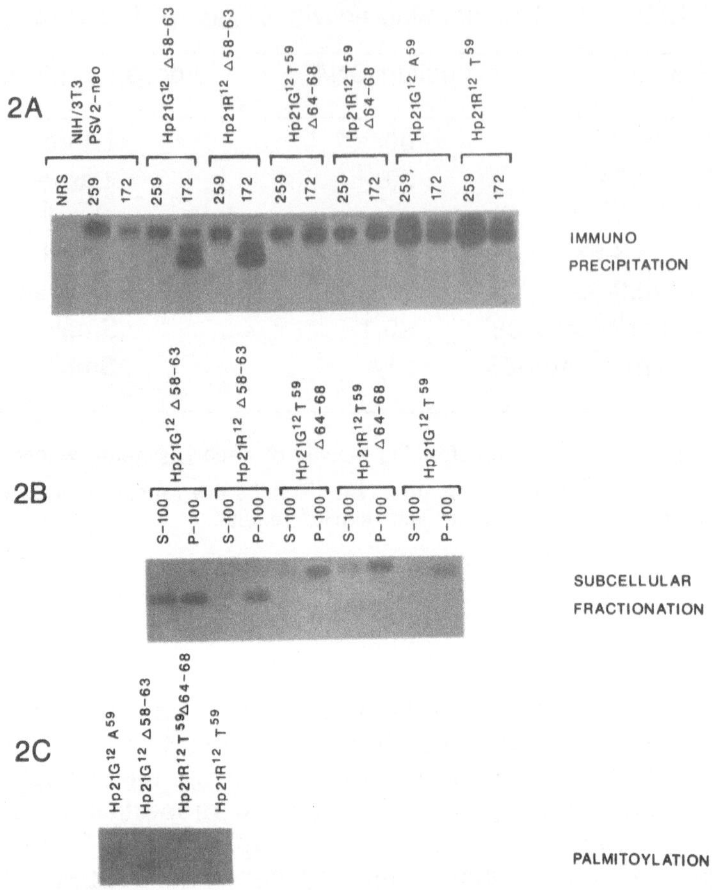

Fig.2 Immunochemical analysis and biogenesis of p21 ras mutants

Analysis of the *ras* proteins was carried out in NIH-3T3 cells transfected with plasmids generated as described in Materials and Methods. (A) Immunoprecipitations were performed utilizing the anti-*ras* p21 monoclonal antibody indicated in each lane, followed by SDS-polyacrylamide gel electrophoresis . (B) Subcellular localization of *ras* p21 proteins was studied by analysis of p21 proteins in cytosolic (S100), and crude membrane (P100) fractions . (C) Fatty acylation of p21 proteins was performed by metabolic labelling of cells with 3H-palmitic acid, followed by immunoprecipitation with anti-*ras* p21 antibody YA6-172.

much lower GTPase activity, which was approximately 1/10th that of the normal protein and similar to that of highly transforming p21 Arg^{12} Thr^{59} protein, but had no transforming activity. Thus, the relative *in vitro* GTPase activities of mutant p21 proteins did not correlate with their transforming potential, in agreement with previous observations (17, 32-34).

There has been evidence that a substitution of Ala^{59} to Thr^{59} increased the off-rate of guanine nucleotides in ras p21 molecules, and that the Y13-259 monoclonal antibody is able to block guanine nucleotide interchange by either normal or mutated ras proteins (13,14). Moreover, an activating substitution of Asp to Ala at position 119 seems to affect the off-rate of GTP as well (35). These results suggested that activation of the ras p21 transforming activity could be achieved by alterations of the interchange ability of the protein. Therefore, we studied the effects of the deletion of residues 58 to 68 on this activity. As shown in Table 2, p21 Arg^{12}(Δ58-63) and p21 Arg^{12} Thr^{59}(Δ64-68) proteins exhibited a slower rate of GTP exchange compared to the mutants with glycine at position 12, p21 Gly^{12} (Δ58-63) and p21 Gly^{12} Thr^{59}(Δ64-68). These mutants exhibited a faster rate of GTP exchange, similar to that of p21 Arg^{12} Thr^{59}. It is evident from these results that the residue at position 12 of p21 proteins lacking either amino acids 58 to 63 or 64 to 68 plays an important role in GTP exchange. This is in agreement with the three-dimensional crystal structure of the c-H-ras p21 (27), where residue 12th is part of a loop in direct contact with the phosphoester bond between ß and phosphates of GTP. However, it was not possible to correlate the *in vitro* GTPase and GTP exchange activities of mutant p21 proteins with their biological activity. For instance,

Table 2.- Biochemical characterization of ras p21 proteins

Protein	GTP-bindind[a]	GTPase[b]	GTP-exchange[c]
Gly12 Ala59	0.60	54	100
Arg12 Thr59	0.37	2	42
Gly12 Δ58-63	0.35	5	35
Arg12 Δ58-63	0.20	20	140
Gly12 Thr59 Δ64-68	0.72	13	38
Arg12 Thr59 Δ64-68	0.70	20	100

a. Apparent K_m for GTP of various ras proteins were calculated as the concentration (nM) at which they exhibit 50% of maximum binding.
b. GTPase activity was analyzed as described before (17), and are expressed as pmoles of phosphate released in 120 min by 10 pmoles of active protein, estimated by their GTP-binding activity.
c. GTP-exchange activity was assayed as described previously (13). Values represent the time in minutes required to display 50% of bound GTP.

mutants containing Gly[12] with very low transforming activity showed GTPase and GTP exchange activities similar to that of potent transforming H-p21 (Arg[12] Thr[59]). Taken together, the results presented here as well as our previous studies characterizing the properties of the interaction of monoclonal Y13-259 with ras p21 proteins in vitro (13) suggest that residues 58 to 68 of the p21 sequence may be important for the interaction of ras proteins with putative effector or modulator molecules, which in turn could control the biological activity of the ras p21 protein.

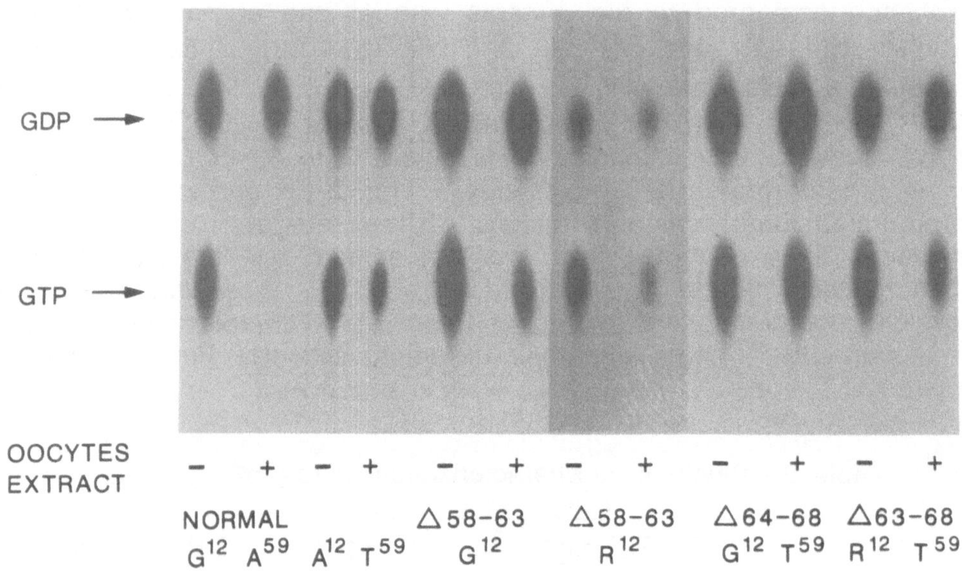

Fig. 3 Stimulation of GTPase activity of p21 deletion mutants by GAP.

Mutant and wild type p21 proteins were analyzed for their GTPase activity in the presence (+) or absence (-) of extract from Xenopus laevis oocytes. The result shown here is a representative of three independent experiments with identical results.

Recently, it has been reported that extracts of Xenopus laevis oocytes contain a factor which is able to markedly increase the in vivo and in vitro GTPase activity of the normal N- or H-ras p21 protein (3,37). This effect was not observed with p21 proteins carrying oncogenic mutations at position 12, 59 or 61. (36,37). This factor, designated GAP for GTPase Activating Protein, was also detected in human lymphocytes , NIH-3T3 cells, and a variety of established cell lines. Furthermore, recent reports have provided evidence that mutations in the putative "effector domain" of the p21 protein (residues 32 to 40) abolish stimulation of the GTPase activity of the normal ras p21 induced by GAP (36,37). Since alterations at the biochemical level found in our deletion mutants did not explain their markedly decreased biological activity, we investigated the effects of the GAP factor on their GTPase activities. As shown in Fig 3, the p21 Gly12(Δ58-63) mutant exhibited a significant reduction in GAP-stimulated GTPase activity, suggesting that this region of the p21 protein was important for its interaction with GAP. In agreement with recent observations (36,37), other p21 deletion mutants carrying oncogenic substitutions at either position 12 or position 59 did not show any significant effect of GAP on their GTPase activities (Fig 3). It is difficult to interpret the results obtained with the deletion mutants carrying oncogenic substitutions, since it has been shown that single point mutations which efficiently activate the transforming potential of the ras protein, abolish GAP mediated stimulation of GTPase activity. Based on the results obtained with deletion of residues 58 to 63 in the normal protein , it is most likely that the deletion mutants carrying oncogenic mutations at position 12 or 59 are also impaired for the interaction with GAP. Thus, the impaired interaction of GAP with each p21 mutant described here may be a common denominator for their significantly reduced transforming activity.

Previously described point mutations in the region of p21 which forms the antigenic determinant for Y13-259 (11) or a deletion/substitution mutant in which residues 63 to 73 were substituted by Ser-Asp-Gln (15), did not significantly reduced the transforming activity of H-ras oncogene. The apparent discrepancy in the properties of the deletion mutants described in this study and the previously reported mutants with a deletion/substitution in a similar region could be explained by an indirect contribution of this region, in the interaction of the ras protein with GAP. Thus, the effects on GAP-p21 interaction will be reflected on the function of the ras p21 protein. Moreover, our results also support the observations that Y13-259 blocks the in vitro interaction of GAP with ras p21 (3, 37, results not shown), and that Y13-259 is able to neutralize the biological activity of either normal or activated ras proteins (1,2). Thus, the results presented here also strongly suggest that GAP functions either as effector of ras p21 activity or as an assential factor required for the interaction of the ras proteins with their actual effector molecule.

MATERIALS AND METHODS

GENERATION OF DELETION MUTANTS

The construction of plasmid pJCL-41 has been published elsewhere (17,19), and contains the complete, unfused gene encoding the viral Harvey ras p21 protein. Digestion with Hind III released a 850 bp fragment containing the complete gene except for the first five codons at the amino terminus of the encoded protein. Further digestion of the purified fragment with Fsp I generated a 650 bp fragment encoding amino acids 73-189. Synthetic linkers were ligated at the blunt end generated by Fsp 1. Linker 1 was designed to delete residues 58-63 from the coding sequence, whereas linker 2 was designed to delete residues 64-68. The resulting 5' end of each fragment ligated to the linkers contained a Taq I end, which was then ligated to an EcoR I - Taq 1 DNA fragment from plasmid pJCL-11, carrying a modified viral K-ras gene. This Eco RI to Taq I fragment encodes amino acids 5 to 57 of the normal p21 with glycine at position 12, and a convenient Hind III site between the 4th and 5th codons. Ligated DNA was further digested with Hind III to purify the monomer which was then inserted into either an eukaryotic or prokaryotic expression vector for the expression of nonfused ras proteins as previously described (19,20).

Linker 1 was

```
        D    Y    S    A    M    R    D    Q    Y    M
        57   64   65   66   67   68   69   70   71   72
5'  CGAC TAT  AGT  GCC  ATG  CGG  GAC  CAG  TAC  ATG  C  3'
3'    TG ATA  TCA  CGG  TAC  GCC  CTG  GTC  ATG  TAC  G  5'
```

Linker 2 was

```
        D    T    T    G    Q    E    E    D    Q    Y    M
        57   58   59   60   61   62   63   69   70   71   72
5'  CGAC ACA  ACA  GGT  CAA  GAA  GAG  GAC  CAG  TAC  AGT C  3'
3'    TG TGT  TGT  CCA  GTT  CTT  CTC  CTG  GTC  ATG  TAC G  5'
```

The above described ras constructs contained glycine at position 12. The corresponding deletion mutants containing arginine at position 12 were generated by site directed mutagenesis in M13 phage (21). For this purpose an oligonucleotide, 5' GTG GGC GCT AGA GGC GTG GGA 3', was used to substitute codon 12 from Gly to Arg. All of the above described deletions and mutations were confirmed by dideoxy-nucleotide sequence analysis.

PROTEIN ANALYSIS

Each of the generated vector DNAs obtained as described in Fig.1 were purified and cotransfected with pSV2-neo into NIH-3T3 cells utilizing the calcium phosphate coprecipitation method and the resistant colonies were grown to mass culture. Cells were labelled with 300 µCi/ml of L-(35S)-methionine (New England Nuclear, 1100 Ci/mmol), for 4 hrs in DMEM without methionine, or for 6 hrs in same medium containing 10% calf serum and 1 mCi/ml of (3H)palmitic acid. A) Immunoprecipitations were carried out as previously described (23) with normal rat serum (NRS), or monoclonal antibodies Y13-259 or YA6-172 as indicated (24). Subcellular localization of the ras proteins was performed after cell lysis in hypotonic buffer (10 mM Tris-HCl, pH 7.5, 0.1 mM dithiotreitol, 0.01% aprotinin). Removal of cell nuclei and cell debris was achieved by low speed centrifugation, and the cell extracts separated into soluble (S-100) and membrane (P-100) fractions by ultracentrifugation at 100,000xg. Equivalent portions of S-100 and P-100 fractions from each cell line were immunoprecipitated with monoclonal antibody YA6-172 and analyzed as above. Palmitylated p21 proteins were immunoprecipitated with antibody YA6-172 and processed as previously described (23).

ANALYSIS OF GAP ACTIVITY

Normal and mutated p21 proteins were expressed in E. coli and purified to homogeneity as previously described (17). Xenopus laevis oocytes from 3-4 ovaries were homogenized using a Dounce homogenizer in 20 mM MES, pH 7.0, 1 mM MgS04, 200 µg/ml leupeptin, 0.01% aprotinin (buffer A). The homogenate was cleared at 5,000xg for 15 min., and the resulting supernatant was then centrifuged at IOO,OOOxg for 60 min. The final supernatant was utilized as a crude GAP preparation for the experiments. The GTPase assay was carried out in 300µl of buffer B (50 mM Hepes, pH 8.0, 1 mM MgS04, 100 mM NaCl, 0.25 mM ADP(NH)P, 100 µg/ml BSA, 5 mM dithiotreitol, 1 µM (-32P)GTP (3500 Ci/mmol)). Equivalent amounts of each of the p21 ras proteins were then added, and samples were incubated at 23º C for 15 min. to allow complete GTP binding. At this time, 50 µl of the oocyte extract or buffer A were added and the reaction continued at 37º C for 45 min. Samples were then immunoprecipitated with 20 µl of monoclonal antibody YA6-172 as previously described (12). This amount of antibody was first standardized as able to remove all active p21 proteins. The immunocomplexes were resuspended in 0.1 % SDS, 1 mM EDTA, and heated at 90º C for 5 min. Aliquots of 5 µl of solubized material were resolved by thin layer chromatography (TLC) on PEI Cellulose in 1 M LiCl, along with GDP and GTP standards. Autoradiograms were obtained as described, after 2-4 hrs exposure (17).

REFERENCES

1. Mulcahy, L.S., Smith, M.R., and Stacey, D.W. Nature 313, 241-243 (1985).
2. Smith, M.R., DeGudicibus, S.J., and Stacey, D.W. Nature 320, 540-543 (1984).
3. Trahey, M., and McCormick, F. Science 238, 542-545 (1987).
4. Gibbs, J.B., Sigal, I.S., and Scolnick, E.M. Trends Biochem. Sci. 10, 350-353 (1985)
5. Lacal, J.C., and Tronick, S.E. The ras oncogene. In The Oncogene Handbook Reddy, P, Curran, T., and Skalka, A. edts. Elsvier, Holland (in press).
6. Tabin, C.J., Bradley, S.M.,Bargmann,C.I., Weinberg, R.A., Papageorge, A.G. Scolnick, E.M., Dhar, R., Lowy, D.R., and Chang, E.H. Nature 300, 143-149 (1982).
7. Reddy, E.P., Reynolds, R.K., Santos, E., and Barbacid, M. Nature 300, 149-152 (1982).
8. Yuasa, Y., Srivastava, S.K., Dunn, C.Y., Rhim,J.S., Reddy, E.P., and Aaronson S.A. Nature 303, 775-779 (1983).
9. Shimizu, K., Birnbaum, D., Ruley, M.A., Fasano, 0., Suard, Y., Edlund, L.

Taparowsky, E., Goldfarb, M., and Wigler, M. Nature 304, 497-500 (1983).
10. Barbacid, M. Ann. Rev. Biochem. 56, 779-827 (1987)
11. Sigal, I.S., Gibbs, J.B., Alonzo, J.S., and Scolnick, E.M. Proc. Natl. Acad. Sci. USA 83, 4725-4729 (1986).
12. Lacal, J.C., and Aaronson, S.A. Mol. Cell. Biol. 6,1002- 1009 (1986).
13. Lacal, J.C., and Aaronson, S.A. Mol. Cell. Biol. 6,4214-4220 (1986).
14. Hattori, S., Clanton, D.J., Satoh, T., Nakamura, S., Kaziro, Y., Kawakita, M. Shih, T.Y. Mol. Cell. Biol. 7, 1999-2002 (1987)
15. Willumsen, B.W., Papageorge, A.G., Kung, H., Bekesi, G., Robins, T., Johnsen, M. Vass, W.C., and Lowy, D.R. Mol. Cell. Biol. 6,2646-2654 (1986).
16. Seeburg, P.H., Colby, W.W., Capon, D.J., Goeddel, D.V., and Levinson, A.D. Nature 312, 71-75 (1984)
17. Lacal, J.C., Srivastava, S.K., Anderson, P.S., and Aaronson, S.A. Cell 44, 609-617 (1986)
18. Fasano, 0., Aldrich, T., Tamamoi, F., Taparowsky, E., Furth, M., and Wigler, M. Proc. Natl. Acad. Sci. USA 81, 4008-4012 (1984).
19. Lacal, J.C., and Aaronson, S.A. EMBO J. 5, 679-687 (1986).
20. Lacal, J.C., Santos, E., Notario, V., Barbacid, M., Yamazaki, S., Kung, H., Seamans, C., McAndrew, S., and Crowl, R. Proc. Natl. Acad. Sci. USA 81, 5305-5309 (1984).
21. Zoller, M.J., and Smith, M. Methods in Enzymol. 100, 468-500 (1983)
22. Graham, F.L., and van der Eb, A.J. Virology 52, 456-467 (1973)
23. Srivastava, S.K., Yuasa, Y., Reynolds, S.H., and Aaronson, S.A. Proc. Natl. Acad. Sci. USA 82, 38-42 (1985).
24. Furth, M., Davis, L.J., Fleurdelys, B., and Scolnick, E.M. J. Virol. 43, 294-304 (1982)
25. Willumsen, B.M., Christensen, A., Hubbert, N.L., Papageorge, A.G., and Lowy D.R. Nature 310, 583-586 (1984)
26. Willumsen, B.M., Norris, K., Papageorge, A.G., Hubbert, N.L., and Lowy, D.R. EMBO J. 3, 2581-2585 (1984).
27. De Vos, A.M., Tong, L., Milburn, M.V., Matias, P.M., Jancarik, J., Noguchi, s., Nishimura, S., Miura, K., Ohtsuka, E., and Kim, S. Science 239, 888-893 (1988)
28. Gibbs, J.B., Sigal, I.S., Poe, M., and Scolnick, E.M. Proc. Natl. Acad. Sci. USA 81, 5704-5708 (1984)
29. McGrath, J.P., Capon, D.J., Goeddel, D.V., and Levinson, A.D. Nature 310, 644-649 (1984).
30. Sweet, R.W., Yokoyama, S., Kamata, T., Feramisco, J.R., Rosenberg, M., and Gross, M. Nature 311, 273-275 (1984).
31. Manne, V., Bekesi, V., and Kung H. Proc. Natl. Acad. Sci. USA 82, 376-380 (1984)
32. Der, C.J., Finkel, T., and Cooper, G.M. Cell 44, 167-176 (1986).
33. Colby, W.W., Hayflick, J.S., Clark, S.G., and Levinson, A.D. Mol. Cell. Bio1.6, 730-734 (1986).
34. Trahey, M, Milley, R.J., Cole, G.E., Innis, M., Paterson, H., Marshall, C.J. Hall, A., and McCormick, F. Mol. Cell. Biol. 7, 541-544 (1987).
35. Sigal, I.S., Gibbs, J.B., D'Alonzo, J.S., Temeles, G.T., Wolanski, B.S., Socher S.H., and Scolnick, E.M. Proc. Natl. Acad. Sci. USA 83, 952-956 (1986).
36. Cales, C., Hancock, J., Marshall, C.J., and Hall, A. Nature 332, 548-551 (1988)
37. Adari, H., Lowy, D.R., Willumsen, B.M., Der, C.J., and McCormick, F. Science 240,518-521 (1988).

PURIFICATION AND MOLECULAR CLONING OF BOVINE GAP

J.B. Gibbs, U.S. Vogel, M.D. Schaber, M.S. Marshall, R. E. Diehl, E.M. Scolnick, R.A.F. Dixon, and I.S. Sigal

Department of Molecular Biology, Merck, Sharp, and Dohme Research Laboratories, West Point, PA 19486, USA

INTRODUCTION

The ras oncogenes encode 3 highly homologous proteins: Harvey (Ha), Kirsten (Ki), and N-ras (Barbacid, 1987). These proteins are 21 kDa (p21) and have the biochemical properties of GTP and GDP binding, GTPase activity, and membrane localization. The ras-encoded proteins are present in normal mammalian cells and in cells of diverse evolutionary origin such as yeast, *Drosophila*, *Dictyostelium*, and *Xenopus*. Up to 40% of human tumors have been identified as having a biologically activated ras gene. Activation can occur by point mutations that inhibit GTPase activity or facilitate GTP for GDP nucleotide exhange. By analogy to the GTP/GDP cycle of known guanine nucleotide-binding proteins, both of these activation mechanisms are predicted to promote formation of the biologically active ras p21-GTP complex. Although the target of ras protein in the yeast *S. cerevisiae* has been identified as adenylyl cyclase (Toda et al., 1985), the ras p21 target in mammalian and other eucaryotic cells has not yet been discovered.

Our structure-function studies of ras p21 have focused on ras interaction with other proteins and on the regulation of guanine nucleotides bound to ras. Work from our laboratory (Sigal et al., 1986; Marshall et al., 1988) and the work of Willumsen et al. (1986) previously identified a region of ras encompassing residues 30-40 that is required for biological activity. Deletion of this region or single amino acid substitutions at residues 33, 35, 36, 38 or 40 dramatically impaired the ability of ras proteins to transform cells or stimulate *S. cerevisiae* adenylyl cyclase. The mutant proteins were able to bind GTP and GDP and become localized to the membrane. These results suggested that residues 30-40 were essential for ras interaction with its target. Exploiting the properties of these mutants, we isolated in a yeast genetic screen a suppressor mutation in the proposed target of ras, adenylyl cyclase, that was able to interact with ras proteins having mutations in the ras 30-40 region (Marshall et al., 1988).

In order to study the regulation of ras p21-GTP, we quantitated the guanine nucleotides bound to ras proteins in living cells. We observed a paradox in that normal mammalian ras p21 expressed in yeast was bound to equimolor GTP and GDP whereas the same protein upon microinjection into

Xenopus oocytes was complexed only to GDP (Gibbs et al., 1987; Sigal et al., 1988). A solution to this paradox was provided by the discovery of the ras GTPase activating protein (GAP) by Trahey and McCormick (1987). GAP was identified as a protein factor in the cytosolic extracts of *Xenopus* oocytes and mammalian cells that could markedly stimulate the GTP hydrolytic rate of normal ras p21 without influencing the impaired GTPase of activated ras proteins. In order to characterize GAP function, we have purified the protein and cloned the cDNA for GAP.

PURIFICATION OF GAP AND RAS/GAP INTERACTION

A survey of rat tissues indicated that GAP activity was highest in brain and testes. Less activity was present in lung, liver, spleen, and heart. For the purification of GAP, the cerebrum of bovine brain was chosen because large amounts of starting material could be obtained. GAP was purified approximately 15,000-fold by a five-column procedure involving DEAE-Sephacel, Sepharose 6B, orange and green dye matrices, and Mono Q resins. Details have been described elsewhere (Gibbs et al., 1988). The fold-purification of GAP indicates that it constitutes <0.01% of total soluble protein in bovine cerebra. GAP eluted from each column as a single symetrical peak suggesting a single molecular form. The critical steps in the purification were chromatography on the dye matrix columns. Elution from the dye columns was only effected by salt (NaCl) and not by ATP, GTP, NAD, or EDTA. This result suggested that the interaction was ionic or hydrophobic and was not due to an affinity interaction with the nucleotide-like dyes. When the purified preparation was examined by SDS/PAGE, a single major protein of 125 kDa was observed by staining with Coomassie Blue or silver (Fig. 1). This band was observed to coelute with GAP activity from the green dye matrix and Mono Q columns (not shown). The size for GAP as determined by gel filtration was 130 kDa indicating that GAP existed as a monomeric polypeptide.

Purified GAP was tested with various ras proteins (Gibbs et al., 1988; Sigal et al., 1988), and the results are summarized in Table 1. GAP readily stimulated the GTP hydrolytic acitivity activity of mammalian Ha p21, the yeast RAS1 N-terminal domain (residues 1-185), and full-length yeast RAS2. The products of the reaction were ras-GDP and free orthophosphate. The effect of GAP was catalytic because we have observed up to 10 turnovers and an apparent second-order rate constant of $5 \times 10^6\ M^{-1}\ s^{-1}$ which is close to diffusion-controlled (Gibbs et al., 1988). The oncogenic variants [Val-12]Ha, [Arg-12]Ha, and [Leu-61]Ha, which have an impaired GTPase were insensitive to GAP. [Pro-12]Ha, which has an intrinsic GTPase activity that is 2-fold greater than that of normal Ha (Gly-12), did not respond to GAP. We speculate that the rigidity of Pro at position 12 prevents adaptation of a conformation required for GAP-stimulated GTP hydrolysis.

In order to test whether the inability of the oncogenic variants to respond to GAP was due to impaired intrinsic GTPase or due to an inability to physically interact with GAP, we developed a competition assay to evaluate ras p21/GAP interaction (Sigal et al., 1988; Vogel et al. 1988). As observed by Vogel et al., (1988), Ha ras p21-GTP competed with an IC_{50} value of 110 µM. The GDP complex of Ha p21 did not compete when tested up to 1.5 mM. The results in Table 1 summarize the competition data from Sigal et al. (1988) and Vogel et al. (1988) for ras proteins tested at 0.5 mM. The GTP complexes of oncogenic ras variants and [Pro-12]Ha effectively competed for GAP binding. The results demonstrate that the insensitivity to GAP by these proteins is not due to an inability to bind GAP. Furthermore, GAP preferentially interacts with the biologically active GTP complexes of ras p21 as compared to the GDP complexes.

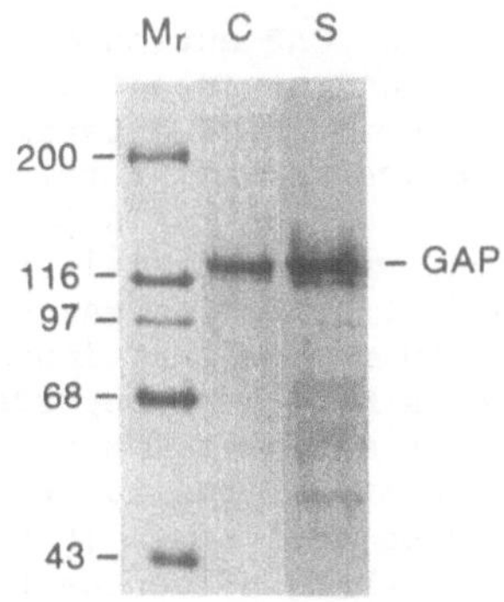

Fig. 1. Analysis of GAP protein. Purified GAP (2 ug) was subjected to SDS/polyacrylamide gel electrophoresis (SDS/PAGE) on 7.5% acrylamide gels. Molecular size standards (M_r) were stained with Coomassie Blue R-250. The apparent M_r values are indicated in kilodaltons. Purified GAP was stained with Coomassie Blue R-250 (C) or silver (S).

Table 1 ras/GAP interaction

ras Protein	Relative GTP Hydrolytic Rate - GAP	+ GAP	Competition
Ha	1	100	+
RAS1 (term. 185)	2	150	ND
RAS2	2	150	ND
[Arg-12]Ha	<0.02	<0.02	+
[Val-12]Ha	<0.02	<0.02	+
[Leu-61]Ha	<0.02	<0.02	+
[Pro-12]Ha	2	2	+
[Asn-33]Ha	2	2	-
[Ala-35]Ha	<0.02	<0.02	-
[Ala-38]Ha	<0.02	<0.02	-
[Asn-38]Ha	2	2	-

The GTP hydrolytic activity of purified ras proteins was measured by the nitrocellulose filtration method (Gibbs et al., 1988) in the absence or presence of 10 ng purified bovine GAP under linear assay conditions. A value of 1 represents a rate of 0.002 min^{-1} at 24°C. To test for physical interaction between GAP and ras proteins, competition assays were performed as described by Sigal et al. (1988) and Vogel et al. (1988). Competition was scored + or - by the ability of nonradioactive ras p21-GTP complexes (0.5mM) to inhibit GAP-stimulated Ha ras p21 [γ-^{32}P]GTP hydrolysis. ND, not determined. The results in this table are a compilation of data found in Gibbs et al. (1988), Sigal et al. (1988), and Vogel et al. (1988).

Two models could account for GAP action on ras GTPase activity (Sigal, 1988). In the upstream model, GAP would act as a negative regulatory element to downregulate ras p21 interaction with a target protein. In the downstream model, GAP would be the immediate target protein of ras p21. In the downstream model, the effect of GAP on ras GTPase activity would be analogous to EF-Tu/ribosome interaction. As an initial attempt to distinguish these two models, we tested the effect of GAP on ras proteins having single amino acid substitutions in the 30-40 region that impair biological activity (Sigal et al., 1988). Similar studies have been reported by Adari et al. (1988) and Cales et al. (1988). First, as shown in Table 1, it is evident that mutations such as a Thr to Ala at residue 35 or an Asp to Ala at residue 38 can inhibit ras intrinsic GTPase activity. This effect is probably due to an effect on ras p21 conformation that influences regions essential for GTP interaction. Second, the Asp to Asn changes at residues 33 or 38 do not inhibit ras p21 intrinsic GTPase. Yet, these mutant proteins do not respond to GAP. Furthermore, the GTP complexes of these proteins when tested at 0.5 mM do not compete with normal Ha for binding to GAP (Sigal et al., 1988; Vogel et al., 1988). The results imply that residues 33 and 38 are involved either directly or indirectly in the interaction between ras p21 and GAP.

CLONING OF GAP

To more fully evaluate GAP function and ras/GAP interaction, we proceeded to clone the gene encoding GAP by first obtaining peptide amino acid sequence of the purified protein (Vogel et al., 1988). Screening of a bovine brain cDNA library with oligonucleotides predicted to be complementary to the GAP mRNA sequence resulted in several clones that had overlapping restriction maps. The open reading frame of 3135 bp encoded a protein of 115,742 daltons, and the predicted amino acid composition was very similar to the experimentally determined amino acid composition of purified GAP (Vogel et al, 1988). Furthermore, all peptide fragments derived from purified GAP were present in the predicted protein sequence. The apparent discrepancy between the molecular size of GAP as calculated from the cDNA sequence (116 kDa) versus that determined for pure protein by SDS/PAGE analysis (125 kDa) may reflect anomalous mobility of GAP and/or that of the molecular size standards on the polyacrylamide gels or a possible post-translational modification of GAP.

Proof for the authenticity of the GAP clones was obtained by functional expression of the GAP cDNA in E. coli HB101 cells using a pUC-derived vector (Vogel et al., 1988). The construction encodes a fusion protein with a predicted molecular size of 107 kDa consisting of the N-terminal 23 residues from β-galactosidase fused to amino acids 128-1044 of GAP. As shown in Fig. 2, a protein of the expected molecular size was detected by immunoblot analysis of the soluble fraction of extracts from E. coli cells transformed with the expression plasmid in the correct orientation (pGAP21 and pGAP26). No immune-specific protein was detected in a lysate derived from cells transformed with the cDNA insert in the reverse orientation (pCON). Biochemical analysis demonstrated GAP activity only in cell lysates expressing GAP protein (Vogel et al., 1988). The detection of GAP activity with clone GAP1 indicates that residues 128-1044 of the protein are sufficient for GAP-stimulated p21 GTPase activity. Future structure-function studies of GAP will define the catalytic domain in more detail.

Initial characterization of genomic DNA by Vogel et al. (1988) indicated that the gene for GAP was probably a single copy as determined by blot hybidization. Blot hybridization of RNA derived from rat C6 glioma cells identified a single 6.0 kb transcript (Fig. 3). Consistent with the RNA data, immunoblot analysis of mammalian tissues using a GAP specific

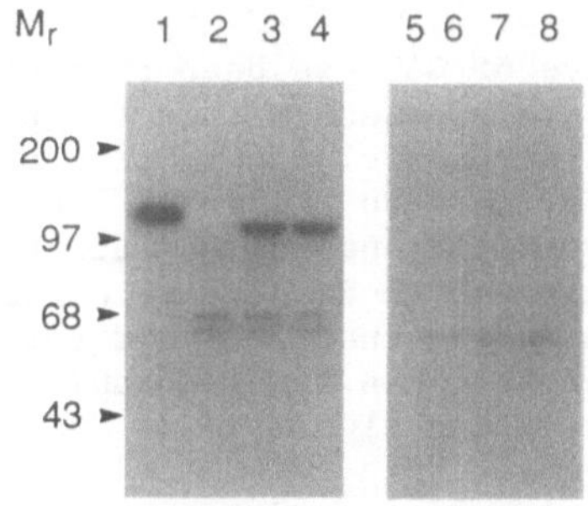

Fig. 2. Expression of bovine GAP in E. coli. Clone GAP1 encoding GAP amino acids 128-1044 (Vogel et al., 1988) was inserted into pUC13 either in the correct orientation with β-galactosidase (pGAP21, pGAP26) or in the opposite orientation (pCON) and was used to express GAP protein in E. coli cells as described previously (Vogel et al., 1988). The soluble fraction of cell lysates were subjected to SDS/PAGE, transferred onto nitrocellulose, and probed sequentially with 1:500 dilutions of rabbit anti-GAP peptide 139-152 serum (lanes 1-4) or preimmune serum (lanes 5-8) and ^{125}I-labelled anti-rabbit immunoglobulin (0.4 uCi/ml). Lanes: 1 and 5, pure bovine GAP; 2 and 6 pCON; 3 and 7, pGAP21; 4 and 8, pGAP26. M_r, Molecular size in kilodaltons.

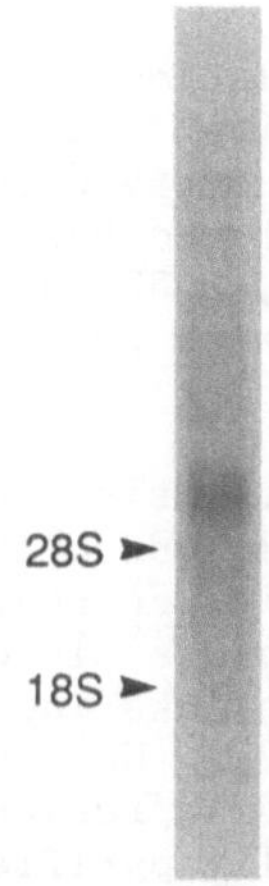

Fig. 3. Blot hybridization of GAP mRNA. Poly (A)+ RNA from rat C6 glioma cells (4 μg) was electrophoresed on a formaldehyde agarose gel, blotted onto nitrocellulose, and probed with ^{32}P-labelled pGAP1 insert. Filter hybridization was carried out at 60°C in 5x SSC/0.1% SDS/1x Denhardt's solution/0.1% NaPPi. The filter was then washed 3x in 2x SSC/0.1% SDS at 60°C. Positions of the 28S and 18S ribosomal bands are indicated.

anti-peptide serum indicated that only a single molecular species of 125 kDa was present (not shown).

SEQUENCE ANALYSIS OF GAP

The complete sequence of GAP has been presented elsewhere (Vogel et al., 1988). Several domains are apparent within the protein. The N-terminal 160 residues are extremely hydrophobic. In particular, there is a very high concentration of Gly, Ala, and Pro residues. The region having the next 700 amino acids is more hydrophilic, has a periodicity of turns approximately every 50 residues and has a high density of surface regions as predicted by the University of Wisconsin sequence analysis software package. The C-terminal half of the protein has high probability of several alpha helixes as predicted by the Chou-Fasman algorithm.

Upon searching protein sequence databases, it was apparent that GAP was a new protein; it was not identical to any other reported sequence. Therefore, a search of the database for protein sequences similar to GAP was performed using the algorithm defined by Wilbur and Lipman (1983). As observed by Vogel et al. (1988), a significant similarity was scored between GAP and the noncatalytic domain of *S. cerevisiae* adenylyl cyclase. The homology consisted of 16% identical and 34% conservative residues between the entire sequence of GAP and residues 425-1462 of adenylyl cyclase from *S. cerevisiae*. By inspection of published sequences, Vogel et al. (1988) also noted a striking homology of GAP residues 178-156 and 348-426 with regions conserved among phospholipase C-148, the *crk* oncogene product, and the nonreceptor tyrosine kinases (Stahl et al., 1988; Mayer et al., 1988).

PHOSPHORYLATION OF GAP

Phosphorylation-dephosphorylation is a common mechanism to regulate the activity of many different proteins. We therefore tested purified bovine GAP as a potential phosphoacceptor substrate for protein kinases. Although two regions with amino acid similarity to known substrates of the cAMP-dependent protein kinase (Feramisco et al., 1980) are predicted at Ser-354 and Ser-662, no phosphorylation was observed with the catalytic subunit of the cAMP-dependent protein kinase.

Protein phosphorylation on tyrosine residues is most commonly observed with cell proliferation mechanisms associated with oncogenes or growth factors. There is biological evidence that ras functions in the same pathway as insulin-mediated responses because microinjection of neutralizing ras antibodies into *Xenopus* oocytes blocks insulin-induced maturation (Deshpande and Kung, 1987; Korn et al., 1987). We therefore, tested whether GAP could serve as a substrate for the insulin receptor kinase. As shown in Fig. 4, GAP was phosphorylated by partially purified rat liver insulin receptor kinase in a ligand-dependent manner. The amount of incorporated phosphate correlated with GAP protein in successive fractions from a Mono Q column. The incorporated phosphate was alkali-stable which is consistent with the presence of phosphotyrosine. Furthermore, GAP phosphorylated by the insulin receptor kinase was immunoreactive with anti-phosphotyrosine antibodies (not shown). The stoichiometry of phosphorylation was low (2%), and no changes in GAP-stimulated ras GTPase activity were observed. Although the site of phosphorylation has not been determined, two potential tyrosine kinase phosphorylation sites are predicted at Tyr-161 and Tyr-720 based on homology to known tyrosine phosphorylation substrates (Pike et al., 1982). It remains to be established whether the observed phosphorylation of GAP occurs in vivo and whether it alters GAP function.

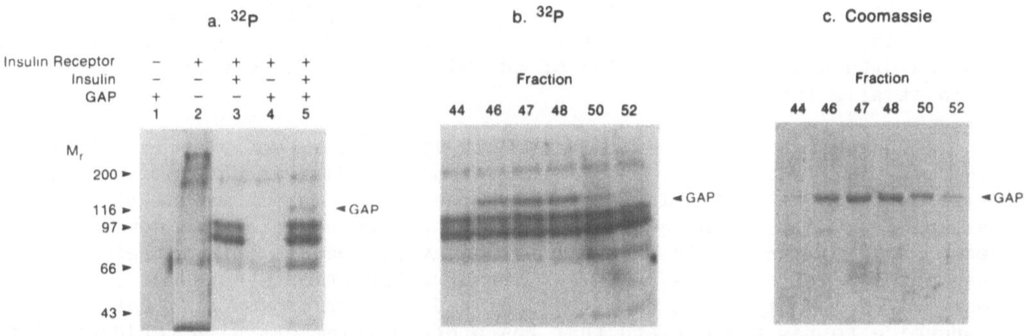

Fig. 4. Phosphorylation of GAP by the insulin receptor kinase. a) Purified bovine GAP (160 ng) was incubated for 1 hour at 24°C in buffer containing 50 mM Na Hepes, pH 7.5/ 10 mM $MgCl_2$/ 1 mM dithiothreitol/ 1 mM Na vanadate/ 50 μM ATP/ 2 x 10^6 cpm [γ-^{32}P]ATP/ rat liver insulin receptor kinase (2 fmol of insulin binding activity). Where indicated, insulin was included at 0.1 μM. Samples were subjected to SDS/PAGE and visualized by autoradiography. In the absence of insulin receptor kinase, insulin did not stimulate phosphorylation of GAP (not shown). b) The indicated fractions containing GAP from the final Mono Q column elution profile were incubated in the insulin receptor kinase reaction with insulin as above. c) Coomassie Blue-stained polyacrylamide gel of the Mono Q fractions used in b. M_r, molecular size in kilodaltons.

DISCUSSION

GAP is the first protein identified in a mammalian cell that interacts with ras p21. The effect of GAP on a biochemical activity of ras p21 that is central to ras function (GTPase) is consistent with the G-protein hypothesis of ras action. Ras p21 is localized at the plasma membrane, and biochemical activity measurements indicate that GAP is cytosolic. Preliminary immunological studies also suggest that GAP is present in the cytosol. More precise cell fractionation studies as well as whole cell immunofluorescence studies should resolve whether GAP can associate with membranes.

Critical to our understanding of GAP function is whether it is upstream or downstream of ras. The analysis of ras/GAP interaction is consistent with, but not proof of, the model of GAP as the immediate downstream target of ras p21. A strong correlation is evident between biological activity of ras proteins and the ability of these ras proteins to interact with GAP as determined in the competition assay. Furthermore, the interaction between GAP and either normal or oncogenic ras proteins is dependent on GTP complexed to ras p21. A similar situation is observed with yeast adenylyl cyclase. RAS-GTP is the biochemically active species, and RAS-GDP apparently does not compete with RAS-GTP stimulation of adenylyl cyclase (Broek et al., 1985; Field et al., 1988).

Although ras protein stimulates adenylyl cyclase activity in the yeast *S.cerevisiae* (Toda et al., 1985; Broek et al., 1985; Field et al., 1988), all evidence suggests that this interaction does not occur in mammalian cells. The homology between the sequences of GAP and the noncatalytic domain of yeast adenylyl cyclase is interesting, however, because both proteins interact selectively with RAS-GTP. Structure-function studies of yeast adenylyl cyclase indicate that the isolated catalytic domain is not responsive to ras protein and that ras protein interacts with residues within the noncatalytic domain (Kataoka et al., 1986; Uno et al., 1987). The homology observed between GAP and the conserved residues of phospholipase C-148, crk, and the nonreceptor tyrosine kinases may reflect a common funtional feature. The function of this conserved region is not known for any of these proteins, although in the nonreceptor tyrosine kinases, it is located in the noncatalytic domain and may regulate biological activity.

The discovery of GAP may have implications beyond the ras pathway. Many G-proteins of approximate M_r 20 kDa have been identified (YPT1, ral, rho, SEC4, rab, and ARF as examples). Unlike the M_r 40 kDa G-proteins (G_s, G_i, and G_o), the 20 kDa G-proteins in general are not associated as heteromeric complexes. This observation suggests that 20 kDa G-proteins are regulated by mechanisms distinct from those described for the 40 kDa G-proteins. It will be interesting whether GAP-like proteins exist for other members of the 20 kDa G-proteins.

ACKNOWLEDGEMENTS

We thank William Snyder for performing the immunoblot analyses and Vincent Strout for genorously providing the insulin receptor preparation.

REFERENCES

Adari, H., Lowy, D. R., Willumsen, B. M., Der, C. J., and McCormick, F., 1988, Guanosine triphosphatase activating protein (GAP) interacts with the p21 *ras* effector binding domain, *Science*, 240:518.
Barbacid, M., 1987, *ras* genes, *Ann. Rev. Biochem.*, 56:779.

Broek, D., Samiy, N., Fasano, O., Fujiyama, A., Tamanoi, F., Northrup, J., and Wigler, M., 1985, Differential activation of yeast adenylate cyclase by wild-type and mutant RAS proteins, Cell, 41:763.

Cales, C. J., Hancock, J. F., Marshall, C. J., and Hall, A., 1988, The cytoplasmic protein GAP is implicated as the target for regulation by the ras gene product, Nature, 332:548.

Defeo-Jones, D., Tatchell, K., Robinson, L. C., Sigal, I. S., Vass, W., Lowy, D. R., and Scolnick, E. M., 1985, Mammalian and yeast ras gene products: biological function in their heterologous systems, Science, 228:179.

Deshpande, A. K., and Kung, H. F., 1987, Insulin induction of Xenopus leavis oocyte maturation is inhibited by monoclonal antibody against ras p21 proteins. Mol. Cell Biol., 7:1285.

Feramisco, J. R., Glass, D. B., and Krebs, E. G., 1980, Optimal spatial requirements for the location of basic residues in peptide substrates for the cyclic AMP-dependent protein kinase, J. Biol. Chem., 255:4240.

Field, J., Broek, D., Kataoka, T., and Wigler, M., 1987, Guanine nucleotide activation of, and competition between, RAS proteins from Saccharomyces cerevisiae, Mol. Cell. Biol., 7:2128.

Gibbs, J. B., Schaber, M. D., Marshall, M. S., Scolnick, E. M., and Sigal, I. S., 1987, Identification of guanine nucleotides bound to ras-encoded proteins in growing yeast cells, J. Biol. Chem., 262:10426.

Gibbs, J. B., Schaber, M. D., Allard, W. J., Scolnick, E. M., and Sigal, I. S., 1988, Purification of ras GTPase activating protein from bovine brain, Proc. Natl. Acad. Sci. U.S.A., 85:5026.

Kataoka, T., Powers, S., Cameron, S., Fasano, O., Goldfarb, M., Broach, J., and Wigler, M., 1985, Functional homology of mammalian and yeast RAS genes, Cell, 40:19.

Kataoka, T., Broek, D., and Wigler, M., 1985, DNA sequence and characterization of the S. cerevisiae gene encoding adenylate cyclase, Cell, 43:493.

Korn, L. J., Siebel, C. W., McCormick, F., and Roth, R. A., 1987, ras p21 as a potential mediator of insulin action in Xenopus oocytes, Science, 236:840.

Marshall, M. S., Gibbs, J. B., Scolnick, E. M., and Sigal, I. S., 1988, An adenylate cyclase from Saccharomyces cerevisiae that is stimulated by ras proteins with effector mutations, Mol. Cell. Biol., 8:52.

Mayer, B. J., Hamaguchi, M., and Hanafusa, H., 1988, A novel viral oncogene with strutural similarity to phospholipase C, Nature, 332:272.

Pike, L. J., Gallis, B., Canellie, J. E., Bornstein, P., and Krebs, E. G., 1982, Epidermal growth factor stimulates the phophorylation of synthetic tyrosine-containing peptides by A-431 cell membranes, Proc. Natl. Acad. Sci. U.S.A., 79:1443.

Sigal, I. S., Gibbs, J. B., D'Alonzo, J. S., and Scolnick, E. M., 1986, Identification of effector residues and a neutralizing epitode of Ha-ras encoded p21, Proc. Natl. Acad. Sci. U.S.A., 83:4725.

Sigal, I. S., 1988, The ras oncogene, a structure and some function, Nature, 332:485.

Sigal, I. S., Marshall, M. S., Schaber, M. D., Vogel, U. S., Scolnick, E. M., and Gibbs, J. B., 1988, Structure-function studies of the ras protein, LIII Cold Spring Harbor Symposium on Quantitative Biology, in press.

Stahl, M. L., Ferenz, C. R., Kelleher, K. L., Kriz, R. W., and Knopf, J. L., 1988, Sequence similarity of phospholipase C with the non-catalytic domain of src, Nature, 332:269.

Toda, T., Uno, I., Ishikawa, T., Powers, S., Kataoka, T., Broek, D., Cameron, S., Broach, J., Matsumoto, K., and Wigler, M., 1985, In yeast, RAS proteins are controlling elements of adenylate cyclase, Cell, 40:27.

Trahey, M., and McCormick, F., 1987, A cytoplasmic protein stimulates normal N-ras GTPase, but does not affect oncogenic mutants, Science, 238:542.

Uno, I., Mitsuzawa, H., Tanaka, K., Oshima, T., and Ishikawa, T., 1987, Identification of the domain of Saccharomyces cerevisiae adenylate

cyclase associated with the regulatory function of RAS products, Mol. Gen. Genet., 210:187.
Vogel, U. S., Dixon, R. A. F., Schaber, M. D., Diehl, R. E., Marshall, M. S., Scolnick, E. M., Sigal, I. S., and Gibbs, J. B., 1988, Cloning of bovine GAP: a protein that interacts with oncogenic *ras* p21, *Nature*, in press.
Wilbur, W. J., and Lipman, D. J., 1983, Rapid similarity searches of nucleic acid and protein data banks, *Proc. Natl. Acad. Sci. U.S.A.*, 80:726.
Willumsen, B. M., Papageorge, A. G., Kung, H., Bekesi, E., Robins, T., Johnsen, M., Vass, W. C., and Lowy, D. R., 1986, Mutational analysis of a *ras* catalytic domain, *Mol. Cell. Biol.*, 6:2646.

THE FUNCTION OF THE MAMMALIAN RAS PROTEINS

Alan Hall, Jonathan D.H. Morris, Brendan Price, John F. Hancock, Sandra Gardener+, Miles D. Houslay+, Michael J.O. Wakelam+ and Christopher J. Marshall

Institute of Cancer Research, Chester Beatty Laboratories, Fulham Road, London SW3 6JB, U.K.

+Molecular Pharmacology Group, Department of Biochemistry University of Glasgow, Glasgow G12 8QQ, U.K.

INTRODUCTION

Single amino acid alterations in one of the three ras proteins have been detected in 25-50% of human cancers and it is believed that the somatic mutational event which generated these amino acid substitutions was an important step in the development of these malignancies (Barbacid, 1986; Bos et al., 1987; Paterson et al., 1987).

The ras proteins, p21$^{\underline{ras}}$, are expressed in most if not all cells. They are located on the cytoplasmic side of the plasma membrane, they bind GTP and GDP and have an intrinsic GTPase activity. This, together with some sequence similarities to other proteins, has suggested that p21$^{\underline{ras}}$ functions as a regulatory guanine nucleotide binding protein (see Barbacid, 1986 for review). It is still not clear, however, which processes are being regulated and with which molecules p21 interacts.

It has been observed that some of the oncogenic mutations of p21$^{\underline{ras}}$ detected in human malignancies (located at codons 12,13 and 61) lead to a decrease in the intrinsic GTPase activity of the protein. This has led to the proposal that increased levels of the active GTP form account for its transforming properties (McGrath et al., 1984). However, there is not always a correlation between the observed reduction in GTPase activity of a particular mutant and the strength of its transforming activity (Trahey et al., 1987). Recently a cytoplasmic protein, GAP, has been identified that increases the GTPase activity of normal but not val12 or asp12 p21$^{\underline{ras}}$ (Trahey and McCormick, 1987) and it is now clear that all oncogenic mutations have a dramatic effect on the *in vivo* GTPase activity of the protein.

The biochemical function of p21$^{\underline{ras}}$ has been very difficult to study. Microinjection of *E. coli* produced oncogenic p21 into quiescent cells induces DNA synthesis and cell division (Stacy and Kung, 1984; Feramisco et al., 1984; Trahey et al., 1987), whilst microinjection of neutralising anti-p21$^{\underline{ras}}$ antibodies into quiescent fibroblasts blocks proliferation after addition of growth factors (Mulcahy et al., 1985). This strongly suggests that ras is involved in regulating growth factor induced cell proliferation and that the oncogenic mutations lead to a breakdown in this regulatory process. In S. cerevisae, the yeast ras proteins directly regulate adenylate cyclase activity (Toda et al., 1985), but this seems not to be the case in mammalian cells and xenopus oocytes (Birchmeier et al., 1985; Beckner et al., 1985). We report here some of our efforts to look for a possible target for $p21^{ras}$ regulation, and in particular for changes in another known second messenger system, namely the breakdown of phosphatidyl inositol lipids by a phospholipase C.

RESULTS

Analysis of stably transfected cell lines

In order to see if $p21^{ras}$ can affect the inositol phospholipid (PI) breakdown pathway, cells expressing a variety of ras gene constructs were labelled overnight in ^{3}[H]inositol, washed extensively and the rate of generation of inositol phosphates measured in the presence of Li^+. Since these measurements are made in the absence of added growth factors, we refer to these conditions as basal rates. The results, using clones constitutively expressing mutant N-ras genes or clones containing N-ras linked to the heterologous inducible mouse mammary tumour virus promoter, are shown in Fig. 1. Induction of the T15 cell line leads to the expression of around 10^6 molecules of normal $p21^{ras}$ (McKay et al., 1986), but no change in the basal rate of

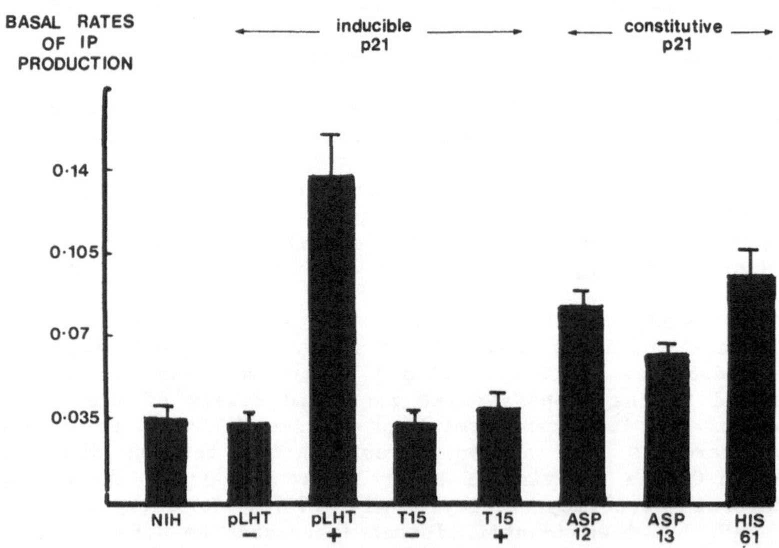

Fig. 1. Basal rates of inositol phosphate (IP) production in N-ras containing NIH-3T3 clones. Clones transfected with a complete N-ras gene with Asp12, Asp13 or His61 oncogenic mutations and which constitutively express the mutant ras protein are shown. Also shown are inducible normal N-ras containing cells (T15) and inducible mutant (lys61) ras containing cells (pLHT) grown in the presence (+) or absence (−) of 80 nM dexamethasone. Rates are expressed in (cpm in water soluble inositol phosphates/cpm in lipids x 100)% min^{-1}.

production of inositol phosphates is observed compared to uninduced cells or parental NIH-3T3 cells (Wakelam et al., 1986). In contrast, induction of the cell line pLHT leads to the expression of around 10^5 molecules of oncogenic p21^{N-ras}, and we observe a 3-4 fold increase in the rate of inositol phosphate production (Hancock et al., 1988). Also shown in Fig. 1 are the results obtained with a variety of clones constitutively expressing activated ras genes; all have a significantly higher basal rate of PI metabolism than parental NIH-3T3 cells. We have observed the same results with NIH-3T3 cells constitutively expressing oncogenically activated Ha- and Ki- ras genes (Hancock et al., 1988).

Although we observed no change in PI metabolism in fully induced T15 cells, we did find a striking increase in bombesin stimulated PI breakdown (Wakelam et al., 1986). Addition of bombesin led to a small (20%) increase in the rate of production of inositol phosphates in uninduced T15 cells and in NIH-3T3 cells. However, addition of bombesin to induced T15 cells gave a 2-3 fold stimulation. Measurement of the binding of radiolabelled bombesin to induced and uninduced T15 cells showed no change in bombesin receptor number. We have recently confirmed that IP_3 is the primary breakdown product of this synergistic reaction and that in induced T15 cells bombesin gives a 2-3 fold stimulation of intracellular Ca^{2+} release compared to NIH-3T3 cells (A. Lloyd, CJM, M Whittaker, SG, MJOW, unpublished results).

An analysis of two independent clones constitutively expressing high levels of normal p21^{N-ras}, however, showed no increase in bombesin stimulated PI breakdown. It is unclear to us whether some other change has occurred in T15 cells or whether the use of an inducible clone allows subtle changes in PI metabolism to be analysed, which are not detectable in cell lines stably transformed by over-expression of normal N-ras protein.

We have also isolated transformed cell lines constitutively expressing normal Ha and normal Ki p21^{ras}. Clones expressing normal Ha are not significantly stimulated by any of the ligands we have tried. Although preliminary work suggested an increased PDGF response in normal Ha-ras overexpressing clones, subsequent isolation of other clones, including an inducible p21^{Ha-ras} containing clone, has not confirmed these original findings. In agreement with others (Parries et al., 1987) we have observed an increased bradykinin stimulated response in inositol phosphate formation in clones overexpressing Ki-ras. However, these effects probably arise through an increase in bradykinin receptor numbers on the cells (Parries et al., 1987). The mechanism by which normal (and mutant) Ki-ras increases bradykinin receptor numbers is unknown.

Analysis of transient ras expression in COS-1 cells

COS-1 cells were transfected with plasmids containing mutant or normal Ha-ras cDNAs under the control of the SV40 early promoter. Since the plasmid used (pEXV) also contains the SV40 origin of replication, a large number of plasmid copies are produced. This coupled with the high level expression from the strong promoter generates large amounts of p21^{Ha-ras} and its effect on PI turnover can be examined (Hancock et al., 1988). Figure 2 shows the results obtained, it is clear that oncogenic but not normal ras leads to a significant increase in the basal rate of PI turnover. Although the effect does not appear to be as large as that obtained with the stably transfected cell lines (see Fig. 1), only around 20% of the COS-1 cells actually take up DNA after transfection. When this is taken into account and assuming that both transfected and untransfected cells contribute to the background basal rate, then the increases observed in the two systems are comparable.

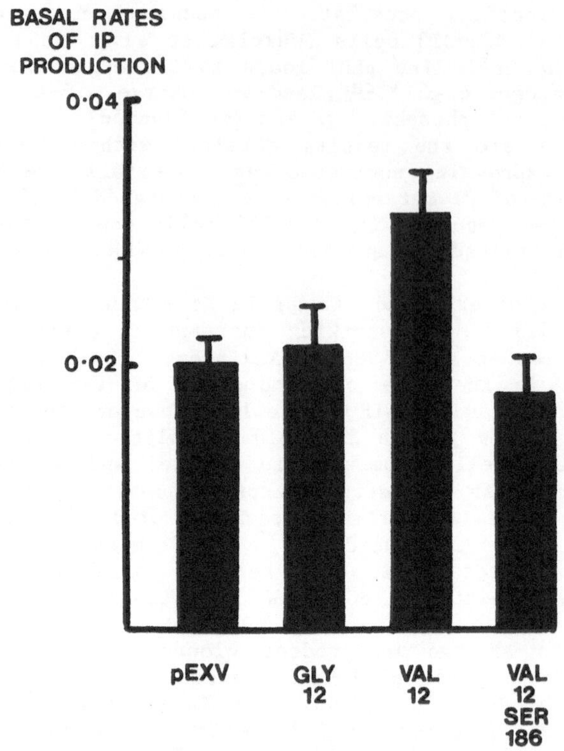

Fig. 2. Basal rates of inositol phosphate (IP) production in COS-1 cells transiently expressing Ha-ras cDNAs. The vector alone (pEXV) or the vector containing normal (gly12) oncogenic (val12) or a biologically inactive (val12ser186) cDNA have been used to transfect COS-1 cells (Hancock et al., 1988).

Analysis of scrape-loaded cells

The experiments described so far have made use of ras DNA introduced into cells to generate ras protein. It is not possible, therefore, to look for very early ras-induced events. Although microinjection of *E. coli* purified ras protein would overcome this problem, it is difficult if not impossible to inject sufficient cells to look at biochemical changes. We have made use of the scrape-loading technique (McNeil et al., 1984) to introduce *E. coli* produced proteins into large numbers (10^5) of Swiss 3T3 cells. Cells, scrape-loaded with oncogenic $p21^{ras}$ become morphologically transformed after 15h (see Fig. 3) and, if quiescent cells are scrape-loaded, oncogenic ras in the presence of insulin stimulates DNA synthesis (Morris et al., 1988).

We have used this approach to look for changes in second messengers induced by oncogenic $p21^{ras}$. 1,2-diacylglycerol (DAG) formation was monitored by looking for phosphorylation of an 80kD substrate of protein kinase C. We find that phosphorylation of the 80kD protein occurs 5 min after scrape-loading val12 $p21^{ras}$. Scrape-loading the biologically inactive val12ser186 ras protein does not produce this effect. When cells, prelabelled with ^3H inositol, were scrape-loaded with val12 $p21^{ras}$, we could find no increase in inositol phosphate production, whereas PDGF stimulated a 4-fold increase in inositol phosphates over 20 minutes (Morris et al., 1988).

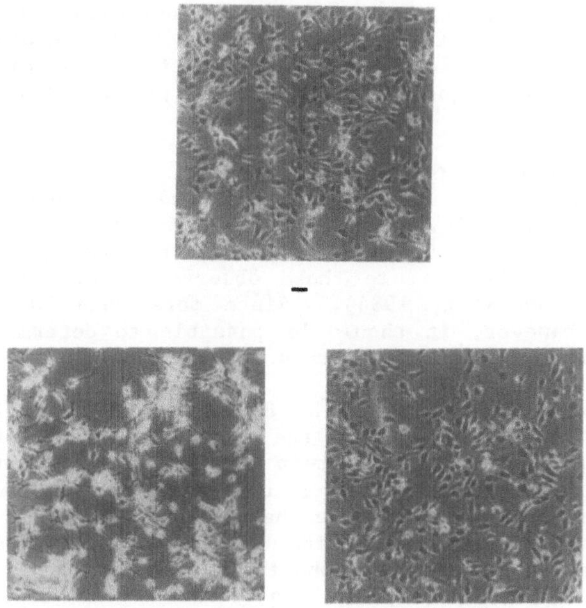

Fig. 3. Quiescent Swiss 3T3 cells were scrape-loaded in buffer (-) or buffer containing val12 or val12ser186 $p21^{ras}$ purified from *E. coli* expression systems. The morphology of the cells was observed 15h after scrape-loading (Morris et al., 1988).

DISCUSSION

It is clear that normal ras proteins play a crucial role in regulating cell proliferation and that oncogenic mutations lead to a breakdown in this control. However, the biochemical processes regulated by mammalian $p21^{ras}$ are unknown and much effort has been expended looking for the effects of ras on second messenger signalling systems.

We have found, in agreement with others (Fleischman et al., 1986), that oncogenically activated *ras* increases the basal rate of inositol phosphate production in NIH-3T3 cells (Hancock et al., 1988). Normal ras, even when expressed at 10 fold higher levels, does not increase this basal rate (Wakelam et al., 1986; Hancock et al., 1988). Since it is likely that oncogenic mutations in ras lead to constitutive activation of the process(es) normally regulated by $p21^{ras}$, we have also looked for effects of normal p21 on growth factor stimulated PI metabolism. We have isolated one cell line, T15, which contains an inducible normal N-*ras* gene in which the ras protein is 50-fold over-expressed. Induction of $p21^{N\text{-}ras}$ leads to an increase in bombesin stimulated inositol phosphate production (Wakelam et al., 1986). This effect has not, however, been observed in clones overexpressing normal N-*ras* constitutively, making it difficult to draw any firm conclusions concerning the role of the normal ras protein, if any, in PI metabolism. We do not know at what level the co-operation between the activated bombesin receptor and normal $p21^{N\text{-}ras}$ occurs in the T15 cells or why this effect is only observed in this inducible cell line.

Indeed, using this kind of approach with transfected cell lines, it is very difficult to determine whether changes in PI metabolism, or any other signalling system, are an immediate effect of ras function or whether they are a consequence of the chronic production of abnormal ras proteins.

In an attempt to overcome these problems, we have used a technique, scrape-loading, capable of introducing purified ras protein into a large number of cells quickly. The effects of oncogenic ras protein, introduced into quiescent Swiss 3T3 cells in this way, on morphology and DNA synthesis are identical to those observed after microinjection of protein (Stacey and King, 1984). Since more than 10^5 cells can be scrape-loaded, however, it should be possible to determine early ras-induced biochemical changes using this technique.

We have found activation of an 80kD protein 5 min after scrape-loading val12 $p21^{ras}$ indicating the rapid formation of 1,2-diacylglycerol and the activation of protein kinase C. Surprisingly, however, we could find no indication for a concomitant increase in inositol phosphates. We are led to conclude that the DAG is coming from a source other than phosphatidyl inositoside breakdown and we are currently looking for changes in other phospholipids. Furthermore, since changes in inositol phosphate production has clearly been observed by many groups in cell lines expressing ras genes, we suppose that this increase in phospholipase C activity is a consequence of the chronic expression of oncogenic $p21^{ras}$ in cells and not of a direct activation of this secondary messenger system.

ACKNOWLEDGEMENTS

The work described in this paper was supported by the Cancer Research Campaign (UK) and the Medical Research Council (UK).

REFERENCES

Barbacid, M. (1986) ras Genes. Ann. Rev. Biochem. 56: 779.

Beckner, S.K., Hattori, S., and Shih, T.Y., 1985, The ras oncogene product is not a regulatory component of adenylate cyclase, Nature, 317:71.

Birchmeier, C., Broek, D., and Wigler, M., 1985, Ras proteins can induce meiosis in xenopus oocytes, Cell, 43:615.

Bos, J.L., Fearon, E.R., Hamilton, S.R., deVries, M.W., van Bloom, J.M., Van der Eb, A.J., and Vogelstein, B., 1987, Prevalence of ras mutations in human colorectal cancer, Nature, 327:293.

Feramisco, J.R., Gross, M., Kamata, T., Rosenberg, M., and Sweet, R.W., 1984, Microinjection of the oncogene form of the human H-ras (T24) protein results in rapid proliferation of quiescent cells. Cell, 38:109.

Fleischman, L.F., Chawala, S.B., and Cantley, L., 1986, Ras transformed cells: Altered levels of phosphatidylinositol 4,5 bisphosphate and catobolites, Science, 231:407.

Hancock, J.F., Marshall, C.J., McKay, I.A., Gardener, S., Houslay, M.D., Hall, A. and Wakelam, M.J.O., 1988, Mutant but not normal $p21^{ras}$ elevates inositol phospholipid breakdown in two different cell systems, Oncogene, in press.

McGrath, J.P., Capon, D.J., Goeddel, D.V., and Levinson, A.D., 1984, Comparative biochemical properties of normal and activated human ras p21 protein, Nature, 310:644.

McKay, I.A., Marshall, C.J., Cales, C., and Hall, A., 1986, Transformation and stimulation of DNA synthesis in NIH-3T3 cells are a titratable function of p21^{N-ras} expression, The EMBO J., 5: 2617.

McNeil, P.L., Murphy, R.F., Lanni, F., and Taylor, D.L., 1984, A method for incorporating macromolecules into adherent cells. J. Cell. Biol., 98:1556.

Morris, J.D., Price, B., Lloyd, A., Self, A.J., Marshall, C.J. and Hall, A., 1988, Scrape-loading of Swiss 3T3 cells with ras protein induces rapid activation of protein kinase C followed by DNA synthesis, in press.

Mulcahy, L.S., Smith, M.P., and Stacey, D.W., 1985, Requirement for ras proto-oncogene function during serum stimulated growth of NIH-3T3 cells, Nature, 313:241.

Parries, G., Hoebel, R., and Racker, I., 1987, Opposing effects of a ras oncogene on growth factor stimulated phosphoinositide hydrolysis: Desensitization to PDGF and enhanced sensitivity to bradykinin, Proc. Natl. Acad. Sci. USA, 84:2648.

Paterson, H., Reeves, B., Brown, R., Hall, A., Furth, M., Bos, J., Jones, P., and Marshall, C.J., 1987, Activated N-ras controls the transformed phenotype of HT1080 human fibrosarcoma cells, Cell, 51:803.

Stacy, D.N. and Kung, H.F., 1984, Transformation of NIH-3T3 cells by microinjection of Ha-ras p21 protein, Nature, 310:508.

Toda, T., Uno, I., Ishikawa, T., Powers, S., Kataoka, T., Broek, D., Broach, J., Matsumoto, K., and Wigler, M., 1985, Yeast RAS proteins are controlling elements in the cyclic AMP pathway, Cell, 40:27.

Trahey, M. and McCormick, F., 1987, A cytoplasmic protein stimulates normal N-ras p21 GTPase, but does not affect oncogenic mutants, Science, 238:542.

Trahey, M., Milley, R.J., Cole, G.E., Innis, M., Paterson, H., Marshall, C.J., Hall, A., and McCormick, F., 1987, Biochemical and biological properties of the human N-ras p21 protein, Molec. Cell. Biol., 7:556.

Wakelam, M.J.O., Davies, S.A., Houslay, M.D., McKay, I., Marshall, C.J., and Hall, A., 1986, Normal p21^{N-ras} couples bombesin and growth factor receptors to inositol phosphate production, Nature, 323:173.

BIOCHEMICAL PROPERTIES OF HA-RAS ENCODED P21 MUTANTS

Jacob John, Mathias Frech, Jürgen Feuerstein,
Roger S.Goody and Fred Wittinghofer

Max-Planck-Institut für medizinische Forschung
Abteilung Biophysik
6900 Heidelberg

INTRODUCTION

Mutations of the ras proto-oncogenes ras^H, ras^K and ras^N have been found in many human tumours and in the ras genes of the acutely transforming animal retroviruses HaMuSV and KiMuSV. (for reviews see Gibbs et al., 1985; Barbacid, 1987). These mutations were found to be single or double point mutations and could be localized in all cases to two different regions of the amino acid sequence of the ras encoded p21 proteins. Thus, either amino acids 12/13 or amino acids 59/61 were found to be mutated. The retroviral p21 proteins of the Kirsten and Harvey strain of MuSV contain a double mutation involving $glycine_{12}$ and $alanine_{59}$, where alanine59 is changed to threonine in both cases (Dhar et al., 1982; Tsuchida et al., 1982; Yasuda et al., 1984).

How these mutations effect the biochemistry and structure of the p21 protein it not known except that they activate the ability of the protein to transform fibroblasts cells in culture and make the protein tumorigenic. The threedimensional structure of the catalytic domain of the cellular p21-GDP complex has recently been obtained by S.-H. Kim and coworkers and activated forms of this protein have been crystallized (deVos et al., 1988). Proton NMR studies (Schlichting, Wittinghofer and Rösch, unpublished)) on p21 show that there are no significant conformational differences between cellular p21 and transforming mutants thereof. We have therefore started to characterize by biochemical means the differences between normal and mutant proteins. Here we show that the effect of mutations in position 12, 59, 61 are different from, and sometimes opposed to each other.

RESULTS

Table 1 shows the results obtained by measuring the kinetics of association and dissociation of cellular p21 with GDP by the nitrocellulose filter binding technique. As we reported earlier (Feuerstein et al., 1987) the ratio of the association and dissociation rate constants lead to a binding constant for the interaction of p21 with GDP, which is $0.87 \times 10^{10} Mol^{-1}$ in the absence and $5.7 \times 10^{10} Mol^{-1}$ in the presence of excess magnesium at 4°C, much higher than has previously been reported. To confirm the high affinity betweeen guanosine nucleotides and

Table 1. Kinetic constants for the binding of GDP and analogue to p21. They were measured by the retention of complexes of radiolabelled guanine nucleotide complexes with p21 on nitrocellulose filters as described (Feuerstein et al., 1987) or by using the change in the fluoresecence emission signal at 455nm (see Fig.2) produced when 3`-mant-2`-deoxy-GDP binds to p21.

	Filterbinding (4°C)		Fluorescence (25°C)
	$+Mg^{2+}$	$-Mg^{2+}$	$+Mg^{2+}$
$k_{+1} \times 10^{-6} [M^{-1}, sec^{-1}]$	1.7	5.2	3.4
$k_{-1} \times 10^{4} [sec^{-1}]$	0.3	6.0	4.5
$K \times 10^{10} [M^{-1}]$	5.7	0.87	0.75

p21 proteins by an independant method we synthesized derivatives of guanine nucleotides, containing the fluorescent methylanthraniloyl-(mant-) group (Hiratsuka, 1983). Since with normal ribo-guanosine nucleotides one obtains a mixture of 2`- and 3` methylanthraniloylesters, which could possibly be problematic to use in fluorescent measurements we used 2`deoxy-GDP and 2`-deoxy-GTP to obtain 2`deoxy-3`-methylanthraniloyl-GDP and -GTP, as shown in Fig.1, which are called mant-GDP and mant-GTP hereafter.

2`-deoxy-GDP and GTP and their derivatized fluorescent analogues have a high affinity for p21, which is only slightly lower than that for GDP or GTP. This indicates that p21, just as EF-Tu, can tolerate large substituents on the ribose ring of the nucleotide. The binding of mant-GDP (and the GTP analogue) to p21 produce a large fluorescence intensity change. Fig.2 shows that on binding to cellular p21 the emission maximum of the fluorophore changes from 443nm to 435nm with an increase in the relative fluorescence intensity of 92% in the absence and 105% in the presence of magnesium ions. The fluorescent analogues thus allow one to measure various aspects of the interaction of GDP/GTP with p21. Furthermore, we find that the fluorescence changes observed on binding of the analogues are different for the different p21 proteins. This shows that these analogues are sensitive probes to measure the association and dissociation kinetics of p21 proteins.

As an example Table 1 shows the results obtained for the association and dissociation of p21c with 3`-mant-2`-deoxy-GDP in the absence of Mg^{2+}. The rate constants obtained are in fairly good agreement with the

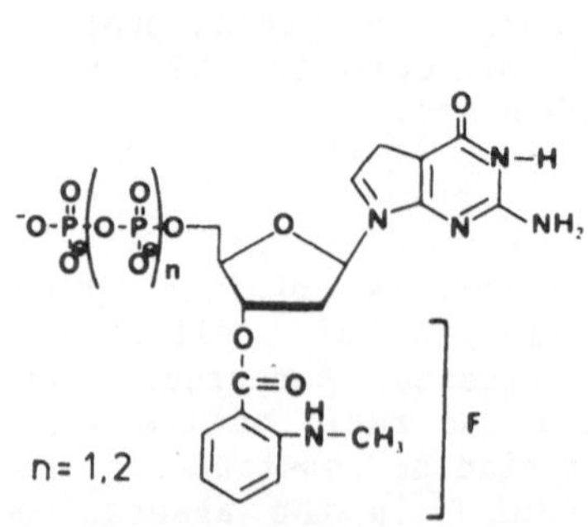

Fig.1. Structure of the fluorescent derivative of GDP and GTP, 3`-methylanthraniloyl-2`-deoxy-GDP (and -GTP)

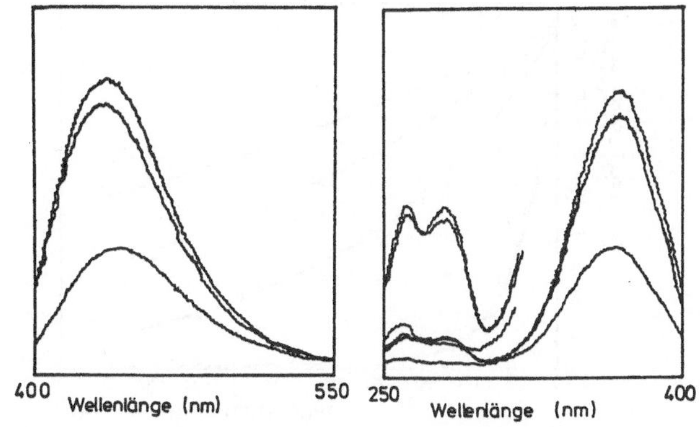

Fig.2. Fluorescence spectra of mant-GDP under different conditions. The spectra were taken with a 1uM solution of the fluorescent analogue alone (lower curve), in the presence of 1uM p21c (middle curve) and in the presence of 1uM p21 and 5mM Mg^{2+}. Left: Emission spectrum (excitation wavelength 370nm); Right: Excitation spectrum (emission wavelength 440nm).

data obtained by the filter binding studies with radioactively labelled nucleotides. This confirms that the binding constants between p21 and guanosine nucleotides are indeed in the order of 10^{10}. We are using the fluorescent analogues to characterize the interaction with guanosine nucleotides and metal ion of the normal and mutant tranforming p21 proteins in order to get an insight into the structural variation caused by the mutations.

We and others (Feuerstein et al., 1986; Hoshino et al., 1987; Lacal and Aaronsen, 1986) have noticed earlier that the guanine nucleotide exchange rates of viral p21 are increased as compared to the cellular form. Since viral p21 has a double mutation involving $alanine_{59}$ and $glycine_{12}$ we wanted to find out the contribution of each single point mutation. Fig.3 shows the mutants we constructed by "in vitro recombination" of various restriction fragments of the p21 expression vectors described earlier (Tucker et al., 1986). We thus obtained expression systems which produce p21 proteins with single point mutations in positions 12, 59 and 61. We also constructed plasmids expressing proteins with the combinations of these mutations, which are also listed in Fig.3. After purification of these proteins we measured the

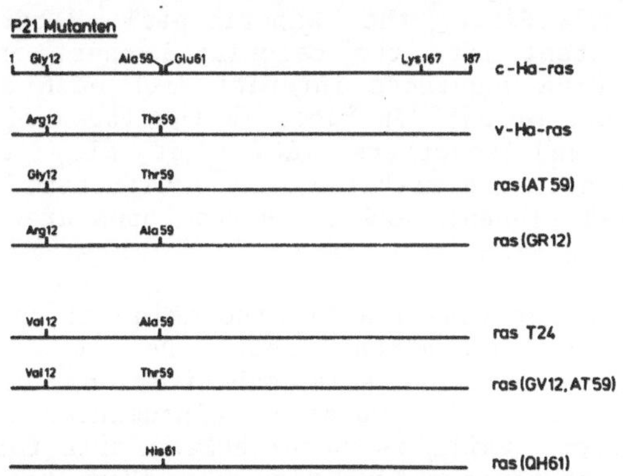

Fig.3. Schematic representation of the primary structure of p21 wild type and mutant proteins whose dissociation kinetics were investigated.

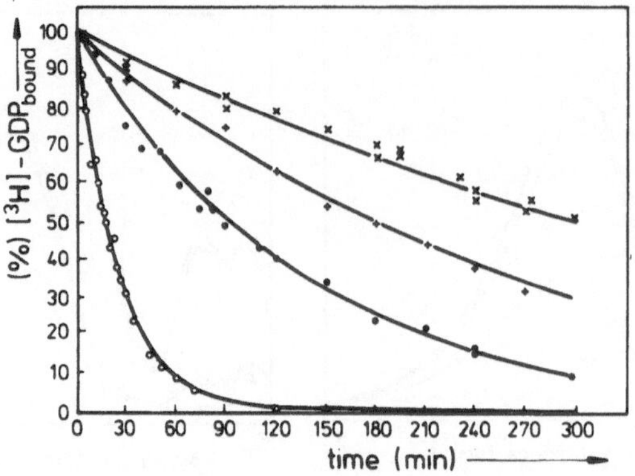

Fig.4. Kinetics of the dissociation of GDP from $p21^H$ mutants. GDP dissociation rates were measured at 37°C in the presence of 10mM Mg^{2+}. (.), cellular p21($Gly_{12}Ala_{59}$); (∘), viral p21($Arg_{12}Thr_{59}$); (+), p21($Val_{12}Ala_{59}$) and (x), p21($Arg_{12}Ala_{59}$). 2uM p21 were equilibrated with [^{3}H]GDP for 30min at 21°C. Mg^{2+} and excess unlabelled GDP were added and the displacement of [^{3}H]GDP from p21 was measured by filtration of aliquots through nitrocellulose filters.

dissociation rate constants of guanine nucleotides in the presence and absence of Mg^{2+}. Fig.4 shows that mutations at position 12 and 59 have opposite effects on the dissociation rate constants of GDP. The Gly_{12}-Val mutation reduces the rate constant by a factor of three, whereas the Ala_{59}-Thr mutation increases it by a factor of 4.5. The mutation of $glycine_{12}$ to arginine again reduces the dissociation rate. The results of these measurements are shown schematically in Fig.5. The double mutants, which have combinations of the above mentionned single mutations, show a behaviour which is intermediate between the increasing and the decreasing effect of the single point mutations. However a closer examination of the results reveals that the effect of each individual mutation on the double mutant is not strictly additive. On the assumption that the dissociation rate constant k_{-1}^{12} is influenced by the single mutation in an additive manner, one obtains

$$k_{-1}^{12} = (k_{-1}^{1} \times k_{-1}^{2}) / k_{-1}^{0}$$

where the subscripts 0,1 and 2 denote the wildtype and mutant proteins 1 and 2, respectively.Since the experimental determined values are significantly different from the calculated ones, one has to conclude that the corresponding sequences interact with each other in the three-dimensional structure of p21. In fact, in the three-dimensional structure published by Kim and coworkers (deVos et al., 1988) the residues $glycine_{12}$ and $alanine_{59}$ are situated in two different loops which connect secondary structural elements and these two loops are situated very close to each other.

Point mutations at position 61, two amino acids away from position 59 have been found in many human tumours. Der et al. (1986) have shown that most mutants with amino acid substitutions at position 61 have transforming activity. We mutated $glutamine_{61}$ to histidine, a conservative mutation, which is nevertheless transforming (Der et al., 1986) and has been found in a lung carcinoma (Yammamoto et al., 1985) and a rhabdomyosarcoma (Chardin et al., 1984). As shown in Fig.5, this substitution does not affect the dissociation rate constant for GDP. The p21($Val_{12}His_{61}$) double mutant shows the same properties as p21 with a

5

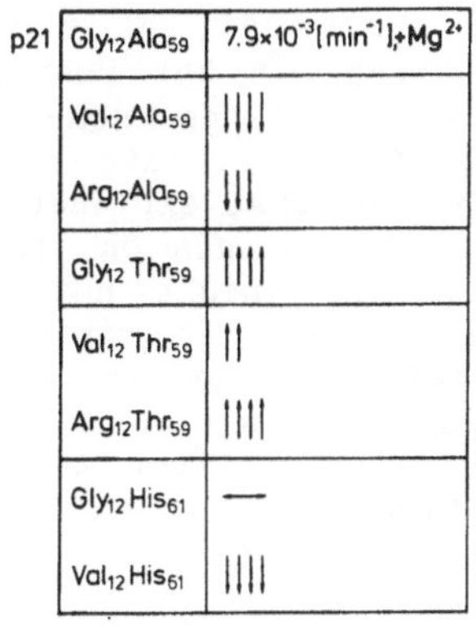

6

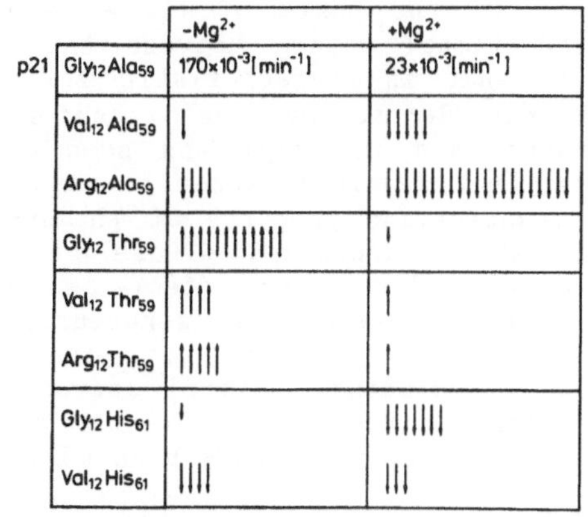

Fig.5. Effect of single and double mutations of $p21^{H}$ on the GDP dissociation rate constants. An upward or downward arrow indicates whether the mutation increases or decreases the dissociation rate. The number of arrows indicates the factor, by which the dissociation rate of wildtype p21 is changed.
Fig.6. Effect of single and double mutations of $p21^{H}$ on the GTP dissociation rate constant.Symbols are as in Fig.5.

single $valine_{12}$ mutation. Thus these two mutations seem to be independant of each other as far as the GDP off rate is concerned. We will show later that this is not the case with the GTP off rates.

We have shown earlier (Feuerstein et al., 1987) that in the active site of EF-Tu and p21, the Mg^{2+}-ion is coordinated as a monodentate complex to a non-bridging oxygen atom of the beta-phosphate group of GDP. In p21 this Mg^{2+} ion is exposed to the solvent, which enables EDTA to rapidly chelate and remove the metal ion from the nucleotide binding site. Thus, as we and others have shown (Hattori et al., 1987; Hall and Self, 1986; Tucker et al., 1986), the release of guanosine nucleotides is accelerated in the presence of EDTA. The dissociation rate constants were also measured in the presence of EDTA (data not shown). Under all circumstances the rate constants are much faster as compared to those in the presence of Mg^{2+}. They are in fact to fast too be measured conveniently at 37°C, and were hence determined at 21°C. The ratios of dissociation rate constants measured in the presence and absence of magnesium vary between 120 and 1000 and are lower for proteins with $threonines_{59}$.

We wanted to find out whether the transforming mutations of p21 have a similar effect on the GTP conformation of the protein. For this we determined the GTP dissociation rate constants using gamma-^{32}P labelled GTP and the results are shown schematically in Fig.6. The GTP off-rate of cellular p21 is about three times faster than the GDP off-rate. Mutations in position 12 again slow down the dissociation rate and the the effect of the glycine-arginine substitution is especially drastic since it decreases the rate constant by a factor of 21. The $alanine_{59}$-threonine exchange again increases the GTP dissociation rate but the effect is much smaller as compared to the GDP dissociation constant. p21(QH61) shows again a different behaviour, but only in the presence of magnesium, where the exchange with histidine slows down the dissociation rate. The results presented in Fig.6 demonstrate that although there are general effects for each type of mutation, one finds that each amino acid exchange has a specific consequence on the structure of p21.

On the basis of these experiments one can also see that the conformational change between the GDP bound state and the GTP bound state, that is characteristic for all G proteins, is influenced to a different degree by the transforming mutations. Since some mutant proteins have increased and some decreased dissociation rate constants these kinetic values cannot be used to provide a general explanation for the transforming property of the mutants. All the mutants on the other hand have a decreased GTPase rate in vitro. This might not be the explanation for their effect on cellular growth but it might be an indication for a general structural change that is common to all these mutant proteins and that is not reflected by the kinetics of guanine nucleotide binding. A final answer might be available after the three-dimensional structure of the GDP and GTP complexes of the cellular and transforming proteins have been solved to high resolution.

REFERENCES

Barbacid,M. (1987), Ann. Rev.Biochem. 56, 779-827

Der,C.J., Finkel,T. and Coper,G.M. (1986), Cell 44, 167-176

Dhar,R., Ellis,R.W., Shiy, T.Y., Oroszlan,S., Shapiro,B., Maizel,J., Lowy,D. and Scolnick,E. (1982), Science 217, 934-936

Chardin,P., Yeramian,P. and Tavitian,A., (1985), Int.J.Cancer 35, 647-652

DeVos,A.M., Tong,L., Milburn,M.V., Matias,P.M., Jancarik,J.,Noguchi,S., Nishimura,S., Miura,K., Ohtsuka,E. and Kim,S.-H. (1988), Science 239,888-893

Feuerstein,J., Kalbitzer,H.R., John,J., Goody,R.S., and Wittinghofer,F. (1987), Eur.J.Biochem. 162, 49-55

Gibbs,J., Sigal,I.S. and Scolnick,E.M. (1985) Trends Biochem.,Sci. 10, 350-353

Hall,A. and Self,A.J. (1986) J.Biol.Chem. 261, 10963-10965

Hattori,S., Clanton,D.H., Satoh,T. Nakamura,S. Kawakita,M. and Shih,T.Y. (1987), Mol.Cell.Biol. 7, 1999-2002

Hiratsuka,T. (1983), Biochim.Biophys.Acta 742, 496-508

Hoshino,M., Clanton,D.H., Shih,T.Y. Kawakitra,M., and Hattori,S. (1987), J.Biochem. (Tokyo) 102, 503-511

Lacal,J.C. ans Aaronsen,S.A. (1986), Mol.Cell.Biol. 6, 4214-4220

Noguchi,S., Nishimura,S., Miura,K., Ohtsuka,E. and Kim, S.-H. (1988) Science, 239, 888-893

Trahey,M. and McCormick,F. (1987) 238, 542-545

Tsuchida, M., Ohtsubo,E. and Ryder,T. (1982), Science 217, 937-939

Tucker,J., Sczakiel,G., Feuerstein,J., John.J., Goody,R.S. and Wittinghofer,F. (1986) EMBO J: 5, 1351-1358

Yammamoto,F. and Perucho,M. (1984), Nucl.Acid Res. 12, 8873-8886

Yasuda,S., Furuichi,M. and Soeda,E., (1984) Nucl.Acid Res.12, 5583-5588

STRUCTURAL AND FUNCTIONAL STUDIES ON c-p21, v-p21 AND THE GENETICALLY ENGINEERED GUANINE NUCLEOTIDE BINDING DOMAIN OF EF-Tu

Alfred Pingoud, Uwe Pieper, Roger Busche, Hans-Jürgen Ehbrecht, Matthias Wehrmann, Frank-Ulrich Gast, Jürgen Feuerstein[1], Alfred Wittinghofer[1], Thomas Jarchau[2], Gert-Wieland Kohring[3] and Frank Mayer[3]

Zentrum Biochemie, Medizinische Hochschule Hannover
D-3000 Hannover, W.-Germany

INTRODUCTION

Guanine nucleotides are intimately involved in a variety of regulatory functions, in particular in protein synthesis (1) and signal transduction (2,3). Common to these functions is that activation of the target proteins occurs by binding of GTP, while deactivation is achieved by GTP hydrolysis. Following GDP dissociation from the guanine nucleotide binding protein reactivation is initiated by GTP binding. Interference with this cycle of events leads to a major disturbance of the metabolic reactions involved.

Two of the best characterized guanine nucleotide binding proteins are the bacterial elongation factor EF-Tu which supplies the protein synthesizing ribosome with aminoacyl-tRNAs (4,5) and the H-ras gene product p21 (6) which presumably is involved in an as yet not identified signal transduction chain. A comparison of sequence data (7,8,9) have suggested that functional similarities of the two proteins are correlated with common structural features. This has prompted us to compare EF-Tu and p21 in structural terms. For this purpose we have chosen two different approaches: namely

I. to investigate whether EF-Tu and p21 cross-react immunologically, as well as to search for ligands, which interact with both proteins
II. to isolate the guanine nucleotide binding domain of EF-Tu, to analyze its secondary structure composition and compare it with that of p21

RESULTS AND DISCUSSION

<u>EF-Tu and p21 cross-react immunologically</u>

In order to find out, whether EF-Tu and the H-ras gene product p21 cross-react immunologically, we have raised polyclonal antibodies against EF-Tu and v-p21 in rabbits. For the immunization procedure we have used homogeneous antigens prepared by polyacrylamide gel electrophoresis in the presence of sodium dodecyl sulfate (SDS-PAGE) starting with antigens already highly purified by chromatographic techniques. The antisera obtained were fractionated to give enriched IgG fractions: anti-EF-Tu and anti-p21. Some of the experiments were performed with these crude antibody preparations. Most of the experiments were carried out with affinity-purified antibodies. For this purpose we have coupled EF-Tu and p21 to vinylsulfone agarose and purified the anti-EF-Tu and anti-p21 preparations on EF-Tu and p21 affinity-columns, such that we obtained the following affinity-purified antibody preparations: anti-

---

[1]Abteilung Biophysik, Max-Planck-Institut für Medizinische Forschung, D-6900 Heidelberg
[2]Institut für Mikrobiologie, Universität Würzburg, D-8700 Würzburg
[3]Institut für Mikrobiologie, Universität Göttingen, D-3400 Göttingen

EF-Tu affinity-purified over an EF-Tu column, anti-EF-Tu affinity-purified over a p21 column, anti-p21 affinity-purified over a p21 column, anti-p21 affinity-purified over an EF-Tu column.

The various antibody preparations were employed in enzyme-linked immunoassays. The results of experiments in which identical IgG concentrations were used are shown in schematic form in Fig. 1. The crude anti-EF-Tu preparation reacts not only with EF-Tu but also with p21. Likewise, the crude anti-p21 reacts with p21 and EF-Tu. Already with these antibody preparations immunological cross-reactivity of EF-Tu and p21 is apparent. At this IgG concentration the effect is small, it increases, however, with increasing concentrations of IgG (not shown). At the same low IgG concentration the affinity-purified antibodies give rise to a much stronger reaction. The immunological cross-reactivity between EF-Tu and p21 is particularly obvious with antibodies affinity-purified over affinity-columns carrying the cross-reacting antigen. For example, affinity-purification of anti-EF-Tu over a p21 column yields an anti-EF-Tu preparation with an identical reactivity towards EF-Tu and p21. This was not unexpected after the immunological cross-reactivity was established with the crude IgG preparations. The results shown in Fig. 1 demonstrate that in the antisera directed against EF-Tu and p21 antibodies are present that cross-react with the two antigens and, therefore, that they can be enriched by the affinity-columns carrying the cross-reacting antigen.

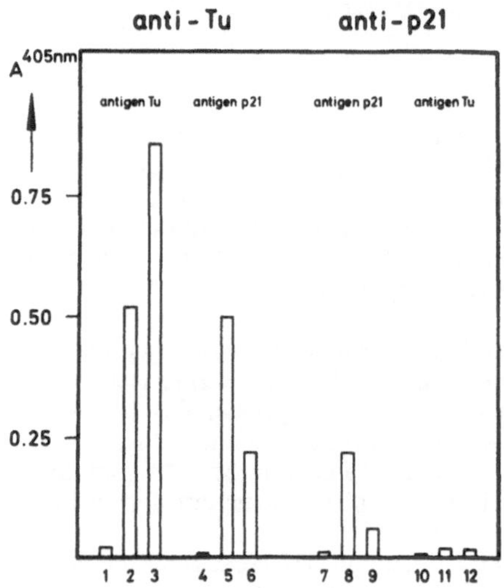

Fig. 1.Enzyme linked immunoassay with EF-Tu and p21 as antigens and various anti-Tu and anti-p21 antibody preparations.
The assay was carried out with EF-Tu and p21 as antigens immobilized in the wells of a microtiter plate. The following antibody preparations were used: 1 and 4 crude anti-Tu, 2 and 5 anti-Tu affinity purified over a p21 column, 3 and 6 anti-Tu affinity purified over an EF-Tu column, 7 and 10 crude anti-p21, 8 and 11 anti-p21 affinity purified over a p21 column, 9 and 13 anti-p21 affinity purified over an EF-Tu column. Pre-immune IgG does not give a positive colour reaction.

Antibodies raised against EF-Tu and p21 interfere with functional activities of both EF-Tu and p21

It was of interest to investigate, whether the binding of antibodies directed against EF-Tu and p21 interferes with the functions of these proteins (cf. 10,11), and of particular interest, whether interference is seen with the cross-reacting antibodies. Fig. 2 shows the effect of the binding of antibodies directed against p21 and affinity-purified over a EF-Tu-column. It can be seen that anti-p21 drastically slows down the GDP exchange with EF-Tu·GDP. Similar experiments were carried out also with p21 and anti-EF-Tu antibodies affinity-purified over a p21-column. There the effect is not as pronounced: while anti-p21 affinity-purified over a p21-column inhibits the GDP exchange with p21 GDP drastically, anti-EF-Tu affinity-purified over a p21-column inhibits the GDP exchange reaction only slightly (Fig. 3). We have in addition looked for an effect of the antibodies on the GTPase reaction. The GTPase activity of EF-Tu and p21 is very low unless stimulated by an appropriate ligand, e.g. in the case of EF-Tu the programmed ribosome or the antibiotic kirromycin. We have now found out that the presence of anti-EF-Tu stimulates the GTPase reaction by approximately 50% in the presence of kirromycin (not shown). It is important to note that this effect is produced also by antibodies directed against p21: Anti-EF-Tu and anti-p21 mimic the effect of physiological effector molecules that stimulate the GTPase activity of EF-Tu. Similar experiments carried out with EF-Tu in the absence of kirromycin and with p21 which have a very low GTPase activity were not sufficiently reproducible to be used for a quantitative evaluation.

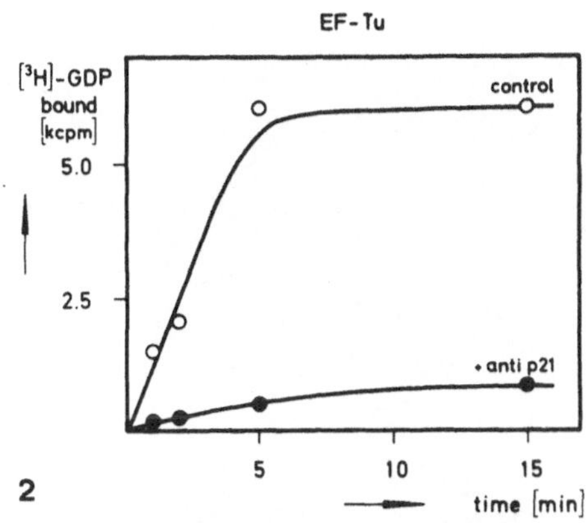

2

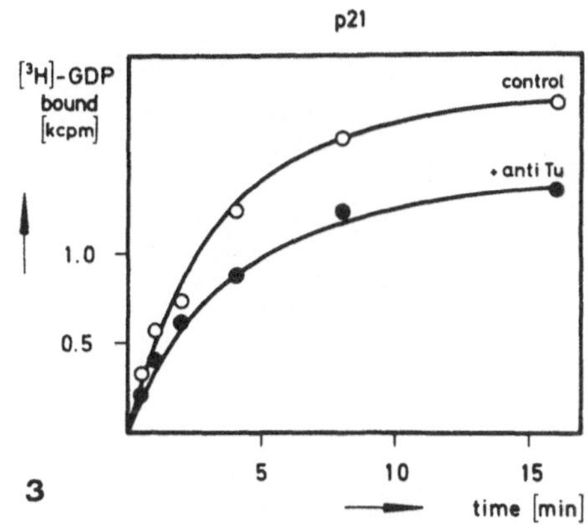

3

Fig. 2. ^{3}*H-GDP binding to EF-Tu in the absence and presence of antibodies directed against p21. 0.5 μM EF-Tu·GDP was incubated for 15 min at 37°C with 0.4* A_{280nm} *units of pre-immune IgG (○) or anti-p21 antibodies affinity purified over an EF-Tu column (●). The GDP exchange reaction was started by addition of 0.8 μM* ^{3}*H-GDP. At the times indicated aliquots were withdrawn and pipetted onto nitrocellulose filters which were then washed and dried. The amount of* ^{3}*H-GDP bound to EF-Tu and retained on the filter was determined in a scintillation counter.*

Fig. 3. ^{3}*H-GDP binding to p21 in the absence and presence of antibodies directed against EF-Tu. 0.4 μM p21·GDPwas incubated for 15 min at 37°C with 2.4* A_{280} *units of pre-immune IgG (○) or anti-Tu antibodies affinity purified over a p21 column (●). The GDP exchange reaction was started and analyzed as described in the legend to fig. 2.*

It might be considered as surprising that the same antibody preparations which inhibit the guanine nucleotide exchange with EF-Tu and p21 stimulate the GTPase activity of EF-Tu. Since we have used polyclonal antibodies these effects could be due to a different set of antibodies contained in this preparation. Alternatively, it could be that the same antibodies elicit both effects, similarly as kirromycin which when bound to EF-Tu changes the guanine nucleotide exchange rate and the GTP hydrolysis rate (12).

Our comparative immunological studies whith EF-Tu and p21 demonstrate that these proteins have common epitopes, some of which are involved in guanine nucleotide binding and GTP hydrolysis.

<u>Other guanine nucleotide binding proteins in E.coli cross-react immunologically with anti-p21 antibodies</u>

Sequence homology with p21 was not only observed for EF-Tu but also for the elongation factor EF-G and other nucleotide binding proteins. We have, therefore, investigated, whether other proteins of E.coli than EF-Tu cross-react immunologically with p21. For this purpose we have analyzed whole cell extracts of E.coli after separation by polyacrylamide gel electrophoresis using an immuno blot procedure. Fig. 4 shows a Western blot of E.coli proteins using a crude anti-p21 antibody preparation. It can be seen that EF-Tu reacts weakly with the antibody. In addition there are strongly cross-reacting proteins of molecular weight 70000 to 80000 and 21000 and weakly cross-reacting proteins of approx. 35000 and 90000 present in the E.coli lysate. The 21000 molecular weight band is not seen with affinity-purified antibodies, while the other bands are still present, in particular the 70000 to 80000 band remains the strongest band in the immuno blot. We conclude that the immunolabelling of the 21000 band is due to a contaminant present in the antigen preparation which was used for immunization. This antigen preparation was obtained by semi-preparative SDS-PAGE in which proteins are separated according to their molecular weight. It is possible, therefore, that the antigen preparation used for the immunisation contained not only p21 but to a very samll extent also other proteins of molecular weight 21000. Since these would be E.coli proteins, they are likely to be very immunogenic in rabbits and will give rise to antibodies. Most of these contaminating antibodies, however, will be eliminated by affinity purification over p21 or EF-Tu-columns.

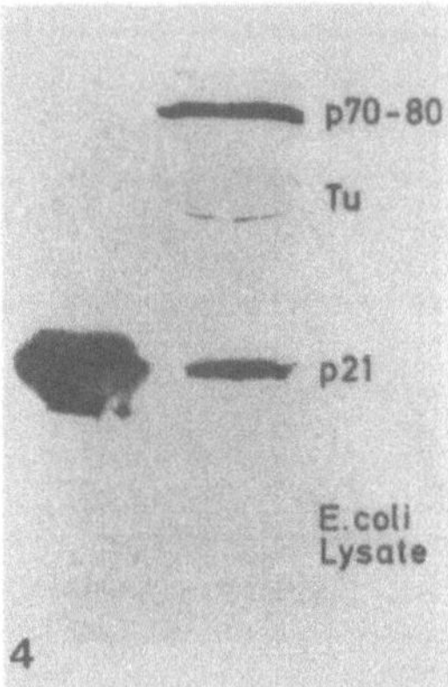

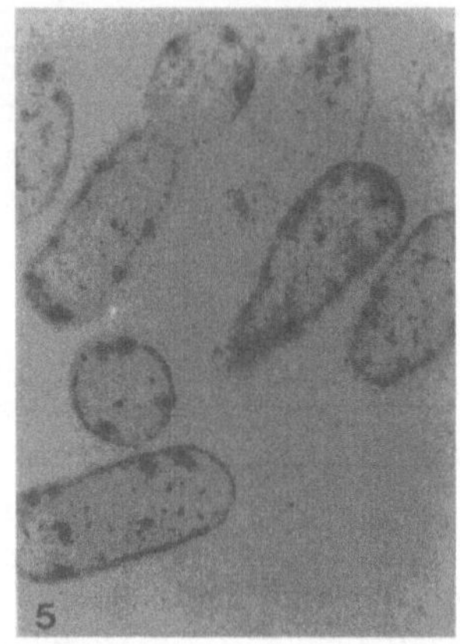

Fig. 4. Western blot of electrophoretically separated E.coli proteins using an anti-p21 antibody preparation.
82 ng of homogeneous p21·GDP (left) and 200 ng of an E.coli extract (right) were separated by electrophoresis on 17.5% polyacrylamide gels in the presence of sodium dodecyl sulfate, blotted onto nitrocellulose sheets and immuno stained with anti-p21.

Fig. 5. Immunolabelling of ultrathin sections of E.coli.
E.coli cells were fixed and embedded in resin. Ultrathin sections were treated with anti-p21 and then with goat anti-rabbit-IgG-gold. 70% of the gold particles are associated with the cell membrane.

In order to identify the E.coli protein of 70000 to 80000 molecular weight which strongly reacts with anti p21 antibodies we have analyzed by immunoblotting homogeneous preparations of the E.coli initiation factor IF-2 (kindly supplied by Dr. C. Gualerzi) and the elongation factor EF-G which have molecular weights in the range of interest and are both guanine nucleotide binding proteins. These two proteins did not react with anti-p21 antibodies to a significant extent. We have also suspected that the lepA protein, a guanine nucleotide binding membrane protein of E.coli with a molecular weight of 76000, and sequence homology with EF-G, EF-Tu and IF (2), represents the strongly stained protein in the immuno blot of E.coli proteins. In order to test this idea we have first tried to analyze the intracellular localization of this protein in E.coli. For this purpose E.coli cells were fixed with paraformaldehyde and glutaraldehyde, embedded in a lowicryl K4M resin and then immunolabelled. The immunolabelling was carried out with ultrathin sections of the embedded cells using anti-p21 antibodies and goat-anti-rabbit-IgG-gold (14). Fig. 5 shows an electron micrograph of immunolabelled E.coli cell sections. A quantitative evaluation reveals that on average 70% of the cross reacting material is present in the cell envelope. This result would agree with the assignment that lepA protein represents the unidentified antigen in E.coli which reacts strongly with anti-p21 antibodies. This identification was confirmed by enzyme linked immunoassays with a homogeneous lepA preparation (kindly supplied by Dr. P. March).

Daunomycin binds to EF-Tu, p21 and other nucleotide binding proteins and inhibits protein synthesis in vitro

A cluster analysis of the protein sequence homology of some nucleotide binding proteins has indicated a distant relationship between P-glycoprotein and among other proteins p21 (15). P-glycoprotein is normally overexpressed in cell lines exhibiting pleiotropic drug resistance. It binds a variety of seemingly unrelated drugs and neutralizes their cytotoxic effects by exporting them across the plasma membrane. We have been interested to find out whether the sequence homology between P-glycoprotein and other nucleotide binding proteins, in particular EF-Tu and p21 is also reflected in their affinity towards certain drugs. We have chosen daunomycin for this investigation, since one of us has recently shown that an analogue of this compound interacts with and can be crosslinked to P-glycoprotein.

Binding of daunomycin to EF-Tu was first demonstrated by photoaffinity labelling with a radioactively labelled daunomycin derivative, iodomycin (Fig. 6). Product analysis of the photo-cross-linking reaction was carried out by SDS-PAGE followed by autoradiography (Fig. 7). The photoaffinity labelling of EF-Tu with iodomycin could be suppressed by excess daunomycin, and also by kirromycin and pulvomycin, two antibiotics known to bind to EF-Tu and change its conformation and activity (data not shown). The effect of kirromycin or pulvomycin, which are structurally not related to daunomycin, is indirect (vide infra). Photo-cross-linking of EF-Tu and daunomycin can also be carried out with commercially available ^{3}H-daunomycin. Product analysis is then carried out by pipetting the reaction mixture on glass fiber filters, which are immersed into cold 10% trichloroacetic acid, washed, dried and subjected to liquid scintillation counting. The cross-linking of daunomycin is specific, since it can be suppressed by excess unlabelled daunomycin (Fig. 8). With ^{3}H-daunomycin the cross-linking yield is approx. 5-10% depending on the reaction conditions.

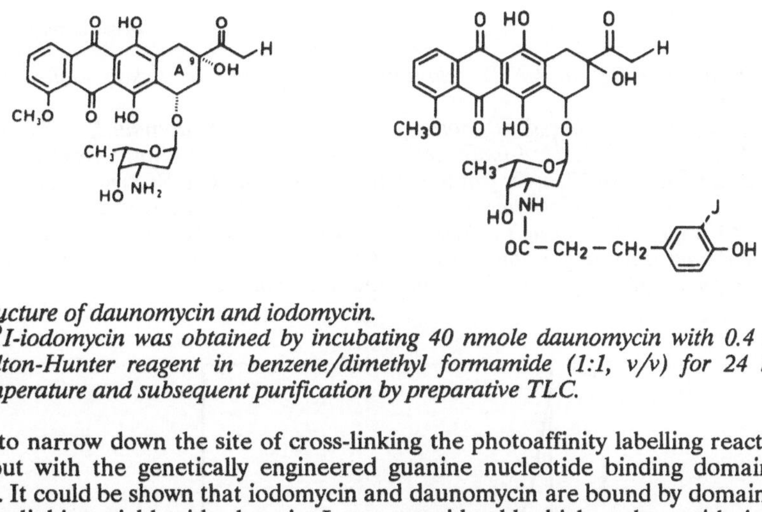

Fig. 6. Structure of daunomycin and iodomycin.
^{125}I-iodomycin was obtained by incubating 40 nmole daunomycin with 0.4 nmole ^{125}I-Bolton-Hunter reagent in benzene/dimethyl formamide (1:1, v/v) for 24 hrs at room temperature and subsequent purification by preparative TLC.

In order to narrow down the site of cross-linking the photoaffinity labelling reaction was also carried out with the genetically engineered guanine nucleotide binding domain of EF-Tu, domain I. It could be shown that iodomycin and daunomycin are bound by domain I (Fig. 7,8). The cross-linking yield with domain I was considerably higher than with intact EF-Tu. Crosslinking could be suppressed in a concentration dependent manner by excess daunomycin, not, however, by kirromycin or GDP, demonstrating that the effect of kirromycin on the cross-linking yield with intact EF-Tu is indirect and that the binding of iodomycin is not to the GDP binding site.

Both UV/VIS (Fig. 9a) as well as CD (Fig. 9b) spectroscopy demonstrate directly that daunomycin interacts with EF-Tu, the change in extinction or ellipticity, however, is small and hardly sufficient for a quantitative assessment of the affinity.

In order to find out whether the binding of daunomycin to EF-Tu interferes with the activity of EF-Tu, we have analyzed polyPhe synthesis on polyU programmed ribosomes in the absence and presence of daunomycin. Fig. 10 shows that daunomycin interferes in a concentration dependent manner with protein synthesis in vitro. Since daunomycin does not change the affinity of EF-Tu for GDP, GTP, and aminoacyl-tRNA (data not shown), its effect on protein synthesis must be due to the impairment of other activities of EF-Tu (GTP hydrolysis, binding to EF-Ts, binding to ribosomes, etc.). This is currently being investigated.

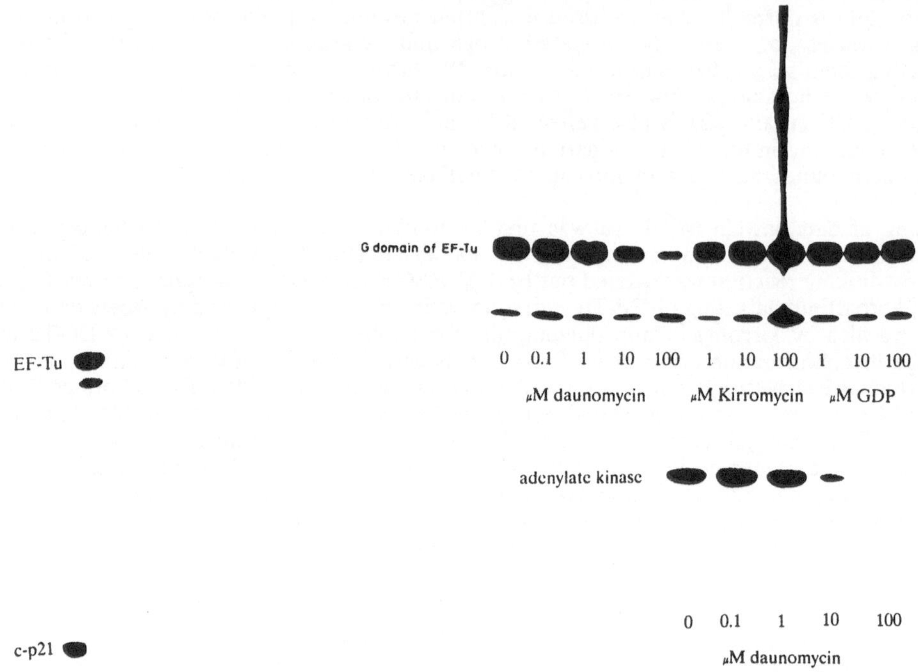

Fig. 7. Photoaffinity labelling of EF-Tu, domain I, c-p21 and adenylate kinase with ^{125}I-iodomycin.

Approx. 1 μM of electrophoretically homogeneous EF-Tu·GDP domain I, c-p21 or adenylate kinase, resp. and 0.2 nM ^{125}I-iodomycin dissolved in 500 μl 10 mM Tris·HCl pH 7.5, 250 mM sucrose, 5 mM $MgCl_2$ were irradiated at 24°C with a 1000 W Xenon lamp whose light was filtered through two cuvettes of 2 cm pathlength, one filled with H_2O, the other with saturated $CuSO_4$ and through a Schott KV 418 filter. Irradiation time was 1 min. After labelling the protein was precipitated with 10% w/v TCA and loaded onto a Laemmli-gel. After electrophoresis the gel was dried and autoradiographed.

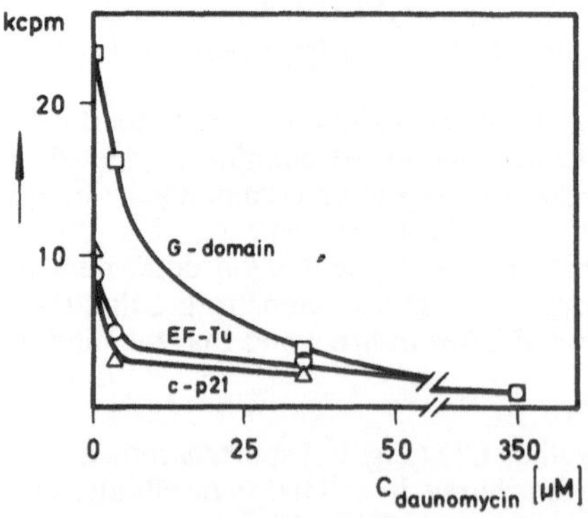

Fig. 8. Photoaffinity labelling of EF-Tu, domain I and c-p21 with ^{3}H-daunomycin.

Approx. 10 μM of EF-Tu, domain I or c-p21, resp., and ^{3}H-daunomycin were irradiated as described in the legend to fig. 7 for 15 sec with increasing amounts of unlabelled daunomycin. The reaction mixture was pipetted onto GF/C filters, immersed into 10% trichloroacetic acid, washed, dried and analyzed for protein bound radioactivity.

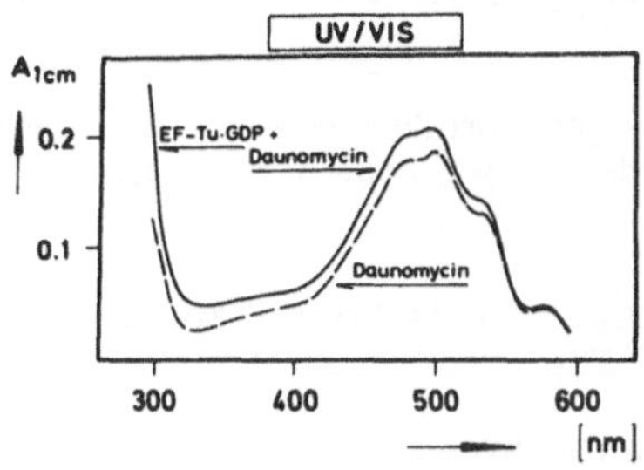

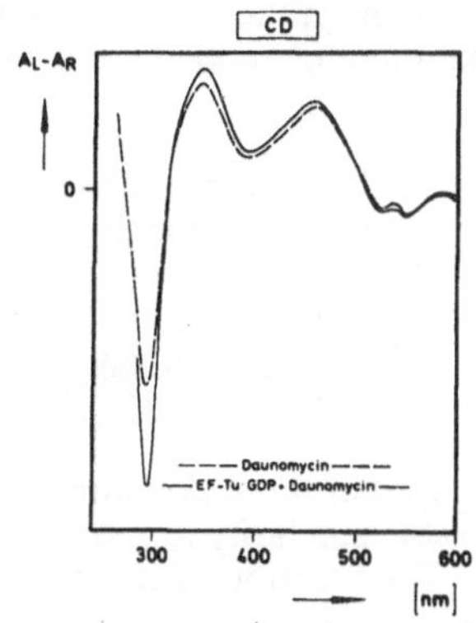

Fig. 9. UV/VIS and CD spectra of daunomycin and the EF-Tu·GDP x daunomycin complex.
The UV/VIS spectrum of 11.5 µMdaunomycin in the absence of presence of 13.4 µMEF-Tu·GDP is shown on the left, the CD spectrum of 115 µMdaunomycin in the absence or presence of 134 µM EF-Tu·GDPin 0.05 M Tris·HCl pH 7.5, 5mM $MgCl_2$ is shown on the right.

Photo-cross-linking experiments demonstrate that also p21 interacts with iodomycin and daunomycin (Fig. 7,8). This interaction enhances the rate of the ^{3}H-GDP/GDP exchange with c-p21 (Fig. 11). It does not seem to affect the affinity of v-p21 towards GDP, since daunomycin, even at 100 µM concentration does not affect the ^{3}H-GDP/GDP exchange with v-p21 (data not shown).

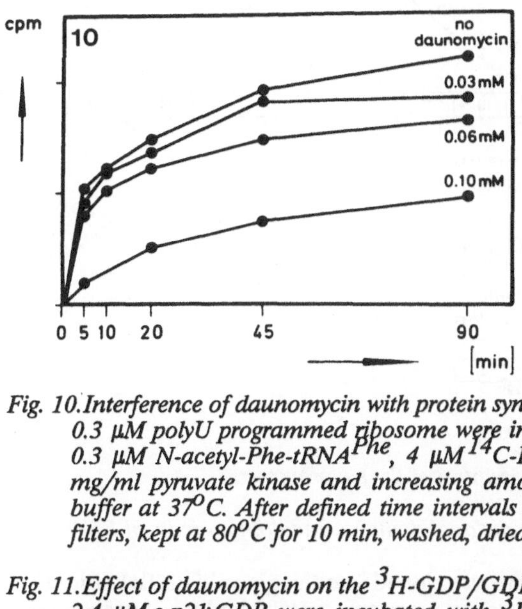

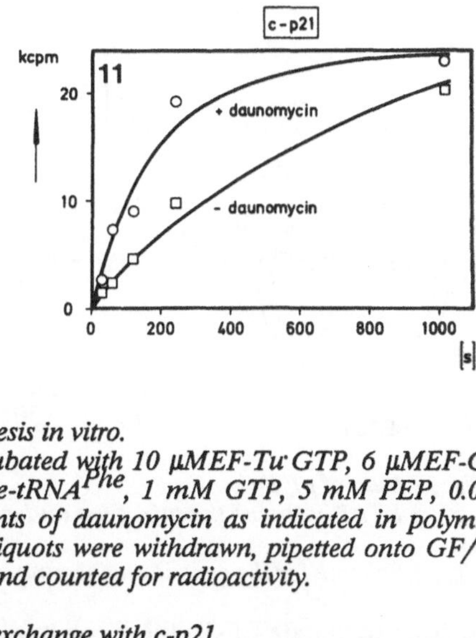

Fig. 10. Interference of daunomycin with protein synthesis in vitro.
0.3 µM polyU programmed ribosome were incubated with 10 µMEF-Tu·GTP, 6 µMEF-G, 0.3 µM N-acetyl-Phe-tRNA^{Phe}, 4 µM ^{14}C-Phe-tRNA^{Phe}, 1 mM GTP, 5 mM PEP, 0.05 mg/ml pyruvate kinase and increasing amounts of daunomycin as indicated in polymix buffer at 37°C. After defined time intervals aliquots were withdrawn, pipetted onto GF/C filters, kept at 80°C for 10 min, washed, dried and counted for radioactivity.

Fig. 11. Effect of daunomycin on the ^{3}H-GDP/GDP exchange with c-p21.
2.4 µM c-p21·GDP were incubated with ^{3}H-GDP at 37°C. After defined time intervals aliquots were withdrawn, pipetted onto nitrocellulose filters, washed, dried and analyzed for protein bound radioactivity.

Daunomycin is also bound specifically by adenylate kinase (Fig. 7). Whether affinity for anthracycline antibiotics is a common property of many nucleotide binding proteins remains to be established.

The identification of the site of cross-linking of daunomycin to EF-Tu and p21 is currently pursued by us.

Our results suggest that the cytotoxic effect of daunomycin and other anthracycline antibiotics may at least in part be due to their interference with the function of nucleotide binding proteins, e.g. in protein synthesis and signal transduction, a conclusion that is supported by the finding that the inhibitory activity of some anthracycline antibiotics is not due to its binding to DNA (16).

Furthermore, our demonstration that the ras-gene product p21 interacts with daunomycin may provide a starting point for the rational design of a drug directed against the oncogenic version of p21.

The isolated guanine nucleotide binding domain of EF-Tu preserves some of the activities of intact EF-Tu but does not productively interact with the other two domains

EF-Tu is composed of three domains. Domain I is involved in GTP binding and hydrolysis, it shows considerable similarity in sequence and structure with G-proteins, in particular the ras-gene products. The function of the other two domains is not well understood. We would like to know

- whether the individual domains and pairs of domains can form stable structures by themselves
- whether they interact with each other
- whether they show partial activities of the intact EF-Tu.

To this end we have begun to produce by genetic engineering individual domains and pairs of domains and to study their structural and functional properties.

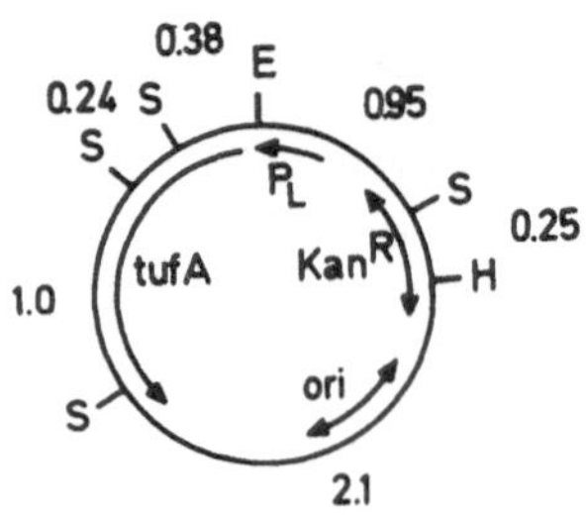

12

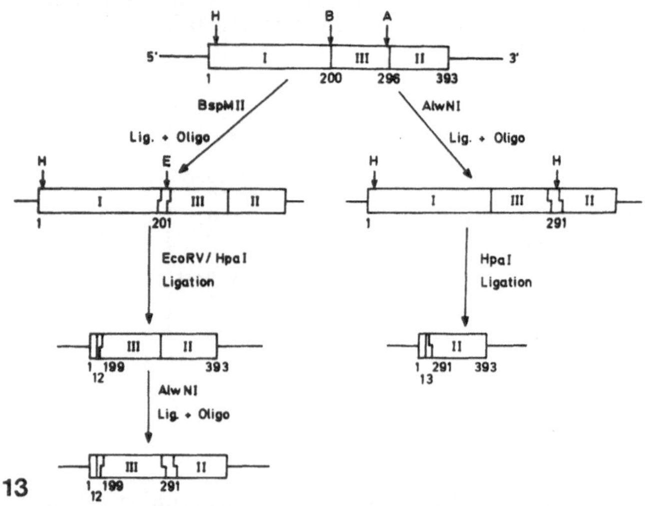

13

Fig. 12:Map of pETu1.
pETu1 harbours the tufA gene under the control of the p_L promoter.

Fig. 13.Scheme for the genetic engineering of individual domains and domain pairs of EF-Tu.

In order to mutate EF-Tu the E.coli tufA gene was cloned into the plasmid pETu1 (Fig. 12) in which the tufA gene is under control of the strong p_L promoter which allows overexpression of mutated EF-Tu in E.coli strains carrying the thermosensitive cI 857 repressor by a temperature shift from 28 to 42°C. For the production of isolated domains or pairs of domains of EF-Tu pETu1 was cleaved with restriction enzymes and religated in the presence of appropriate linkers (Fig. 13). So far we have genetically engineered and expressed domain I, domain pair III+II and domain pair I+III. Domain I and domain pair III+II were purified to near homogeneity.

Circular dichroism spectroscopy (CD) was employed to determine structural features of EF-Tu and domain mutants of EF-Tu, since CD spectra can be analyzed to obtain information on the α-helix and ß-pleated sheet structure content of proteins in solution. Fig. 14 shows the CD spectra of EF-Tu·GDP and of an equimolar mixture of domain I·GDP and the domain pair III+II. According to the secondary structure analysis and using results from the X-ray structure analysis of EF-Tu the isolated domain I has the same secondary structure as in the intact EF-Tu. Furthermore, since the equimolar mixture of domain I and domain pair III+II gives a similar CD spectrum as intact EF-Tu, we infer that the isolated domain pair III+II has a similar secondary structure as in the intact EF-Tu.

CD-Spectra of intact EF-Tu and domain I + domain pair II+III

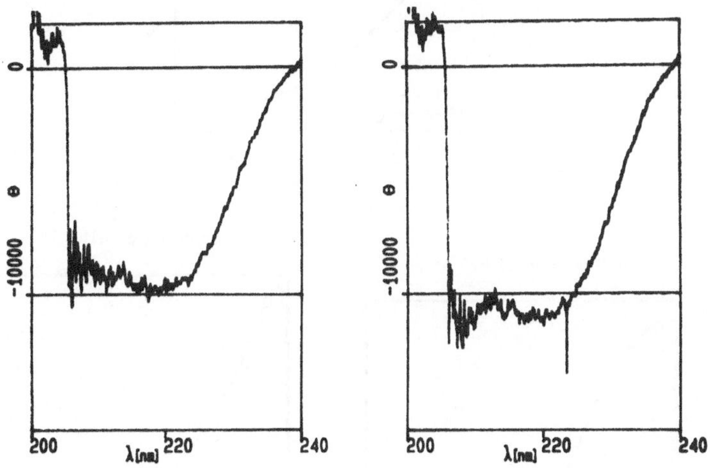

Fig. 14. Circular dichroism of EF-Tu and the equimolar mixture of domain I and domains III+II. CD spectra were recorded with 10 μM protein solutions in 0.05 M Tris HCl pH 7.5, 10 mM $MgCl_2$, 5 μM GDP, 0.1% lubrol, 15% glycerol. The spectra are represented in the conventional form for proteins, i.e. mean residue ellipticity vs. wavelength.

We have analyzed the temperature induced denaturation of EF-Tu·GDP, domain I·GDP and domain pair III+II by recording the ellipticity at 220 nm as a function of temperature. Under the conditions used all proteins denature in an irreversible manner. But while intact EF-Tu precipitates at temperatures above 50°C, domain I and domain pair III+II stay in solution up to 70°C, with only a partial loss in secondary structure (data not shown).

Guanidinium chloride induced denaturation of EF-Tu·GDP, domain I·GDP and domain III+II leads to a total unfolding of the proteins as demonstrated by CD spectroscopy (Fig. 15). While intact EF-Tu denatures in a biphasic manner, the denaturation of domain I and domain pair III+II follows apparently a monophasic course. The denaturation curve for intact EF-Tu is not exactly the same as the sum of the denaturation curves of domain I and domain pair III+II. In particular the first phase of the EF-Tu denaturation is not seen in the denaturation of either domain I or domain pair III+II. It might represent a partial separation of the individual domains. The second phase of the EF-Tu denaturation occurs at similar guanidinium chloride concentrations as that of domain I and domain pair III+II. GDP binding to EF-Tu and domain I is lost before unfolding occurs indicating a direct interference of the denaturating agent with GDP binding.

The guanine nucleotide binding as well as the GTPase activity of domain I has been investigated by Parmeggiani and coworkers (17) and their results have been confirmed by us (Table 1). As shown in Fig. 16 domain pair III+II does not alter the GDP/^{3}H-GDP exchange rate of domain I which is by at least a factor of 5 faster than that of intact EF-Tu indicating that the covalent connection of the three domains is necessary for the high affinity of GDP for EF-Tu. Protection experiments against non-enzymatic hydrolysis of aminoacyl-tRNA (18) indicate that domain I·GTP in μM concentrations does not detectably interact with aminoacyl tRNAs.

Domain II presumably harbours the binding site for kirromycin (19,20,21). Since the binding of kirromycin to EF-Tu can be monitored spectroscopically, we have investigated whether kirromycin is bound to domain pair III+II. No binding could be detected at μM concentrations (data not shown).

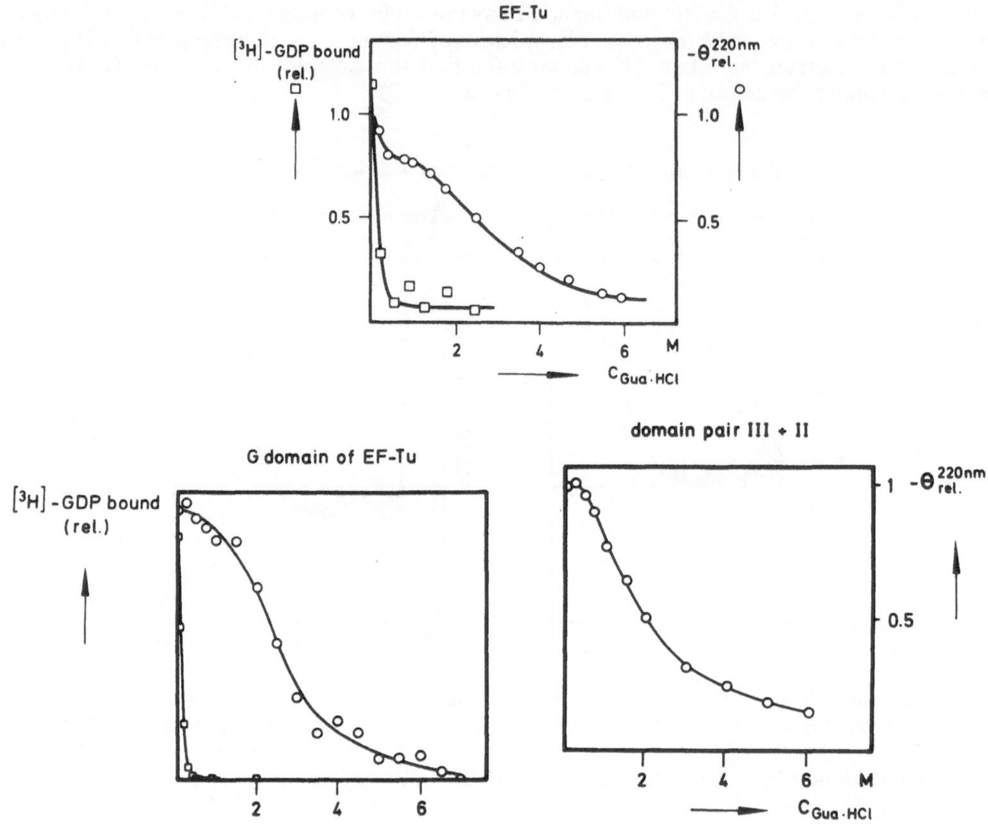

Fig. 15. Guanidinium chloride induced denaturation of EF-Tu, domain I and domain pair III+II.
A 0.6 μM solution of each protein was incubated for 30 min at room temperature at various guanidinium chloride concentrations in the same buffer as given in the legend to fig. 14 but without glycerol. The unfolding was measured by recording the ellipticity at 220 nm, the GDP binding activity was determined by measuring the GDP/³H-GDP exchange.

Table 1. Functional properties of domain I

	Domain I		EF-Tu
	This work	Parmeggiani et al.	
GTPase activity at 30°C***	0.08 *	0.1	0.06
	0.21 **		
K_{Ass} [M^{-1}] for GDP [30°C]	$0.2\cdot10^6$	$0.33\cdot10^6$	$170\cdot10^6$
K_{Ass} [M^{-1}] for GTP [30°C]	$0.16\cdot10^6$	$0.25\cdot10^6$	$1.7\cdot10^6$
Half life of GDP complex			
[min] at 0°C	10	2.8	52
at 25°C	1.7		

* freshly prepared
** after 1 month at 0°C
*** [mmol GTP hydrol./s/mol prot.]

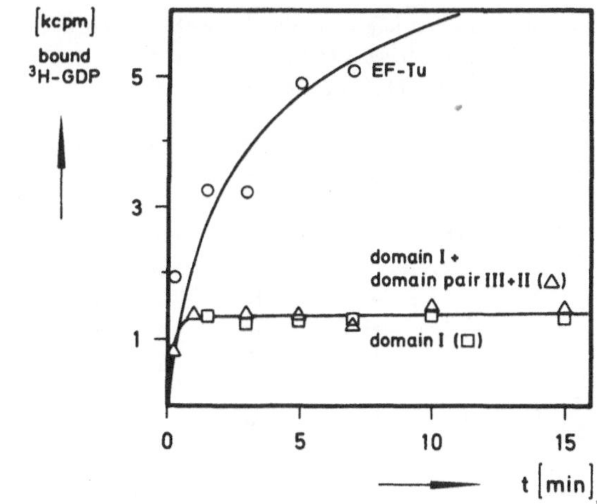

Fig. 16. GDP/^{3}H-GDP exchange of EF-Tu, domain I and of a mixture of domain I and domain pair III+II.
A 1.9 μM solution of each protein in 50 mM imidazole acetate pH 7.6, 50 mM NH_4Cl, 10 mM $MgCl_2$, 1 mM DTE was incubated at 25°C with 50 μM ^{3}H-GDP. After the times indicated aliquots were withdrawn and analyzed for protein bound ^{3}H-GDP.

<u>Circular dichroism spectroscopy reveals structural differences between the guanine nucleotide binding domain of EF-Tu, normal and transforming H-ras gene products</u>

We have recorded the circular dichroism spectra of domain I of EF-Tu, as well as of c-p21 and v-p21. The experiments with domain I were carried out only in the presence of GDP (Fig. 17a) while those with c-p21 (Fig. 17b) and v-p21 (Fig. 17c) were performed in the absence and in the presence of GDP and GTP, resp. These spectra are not identical, indicating that the secondary structure composition of these proteins is different. Table 2 gives a summary of the secondary structure analysis which is based on the structure analysis of five reference proteins (22). It is evident from this analysis that major structural differences must exist between the domain I of EF-Tu on one hand and v-p21 and c-p21 on the other hand, a conclusion which is evident from a comparison of the X-ray structure of EF-Tu and p21 (23,24,25). There are also slight differences in the secondary structure composition between c-p21 and v-p21, as well as between the GTP- and GDP-complexes of these two proteins. The differences in structure between c-p21 and v-p21 are also apparent in hydrodynamic (26) and gel-electrophoretic studies (27) which indicate that c-p21 is somewhat more compact than v-p21.

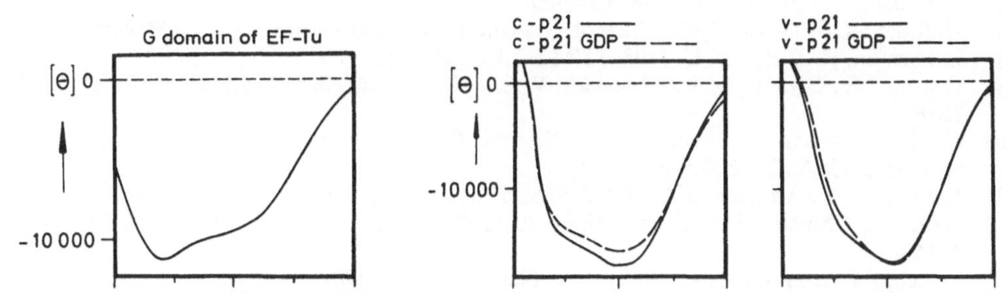

Fig. 17. Circular dichroism spectra of domain I, c-p21 and v-p21.
CD spectra of domain I, c-p21 and v-p21 were recorded at a concentration of approx. 0.2 mg/ml. Experimental details are given in (26).

Table 2. Secondary structure composition of c-p21 and v-p21 in the presence and absence of GDP or GTP, and for comparison of the G-domain of EF-Tu (domain I)

	%α-HELIX	%ß-SHEET	%RES. STRUC.
c-p21	56	25	19
c-p21xGTP	54	23	23
c-p21xGDP	47	22	31
v-p21	52	18	29
v-p21xGTP	52	18	29
v-p21xGDP	39	18	41
G-DOMAIN	30	16	54

CONCLUDING REMARKS

The elongation-factor EF-Tu and the H-ras gene product p21 are involved in very different metabolic pathways, but have one function in common: guanine nucleotide binding and GTP hydrolysis. This common function is reflected first of all in similarities in primary structure which has led to the definition of a family of proteins (G-proteins). Similarities in primary structure usually correlate with similarities in tertiary structure. In the case of the guanine nucleotide binding domain of EF-Tu and p21 our secondary structure analysis suggests that there must be major differences in the overall structure of these proteins. This does not exclude that the topology is similar, i.e. the way in which α-helices, ß-sheets and loops are arranged in this protein. The fact that we observe an immunological cross-reaction between EF-Tu and p21, as well as our finding that both proteins interact with anthracycline antibiotics demonstrates that there are similarities in structural details presumably also outside of the guanine nucleotide binding site.

REFERENCES

(1) Kaziro, Y. (1978) Biochim. Biophys. Acta 505, 95-127
(2) Stryer, L. & Bourne, H.R. (1986) Annu. Rev. Cell Biol. 2, 391-420
(3) Gilman, A.G. (1987) Annu. Rev. Biochem. 56, 615-649
(4) Bosch, L., Kraal, B., van der Meide, P.H., Duisterwinkel, F.J. & van Noort, J.M. (1983) Progr. Nucl. Acid Res. Mol. Biol. 30, 91-126
(5) Parmeggiani, A. & Swart, G.W.M. (1985) Annu. Rev. Microbiol. 39, 557-577
(6) Barbacid, M. (1987) Annu. Rev. Biochem. 56, 779-827
(7) Halliday, K. (1984) J. Cyclic Nucleotide Protein Phosphorylation Res. 9, 435-448
(8) Leberman, R. & Egner, U. (1984) EMBO J. 3, 339-341
(9) Dever, T.E., Glynias, M.J. & Merrick, W.C. (1987) Proc. Natl. Acad. Sci. USA 84, 1814, 1818
(10) Clark, R., Wong, G., Arnheim, N., Nitecki, D. & McCormick, F. (1985) Proc. Natl. Acad. Sci. USA 82, 5280-5284
(11) Lacal, J.C. & Aaronson, S.A. (1986) Proc. Natl. Acad. Sci. USA 83, 5400-5404
(12) Wolf, H., Chinali, G. & Parmeggiani, A. (1974) Proc. Natl. Acad. Sci. USA 71, 4910-4914
(13) March, P.E. & Inouye, M. (1985) Proc. Natl. Acad. Sci. USA 82, 7500-7504
(14) Kohring, G.-W., Mayer, F. & Mayer, H. (1985) Eur. J. Cell Biol. 37, 1-6
(15) Gerlach, J.H., Endicott, J.A., Juranka, P.F., Henderson, G., Sarangi, F., Deuchars, K.L. & Ling, V. (1986) Nature 324, 485-489
(16) Israel, M., Idriss, J.M., Koseki, Y. & Khetarpal, V.K. (1987) Cancers Chemother. Pharmacol. 20, 277-284
(17) Parmeggiani, A., Swart, G.W.M., Mortensen, K.K., Jensen, M., Clark, B.F.C., Dente, L. & Cortese, R. (1987) Proc. Natl. Acad. Sci. USA 84, 3141-3145

(18) Pingoud, A., Urbanke, C., Krauss, G., Peters, F. & Maass, G. (1977) Eur. J. Biochem. 78, 403-409
(19) Duisterwinkel, F.J., de Graaf, J.M., Kraal, B. & Bosch, L. (1981) FEBS Lett. 131, 89-93
(20) Duisterwinkel, F.J., Kraal, B., de Graaf, J.M., Talens, A., Bosch, L., Swart, G.W.M., Parmeggiani, A., LaCour, T.F.M., Nyborg, J. & Clark, B.F.C. (1984), EMBO J. 3, 113-120
(21) van Noort, J.M., Kraal, B. & Bosch, L. (1985) Proc. Natl. Acad. Sci. USA 82, 3212-3216
(22) Chen, Y.-H., Yang, J.T. & Chau, K.H. (1974) Biochemistry 13, 3350-3361
(23) LaCour, T.F.M., Nyborg, J., Thirup, S. & Clark, B.F.C. (1985) EMBO J. 4, 2385-2388
(24) Jurnak, F. (1985) Science 230, 32-36
(25) de Vos, A.M., Tong, L., Milburn, M.V., Matias, P.M., Jancarik, J., Noguchi, S., Nishimura, S., Miura, K., Ohtsuka, E. & Kim, S.-H. (1988), Science 239, 888-893
(26) Pingoud, A., Wehrmann, M., Pieper, U., Gast, F.-U., Urbanke, C., Alves, J., Feuerstein, J. & Wittinghofer, A. (1988) Biochemistry 27, 4735-4740
(27) Tucker, J., Szakiel, G., Feuerstein, J., John, J., Goody, R. & Wittinghofer, A. (1986) EMBO J. 5, 1351-1358

STRUCTURE AND FUNCTION OF *ras* p21: STUDIES BY SITE-DIRECTED MUTAGENESIS

Thomas Y. Shih, David J. Clanton, Pothana Saikumar, Linda S. Ulsh and Seisuke Hattori*

Laboratory of Molecular Oncology, National Cancer Institute-Frederick Cancer Research Facility, Frederick, MD., 21701 U. S. A.

Are studies by analogy fruitful to understanding protein structure and function? With protein sequences accumulating at an astonishing rate through gene cloning and DNA sequencing, a great deal of protein structure and function can be learned by analogy with members of the superfamily whose structures have been determined and whose molecular mechanisms of action are known. The best case in point, perhaps is in unraveling the elusive function of *ras* p21. The molecular model of p21 has been constructed by analogy *with* the crystal structure of the *E. coli* elongation factor, EF-Tu (McCormick et al., 1985; Jurnak, 1985). *This* p21 model is remarkably consistent with the actual three dimensional structure of p21 later determined by X-ray crystallography (De Vos et al., 1988). The recent identification of the GAP protein which stimulates GTPase activity of p21 (Trahey and McCormick, 1987), is a conceptual offspring of similar biochemical mechanism well understood for the function of EF-Tu in protein synthesis (Kaziro, 1978). Furthermore, analogy with the well characterized G-proteins, which regulate transmembrane cell signalling in the adenylate cyclase systems and light transduction in retina, forms the foundation for the current belief that *ras* p21 mediates transmission of growth signals to their intracellular *effectors* that control cell proliferation and differentiation (Bourne and Sullivan, 1986). Oncogenic activation of proto-oncogenes, which have been implicated in various aspects of human carcinogenesis, very often mutates amino acid residues at positions 12, 13 or 61 at the GTP-binding site of *ras* p21 (for review see Barbacid, 1987). These changes all render p21 *constitutively* active in its GTP-bound conformation. One molecular mechanism becomes apparent - that persistent emission of growth control signals is vital to malignant transformation of normal cells. In this paper, we describe our studies using site-directed mutagenesis in understanding the structure and function of *ras* p21, particularly that of the GTP-binding domain.

ras p21 IS A GTP-BINDING PROTEIN

Before the dawn of recombinant DNA technology, p21 proteins encoded by *ras* oncogenes of Harvey and Kirsten murine sarcoma viruses, were *identified* by a somewhat cumbersome and unpredictable method of immunology

*Present address: Department of Pure and Applied Sciences, University of Tokyo, Tokyo 153, Japan

(Shih et al., 1979a). The sarcoma virus p21 is not a structural protein of virus particles and is expressed in minute amounts in transformed cells. Identification of this protein, therefore, had to relied on finding of antisera from Osborne-Mendel rats bearing tumors induced by syngeneic transplatation of NRK cells transformed by the Harvey virus (Shih et al., 1979a). These antisera also cross-reacted with K-ras p21 of the Kirsten virus. Early studies with a temperature-sensitive (ts) mutant of the Kirsten sarcoma virus, suggested that ras p21 was required for maintenance of virus-induced malignant transformation (Shih et al., 1979b). It is telling in recounting the story how this ts mutant has helped in identifying ligands that bind to p21. This ts mutant p21 is thermal labile by heating cell lysates containing this protein at 42° C for 5 min. This property was used to search for possible ligands that might interact with p21. Indeed, it was found that among many classses of chemicals screened, pre-incubation with GTP or GDP stabilized this ts p21 against thermal inactivation, suggesting interaction of p21 with these guanine nucleotides (Scolnick, Papageorge and Shih, 1979). Purified p21 from NIH3T3 cells transformed by the Harvey virus and p21 isolated from E. coli overproducing this protein demonstrated the specific binding with guanine nucleotides (Shih et al, 1980; Hattori et al, 1985). Furthermore, p21 of Harvey and Kirsten viruses having a threonine mutation of Ala-59 residue of the c-ras p21, displays a GTP-specific autokinase activity - the first indication that p21 has enzymic activity associated with GTP-binding (Shih et al., 1980). A GTPase activity which is very important for p21 function was later discovered by many investigators (Gibbs et al., 1984; Manne et al., 1985; McGrath et al., 1984; Sweet et al., 1984). Recently, Trahey and McCormick (1987) have identified a cytoplasmic protein, GAP, which has the ability to stimulate the p21 GTPase activity of normal ras but not that of some oncogenic mutations.

Our studies by Scatchard analysis of purified p21 indicates a single nucleotide binding site per p21 molecule (Hattori et al., 1985). The dissociation constants (K_d) for GDP and GTP are, respectively, 1.0 and 2.6 nM for c-H-ras p21 and 8.9 and 8.2 nM for v-H-ras p21. Removal of divalent cations by EDTA and addition of ammonium sulphate synergistically enhance the dissociation reaction. The dissociation off-rate is significantly higher in p21 of v-H-ras than that of c-H-ras (Hall and Self, 1986; Hoshino et al., 1987).

MUTATIONS OF THE GUANINE BASE BINDING SITE

Regional homology in GTP-binding ras p21 and elongation factors was first noticed by Karen R. Halliday (1984) and elaborated by many others (Leberman and Egner, 1984; Moeller and Amons, 1985; Dever et al., 1987). Three consensus sequence elements were found in most GTP-binding proteins (Figure 1). In this study, we focused our attention to the structural requirments of these sequence elements in GTP-binding and biological activities of ras p21. We first investigated the consensus sequence element of NKXD (Asn^{116}-Lys-X-Asp^{119} in the p21 sequence). We took the cue from Brian Clark and his colleages' early preliminary results of EF-Tu crystal structure (Rubin et al., 1981) that the homologous Asn-116 of p21 might interact with the guanine base. We made systematic mutations in this region with amino acid substitutions by oligonucleotide-directed mutagenesis of v-H-ras oncogene of the proviral DNA clone, pH-1 (Clanton et al., 1986). Asn-116 was changed into either lysine (clone 116K) or tyrosine (clone 116Y) with drastic differences in amino acid side chains. As controls, additional mutations were made that altered the adjacent Lys-117 to a glutamine residue (clone 117Q) or Cys-118 to arginine (clone 118R) or serine (clone 118S). To evaluate the biological activities of mutant ras, the pH-1 DNA was inserted into a pSVneo plasmid to allow selection of G418-resistant cell clones independent of transformation

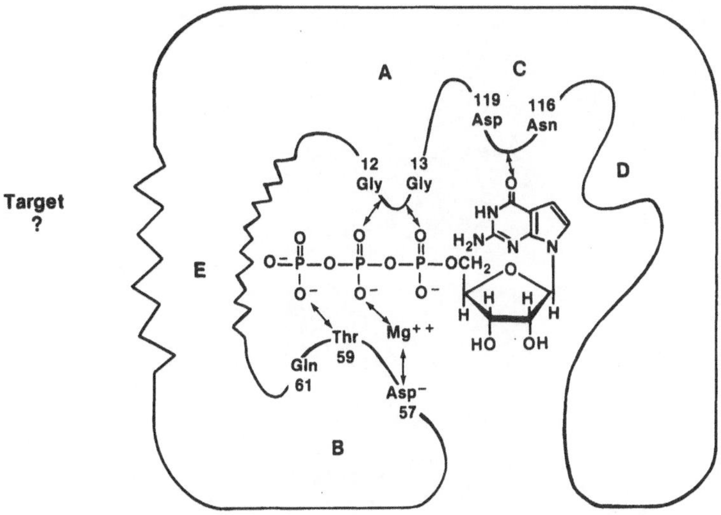

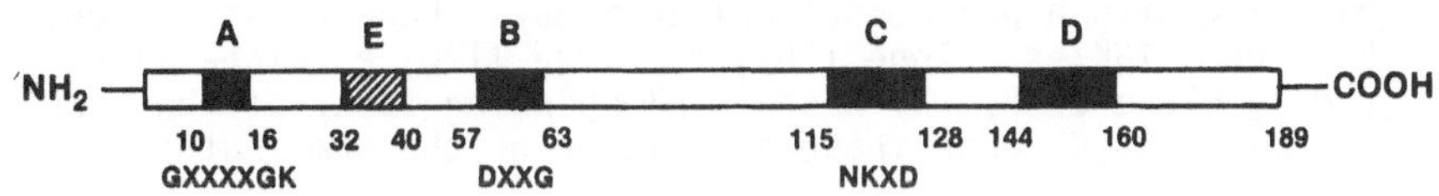

Fig. 1. The GTP-binding domain of _ras_ p21. The three consensus sequence elements (A, B, and C) which characterize G-proteins are indicated in the p21 linear structure. These sequence elements constitute the GTP-binding site. The effector region, E, appears to interact with the p21 target, GAP.

after being transfected into NIH3T3 cells. It should be emphasized here that in the present constructs of proviral DNA clones, the _ras_ gene is under the influence of viral LTR, and normal c-_ras_ is transforming when transfected into NIH3T3 cells. Therefore, transforming activities should be regarded as an assay of biological activities of _ras_ mutants rather then a measurement of oncogenic activation of proto-oncogenes. To study the biochemical properties of p21, mutant _ras_ was inserted into the bacterial expression vector, pJL6, and p21 was isolated from _E. coli_ over-producing this protein (Lautenberger et al., 1983). As shown in Figure 2, mutations of Asn-116 abolish both the GTP-binding and associated autokinase activities of p21, while changes at the adjacent positions at 117 or 118 have little effect. Other investigators have found that mutations at Asp-119 are also critical for GTP-binding (Sigal et al., 1986; Feig et al., 1986). These results are completely consistent with the three dimensional structure later solved by X-ray crystallography for EF-Tu and H-_ras_ p21 (Jurnak, 1985; McCormick et al., 1985; De Vos et al, 1988). Asn-116 and Asp-119, respectively, interact by H-bonding to the keto and amino groups of guanine ring of the bound GDP, and apparently confer the base specificity to the ligand binding. These two mutants of _ras_, 116K and 116Y, which have lost their GTP-binding activity also exhibit reduced transforming ability after transfection into NIH3T3 cells. Intriguingly, other substitutions of the critical Asn-116 and Asp-119 of the c-H-_ras_ proto-oncogene alter its GTP-binding and activate its transforming potential (Walter et al., 1986; Sigal et al., 1986). These results all indicate that interaction of p21 with nucleotides is critical for _ras_ function.

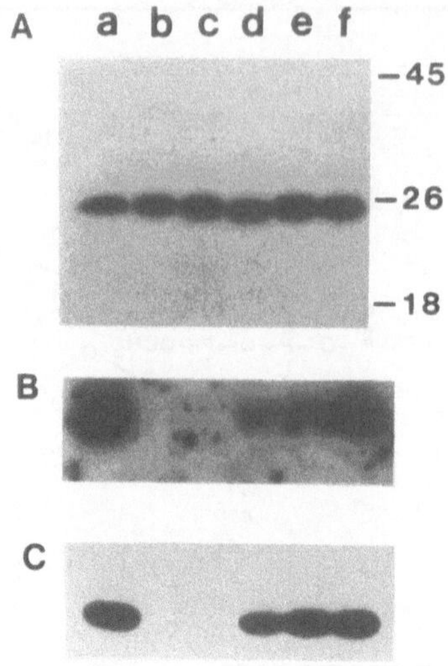

Fig. 2. Mutant p21 overproduced in E. coli. A: Western blotting with a p21 monoclonal antibody. Lane a, wild type; b, clone 116K; c, clone 116Y; d, clone 117Q; e, clone 118R; and f, clone 118S. B: Western GTP-binding assay of the same p21 mutants; C: autokinase activities of the same mutants.

Table 1. Ras p21 mutants: GTP-binding and transformation

Clone	Mutation	GTP-Binding	Autokinase	Transformation
pH-1	wt v-H-ras*	+	+	+
	GXXXXGK Consensus: GDP-phosphoryl binding site			
10V	gly to val	-	-	+
12G(T)	arg to gly	+	+	+
13V	gly to val	-	-	-
15V	gly to val	-	-	-
22K	gln to lys	+	+	+
51S	cys to ser	+	+	+
	DXXG Consensus: autophosphorylation site			
59A(R)	thr to ala	+	-	+
59S(R)	thr to ser	+	+	+
	NKXD Consensus: guanine base binding site			
116K	asn to lys	-	-	-
116Y	asn to tyr	-	-	-
117Q	lys to gln	+	+	+
118R	cys to arg	+	+	+
118S	cys to ser	+	+	+

*v-H-ras: 12R/59T (c-H-ras: 12G/59A)

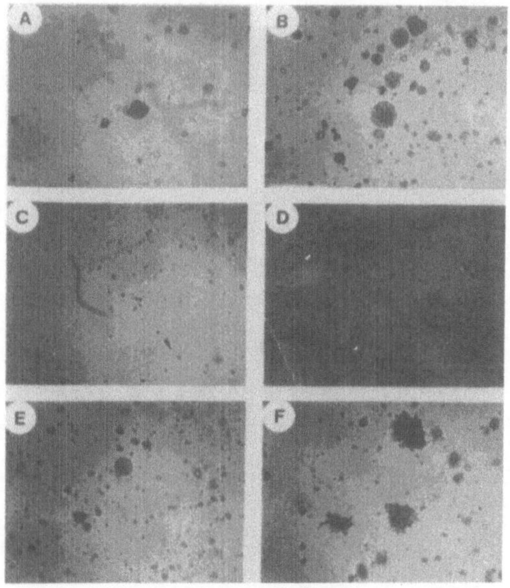

Fig. 3. Colony-forming activities in soft agar of NIH3T3 cells transfected with the following ras mutants: A, pSVneo/ras; B, pSVneo/ras 10V; C, pSVneo/ras 13V; D, pSVneo/ras 15V; E, pSVneo/ras 33H; and F, pSVneo/ras 51S.

MUTATIONS OF THE GDP-PHOSPHORYL BINDING SITE

This site involves the N-terminal glycine-rich consensus sequence for nucleotide binding, GXXXXGK (Gly^{10}-X-Gly^{12}-Gly^{13}-X-Gly^{15}-Lys^{16} in the p21 sequence of the ras proto-oncogenes). Gly-12 is one of the hot spots of mutations that activate ras proto-oncogenes in natural and experimental tumors, and Gly-13 is also the site of oncogenic mutations although found less frequently (reviewed in Barbacid, 1987). Two glycine residues at positions 10 and 15 are invariant in most GTP-binding proteins (Dever et al., 1987). In the three dimensional structure of p21, this glycine-rich sequence is seen as a peptide loop connecting a short α helix to the N-terminal β strand; it apparently straddles the phosphodiester bond between the β and γ phosphates of GTP; and it probably functions as the catalytic site for the GTPase activity (de Vos et al., 1988). We have systematically mutated these glycine residues in the context of the viral H-ras p21 (with Arg-12 and Thr-59 differing from the normal ras proto-oncogenes) (Clanton et al, 1987). When the glycine residue at position 10, 13 or 15 was substituted with a valine residue, the mutant p21 lost most of its GTP-binding and autokinase activities (Table 1). These results suggest that addition of a bulky side chain to any of these glycine residues profoundly distorts the conformation of this peptide loop and affects its nucleotide interactions. Consequently, valine mutations of Gly-13 (clone 13V) and Gly-15 (clone 15V) greatly diminish the transforming ability of p21 (Figure 3C, 3D). Most of the other mutations which do not affect GTP-binding are all transforming. Unexpectedly, ras p21 with a valine mutation of the Gly-10 residue (clone 10V) was found to be highly transforming. These cells grew in soft agar (Fig. 3B), and rapidly formed tumors in nude mice. To eliminate the possibility that these tumors might have arisen from cells unrelated to 10V transformation, explants were taken from tumors induced by the 10V mutant and the wild type viral

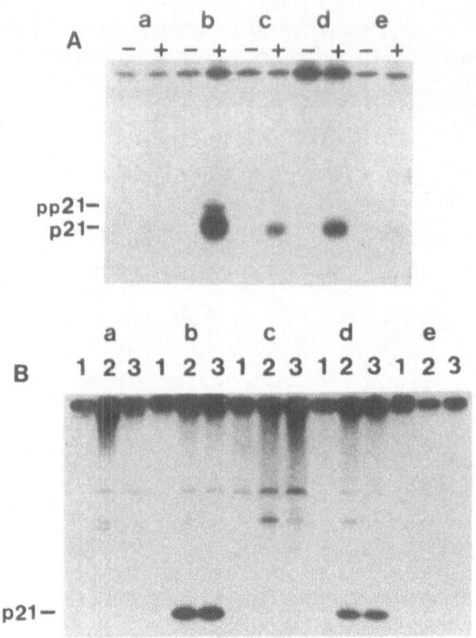

Fig. 4. Lack of autokinase activity of 10V p21 in tumor cells. (A) Cells were labeled with ^{35}S-methionine and p21 was immunoprecipitated with monoclonal antibody against p21 (+) or with IgG (-). a, pSVneo control; b, v-*ras*; c, 10V mass cell culture; d, 10V tumor cell line; and e, 10V cell clone.
(B) Cells were labeled with ^{32}P-orthophosphate and p21 was immunoprecipitated with normal IgG (lanes 1), YA6-172 antibody (lanes 2) and Y13-259 antibody (lanes 3). a, pSVneo control; b, v-*ras* tumor cell line; c, 10V tumor cell line; d, v-*ras* mass cell culture; and e, 10V mass cell culture.

H-*ras*, and were grown as cell lines. The p21 expressed in these cell lines is shown in Fig. 4. The p21 of wild type H-*ras* is characteristically phosphorylated by its autokinase activity (Fig. 4A, lane b and Fig. 4B, lanes b and d), while mutant p21 of clone 10V is seen as a single band in ^{35}S-methionine label (Fig. 4A, lanes c and d), and no phosphorylated form detectable in the ^{32}P-labeling (Fig. 4B, lanes c and e). Thus, 10V tumors are derived from cells containing a *ras* mutation apparently still defective in its autokinase activity and presumably also in its GTP-binding activity. These findings suggest that the glycine-rich consensus sequence is important in controlling p21 activities and that certain mutations may confer to p21 its active conformation without participation of ligand binding. We entertain the idea that substitution with valine of the invariant Gly-10 residue of the viral *ras* p21 may distort the conformation of this peptide loop in such a way that it mimicks GTP binding and locks the p21 molecule in its active conformation. This mutant may be useful in unraveling the mechanism of conformational switching of the p21 molecule.

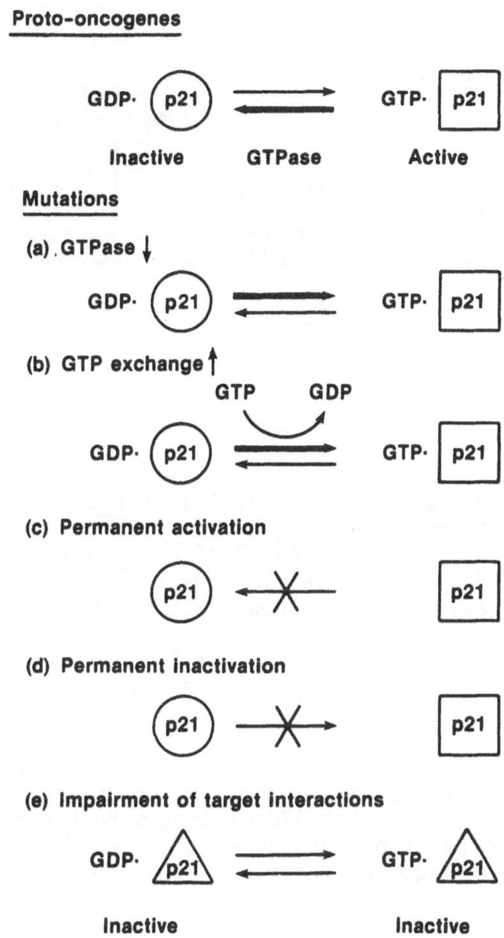

Fig. 5. Guanine nucleotide binding as a molecular switch for p21.

GUANINE NUCLEOTIDE BINDING AS A MOLECULAR SWITCH FOR p21

Biological properties of ras mutants, either occurring in nature or generated in the laboratory, are classified as shown in Fig. 5. The approach we used to study the structure and function of ras p21 is confined to making mutational changes with single amino acid alterations, bearing in mind that the highly conserved sequences of ras genes have structural significance. These mutational studies support the model that p21 exists in two different forms controlled by its binding with guanine nucleotides (Fig. 5). Recent experimental evidence demonstrates that the GTP-bound form is the active species in producing ras biological effects (Gibbs et al., 1987; Trahey and McCormick, 1987; Satoh et al, 1987). Oncogenic mutations found in most tumors occur mostly at the GTP-binding site at positions 12, 13, 59 or 61. These mutations all favor the p21 in its GTP-bound conformation due to a decrease in GTPase activity or possibly to an enhanced GTP/GDP exchange reaction (Fig. 5a, b). The inactivating mutations such as 116K and 116Y that lose their GTP/GDP binding activities may represent the class of mutations which result in permanently inactivated p21 (Fig. 5d). Interestingly, the 10V mutation described here may represent a new class of p21 molecules that are permanently activated without the

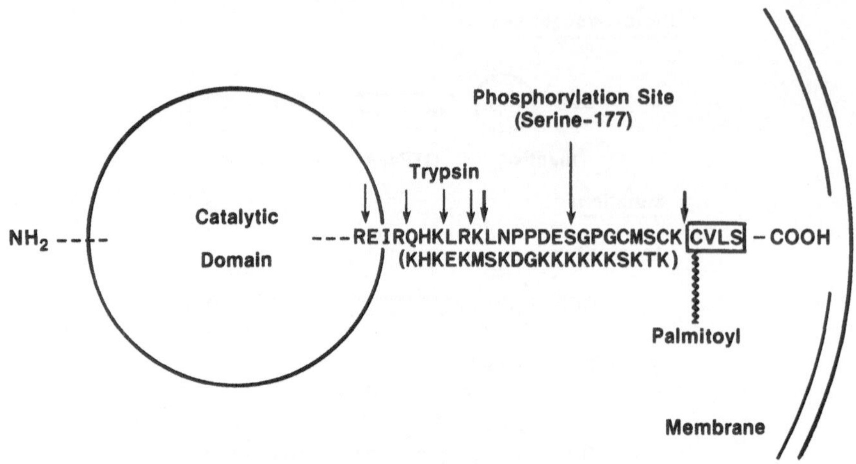

Fig. 6. Novel phosphorylation site(s) of ras p21 common for both protein kinases A and C. The amino acid sequence around the phosphorylation site at serine-177 of H-ras p21 region is shown. Small arrows indicate trypsin cleavage sites. The homologous sequence of K-ras(4B) p21 is shown in parenthesis. This hypervariable region is flanked by a membrane binding site involving the palmitoylated cysteine-186 at the C-terminal region and the large globular catalytic domain comprising the 164 amino acid residues of the p21 N-terminal region.

need of the bound GTP to remain in this conformation (Fig. 5c). Finally, the class of mutants with intact GTP binding site but can no longer interact with target proteins such as GAP, are described by Lowy, Sigal and their colleagues (Fig. 5e) (Adari et al., 1988; Cales et al., 1988; Sigal, 1988).

MUTATIONS OF THE NOVEL PHOSPHORYLATION SITE(S) FOR PROTEIN KINASES

We have recently found a novel phosphorylation of p21 in cells expressing high levels of this product of c-H-ras and c-K-ras genes (Saikumar et al., 1988). Phorbol ester, a protein kinase C activator, and permeable c-AMP derivatives, which activate protein kinase A, stimulated phosphorylation of K-ras p21 3 to 5 fold in 416B cells, an early mouse myeloid cell line expressing high levels of c-K-ras p21. By tryptic peptide mapping, it was found that both protein kinase C and protein kinase A phosphorylated the K-ras p21 at the same site *in vivo* and *in vitro*. Similar experiments with H-ras p21 also suggested a common phosphorylation site for both protein kinases. However, tryptic peptide mapping revealed that the phosphorylation site of H-ras p21 was distinct from that of K-ras p21. The construction of a mutant allowed the identification of serine-177 as the phosphorylation site of H-ras p21 for protein kinases. As depicted in Fig. 6, this novel phosphorylation site lies in the hypervariable region, which links the catalytic globular domain of p21 to the membrane anchoring site at the C-terminus. This flexible region of p21 tail may serve an important but not yet identified function in transmitting signals from interaction of p21 molecules with the plasma membrane. Phosphorylation of this strategically important region of p21 by protein kinases may modulate signal transmission from membrane interactions to the catalytic domain within p21 molecules. Phosphorylation of identical sites of target proteins by multiple protein kinases suggests a potential mechanism of cross-talk between different signal transduction pathways mediated by protein kinases A and C.

SUMMARY

In this study, we have explored the structure-function relationship of H-ras p21 by using site-directed mutagenesis to alter single amino acid residues. Systematic mutations were constructed in the consensus sequence elements of the GTP-binding proteins. Mutations of the guanine base binding site, NKXD, indicate that asn-116 of p21 is critical for GTP-binding and profoundly affects biological properties of ras p21. Mutations of the invariant glycine residues of the N-terminal GDP-phosphoryl binding site, GXXXXGK, reveal the switching function of this consensus sequence element. Replacements of these glycine residues with valine abolish GTP-binding. Interestingly, although the valine-10 mutant has lost its ability to interact with GTP, it is highly transforming. This mutant may represent a new class of mutations in which p21 is permanently locked in its active GTP-bound conformation. Mutations of serine-177 at the C-terminal hypervariable region abolish p21 phosphorylation by both protein kinases A and C, suggesting the role of this phosphorylation in signal transmission from membrane interactions to the catalytic domain of p21 molecules. These studies are remarkably consistent with the three dimensional structure recently determined for H-ras p21 by X-ray crystallography.

ACKNOWLEDGMENTS. We thank Dr. Takis Papas for his support of this study, Drs. James Lautenberger and Takis Papas for the generous use of the pJL6 expression vector, Dr. Donald Blair for help in DNA transfection and Dr. Donald Court for consultation on bacteriology.

REFERENCES

Adari, H., Lowy, D. R., Willumsen, B. M., Der, C. J. and McCormick, F., 1988, Guanosine triphosphatase activating protein (GAP) interacts with the p21 ras effector binding domain, Science, 240:518-521.

Barbacid, M., 1987, ras genes, Annu. Rev. Biochem., 56:779-827.

Bourne, H. R. and Sullivan, K. S., 1986, Mammalian G proteins: model for ras proteins in transmembrane signalling? Cancer Surveys, 5:257-274.

Cales, C., Hancock, J. F., Marshall, C. J. and Hall, A., 1988, The cytoplasmic protein GAP is implicated as the target for regulation by the ras gene product, Nature, 332:548-551.

Clanton, D. J., Hattori, S. and Shih, T. Y., 1986, Mutations of the ras gene product p21 that abolish guanine nucleotide binding, Proc. Nat. Acad. Sci., USA, 83:5076-5080.

Clanton, D. J., Lu, Y., Blair, D. G. and Shih, T. Y., 1987, Structural significance of the GTP-binding domain of ras p21 studied by site-directed mutagenesis, Mol. Cell. Biol., 7:3092-3097.

Dever, T. E., Glynias, M. J. and Merrick, W. C., 1987, GTP-binding domain: three consensus sequence elements with distinct spacing, Proc. Nat. Acad. Sci., USA, 84:1814-1818.

De Vos, A. M., Tong, L., Milburn, M. V., Matias, P. M., Jancarik, J., Noguchi, S., Nishimura, S., Miura, K., Ohtsuka, E. and Kim, S. H., 1988, Three-dimensional structure of an oncogenic protein: catalytic domain of human c-H-ras p21, Science, 239:888-893.

Feig, L. A., Pan, B. T., Roberts, T. M. and Cooper, G. M., 1986, Isolation of ras GTP-binding mutants using an in situ colony-binding assay, Proc. Nat. Acad. Sci., USA, 83:4607-4611.

Gibbs, J. B., Schaber, M. D., Marshall, M. S., Scolnick, E. M. and Sigal, I.

S., 1987, Identification of guanine nucleotides bound to ras-encoded proteins in growing yeast cells, J. Biol. Chem., 262:10426-10429.
Gibbs, J. B., Sigal, I. S., Poe, M. and Scolnick, E. M., 1984, Intrinsic GTPase activity distinguishes normal and oncogenic ras p21 molecules, Proc. Nat. Acad. Sci., USA, 81:5704-5708.
Hall, A. and Self, A. J., 1986, The effect of Mg^{2+} on the guanine nucleotide exchange rate of $p21^{N\text{-}ras}$, J. Biol. Chem., 261:10963-10965.
Halliday, K. R., 1984, Regional homology in GTP-binding proto-oncogene products and elongation factors, J. Cyclic Nucleotide and Protein Phosphorylation Research, 9:435-448.
Hattori, S., Ulsh, L. S., Halliday, K. and Shih, T. Y., 1985, Biochemical properties of a highly purified v-ras^H p21 protein overproduced in E. coli and inhibition of its activities by a monoclonal antibody, Mol. Cell. Biol., 5:1449-1455.
Hattori, S., Yamashita, T., Copeland, T. D., Oroszlan, S. and Shih, T. Y., 1986, Reactivity of a sulfhydryl group of the ras oncogene product p21 modulated by GTP binding, J. Biol. Chem., 261:14582-14586.
Hoshino, M., Clanton, D. J., Shih, T. Y., Kawakita, M. and Hattori, S., 1987, Interaction of ras oncogene product p21 with guanine nucleotides, J. Biochem. (Tokyo), 102:503-511.
Jurnak, F., 1985, Structure of the GDP domain of EF-Tu and location of the amino acids homologous to ras oncogene proteins, Science, 230: 32-36.
Kaziro, Y., 1978, The role of guanosine 5'-triphosphate in polypeptide chain elongation, Biochim. Biophys. Acta, 505:95-127.
Lautenberger, J. A., Ulsh, L., Shih, T. Y. and Papas, T. S., 1983, High level expression in E. coli of enzymatically active Harvey murine sarcoma virus p21 ras protein, Science, 221:858-860.
Leberman, R. and Egner, U., 1984, Homologies in the primary structure of GTP-binding proteins: the nucleotide-binding site of EF-Tu and p21, The EMBO J., 3:339-341.
Manne, V., Bekesi, E. and Kung, H. F., 1985, Ha-ras proteins exhibit GTPase activity: point mutations that activate Ha-ras products result in decreased GTPase activity, Proc. Nat. Acad. Sci., USA, 82:376-380.
McCormick, F., Clark, B. F., La Cour, T. F. M., Kjeldgaard, M., Norskov-Lauritsen, L. and Nyborg, J., 1985, A model for the tertiary structure of p21, the product of the ras oncogene, Science, 230: 78-82.
McGrath, J. P., Capon, D. V., Goeddel, D. V. and Levinson, A. D., 1984, Comparative biochemical properties of normal and activated human ras p21 proteins, Nature, 310:644-649.
Moeller, W. and Amons, R., 1985, Phosphate-binding sequences in nucleotide-binding proteins, FEBS Letters, 186:1-7.
Rubin, J. R., Morikawa, K., Nyborg, J., La Cour, T. F. M., Clark, B. F. C. and Miller, D. L., 1981, Structural features of the GDP binding site of elongation factor Tu from E. coli as determined by X-ray diffraction, FEBS Letters, 129:177-179.
Saikumar, P., Ulsh, L. S., Clanton, D. J., Huang, K. P. and Shih, T. Y., 1988, Novel phosphorylation of c-ras p21 by protein kinases, Oncogene Research, in press.
Satoh, T., Nakamura, S. and Kaziro, Y., 1987, Induction of neurite formation in PC12 cells by microinjection of proto-oncogenic Ha-ras protein preincubated with guanosine-5'-O-(3-thiotriphosphate), Mol. Cell. Biol., 7:4553-4556.
Scolnick, E. M., Papageorge, A. and Shih, T. Y., 1979, Guanine nucleotide-binding activity as an assay for src protein of rat-derived murine sarcoma viruses, Proc. Nat. Acad. Sci., USA, 76:5355-5359.
Shih, T. Y., Papageorge, A. G., Stokes, P. E., Weeks, M. O. and Scolnick, E. M., 1980, Guanine nucleotide-binding and autophosphorylating activities associated with the $p21^{src}$ protein of Harvey murine sarcoma virus, Nature, 287:686-691.

Shih, T. Y., Weeks, M. O., Young, H. A. and Scolnick, E. M., 1979a, Identification of a sarcoma virus coded phosphoprotein in nonproducer cells transformed by Kirsten or Harvey murine sarcoma virus. Virology, 96:64-79.
Shih, T. Y., Weeks, M. O., Young, H. A. and Scolnick, E. M., 1979b, p21 of Kirsten murine sarcoma virus is thermolabile in a viral mutant temperature-sensitive for the maintenance of transformation, J. Virol., 31:546-556.
Sigal, I. S., 1988, A structure and some function, Nature, 332:485-486.
Sigal, I. S., Gibbs, J. B., D'Alonzo, J. S., Temeles, G. L., Wolanski, B. S., Socher, S. and Scolnick, E. S., 1986, Mutant ras-encoded proteins with altered nucleotide binding exert dominant biological effects, Proc. Nat. Acad. Sci., USA, 83:952-956.
Sweet, R. W., Yokoyama, S., Kamata, T., Feramisco, J. R., Rosenberg, M. and Gross, M., 1984, The product of ras is a GTPase and T24 oncogenic mutant is deficient in this activity, Nature, 311:273-275.
Trahey, M. and McCormick, F., 1987, A cytoplasmic protein stimulates normal N-ras p21 GTPase, but does not affect oncogenic mutants, Science, 238:542-545.
Walter, M., Clark, S. G. and Levinson, A. D., 1986, The oncogenic activation of human $p21^{ras}$ by a novel mechanism, Science, 233:649-652.

THE CYCLIC AMP PRODUCING PATHWAY IN *SACCHAROMYCES CEREVISIAE* INVOLVES *CDC25* AND *RAS* GENES PRODUCTS

Michel Jacquet , Jacques Camonis, Emmanuelle Boy-Marcotte,
Faten Damak and Hervé Garreau

Groupe Information Génétique et Développement. Université Paris-Sud
Bât. 400 - 91405 Orsay Cédex, France

INTRODUCTION

In mammalian cells, the adenylate cyclase is regulated by extracellular signaling molecules, hormones and neurotransmitters. This regulation involves specific transmembrane receptors and transducers which are heterotrimer G-proteins. In response to the liganded receptor the G-protein is activated by dissociation of the GTP-bound a subunit which can then activate the adenylate cyclase. Although this complex system which also involves negative regulatory circuit, has been extensively studied, the details of its functioning have not yet been completely elucidated.

The discovery of mutants of the adenylate cyclase gene by Matsumoto, Uno and Ishikawa (1982) in yeast, *Saccharomyces cerevisiae*, has offered a new opportunity to analyse molecular devices involved in cAMP production in eucaryotic cells. As discussed below, the yeast cAMP producing pathway differs from the homologous system in mammalian cells but it seems to present a similar cascade of sequential activation which also involves three elements: the adenylate cyclase itself, the products of the RAS genes which are supposed to act as transducers, and the product of *CDC25* which could activate *RAS* proteins in response to nutritional signals.

The adenylate cyclase is the product of the *CYR1* gene (Matsumoto *et al.*, 1982). This gene is allelic to the *CDC35* gene (Boutelet *et al.*, 1985) which has been shown to be essential for the cell division cycle (Pringle and Hartwell 1981). It has been cloned in several laboratories (Mason *et al.*, 1984, Casperson *et al.*, 1985 ; Kataoka *et al.*, 1985b). It is encoded by a gene which possesses an open reading frame of 2026 amino acids (Kataoka *et al.*, 1985b). The C-terminal end of the ORF appears to be sufficient to perform cAMP synthesis (Mason *et al.*, 1986 ; Kataoka *et al.*, 1985b). This enzyme is membrane bound as in mammals and is activated by GTP binding proteins as demonstrated by the requirement for GTP in Mg^{++} stimulated assays (Casperson *et al.*, 1983).

Two genes homologous to the protooncogene ras (Barbacid, 1987) have been identified as being necessary for cAMP production: *RAS1* and *RAS2* (DeFoe-Jones *et al.*, 1983, Powers *et al.*, 1984). These genes are functionally interchangeable, gene disruption of the two is required to block cAMP production and cell division (Tatchell *et al.*, 1984; Kataoka *et al.*, 1984). An oncogene-like allele of the yeast *RAS2* gene has been obtained by the replacement of a glycine by a valine at position 19 (equivalent to position 12 in the mammalian p21) (Kataoka *et al.*, 1985a ; Sigal *et al.*, 1986) . The mutated $RAS2^{val19}$ gene is able to transform NIH 3T3 mouse fibroblasts (Kataoka *et al.*, 1984). When introduced into yeast, it leads to a phenotype characteristic of a permanently activated cAMP dependent protein kinase : loss of viability in nitrogen depleted medium, inability to sporulate, increase thermosensitivity, lack of glycogen and trehalose accumulation. These effects are dominant.

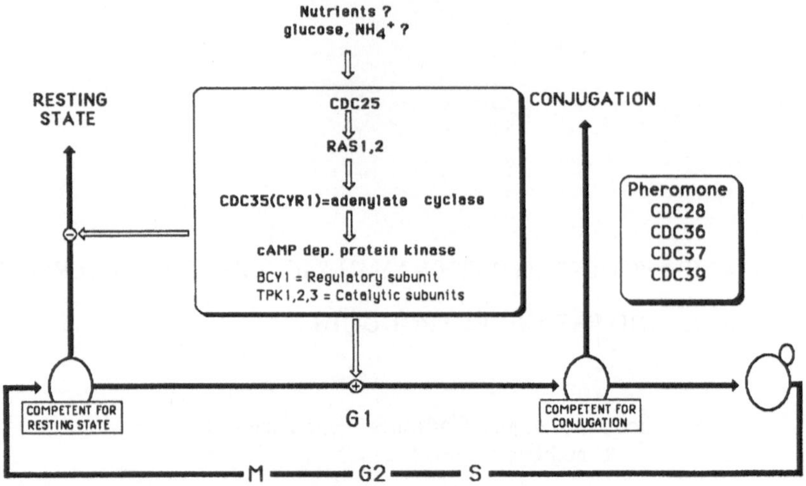

FIGURE 1. Schematic diagram of the implication of cAMP in the control of the cell division cycle in *Saccharomyces cerevisiae*.

The third element necessary for cAMP production is the *CDC25* gene. Its analysis is presented below.

From analysis of the phenotype of the mutants of the adenylate cyclase gene it can be deduced, that cAMP is required for the onset of a new cell division cycle (figure 1). Mutations in the adenylate cyclase gene (*CDC35*) or in *RAS1* and *RAS2* or in *CDC25* arrest the cell in the unbudded state prior the S phase and before the acquisition of the competent state for conjugation in haploid cells. Then the cells enter the resting state as nutritionally deprived cells. In contrast, when excess cAMP is added to responsive cells or if cell is deficient in the regulatory subunit of the cAMP dependent protein kinase (*bcy1* mutants) then the cell is unable to enter the resting state even in absence of essential nutrient such as nitrogen (Boy-Marcotte *et al.*, 1987). Therefore, in *Saccharomyces cerevisiae*, cAMP appears to be an internal signal for the switch between resting state and cell division in addition to its effects on metabolism control such as glycogen or trehalose accumulation. The step of the cell division cycle controlled by the cAMP pathway precedes the step which requires *CDC28* and is controlled by the pheromones (Pringle and Hartwell 1981; Jacquet and Camonis 1985). cAMP is most likely modulated by the availability of nutrients but the nature of any modulator is unknown. The best characterised variation in cAMP level is obtained by adding glucose to cells grown without glucose (Thevelein 1984).

THE *CDC25* GENE PRODUCT ACTS UPSTREAM OF THE RAS PROTEINS

We were first interested in the study of *CDC25*, because mutations of this gene give the same phenotype as mutations of the adenylate cyclase gene. In collaboration with F. Hilger we have shown that, as with the *cdc35* mutation, the growth defect of *cdc25* mutants at restrictive temperature can be overcome by adding cAMP to responsive cells. Then, the cAMP level in *cdc25-5* mutants fails dramatically in the few minutes following the transfer of the mutant cells to restrictive temperature (Camonis *et al.*, 1986). The epistatic relationship between *CDC25*, *RAS2* and *CDC35* has been examined as summarized in table I. *cdc25* and *cdc35* mutants fail to grow at 36°C and when diploids, sporulate in presence of ammonium whereas the wild type strain does not. The addition of cAMP to cells containing an allele which allows the response to exogenously added cAMP, *rca1* (Boy-Marcotte *et al.*, 1987), restores the growth at 36°C but does not prevent hypersporulation suggesting that the two phenomenons are regulated separately. Similarly, the addition of a truncated adenylate cyclase gene (Masson *et al.*, 1984) which allows unregulated production of cAMP to *cdc25* mutant cell restores growth at 36°C The *RAS2*^{val19} allele which is known to permanently activate the adenylate cyclase, prevents

TABLE I. Epistatic relationships between *cdc25*, *cdc35* and *RAS2*^{val19} mutations

Relevant genotype	cell division at 36°C	Glycogen	Sporulation	Hypersporulation
cdc25-5	-	+	+	+
cdc35-10	-	+	+	+
wild type	+	+	+	-
rca1 + cAMP	+	-	+	-
RAS2^{val19}	+	-	-	-
cdc25-5, rca1 + cAMP	+	-	+	+
cdc25-5, pCDC35	+	+	nd	nd
cdc25-5, RAS2^{val19}	+	-	-	-
cdc25-5, cdc35-10	-	+	+	+
cdc35-10, rca1+cAMP	+	-	+	+
cdc35-10, pCDC25	-	+	nd	nd
cdc35-10, RAS2^{val19}	-	+/-	+	-

Cell division was followed on YEPD plates at 36°C and 24°C. Glycogen accumulation was tested by iodine staining, Sporulation was followed microscopically five days after plating homozygous diploids on sporulation medium. Hypersporulation means sporulation in the presence of 5mM ammonium sulfate.

the accumulation of glycogen and inhibits sporulation. *RAS2*^{val19} is epistatic to *cdc25-5*: it allows the double mutant to grow at 36°C and prevents glycogen accumulation and sporulation. In contrast, the *cdc35* mutation is epistatic to the *RAS2*^{val19} mutation. Similar results have been obtained in other groups (Robinson *et al.*, 1987, Broek *et al.*, 1987). These epistatic relationships suggest that the functioning of the adenylate cyclase is positively regulated by the product of the *RAS2* gene which is itself controlled by the product of the *cdc25* gene.

We have cloned the *CDC25* gene by complementation of *cdc25* mutants (Camonis *et al* 1986). It is transcribed as a 5200 nt mRNA and contains an ORF of 1589 codons. Similar results were obtained by other groups (Daniel and Simchen, 1986; Martegani *et al.*, 1986; Robinson *et al.*, 1987; Broek *et al.*, 1987). This sequence contains several putative sites of phosphorylation by cAMP dependent protein kinase as well as putative sites for glycosylation. The structure of the protein and its cellular localization are still unknown. The 3' part of the gene (figure 4), is capable, when inserted in a plasmid, to complement the *cdc25* mutations. The same minimal fragment is able to suppress different *cdc25* mutations as well as a *CDC25* disruption (Robinson *et al.*, 1987, Broek *et al.*, 1987). Thus it seems that the C-terminal part of the ORF is capable by itself to ensure the function of *CDC25* for growth.

RAS2^{ile152} A SPONTANEOUS SUPPRESSOR FOR *CDC25*

Spontaneous revertants of the growth arrest by *cdc25-5* mutants at restrictive temperature frequently appear. Several of these suppressors have been genetically analysed in our laboratory. One of them, named *25SU3* which presents a dominant phenotype was further characterized (Camonis and Jacquet 1988). This extragenic suppressor of *cdc25-5* mutation does not suppress the *cdc35* mutation and thus corresponds to a mutation in a element probably involved in the activation of the adenylate cyclase between the *CDC25* product and the adenylate cyclase itself. Genetic linkage analysis of this gene with other genes of the system reveals a strong association with the *RAS2* gene. No recombinants of *25SU3* and *RAS2::URA3* were found among 118 tetrads analysed. This result suggested that suppression was due to a mutation in the *RAS2* gene. This hypothesis was confirmed by cloning and sequencing the mutated gene. It was cloned on the basis of its ability to suppress the *cdc25* mutation. The complementing piece of the cloned DNA shows the same restriction map as the *RAS2* gene with an additional EcoRV site. The nucleotide sequence of a DNA fragment encompassing the *RAS2* gene differs from the sequence of the wild type gene DNA by one point mutation at position 455 of the coding region. The transition C/G to A/T leads to the creation of a new EcoRV site and to the change of codon ACA to ATA replacing a threonine by a isoleucine at position 152 of the ORF.

The experiment presented in figure 2 demonstrates that this mutation is sufficient for the suppression of the *cdc25-5* growth arrest. The *cdc25-5* mutant does not grow at 36°C . When the cell is transformed with a multicopy plasmid containing the mutated $RAS2^{ile152}$ or when the $RAS2^{ile152}$ has replaced the normal gene by gene replacement then growth is restored at 36°C. As a control the normal *RAS2* gene, even on a multicopy plasmid, is unable to suppress the *cdc25-5* mutation. Positive control for suppression is also shown with the gene encoding the catalytic subunit of the cAMP dependant protein kinase on a multicopy plasmid. To test whether or not a second mutation in the 5' region of the gene could be responsible for suppression, chimeric plasmids were constructed with part of the normal gene and part of the mutated gene; these plasmids were inserted at the *RAS2* locus using integrative vectors. As shown in figure 2 suppression is obtained only with the plasmid containing the mutation in the ORF.

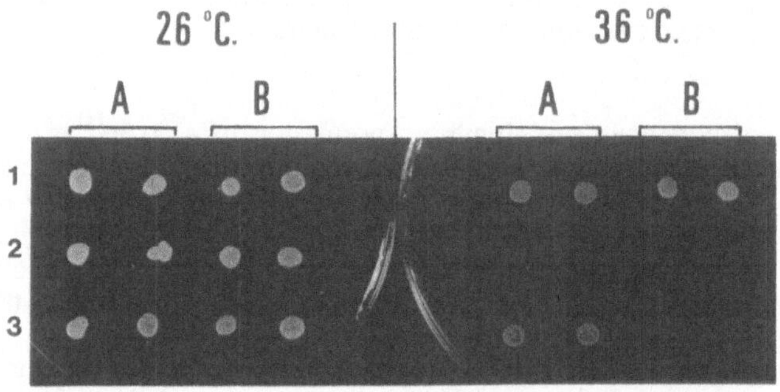

FIGURE 2. Suppression of *cdc25-5* by $RAS2^{ile152}$

Different clones containing the *cdc25-5* allele (Camonis and Jacquet 1988) were spotted on YEPD plates at 26°C and 36°C respectively as indicated.: A1 cells containing the integrated chimeric *RAS2* gene composed of the 5' half of the *RAS2* gene and the 3' half of the $RAS2^{ile152}$(with the mutation), A2- cells containing the integrated chimeric plasmid composed of the 5' half of the mutated ras2 gene (without the mutation) and the 3' half of the wild type *RAS2* gene, A3- transformed cell with a plasmid containing the gene encoding the catalytic subunit of the cAMP dependent protein kinase *SRA3*, B1- cells containing the integrated complete $RAS2^{ile152}$ gene, B2- cell containing the integrated complete wild type gene, B3 cell transformed with a control plasmid: YCP50.

Position 152 of the yeast *RAS2* gene corresponds to position 144 of the ras genes product: p21. This region of p21 is highly conserved among ras proteins. This part of the sequence appears to be involved in the binding of the guanosine nucleotide. Mutations in this region have not been involved in oncogenesis. However, the same exchange of a threonine by an isoleucine at position 144 has been described in a mutant of the Ha-ras gene selected on its reduced capacity to bind guanosine nucleotides (Feig *et al.*, 1986).

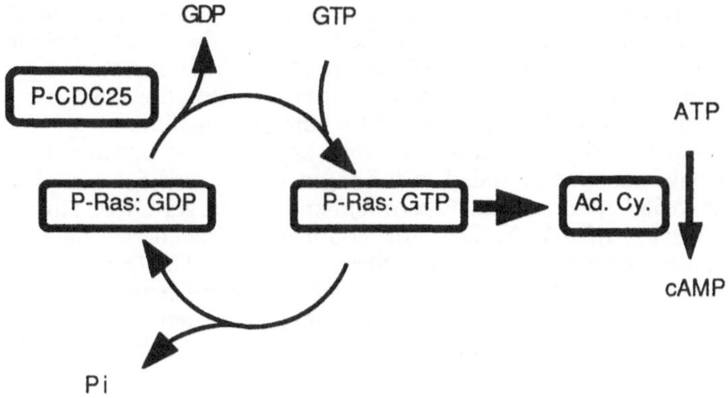

FIGURE 3 . Schematic representation of the recycling of ras protein

A MODEL FOR RAS ACTIVATION

A model for the activation of the ras protein leading to stimulation of the adenylate cyclase is presented in figure 3. This model gives the *CDC25* gene product a role in the recycling of the GTP bound form of the ras protein As discussed below it is supported by the existence of the *RAS2*^{ile152} gene which suppresses *cdc25* mutations.

If the *RAS2*^{ile152} mutation abolishes the cell's need of a functional *CDC25* gene product, then it can be supposed that the modified ras protein is capable of being activated by itself. The same reasoning applies to the *RAS*^{val19} mutation which is also epistatic to *cdc25*. In this latter case, permanent activation of the ras protein is explained by the block of the GTPase activity and then the accumulation of the GTP-bound form of the ras protein within the cell. In the former case, if we assume that the modification of the position 152 confers a reduced affinity for guanine nucleotides as described for the Ha-ras^{ile144} gene, then the increase amount of the activated ras protein can be accounted for by an increase in the exchange rate between GTP and GDP. Indeed, the intracellular concentration of guanine nucleotides is in the range of µM and the K_D of ras protein for these nucleotides is in the nM range, therefore it is expected that most of the ras protein are in a bound form within the normal cell. The cellular excess of GTP over GDP favours the GTP bound form while the intrinsic GTPase activity of the ras protein leads to a higher proportion of the GDP bound form. Since the dissociation rate is very low in the absence of any GDP exchange factor, most of the ras protein should remain bound to GDP. Activation, or exchange of GDP by GTP can then be achieved by an increased rate of dissociation (and thus a lower affinity). This is expected to occur as a consequence of the *RAS2*^{ile152} mutation. In normal cell a specific exchange factor should be required. The product of *CDC25* could be (or closely controls) this factor allowing the recycling of the activated ras protein in response to some other signal.

SCD25 A GENE PARTIALLY HOMOLOGOUS TO *CDC25*

The identification of elements involved in a multicomponent system can be carried out by analysis of extragenic suppressors or by isolation of normal genes cloned by their capacity to suppress defective mutations. The *cdc25* mutation which prevents the functioning of the cAMP cascade at the more upstream level can be bypassed by transfer into the cell of normal genes such as the genes encoding the catalytic subunit of the cAMP dependant protein kinase (Toda *et al.*, 1987; Lisszieiwiecz *et al.*, 1987)). It can also be suppressed by fragmented genes such as the catalytic part of the adenylate cyclase which has lost its regulatory part (table I).

In the analysis of *cdc25-5* transformed cells with DNA inserted on low copy number plasmid (YRp7), we have isolated a piece of DNA containing the C-terminal of a new gene.

Yeast genomic DNA was isolated from a strain containing the two thermosensitive mutations *cdc25-5* and *cdc35-10*. This avoids the cloning of the wild type allele of these genes in a *cdc25-5* mutant strain (OL86). A DNA fragment of 3,5 Kb inserted in a complementing plasmid was isolated from the transformed yeast strain. Deletion analysis of this plasmid reveals that only part of this DNA was required for complementation.

The DNA sequence of 2940 bp of this DNA segment was performed. Two large ORF were found, but only one was included in the segment required for complementation. This ORF corresponds to 583 codons of the C-terminal end of a larger ORF. This ORF was named *SCD25* for suppression of *cdc25*. The stop codon is followed by putative transcription termination consensus sequences (Zarret and Sherman, 1986). The DNA region corresponding to the N-terminal part of the ORF was cloned as an overlapping segment inserted in a plasmid. It was detected by cross hybridization with the previously isolated DNA fragment. Sequencing of this new DNA fragment was performed and reveals that the complete ORF contains 1251 codons.

The deduced amino acid sequence from this ORF shares large regions of homology with the *CDC25* ORF. As illustrated (figure 4) the homology extends on the complete sequence but is more pronounced in the C-terminal part. The C-terminal part ,in both cases, is sufficient for complemention of the *cdc25* mutation. Deletion analysis of the complementing region of the *CDC25* gene has shown that the minimal complementing fragment is included between the 355 and the 577 C-terminal amino acids. The same analysis performed with *SCD25* reveals that the last C-terminal 377 amino acids are sufficient to bring about complementation of the *cdc25* mutation. These results strongly suggest that the two homologous DNA fragments encode polypeptides capable of assuming the same biochemical function. This function is sufficient to complement the defect in the *CDC25* gene. It has been shown by others that he C-terminal part of the *CDC25* ORF complements different *cdc25* mutations and even complete disruption of the gene (Robinson *et al.*, 1987; Broek *et al.*, 1987). Thus only the C-terminal part of the molecule seems to be responsible for the function of the gene required in production of cAMP and for growth.

Preliminary experiments seems to indicate that the reconstituted *SCD25* complete gene is unable to complement the *cdc25* mutation thus the remaining N-terminal part of the gene could be involved in specification of the target and the target of *SCD25* could be different from the target of *CDC25*.

If we assume that the role of the *CDC25* gene product is to increase the exchange rate of guanosine nucleotides for RAS protein then it can be imagined that the *SCD25* gene product has a similar role with another GTP binding protein. Several GTP binding proteins are known to exist in yeast as in other eucaryotes and the existence of various exchange proteins is not excluded.

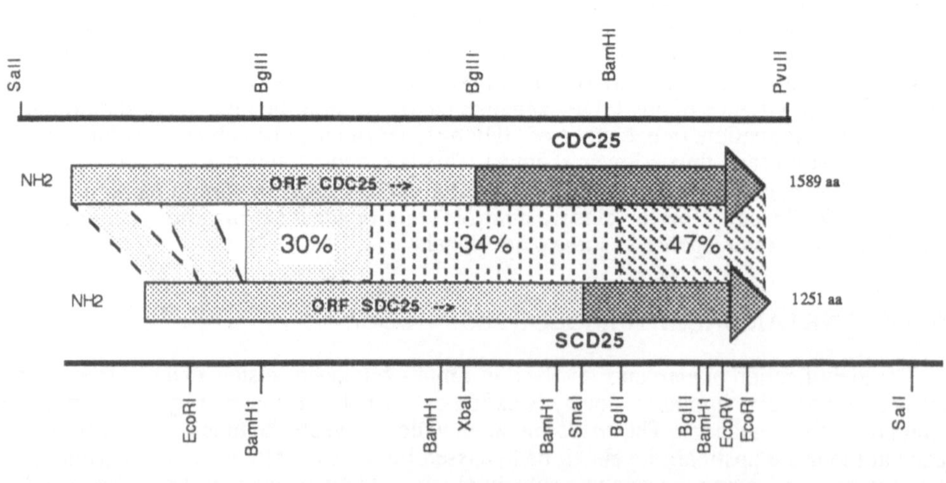

FIGURE 4 . Comparison between the two ORFs: CDC25 and SCD25
Restriction map of the DNA fragments containing respectively CDC25 and SCD25 gene are given respectively at the top and the bottom. The ORF are schematised by large arrows, the darker part corresponds to the region which is sufficient to complement cdc25 mutations.

CONCLUSIONS

The adenylate cyclase of the yeast *Saccharomyces cerevisiae* is activated by a member of the ras protein as a transducer rather than by a G protein as in mammalian cells. This peculiarity is interesting because it provides a defined system for the study of transduction by a ras protein. The product of the *CDC25* gene is the only identified element in yeast for the activation of ras. Although, the biochemical role of the *CDC25* gene product remains to be established, it can be proposed as a working hypothesis that it corresponds to the exchange factor in the recycling of a GTP bound ras protein. This hypothesis is based on several genetic observations such as our discovery of a new mutation of *RAS2* which abolishes the need for a functional *CDC25* gene. Finally complementation experiments led us to discover a new gene which possesses an ORF partially homologous to the ORF of *CDC25*. Therefore the active domain of *CDC25* seems to belong to a family of proteins which could share similar function.

ACKNOWLEDGEMENTS

We thank Rolande Guilbaud and Danièle Le Roscouet for their skillful technical assistance . Faten Damak was a recipient of a fellowship from the C.I.E.S (Centre International des Etudiants Stagiaires). This work was supported by grants from INSERM (Institut Nationale de la Santé et de la Recherche Médicale), ARC (Association pour la Recherche contre le Cancer) , and Ligue Nationale Française de la Recherche contre le Cancer

REFERENCES

Barbacid, M., 1987, ras genes *Ann. Rev. Biochem.* 56, 779-827

Boutelet, F., PetitJean, A., and Hilger, F., 1985, *CDC35* mutants are defective in adenylate cyclase and are allelic with *cyr1* mutants while *cas1*, a new gene , is involved in the regulation of adenylate cyclase . *Embo J.* , 4 , 2635-2642.

Boy-Marcotte, E., Garreau, H., and Jacquet, M., 1987, Cyclic AMP controls the switch between division cycle and resting state programs in response to ammonium availability in *Saccharomyces cerevisiae. Yeast* 3, 85-93

Broek, D., Toda, T., Michaeli, T., Levin, L., Birchmeier, C., Zoller, M., Powers, S., and Wigler, M., 1987, The *S. cerevisiae CDC25* gene product regulates the RAS/ adenylate cyclase pathway. *Cell* , 48 , 789-799

Camonis, J., Kalekine, M., Gondré, B., Garreau, H., Boy-Marcotte, E., and Jacquet, M., 1986, Characterization, cloning and sequence analysis of *CDC25* gene which controls the cyclic AMP level of *Saccharomyces cerevisiae. The EMBO J.* , 5, 375-380 .

Camonis, J., and Jacquet, M., 1988, A new *RAS* mutation which suppresses the *CDC25* gene requirement for growth in *Saccharomyces cerevisiae. Mol. Cell Biol.* 8, 2980-2983.

Casperson, G.F., Walker, N., Brassier, A.R., and Bourne, H.R., 1983, A guanine nucleotide sensitive adenylate cyclase in the yeast *Saccharomyces cerevisiae. J. Biol. Chem.* 158, 7911-7914

Casperson, G.F., Walker, N., and Bourne, R.H., 1985, Isolation of the gene encoding adenylate cyclase in *Saccharomyces cerevisiae. Proc. Natl. Acad. Sci.* USA 82, 5060-5063

Daniel, J., and Simchen, C., 1986, Clones from two different genomic regions complement the *cdc25* start mutation of *Saccharomyces cerevisiae Curr. Genet.*, 10, 643-646.

DeFeo-Jones, D., Scolnick, E., Koller, R., and Dhar, R. 1983 *ras*-related gene sequence identified from *Saccharomyces cerevisiae. Nature*, 306, 707-709

Feig, L.A.,Pan, B., Roberts, T.M., and Cooper, G.M., 1986, Isolation of ras GTP-binding mutants using an *in situ* colony binding assay. *Proc. Natl. Acad. Sci.* USA 83, 4607-4611

Jacquet, M., and Camonis, J., 1985, Contrôle du cycle de division cellulaire et de la sporulation chez *Saccharomyces cerevisiae* par le système de l'AMPc. *Biochimie*, 67, 35-43.

Kataoka, T., Powers, S.,McGill, C., Fasano, O., Strathern, J., Broach, J.R., and Wigler, M., 1984, Genetic analysis of yeast RAS1 and RAS2 genes. *Cell*, 37, 437-445.

Kataoka, T., Powers, S., Cameron, S., Fasano, O., Golfarb, M., Broach, J., and Wigler, M., 1985a, Functional homology of mammalian and yeast ras genes *Cell* 40, 19-26

Kataoka, T., Broek, D., and Wigler, M., 1985b, DNA sequence and characterization of the *S. cerevisiae* gene encoding adenylate cyclase *Cell* 43, 493-505

Lissziewiecz, J., Godany, A., Förster, H.H., Küntzel, H., 1987, Isolation and nucleotide sequence of a *Saccharomyces cerevisiae* protein kinase gene suppressing the cell cycle start mutation *cdc25*. *J. Biol. Chem* 262, 2549-2553

Martegani, E., Baroni, M. D., Frascotti, G., and Alberghina, L., 1986, Molecular cloning and transcriptional analysis of the start gene *CDC25* of *Saccharomyces cerevisiae*. *EMBO J.* 5, 2363-2369.

Masson, P., Jacquemin, J. M., and Culot, M., 1984, Molecular cloning of the tsm0185 gene responsible for adenylate cyclase activity in *Saccharomyces cerevisiae. Ann. Microbiol.* (Inst. Pasteur),135, 344-351

Masson, P., Lenzen, G., Jacquemin, J.M., and Danchin, A., 1986, Yeast adenylate cyclase catalytic domain is carboxy terminal *Currr. Genet.* 10, 343-352

Matsumoto, K., Uno, I., Oshima Y., and Ishikawa, T., 1982, Isolation and characterization of yeast mutants deficient in adenylate cyclase and cAMP -dependent protein kinase. *Proc. Natl. Acad. Sci.* USA 79, 2355-2359

Powers, S., Kataoka, T., Fasano, O., Goldfarb, M., Strathern, J.N., Broach, J.R. and Wigler, M., 1984, Genes in *S. cerevisiae* encoding proteins with domains homologous to the mammalian ras proteins. *Cell*, 36, 607-612.

Pringle, J.R., and Hartwell, L.W., 1981, The *Saccharomyces cerevisiae* cell cycle. In Strathern, J.N., Jones, E.W., and Broach, J.R. (Eds)" The molecular biology of the yeast *Saccharomyces cerevisiae* : life cycle and inheritance. Cold Spring Harbor laboratories, N.Y. 97-142.

Robinson, L.C., Gibbs, J.B., Marshall, M.S., Sigal, I.S., and Tatchell, K., 1987, *CDC25* : a component of the *RAS*-adenylate cyclase pathway in *Saccharomyces cerevisiae*. *Science*, 235, 1218-1221.

Sigal, I.S., Gibbs, J.B., D'Alonzo, J.S., Temeles, G.L., Wolanski, B.S., Socher, S.H., and Scolnick, E.M., 1986, Mutant ras-encoded proteins with altered nucleotide binding exert dominant biological effect. *Proc. Natl. Acad. Sci.* USA 83, 952-956

Tatchell, K., Chaleff, D., DeFoe-Jones, D., and Scolnick, E.M., 1984, Requirement of either of a pair of ras-related genes of *Saccharomyces cerevisiae* for spore viability *Nature* 309, 523-527

Thevelein, J.M., 1984, Cyclic-AMP content and trehalase activation in vegetative cells and ascospores of yeast. *Arch. Microbiol.* 138, 64-67

Toda, T., Cameron, S., Sass, P., Zoller, M., and Wigler, M., 1987, Three different genes in *S. cerevisiae* encode the catalytic subunits of the cAMP dependent protein kinase. *Cell.*, 5O , 277-287 .

Zarret, S., and Sherman, F., 1986, DNA sequence required for efficient transcription termination in yeast . *Cell* , 28 , 563-573 .

MUTATIONS AT THE *RAS2* LOCUS THAT, IN A *ras1*⁻ BACKGROUND, IMPAIR THE GROWTH OF YEAST ON NONFERMENTABLE CARBON SOURCES

Ottavio Fasano, Jean Bernard Crechet*, Emmanuele De Vendittis, Regina Zahn, Georg Feger, Alessandra Vitelli, and Andrea Parmeggiani*

Differentiation Programme, European Molecular Biology Laboratory, Postfach 10.2209, D-6900 Heidelberg, FRG, and (*) Laboratoire de Biochimie, Ecole Polytechnique, F-91128 Palaiseau-Cedex, France

INTRODUCTION

The proteins encoded by the *RAS1* and *RAS2* genes (collectively denominated *RAS*) are important for progression of *S. cerevisiae* through the cell cycle. In fact, yeast cells in which the function of the *RAS2* protein is constitutively activated by a point mutation cannot properly arrest in G1 (Toda et al., 1985). On the other hand, cells with a disrupted *RAS1* gene and with a temperature-sensitive RAS2 gene product are found prevalently as unbudded, G1-arrested cells upon temperature shift-up (De Vendittis et al., 1986).

To test whether other cellular functions, besides growth, are controlled by the RAS2 gene product, it would be interesting to compare the properties of *ras1⁻RAS2* strains to those of otherwise isogenic *ras1⁻ras2⁻* strains. However, the lethality of *ras1⁻ras2⁻* double mutants prevents the detection of the effects of a disrupted *RAS* function on cellular functions other than growth.

To overcome this difficulty, we have constructed yeast strains in which a partially deficient *RAS* function results from the deletion of the *RAS1* gene, and from the presence of a mutated chromosomal *RAS2* gene. By comparing the phenotype of these mutants with those of isogenic *ras1⁻RAS2* strains, we have explored the requirement for the *RAS2* gene function in yeast cells growing in different conditions. The choice of a *ras1⁻*background was important, because the absence of the *RAS1* gene eliminated the possible interference caused by the differential expression of the *RAS1* transcript during growth on different substrates (Breviario et al., 1986). The validity of our conclusions has been verified using distinct mutated alleles of *RAS2*.

METHODS

Yeast strains, growth media and trasformation

The strain JR26-19D (*MATα ade2-1 can1-100 his3 leu2-3,112 lys1-1 ura3-52 ras1::URA3*) was used as a recipient for transformation with the purified large EcoRI-HindIII fragment of the plasmid pR2S (Figure 1) harboring either wild-type or mutated *RAS2* gene sequences with their 5' and 3' flanking regions, and the selectable marker *SUP16*. Nomenclature for genotypes is according to standard conventions: dominant alleles are in capital letters, recessive alleles are lower case and genetic marker *X* integrated at locus *Y* is designated *Y::X*. Sporulation of diploids was performed as previously described (Kataoka et al., 1984). The standard rich medium used was YEPD (2% bacto-peptone, 1% yeast extract, and 2% glucose). Standard media with carbon

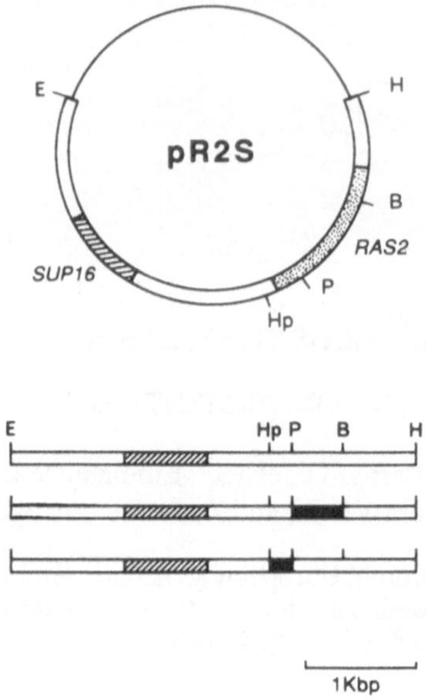

RAS2 genotype	Nucleotide and amino acid sequence at position 40	82	84	Plasmid clone
RAS2 (wild type)	GAT asp	GGC gly	GGA gly	pR2S
ras2-ts1	GAT asp	AGC ser	AGA arg	pR4
ras2-3	AAT asn	GGC gly	GGA gly	pR6

Figure 1. Schematic structure of the plasmids used to direct integration of cloned sequences at the chromosomal *RAS2 locus* of yeast recipient strains. pR2S carries wild-type *RAS2* sequences, while pR4 and pR6 are chimaeric constructions in which the filled regions represent mutated *RAS2* sequences cloned from the original yeast mutants. Differences in nucleotide sequence are indicated in the lower part of the figure. The plasmids have otherwise the same structure. E=EcoRI, Hp=HpaI, P=PstI, B=BalI, H=HindIII.

sources other than glucose were obtained by using 2% bacto-peptone and 1% yeast extract, supplemented with 2% galactose (YEPGA), 3% glycerol (YEPGL), or 3% pyruvate (YEPPY). Complete synthetic medium contained 0.67% yeast nitrogen base without amino acids (Difco), and amino acids and nucleic acid bases as indicated by the yeast manual (Sherman et al., 1986). The carbon sources used to supplement the complete synthetic medium were 2% glucose (SDC), 2% galactose (SGAC), 3% glycerol (SGLC), 3% pyruvate (SPYC).

Yeast transformations were carried out after treatment of yeast cells with lithium acetate, according to Ito et al. (1983).

RESULTS

In *ras1*⁻ strains, the replacement of glycines 82 and 84 of the RAS2 protein by serine and arginine, respectively, results in temperature-sensitive growth on nonfermentable carbon sources and galactose

A yeast strain with a disrupted *RAS1* gene and with the chromosomal *RAS2* gene replaced by a mutant allele (*ras1⁻ras2-ts1*) was obtained by replacing the chromosomal *RAS2* gene of a *ras1⁻*strain with randomly mutagenized *RAS2* sequences (De Vendittis et al., 1986). This mutant, that we called TS1, was unable to grow at 37 °C on rich medium using either galactose or glycerol as a carbon source, while the growth on glucose at 37 °C was reduced as compared to an isogenic *ras1⁻RAS2* strain. After cloning and sequencing the *ras2-ts1* gene from TS1 cells, using the wild-type *RAS2* gene as a probe, we identified two point mutations that predicted the replacement of glycines 82 and 84 of the RAS2 protein by serine and arginine, respectively (Fasano et al., EMBO J., in press).

We verified if both the mutations found in the *ras2-ts1* mutant allele were necessary for temperature-sensitive growth on nonfermentable carbon sources. Using oligonucleotide directed mutagenesis, we constructed plasmid clones of the *RAS2* gene that separately harbored one of the two mutations. The structure of these plasmid was otherwise identical to that of pR2S (Figure 1). After the isolation of the large insert from each plasmid (Figure 1), we transformed the yeast strains JR26-19D (relevant genotype *ras1⁻ RAS2 lys1-1*). Transformants were selected after incubation at 30 °C on synthetic medium with glucose and without lysine, taking advantage of the ability of the *SUP16* marker to suppress the *lys1-1* mutation carried by the recipient strain JR26-19D (De Vendittis et al., 1986). Growth of the tranformants on different carbon sources was evaluated by replica plating. We found that the *RAS2* gene clones harboring a single mutation at either position 82 or 84, similarly to the wild-type *RAS2* gene, were unable to transfer the TS1 phenotype (no growth at 37 °C on either galactose or nonfermentable carbon sources, slow growth on glucose at the same temperature) to the recipient cells. The majority of the transformants had a wild-type phenotype. Only by using a purified fragment of the cloned *ras2-ts1* gene, we obtained with high frequency transformants unable to grow at 37 °C on either galactose or nonfermentable carbon sources. Southern blot analysis of the *DNA* of randomly selected transformants showed that in most cases integration of the *SUP16* marker had taken place correctly at the *RAS2 locus* (not shown). On the basis of the similarity with the growth phenotype of the original TS1 mutant, we tentatively assigned to the temperature-sensitive transformants a *ras1⁻ ras2-ts1::SUP16* genotype. The presence of both mutations (Figure 1, lower panel) was confirmed by Southern blot analysis of the DNA of the transformants, using differential hybridization with radioactively labelled oligonucleotide probes. The oligonucleotides (18-mer) carried either a wild-type or a mutated sequence (single or double mutation), and allowed to discriminate between wild-type and mutated chromosomal *RAS2* sequences, after treating the blots with washing solutions at increasing temperatures (not shown).

In *ras1*⁻ strains, the replacement of aspartic acid at position 40 of RAS2 protein by asparagine impairs growth on nonfermentable carbon sources and galactose, both at 30 and 37 °C

Using the random mutagenesis procedure described by De Vendittis et al. (1986) to mutagenize the *RAS2* locus of *ras1⁻ RAS2* cells, we isolated a yeast strain, that we called TX3, which could not grow at either 30 or 37 °C on nonfermentable carbon sources. TX3 was not a *rho* ⁻ induced by the transformation procedure. In fact, after crossing TX3 with haploid *RAS1 RAS2 rho* ⁻ yeast cells, we obtained diploids which were able to grow on nonfermentable carbon sources at 30 and 37 °C.

Sequence analysis of the cloned mutant gene, that we called *ras2-3*, revealed a single mutation, that predicted the replacement of aspartic at position 40 of the *RAS2* gene product by asparagine. The plasmid carrying the *ras2-3* allele was called pR6 (Figure 1).

To verify that the replacement of aspartic 40 of the RAS2 protein by asparagine resulted in inability to grow on nonfermentable carbon sources and galactose, we attempted to transfer the mutation to the chromosomal *RAS2* locus of *ras1⁻RAS2* recipient cells. Therefore, we used the purified insert of the plasmid pR6, carrying the *ras2-3* gene and the *SUP16* marker, to transform LiCl-treated *ras1⁻RAS2* cells. After selection on synthetic medium with glucose and without lysine, the growth properties of randomly selected transformants were evaluated by replica plating. A high percentage of transformants had the same growth characteristics of the original TX3 strain (no growth at either 30 or 37 °C on nonfermentable carbon sources and galactose). Among these transformants, a small percentage were *rho* ⁻, as revealed by their inability to produce diploids able to grow on nonfermentable carbon sources, after crossing to haploid *RAS1 RAS2 rho* ⁻ strains. However, by the same procedure we found that most of the remaining transformants were *RHO* ⁺. The structure of the chromosomal *RAS2* locus of randomly selected transformants was investigated by Southern blot. We used DNA cleaved with EcoRI and HindIII to verify the correct integration of the *SUP16* marker (De Vendittis et al., 1986). Moreover, since the presence of a mutated *ras2-3* allele leads to the disappearance of a Sau3A site (for the sequence of the *RAS2* gene see Powers et al., 1984), we used Southern blot analysis of DNA cleaved with this enzyme to discriminate between *ras2-3::SUP16* and *RAS2::SUP16* chromosomal structures. Four out of four tranformants showed a *ras1⁻ras2-3::SUP16* genotype (not shown).

We conclude that the replacement of aspartic 40 of the RAS2 gene product by asparagine leads to inability to grow on nonfermentable carbon sources. The finding of a *RHO* ⁺ genotype in most of the transformants, and their ability to give rise to spontaneous revertants, indicates that the mitochondrial genome was intact in these transformants and that the inability to grow on nonfermentable carbon sources was a consequence of the mutation at the *RAS2* locus.

Impaired *RAS*-dependent adenylate cyclase activity of *ras2-ts1* and *ras2-3* mutants

The ability of the ras2-ts1 and ras2-3 protein to stimulate yeast adenylate cyclase was evaluated by measuring cyclic AMP formation *in vitro*, using total membrane preparations as a source of ras proteins and adenylate cyclase. The results, shown in Table I, indicate that the Mg- and GppNHp-dependent adenylate cyclase activity of

Table I. Adenylate cyclase activity of yeast membranes[a].

	adenylate cyclase activity (pmol cAMP produced/min/mg membrane 30 °C)		
RAS genotype	1 mM Mn	10 mM Mg	10mMMg+GppNHp
ras1⁻RAS2	22	8	16
ras1⁻ras2-ts1	10	<0.5	<0.5
ras1⁻ras2-3	24	1	3

(a) Membranes were prepared from cells grown at 30 °C.

membranes from *ras1*$^-$*ras2-ts1* and *ras1*$^-$*ras2-3* strains was either undetectable *(ras2-ts1)* or strongly depressed *(ras2-3)*. The Mn-dependent activity was changed by a factor two in *ras2-ts1* strains, thus indicating that the amount of the catalytic unit of the yeast adenylate cyclase present into the membranes was relatively unaffected by the mutations.

DISCUSSION

We have constructed, by random mutagenesis, isogenic yeast strains carrying, in a *ras1*$^-$ background, distinct mutated chromosomal *RAS2* alleles. The absence of a functional *RAS1* gene was important to eliminate any possible interference deriving from the differential expression of this locus (Breviario et al., 1986) in different carbon sources. We have carried out a detailed characterization of these mutants. The Mg- and GTP-dependent in vitro adenylate cyclase activity of membranes prepared from mutant strains was either absent *(ras2-ts1)* or strongly depressed *(ras2-3)*. This suggests that the mutations prevalently affect the ability of the RAS2 protein to stimulate adenylate cyclase. The impaired adenylate cyclase activity of the mutant *ras2-3* (position 40), is in line with the "effector" role attributed to amino acid residues 40-47 of the RAS proteins (Marshall et al., 1988; the coordinates refer to the yeast RAS2 protein). However, the absence of Mg-dependent adenylate cyclase activity at 30 °C in strains expressing the ras2-ts1 protein was not predicted by this model. Our results raise the possibility that, in addition to residues 40-47, also residues 82 and 84 might be important for the regulation of yeast adenylate cyclase by the RAS2 protein. It is possible that these residues directly interact with adenylate cyclase. Alternatively, since different conformational states of the GTP- and the GDP-bound form of the RAS2 protein need to be postulated to explain the different biological properties of the two complexes (De Vendittis et al., 1986), we might envisage that glycines 82 and 84 are situated in a critical region, whose flexibility is essential to allow transition between the two states. If this hypothesis is correct, the replacement of both glycines by amino acid residues with larger side chains might increase the rigidity of this region, thereby "freezing" the protein in the inactive state. A third possibible explanation of our data, perhaps not excluding other models, is that the ras2-ts1 protein might be particularly prone to thermal denaturation. This hypothesis could explain the temperature-sensitive growth of *ras2-ts1* cells, but not the phenotypic differences observed at 30 °C. In fact, the growth of these mutants at 30 °C on glucose shows that the mutant protein retains most of its function in these conditions (note that *ras1*$^-$*ras2*$^-$mutants are not viable), yet the slow growth in nonfermentable carbon sources and the lack of GTP-dependent adenylate cyclase activity point to a more severe functional lesion. Therefore, it is likely that the altered function of the protein at 30 °C is not caused by denaturation in vivo, but rather by a more localized lesion. The tertiary structure of the H-*ras* p21 (De Vos et al., 1988) indicates that glycines 75 and 77 (corresponding to the conserved glycines 82 and 84 of the yeast RAS proteins) are located at the end of a large loop. Some residues in the loop (not comprising glycines 82 and 84) are involved in the binding of the neutralizing antibody Y13-259 (Barbacid, 1987). Indeed, we have found that the ras2-ts1 protein expressed in *E. coli* was recognized by this antibody (O. Fasano, unpublished results). In the future, the three-dimensional structure of the mutated protein might help discriminating between the various possibilities.

It is surprising that strains expressing the ras2-ts1 or the ras2-3 protein could efficiently grow at 30 °C on glucose, yet the adenylate cyclase activity of membranes prepared from cells growing in these conditions was practically absent. It should be noted that the adenylate cyclase catalytic unit was functional, as shown by the detectable Mn-dependent activity. This contradiction could be explained by assuming that a cyclic AMP-independent pathway, postulated by Cameron et al. (1988), might prevalently mediate the effects of *RAS* during growth of yeast on glucose, while a cyclic AMP-dependent pathway might be more important during respiratory growth. If this model is correct, an inefficient interaction of the mutated RAS2 proteins with adenylate cyclase might explain the poor growth of the mutants described in this paper during respiratory metabolims, both at 30 and 37 °C. On the other hand, a conserved and efficient interaction of the mutated RAS2 proteins with still unidentified cellular elements might explain the better growth of the same strains in glucose. As expected, the mutations affected *RAS*-and cyclic AMP-dependent cellular functions such as glycogen and threalose levels (revealed by iodine/iodide staining), as well as sporulation frequency (De Vendittis et al., 1986). The direction of the changes was opposite to that observed in yeast strains with a *RAS2* gene activated by a point mutation at position 19 (Kataoka et al., 1984; Toda et al., 1985). This confirmed that the mutated RAS2 proteins do not interact very efficiently with adenylate cyclase *in vivo*, thus leading to a generalized depression of cyclic AMP-dependent cellular functions.

REFERENCES

Barbacid, M., 1987, *ras* Genes. Annu. Rev. Biochem., 56:779.

Breviario, D., Hinnebush, A., Cannon, J., Tatchell, K. and Dhar, R., 1986, Carbon Source Regulation of *RAS1* expression in *Saccharomyces cerevisiae* and the Phenotype of *ras2⁻*cells. Proc. Natl. Acad. Sci. USA, 83:4152.

Cameron, S., Levin, L., Zoller, M. and Wigler, M., 1988, cAMP-Independent Control of Sporulation, Glycogen metabolism, and Heat Shock Resistance in *S. cerevisiae*. Cell, 53:555.

De Vendittis, E., Vitelli, A., Zahn, R. and Fasano, O., 1986, Suppression of Defective *RAS1* and *RAS2* Functions in Yeast by an Adenylate Cyclase Activated by a Single Amino Acid Change EMBO J., 5:3657.

De Vos, A.M., Tong, L., Milburn, M.V., Matias, P.M., Jancarik, J., Noguchi, S., Nishimura, S., Miura, K., Ohtsuka, E. and Kim, S., 1988, Three-Dimensional Structure od an Oncogene Protein: Catalytic Domain of Human c-H-*ras* p21. Science, 239:888.

Ito, H., Fukuda,, Y., Murata, K. and Kimura, A., 1983, Transformation of Intact Cells Treated with Alkali Cations. J. Bacteriol., 153:163.

Kataoka, T., Powers, S., McGill, C., Fasano, O., Stratern, J., Broach, J. and Wigler, M., 1984, Genetic Analysis of Yeast *RAS1* and *RAS2* Genes Cell, 37:437.

Marshall, M.S., Gibbs, J., Scolnick, E.M. and Sigal, I.S., 1988, An Adenylate Cyclase from *Saccharomyces cerevisiae* That Is Stimulated by *RAS* Proteins with Effector Mutations. Mol. Cell. Biol., 8:52.

Powers, S., Kataoka, T., Fasano, O., Goldfarb, M., Stratern, J., Broach, J. and Wigler, M., 1984, Genes in *Saccharomyces cerevisiae* Encoding Proteins with Domains Homologous to the Mammalian *ras* Proteins Cell, 36:607.

Sherman. F., Fink, G.R. and Hicks, J., 1983, in Methods in Yeast Genetics, Cold Spring Harbor Laboratory, Cold Spring Harbor, New York.

Toda, T., Uno, I., Ishikawa, T., Powers, S., Kataoka, T., Broek, D., Cameron, S., Broach, J., Matsumoto, K. and Wigler, M., 1985, In Yeast, *RAS* Proteins Are Controlling Elements of Adenylate Cyclase. Cell, 40:27.

STRUCTURAL AND FUNCTIONAL ANALYSIS OF *ypt* PROTEINS, A FAMILY OF *ras*-RELATED NUCLEOTIDE-BINDING PROTEINS IN EUKARYOTIC CELLS

Dieter Gallwitz, Heinz Haubruck, Constance Molenaar, Reinhild Prange, Mechthild Puzicha, Hans Dieter Schmitt, Constantin Vorgias and Peter Wagner

Max-Planck-Institute for Biophysical Chemistry
P.O. Box 2841,
D-3400 Göttingen, FRG

INTRODUCTION

The advance in cloning and sequencing of DNA has led in recent years to the discovery of a large number of genes encoding guanine nucleotide-binding proteins that are related to the potentially oncogenic *ras* proteins. *Ras* proteins (for review: Barbacid, 1987) and various *ras*-related proteins are, as far as this has been studied, evolutionary highly conserved. These proteins share similar biochemical properties, they are membrane-bound and likely to be involved in different regulatory pathways in all eukaryotic cells.

The first of the *ras*-like proteins discovered was the *Ypt1* protein of the budding yeast *Saccharomyces cerevisiae* (Gallwitz et al., 1983). Besides the genes of very diverse eukaryotic species coding for proteins that exhibit extensive homology to the mammalian H-, K- and N-*ras* proteins, genes encoding related proteins were subsequently identified in the mollusc *Aplysia* and in human cells, *rho* proteins (Madaule and Axel, 1985), and in mammalian cells, *ral* (Chardin and Tavitian, 1986) and R-*ras* proteins (Lowe et al., 1987). Mammals were also shown to express proteins highly homologous to the yeast *YPT1* gene product (Haubruck et al., 1987; Touchot et al., 1987), and a yeast gene known to be involved in protein secretion, *SEC4*, was found to encode a GTP-binding protein related to the *ras* gene products (Salminen and Novick, 1987).

All the sequence data presently available, together with the still very limited functional studies that have been performed, allow to distinguish between three families of *ras*-related proteins, *ras*, *rho* and *ypt* proteins (Haubruck et al., 1987).

This paper summarizes the present knowledge of the biochemical properties and the cellular functions of *ypt* proteins with special emphasis of the *S. cerevisiae YPT1* gene product.

BIOCHEMICAL PROPERTIES OF *ypt* PROTEINS

There are four domains of *ras* and *ras*-like proteins that are highly conserved and that have been shown by biochemical studies of wild-type and mutant proteins (for review: Barbacid, 1987; Wagner et al., 1987) and by a recent crystallographic investigation (De Vos et al., 1988) to be of importance for GTP binding and hydrolysis. These domains, residues 5-17, 57-62, 116-119 and 143-147 (with respect to mammalian *ras* proteins) are similarly spaced with respect to the primary structures suggesting rather similar tertiary structures of all of these proteins build of roughly 200 amino acid residues.

GTP and GDP binding of *ypt* proteins from different eukaryotic species has been shown by protein blotting procedures and with the purified yeast *Ypt1* protein and the mouse *ypt1* protein in solution. The binding affinity for GDP is high (2.2 x 10^9 M^{-1} measured for the wild-type yeast *Ypt1* protein at 30°C), and different mutant forms of the protein exhibiting altered GTPase activity (substitutions Ser-17 → Gly, in short S17G, or A65T) had only slightly impaired affinities for the nucleotide (P.W., C.V. and D.G., unpublished). As is the case for mammalian *ras* proteins, the A65T substitution in the yeast *Ypt1* protein, corresponding to the same change in position 59 of *ras* proteins, led to a significant decrease in GTPase activity and to an autophosphorylation of the protein (Wagner et al., 1987). The intrinsic GTPase activity of the yeast *Ypt1* protein was found to be lower than that of *ras* proteins and the S17G substitution resulted in an expected increase of the GTP hydrolyzing activity (Wagner et al., 1987). Whereas the two mutations mentioned did not affect the biological activity of the yeast protein, investigated in haploid cells after exchanging the wild-type *YPT1* gene with the respective mutant genes, the substitutions K21M of the first and N121I of the third nucleotide binding domain led to a significant impairment of GTP binding and a loss of cellular function of the *Ypt1* protein (Schmitt et al., 1986; Wagner et al., 1987). It is clear, therefore, that *ypt* and *ras* proteins are biochemically very similar.

Like the *ras* proteins that are bound to the inner surface of the plasma membrane via palmitic acid covalently attached to a cysteine four residues from the carboxyl terminal end (Willumsen et al., 1984; Deschenes and Broach, 1987), the yeast *Ypt1* protein requires one of its two C-terminal cysteine residues for fatty acid binding and membrane attachment in order to exert its cellular function (Molenaar et al., 1988).

A summary of amino acid substitutions and their effects on the biological function of the *S. cerevisiae Ypt1* protein is given in Figure 1.

TOWARDS AN UNDERSTANDING OF *ypt* FUNCTION

The very efficient system of homologous recombination in the yeast *S. cerevisiae* allows to easily exchange a wild-type gene with a mutant gene or to precisely inactivate a gene on the chromosome. These techniques have been successfully employed to get insight into the possible functioning of the *YPT1* gene. Gene disruption and the replacement of the wild-type gene with mutant genes encoding proteins with significantly impaired guanine nucleotide-binding capacity proved to be lethal (Schmitt et al., 1986; Wagner et al., 1987). The essential role of the *Ypt1* protein was further investigated by placing the gene under an inducible yeast promoter, *GAL10*, allowing to switch on and off its expression. Growth arrest following *Ypt1* protein depletion was accompanied by a conspicious disorganization of microtubules (Schmitt et al., 1986) and by an accumulation of intracellular membranes and vesicles (see Figure 2), by an increase of intracellular calcium and by a severe inhibition of protein secretion (Schmitt et al., 1988). Similar observations were made with a temperature-sensitive (Schmitt et al., 1988) and a cold-sensitive mutant (Segev et al., 1988) of the *Ypt1* protein at the nonpermissive temperature. It was shown in addition that with high concentrations of extracellular calcium the phenotypic alterations of the ts mutant at the restrictive temperature could be prevented partially (Schmitt et al., 1988).

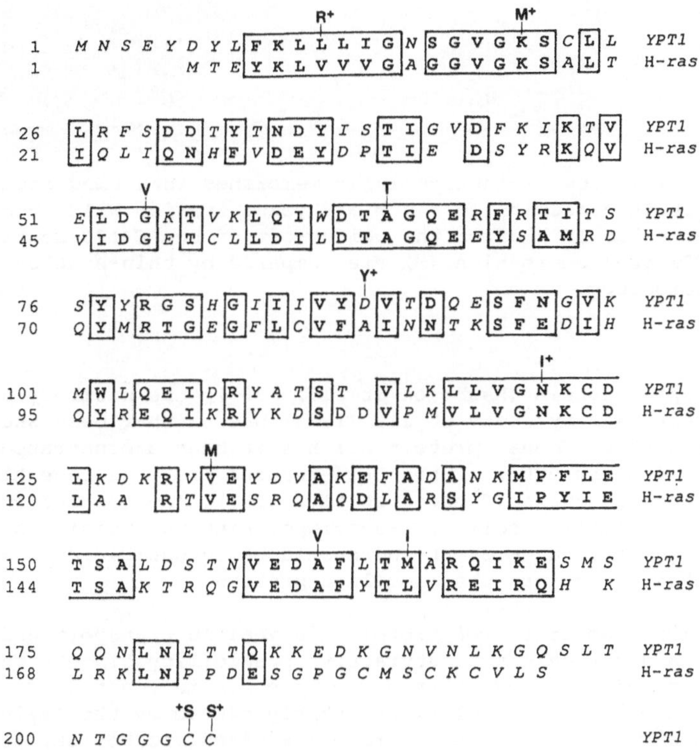

Fig. 1. Comparison of primary sequences of the *S. cerevisiae Ypt1* protein and the human H-*ras* protein. Identical residues and preferred substitutions are boxed. Amino acid substitutions of the *Ypt1* protein generated by in vitro mutagenesis of the gene (Wagner et al., 1987; Molenaar et al., 1988) or found by screening for second-site mutations suppressing the dominant-lethal N121I mutation (Schmitt et al., 1988) are shown above the *Ypt1* sequence. Substitions marked by asterisks resulted in a loss of biological activity and caused lethality, other mutations were neutral with respect to the protein's cellular function.

These observations have led to somewhat different conclusions as to the primary action of the *Ypt1* protein. Whereas Segev et al. (1988) suggest that the protein functions as a membrane label in vesicular transport, we consider the possibility that the regulation of intracellular calcium, known to affect different cellular activities including the cytoskeletal network, mitosis and protein secretion, might be the principle function exerted by the *Ypt1* protein (Schmitt et al., 1988).

In favor for a role of the *Ypt1* protein in the secretory pathway could be the finding by Segev et al. (1988) that an antibody directed against the yeast protein stained Golgi structures in mouse fibroblasts. Although we have recently identified a gene in mouse, called *ypt1*, encoding a protein with more than 70 % amino acid identities compared to

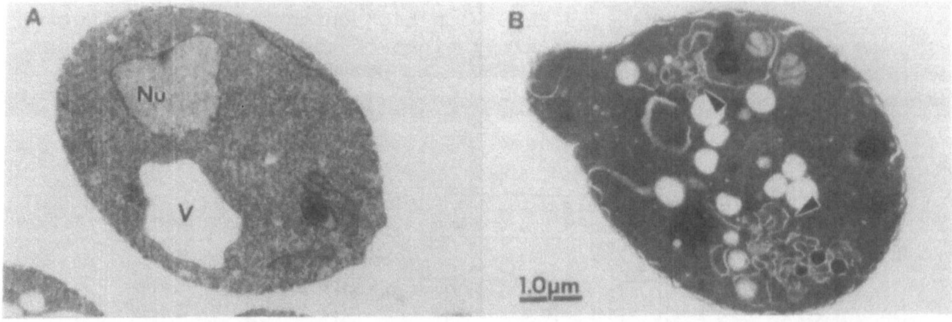

Fig. 2. Accumulation of intracellular membranes resulting from *Ypt1* protein depletion in a haploid yeast strain. Wild-type cells (A) and cells 15 h after the switch-off of the *GAL10*-controlled *YPT1* gene expression (B) are compared by thin-section electron microscopy.

the yeast *Ypt1* protein (Haubruck et al., 1987), the antibody against the yeast protein used by Segev et al. (1988) has neither been shown to crossreact with the mouse protein nor has it been demonstrated to not crossreact with any other of the *Ypt1*-related proteins recently identified in mammalian cells (Touchot et al., 1987). The conclusion that a *ypt* homologue in mammalian cells is associated with the Golgi apparatus is therefore premature at present and should be investigated with monospecific antibodies against the mammalian protein.

The involvement of *ypt* proteins in vesicle transport and protein secretion is nevertheless an attractive possibility since a conditional lethal mutant of the *Ypt1*-related *Sec4* protein in yeast leads to the accumulation of small vesicles, presumably caused by the failure of vesicles to fuse with the plasma membrane (Salminen and Novick, 1987; Goud et al., 1988). This led to the speculation that *ras*-related, GTP-binding proteins, like the *Sec4* and the *Ypt1* protein, might recognize specific proteins of different membrane compartments and thereby direct vesicle transport for fusion with different acceptor membranes (Bourne, 1988). The identification of proteins with which the *Ypt1* protein interacts within the yeast cell is now being tackled with the help of molecular genetic and biochemical methods.

Proteins with extensive homology to the *S. cerevisiae Ypt1* protein are likely to serve a very basic function in all eukaryotic cells. This is suggested by the essential function of a protein from the fission yeast *Schizosaccharomyces prombe*, called *ypt1* (M. Yamamoto, personal communication), and by the expression of the mouse *ypt1* gene in all tissues examined and in differentiating and non-differentiating F9 cells (Haubruck et al., 1987). Both the fission yeast and the mouse *ypt1* protein share more than 70 % amino acid identities with the *S. cerevisiae Ypt1* protein.

CONSERVED STRUCTURAL FEATURES OF *ypt* PROTEINS

As already pointed out above, four domains involved in guanosine nucleotide binding of *ras* and their related proteins are highly conserved and similarly contained in other GTP-binding proteins, like elongation factor EF-Tu (Jurnak, 1985) and the α subunits of G-proteins (Tanabe et al., 1985).

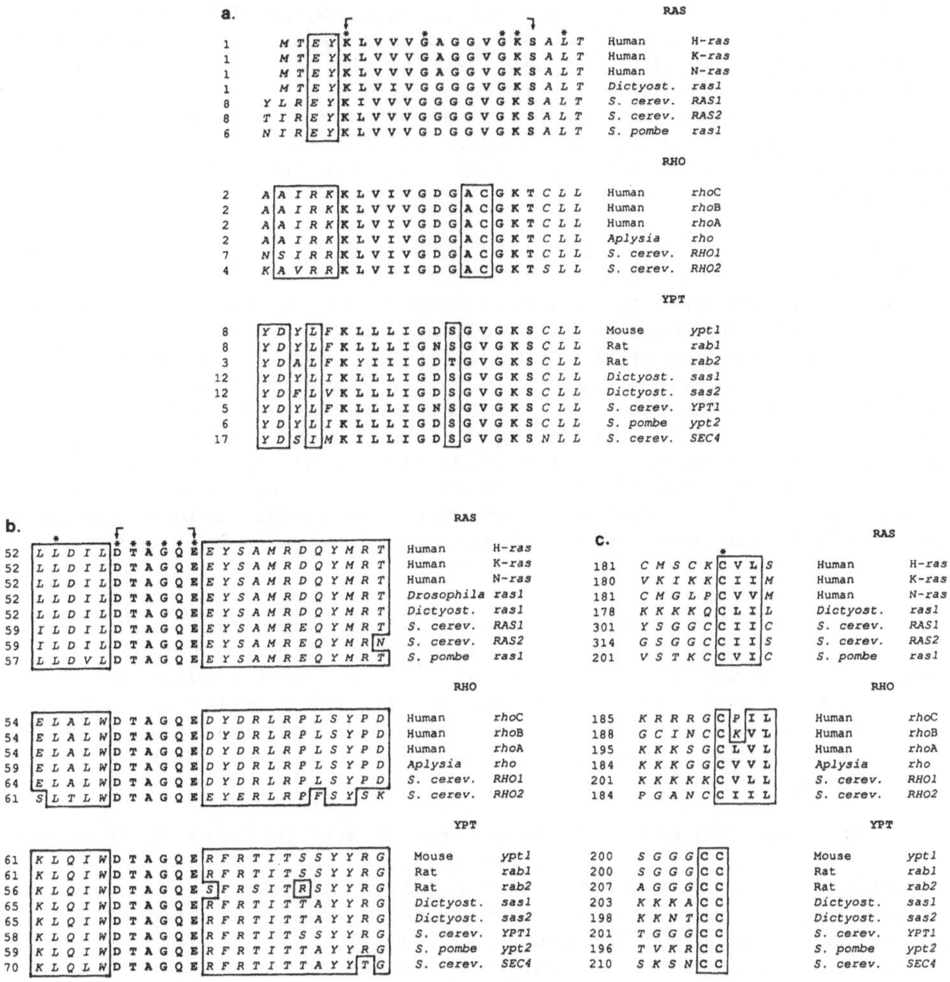

a.
RAS
1 M T E Y K L V V V G A G G V G K S A L T Human H-ras
1 M T E Y K L V V V G A G G V G K S A L T Human K-ras
1 M T E Y K L V V V G A G G V G K S A L T Human N-ras
1 M T E Y K L V I V G G G G V G K S A L T Dictyost. ras1
8 Y L R E Y K I V V V G G G G V G K S A L T S. cerev. RAS1
8 T I R E Y K L V V V G G G G V G K S A L T S. cerev. RAS2
6 N I R E Y K L V V V G D G G V G K S A L T S. pombe ras1
RHO
2 A A I R K K L V I V G D G A C G K T C L L Human rhoC
2 A A I R K K L V V V G D G A C G K T C L L Human rhoB
2 A A I R K K L V I V G D G A C G K T C L L Human rhoA
2 A A I R K K L V I V G D G A C G K T C L L Aplysia rho
7 N S I R R K L V I V G D G A C G K T C L L S. cerev. RHO1
4 K A V R R K L V I I G D G A C G K T S L L S. cerev. RHO2
YPT
8 Y D Y L F K L L L I G D S G V G K S C L L Mouse ypt1
8 Y D Y L F K L L L I G N S G V G K S C L L Rat rab1
3 Y D A L F K Y I I I G D T G V G K S C L L Rat rab2
12 Y D Y L I K L L L I G D S G V G K S C L L Dictyost. sas1
12 Y D F L V K L L L I G D S G V G K S C L L Dictyost. sas2
5 Y D Y L F K L L L I G N S G V G K S C L L S. cerev. YPT1
6 Y D Y L I K L L L I G D S G V G K S C L L S. pombe ypt2
17 Y D S I M K I L L I G D S G V G K S N L L S. cerev. SEC4
b.
RAS
52 L L D I L D T A G Q E E Y S A M R D Q Y M R T Human H-ras
52 L L D I L D T A G Q E E Y S A M R D Q Y M R T Human K-ras
52 L L D I L D T A G Q E E Y S A M R D Q Y M R T Human N-ras
52 L L D I L D T A G Q E E Y S A M R D Q Y M R T Drosophila ras1
52 L L D I L D T A G Q E E Y S A M R D Q Y M R T Dictyost. ras1
59 I L D I L D T A G Q E E Y S A M R E Q Y M R T S. cerev. RAS1
59 I L D I L D T A G Q E E Y S A M R E Q Y M R N S. cerev. RAS2
57 L L D V L D T A G Q E E Y S A M R E Q Y M R T S. pombe ras1
RHO
54 E L A L W D T A G Q E D Y D R L R P L S Y P D Human rhoC
54 E L A L W D T A G Q E D Y D R L R P L S Y P D Human rhoB
54 E L A L W D T A G Q E D Y D R L R P L S Y P D Human rhoA
59 E L A L W D T A G Q E D Y D R L R P L S Y P D Aplysia rho
64 E L A L W D T A G Q E D Y D R L R P L S Y P D S. cerev. RHO1
61 S L T L W D T A G Q E E Y E R L R P F S Y S K S. cerev. RHO2
YPT
61 K L Q I W D T A G Q E R F R T I T S S Y Y R G Mouse ypt1
61 K L Q I W D T A G Q E R F R T I T S S Y Y R G Rat rab1
56 K L Q I W D T A G Q E S F R S I T R S Y Y R G Rat rab2
65 K L Q I W D T A G Q E R F R T I T T A Y Y R G Dictyost. sas1
65 K L Q I W D T A G Q E R F R T I T T A Y Y R G Dictyost. sas2
58 K L Q I W D T A G Q E R F R T I T S S Y Y R G S. cerev. YPT1
59 K L Q I W D T A G Q E R F R T I T T A Y Y R G S. pombe ypt2
70 K L Q L W D T A G Q E R F R T I T T A Y Y T G S. cerev. SEC4
c.
RAS
181 C M S C K C V L S Human H-ras
180 V K I K K C I I M Human K-ras
181 C M G L P C V V M Human N-ras
178 K K K K Q C L I L Dictyost. ras1
301 Y S G G C C I I C S. cerev. RAS1
314 G S G G C C I I S S. cerev. RAS2
201 V S T K C C V I C S. pombe ras1
RHO
185 K R R R G C P I L Human rhoC
188 G C I N C C K V L Human rhoB
195 K K K S G C L V L Human rhoA
184 K K K G G C V V L Aplysia rho
201 K K K K K C V L L S. cerev. RHO1
184 P G A N C C I I L S. cerev. RHO2
YPT
200 S G G G C C Mouse ypt1
200 S G G G C C Rat rab1
207 A G G G C C Rat rab2
203 K K K A C C Dictyost. sas1
198 K K N T C C Dictyost. sas2
201 T G G G C C S. cerev. YPT1
196 T V K R C C S. pombe ypt2
210 S K S N C C S. cerev. SEC4

Fig. 3. Sequence comparison of segments of *ras*, *rho* and *ypt* proteins from different species. Sequences diagnostic for the three families of *ras*-related proteins that reside within or adjacent to the GTP-binding domains I (a) and II (b) or at the C-terminal end are boxed. Residues identical in all proteins are indicated by an asterisk above the H-*ras* sequence. Sequence data are from Barbacid (1987), *ras* proteins; Madaule et al. (1987), *Aplysia* rho and yeast *RHO* proteins; Chardin et al. (1988) and Yeramian et al. (1987), human *rho* proteins; Haubruck et al. (1987), mouse *ypt1* protein; Touchot et al. (1987), rat *rab* proteins; Saxe and Kimmel (personal communication), *Dictyostelium sas* proteins; Gallwitz et al. (1983), *S. cerevisiae Ypt1* protein; Salminen and Novick (1987), *S. cerevisiae Sec4* protein; and our unpublished results, *S. pombe ypt2* protein.

A prominent structural feature of *ypt* proteins is a serine instead of glycine-12 found in *ras* proteins, a residue critical for GTPase activity. The corresponding position in *rho* proteins is also occupied by a glycine, but a number of residues within the region forming loop 1 in *ras* proteins (De Vos et al., 1988) are either typical for *ras*, *rho* or *ypt* proteins and so are residues adjacent to this region (Figure 3a). The distinction between these three groups of GTP-binding proteins can also be made by comparing the sequences flanking the absolutely conserved region comprising residues -Asp-Thr-Ala-Gly-Gln-Glu- in position 57 to 62 (with respect to mammalian *ras* proteins) (Figure 3b). In addition, *ras* and *rho* proteins terminate with a cysteine followed by three other residues whereas *ypt* proteins typically end with two consecutive cysteines (Figure 3c).

As the cysteine residues of the C-terminal region in *ras* (Willumsen et al., 1984; Deschenes and Broach, 1987) and *ypt* proteins (Molenaar et al., 1988) are the site for palmitic acid binding which in turn is required for membrane attachment, we already speculated that the distinct structural differences of the C-terminal sequences of *ras* and *rho* proteins on the one hand and *ypt* proteins on the other might be a means to direct these proteins to different membrane compartments within the cell (Molenaar et al., 1988). However, the elongation of the C-terminus of the yeast *Ypt1* protein by three amino acids did not affect the protein's proper biological function (Molenaar et al., 1988) but there are indications from recent experiments that these additional residues are being cleaved off before membrane attachment of the protein (C.M. and D.G., unpublished).

That *ras* and *ypt* are functionally different groups of GTP-binding proteins is not only shown by their sequence differences but also by the finding that mammalian *ras* proteins are able to functionally replace the essential yeast *Ras* proteins (Kataoka et al., 1985; De Feo-Jones et al., 1985) and that, likewise, the mouse *ypt1* protein can perfectly substitute for the loss of the essential *YPT1* gene product in yeast (R.P., H.H. and D.G., in preparation).

The recent discovery of several *ras*-related proteins in different eukaryotes has led to some confusion regarding the nomenclature. For instance, a rat protein identical in primary structure to the mouse *ypt1* protein, the structural and functional homologue of the yeast *Ypt1* protein, has been named *rab1* (Touchot et al., 1987), and two proteins from *Dictyostelium discoideum* with more than 50 % amino acid identities compared to the yeast *Ypt1* protein have been designated *sas1* and *sas2* (A. Kimmel, personal communication). As outlined above, the different *ras*-related proteins are characterized by the highly conserved and identically spaced four nucleotide-binding domains, but other, rather specific sequence features allow the distinction of three families: *ras*, *rho* and *ypt*. Proteins not fitting into either of these groups could be named *ryh* (for ***r**as/**r**ho-**y**pt* **h**omologue). Members of either family in one species might be functionally homologous, like the *Ras1* and the *Ras2* protein of *S. cerevisiae*, or they might serve different and essential functions, like three *ypt* proteins in the fission yeast *S. pombe* that share more than 50 % of identical amino acid residues. Nevertheless, as proteins belonging to either family distinguished have been found in all eukaryotic kingdoms, it seems desirable to us to adopt a simplified nomenclature to avoid confusion.

ACKNOWLEDGEMENT

The work performed in the authors' laboratory has been supported by grants from the Deutsche Forschungsgemeinschaft and the Fonds der Chemischen Industrie. We thank Ingrid Balshüsemann for help in preparing the manuscript.

REFERENCES

Barbacid, M., 1987, *ras* genes,
Annu. Rev. Biochem., 56:779.

Bourne, H. R., 1988, Do GTPases direct membrane traffic in secretion?,
Cell, 53:669.

Chardin, P., Madaule, P., and Tavitian, A., 1988, Coding sequence of human *rho* cDNAs clone 6 and clone 9,
Nucl. Acids Res., 16:2117.

Chardin, P., and Tavitian, A., 1986, The *ral* gene: a new *ras* related gene isolated by the use of a synthetic probe,
EMBO J., 5:2203.

De Feo-Jones, D., Tatchell, K., Robinson, L. C., Sigal, I. S., Vass, W. C., Lowy, D. R., and Scolnick, E. M., 1985, Mammalian and yeast *ras* gene products: biological function in their heterologous systems,
Science, 228:179.

Deschenes, R. J., and Broach, J. R., 1987, Fatty acylation is important but not essential for *Saccharomyces cerevisiae RAS* function,
Mol. Cell Biol., 7:2344.

De Vos, A. M., Tong, L., Milburn, M. V., Matias, P. M., Jangarik, J., Noguchi, S., Nishimura, S., Miura, K., Ohtsuka, E., Kim, S.-H., 1988, Three dimensional structure of an oncogene protein: catalytic domain of human c-H-*ras*,
Sciene, 239:888.

Gallwitz, D., Donath, C., and Sander, C., 1983, A yeast gene encoding a protein homolgous to the human c-has/bas proto-oncogene product,
Nature, 306:704.

Goud, B., Salminen, A., Walworth, N. C., and Novick, P. J., 1988, A GTP-binding protein required for secretion rapidly associates with secretory vesicles and the plasma membrane in yeast,
Cell, 53:753.

Haubruck, H., Disela, C., Wagner, P., and Gallwitz, D., 1987, The *ras*-related *ypt* protein is an ubiquitous eukaryotic protein: isolation and sequence analysis of mouse cDNA clones highly homologous to the yeast *YPT1* gene,
EMBO. J., 6:4049.

Jurnak, F., 1985, Structure of the GDP domain of EF-Tu and location of the amino acids homologous to *ras* oncogene proteins,
Science, 230:32.

Kataoka, T., Powers, S., Cameron, S., Fasano, O., Goldfarb, M., Broach, J., and Wigler, M., 1985, Functional homology of mammalian and yeast *RAS* genes,
Cell, 40:19.

Lowe, D. G., Capon, D. J., Delwart, E., Sakaguchi, A. Y., Naylor, S. L., and Goeddel, D. V., 1987, Structure of the human and murine R-*ras* genes, novel genes closely related to *ras* proto-oncogenes,
Cell, 48:137.

Madaule, P., and Axel, R., 1985, A novel *ras*-related gene family,
Cell, 41:31.

Madaule, P., Axel, R., and Myers, A. M., 1987, Characterization of two members of the *rho* gene family from the yeast *Saccharomyces cerevisiae*,
Proc. Natl. Acad. Sci. USA, 84:779.

Molenaar, C. M. T., Prange, R., and Gallwitz, D., 1988, A carboxyl-terminal cysteine residue is required for palmitic acid binding and biological activity of the *ras*-related yeast *YPT1* protein,
EMBO J., 7:971.

Salminen, A., and Novick, P. J., 1987, A *ras*-like protein is required for a post-Golgi event in yeast secretion,
Cell, 49:527.
Schmitt, H. D., Puzicha, M., and Gallwitz, D., 1988, Study of a temperature-sensitive mutant of the *ras*-related *YPT1* gene product in yeast suggests a role in the regulation of intracellular calcium,
Cell, 53:635.
Schmitt, H. D., Wagner, P., Pfaff, E., and Gallwitz, D., 1986, The *ras*-related *YPT1* gene product in yeast: a GTP-binding protein that might be involved in microtubule organization,
Cell, 47:401.
Segev, N., Mulholland, J., and Botstein, D., 1988, The yeast GTP-binding *YPT1* protein and a mammalian counterpart are associated with the secretion machinery,
Cell, 52:915.
Tanabe, T., Nukada, T., Nishikawa, Y., Sugimoto, K., Suzuki, H., and Takahashi, H., Noda, M., Haga, T., Ichiyama, A., Kangawa, K., Minamino, N., Matsuo, H., and Numa, S., 1985, Primary structure of the α-subunit of transducin and its relationship to *ras* proteins,
Nature, 315:242.
Touchot, N., Chardin, P., and Tavitian, A., 1987, Four additional members of the *ras* gene superfamiliy isolated by an oligonucleotide strategy: molecular cloning of YPT-related cDNAs from a rat brain library,
Proc. Natl. Acad. Sci. USA, 84:8210.
Yeramian, P., Chardin, P., Madaule, P., and Tavitian, A., 1987, Nucleotide sequence of human rho cDNA clone 12,
Nucl. Acids Res., 15:1869.
Wagner, P., Molenaar, C. M. T., Rauh, A. J. G., Brökel, R., Schmitt, H. D., and Gallwitz, D., 1987, Biochemcical properties of the *ras*-related *YPT* protein in yeast: a mutational analysis,
EMBO J., 6:2373.
Willumsen, B. M., Norris, K., Papageorge, A. G., Hubbert, N. L., and Lowy, D. R., 1984, Harvey murine sarcoma virus p21 ras protein: biological signifcance of the cysteine nearest the carboxy terminus.
EMBO J., 3:2581.

ANALYSIS OF THE *RAS* GENE FUNCTION IN *DICTYOSTELIUM DISCOIDEUM*

C.D. Reymond[1], M.E.E. Ludérus[2], G.N. Europe-Finner[3], N.A. Thompson[1], E. Bürki[1], R. Van Driel[2] and P.C. Newell[3]

[1]Institut suisse de recherches expérimentales sur le cancer, CH-1066 Epalinges, Switzerland

[2]Department of Biochemistry, University of Amsterdam Plantage Muidergracht 12, 1018 TV Amsterdam, The Netherlands

[3]Department of Biochemistry, University of Oxford South Park Road, Oxford OX1 3QU, UK

The *Dictyostelium discoideum* genome contains a single *ras* gene encoding a putative protein related to mammalian *ras* (1). This gene is expressed in vegetative cells, growing in the presence of a food source. Upon starvation cell division stops and a specific developmental cycle is induced which leads to the formation of a fruiting body composed of only two parts, stalk and spores. The latter cells can give rise to vegetative cells in the presence of a new food source (for review see 2). Transcripts from the *Dictyostelium ras* gene can be detected in vegetatively growing cells and accumulate during late development only in prestalk cells. At that stage two mRNAs of different size can be detected. The *ras* protein steady state level present in vegetative cells decreases steadily during early development, then more abruptly passed the pseudoplasmodium stage. Studies by others showed that the *ras* protein is mainly synthesized in vegetative cells and during late development, corresponding to the detection of *ras* mRNA (3).

In order to analyze the sequences responsible for the regulation of the *ras* gene, we used a transformation system developed for *Dictyostelium discoideum* (4). DNA sequences, upstream of the gene were introduced into a transformation vector, together with different parts of the coding region (5). All the constructs were expressed upon re-introduction into *Dictyostelium*, thus indicating the presence of promoter sequences within a fragment about 650 bp upstream of the AUG start codon, as well as of a sufficient amount of transcription factor(s) in these cells. This DNA region contains a direct 7 bp repeat (AACACAC), which can also be found as an inverted repeat within the regulatory region of another prestalk specific gene, pst-cathepsin (6). Using synthetic oligonucleotides, either one or both elements of the repeat were specifically deleted. Re-introduction of genes containing such deletions into *Dictyostelium* showed no RNA accumulation at any stage during development, even though another gene present on the same plasmid (neo^R)[1] was expressed. These results indicate the importance of the 7 bp repeat element in the accumulation of *ras* gene transcripts. Preliminary results indicate that this element may be included in the 5' untranslated region of *ras* mRNA, thus suggesting a possible posttranscriptional role. However, we are still unable to distinguish which sequences are responsible for the regulation of the *ras* gene during development.

The previous results showed that *ras* gene constructs modified *in vitro* can be efficiently expressed upon re-introduction into *Dictyostelium discoideum* cells. Since amino acid changes at a restricted number of positions within mammalian *ras* proteins produced a transformed phenotype (7), we introduced a missense mutation within the *Dictyostelium ras* gene at position encoding the amino acid 12 (8). After transformation, cells containing 200 copies of non- mutated gene (*ras*-Gly_{12}) behaved like wild type cells, whereas the presence of a threonine instead of the original glycine at position 12 (*ras*-Thr_{12}) resulted in a morphologically modified development. When *ras*-Thr_{12} cells are starved at low density, they are unable to form streams of cells migrating chemotacticly toward aggregation centres like wild type strains. A further peculiarity of *ras*-Thr_{12} cells is the formation of multiple tips when allowed to develop at high cell density (Fig.1). Both observations at high and low cell density are consistent with the hypothesis that the *ras*

[1]Reymond & Firtel, unpublished.

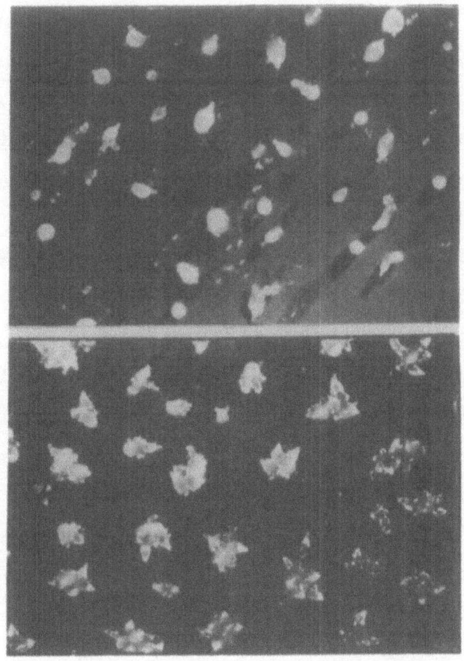

Fig1 *ras*-Thr_{12} phenotype: Ax3 (Kessin)cells (wt, upper panel) and *ras*-Thr_{12} cells (lower panel) were allowed to develop on black nitrocellulose filters saturated with PDF (17). Under these conditions development occurs synchronously and is accomplished in about 27 hours at 21 ^{o}C. The pictures were taken after about 17 hours of development. Ax3 cells formed typical pseudoplasmodiums showing a narrow tip containing the prestalk cells and a mass of prespore cells, whereas *ras*-Thr_{12} cells formed multiple tips.

mutation affects cAMP signalling in *Dictyostelium*.

cAMP is used as chemoattractant during the aggregation phase in *Dictyostelium* (for review see 2). In pseudoplasmodiums, prestalk cells keep their ability to sense cAMP. It was shown that cAMP binds specifically to different classes of receptors on the cell surface according to their kinetics. The signal is then transmitted via G proteins to the adenylate cyclase. A pulse of external cAMP increases adenylate cyclase activity leading to the secretion of cAMP as a signal relay mechanism. Even though the observation of the *ras*-Thr_{12} phenotype indicated an impaired cAMP signaling, the level of cAMP secreted by these cells was indistinguishable from control cells (9). Adenylate cyclase activity as well as steady state receptor level and affinity for cAMP were unchanged (8) in *ras*-Thr_{12} cells.

Previous results indicated the presence in *Dictyostelium* of a second independent pathway linked to cAMP receptors. Starved *Dictyostelium* cells show a stimulation of InsP3 formation after binding of cAMP to cell surface receptors (10), followed by oscillations in its level within the cell. As a result of InsP3 increase, Ca^{2+} is released from internal stores (11) which leads to multiple effects within the cell, one of which is the activation of guanylate cyclase (12). We measured directly inositol polyphosphate levels in the strain containing mutated *ras* genes. At time intervals after a stimulus of cAMP, inositol polyphosphates were separated on Dowex columns. The basal level of InsPX is elevated about threefold in *ras*-Thr_{12} cells as compared to control cells (13). Using HPLC separation techniques, we were able to show that both InsP3 and InsP6 are increased by three to fivefold in this strain. These findings implicate a role for the *ras* protein in the signal transduction pathway leading from the cell surface cAMP receptor to the formation of Inositol polyphosphates.

Other experiments (Collaboration with P. Van Haastert, Leiden) indicate that the presence of mutated *ras* affects cGMP accumulation (9). In *Dictyostelium discoideum*, intracellular cGMP is only transiently raised in response to the binding of cAMP to its receptor (14). Repeated cAMP pulses will lead to desensitization, so that the level of cGMP reached in the cells will be reduced. A phenomenon implicating cAMP receptor, but not adenylate cyclase activity. The maximal level of cGMP reached in *ras*-Thr_{12} cells after one cAMP pulse is reduced by about half as compared to wild type or *ras*-Gly_{12} cells. Furthermore guanylate cyclase in *ras*-Thr_{12} cells seem already desensitized upon repetition of cAMP pulsing (9). Since guanylate cyclase activation has been linked to a cytoplasmic increase in Ca^{2+} concentration, these results are consistent with the implication of *ras* in the inositol polyphosphate pathway. However, one would expect either an increase in cGMP level or an hyper-activation of guanylate cyclase in response to elevated InsP3. A better understanding of the elements of this intricate pathway should lead to the solution of this discrepancy.

Both Ca^{2+} and cGMP have been implicated in the production of coordinate chemotactic movements (for review see 15). Part of the cell shape changes may result from cytoskeletal rearrangements. We used a monoclonal anti- α-actinin antibody (kindly provided by M. Schleicher, Münich) to analyze *Dictyostelium* cytoskeleton (Fig.2). A large proportion of *ras*-Thr_{12} cells showed a patchy staining with this antibody, whereas control cells (*ras*-Gly_{12}) accumulate α-actinin close to the cell membrane. Since α-actinin is known to crosslink actin filaments, these results suggest that the presence of a mutated *ras* influences actin polymerisation, possibly via a change in Ca^{2+} concentration within the cell.

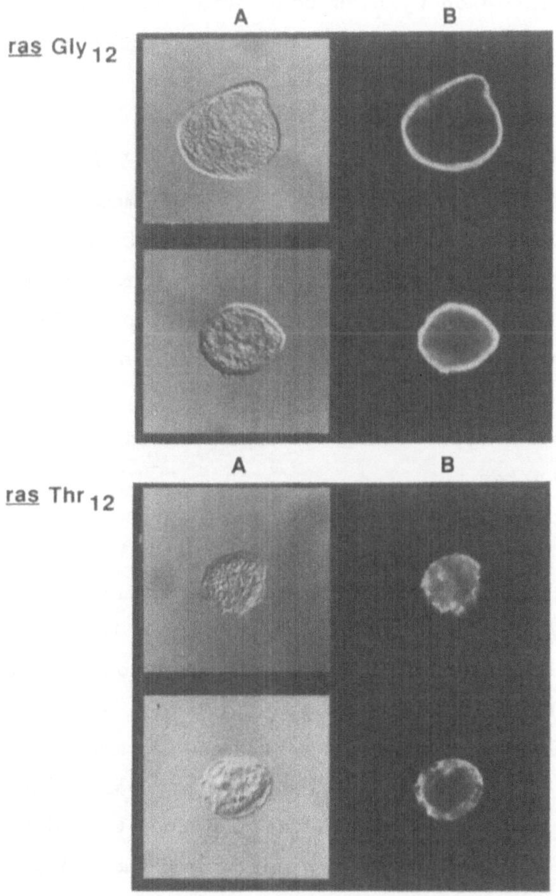

Fig. 2 α-actinin localisation: Cells were allowed to sediment on coverslips for 5 min. and extracted by cold methanol (-80 ^{o}C for 3 min.). After air drying they were incubated for 5 min. in anti α-actinin containing PBS, rinsed twice in PBS and stained with fluorescein conjugated goat anti mouse antibody. After two further rinses in PBS, the preparation was mounted for microscopy in 80% glycerol. α-actinin staining is observed around the cell, close to the cortex both in Ax3 (not shown) and *ras*-Gly_{12} (upper panel). *ras*-Thr_{12} cells (lower panel) show a more patchy localisation throughout the cell. A significant proportion of *ras*-Thr_{12} cells showed a staining resembling wild type cells. Controls omitting the first antibody showed a very weak residual staining evenly distributed over whole cells of both strains.

We further asked whether *ras* may influence cAMP receptor properties themselves. We modified an *in vitro* receptor assay system (16). The number of cAMP receptors on crude membrane preparations can be reduced by addition of ATP and Ca^{2+}. This decrease occurs at a lower Ca^{2+} concentration using preparation of *ras*-Thr_{12} membranes than with *ras*-Gly_{12} or wild type cells. Furthermore, addition of GTP reduced the Ca^{2+} concentration required to decrease receptor number in *ras*-Gly_{12} and wild type cells, but did not affect *ras*-Thr_{12} cAMP receptors. These results indicate that the effect of the *ras* mutation can be mimicked by high concentrations of GTP, but not GDP. The requirement for ATP and calcium to decrease the receptor number in these experiments suggested the involvement of protein kinase C, even though the detection of such enzyme activity in *Dictyostelium* has remained tentative. Addition of PMA (or TPA, a phorbol ester known to activate protein kinase C) to membrane preparations resulted in a effect similar to the addition of GTP. Furthermore the decrease in receptor number is specific to PMA, since it was not observed using a phorbol ester analog 4 alpha PDD. The maximal decrease in the number of cAMP receptors in all these experiments reached only about 50 %. One thus has to postulate the existence of different receptor types half of which would be influenced by a pathway involving *ras*.

In conclusion, the results presented here indicate that *ras* is involved in a cAMP signal transduction pathway in *Dictyostelium*, but has no influence on adenylate cyclase activation. A mutation in *ras* influences InsP3 basal level and probably also diacylglycerol concentration in the cell, since we see a modification of cAMP receptor number which can be mimicked by phorbol ester addition. We thus believe that *ras* participates in the transduction of cAMP signals to modification of cell properties, including cytoskeleton, via a pathway probably involving a protein kinase C in *Dictyostelium*. cAMP receptor number at the cell surface is in turn under the control of this pathway.

In the previous experiments we made use of a *Dictyostelium* strain containing about 200 extra copies of a mutated *ras* gene. The cellular *ras* gene was still present in these cells, thus allowing the detection of dominant effects only. We have attempted to reduce cellular *ras* gene expression using anti-sense approaches, but could not obtain stable cell lines. Further efforts to use a heat shock inducible promoter failed (unpublished). We are currently attempting to use homologous recombination (17) to inactivate the cellular *ras* gene.

REFERENCES

1. Reymond, C. D., Gomer, R. H., Mehdy, M. C. & Firtel, R. A., 1984, Developmental regulation of a Dictyostelium gene encoding a protein homologous to mammalian ras protein, Cell, 39:141.
2. Loomis, W. F., 1982, The development of Dictyostelium discoideum. Academic Press, Inc.
3. Weeks, G. & Pawson, T., 1987, The synthesis and degradation of ras-related gene products during growth and differentiation in Dictyostelium discoideum, Differentiation, 33:207.
4. Nellen, W. & Firtel, R. A., 1985, High-copy number transformants and co-transformation in Dictyostelium, Gene, 39:155.
5. Reymond, C. D., Nellen, W. & Firtel, R.A., 1985, Regulated expression of ras gene constructs in Dictyostelium transformants, Proc. Natl. Acad. Sci. USA, 82:7005.
6. Datta, S. & Firtel, R. A., 1987, Identification of the sequences controlling cyclic AMP regulation and cell-type-specific expression of a prestalk-specific gene in Dictyostelium discoideum, Mol. Cell. Biol., 7:149.
7. Seeburg, P. H., Colby, W. W., Capon, D. J., Goeddel, D. V. & Levinson, A. D., 1984, Biological properties of human c-Ha-ras1 genes mutated at codon 12, Nature, 312:71.
8. Reymond, C. D., Gomer, R. H., Nellen, W., Theibert, A., Devreotes, P. & Firtel, R., 1986, Phenotypic changes induced by a mutated ras gene during the development of Dictyostelium transformants, Nature, 323:340.
9. Van Haastert, P. J. M., Kesbeke, F., Reymond, C. D., Firtel, R. A., Luderus, E. & Van Driel, R., 1987, Aberrant transmembrane signal transduction in Dictyostelium cells expressing a mutated ras gene. Proc. Natl. Acad. Sci. USA, 84:4905.
10. Europe-Finner, G. N. & Newell, P. C., 1987, Cyclic AMP stimulates accumulation of inositol triphosphate in Dictyostelium, J. Sci., 87:221.
11. Europe-Finner, G. N. & Newell, P. C., 1986, Inositol 1,4,5-triphosphate induces calcium release from a non-mitochondrial pool in amoebae of Dictyostelium, Biochim. Biophys. Acta, 887:335.
12. Europe-Finner, G. N. & Newell, P. C., 1985, Inositol 1,4,5-triphosphate induces cyclic GMP formation in Dictyostelium discoideum, Biochem. Biophys. Res. Commun., 130:1115.
13. Europe-Finner, G. N., Ludérus, M. E. E., Small, N. V., Van Driel, R., Reymond, C. D., Firtel, R. A. & Newell, P. C., 1988, Mutant ras gene induces elevated levels of inositol tris- and hexakisphosphates in Dictyostelium, J. Cell. Sci., 89:13.
14. Van Haastert, P. J. M. & Van der Heijden, P. R., 1983, Excitation, adaption, and deadaption of the cAMP-mediated cGMP response in Dictyostelium discoideum, J. Biol., 96:347.
15. Newell, P. C., Europe-Finner, G. N., Small, N. V. & Liu, G., 1988, Inositol phosphate, G-proteins and ras genes involved in chemotactic signal transduction of Dictyostelium, J. Sci., 89:123.
16. Ludérus, M. E. E., Reymond, C. D., Van Haastert, P. J. M. & Van Driel, R., Expression of a mutated ras gene in Dictyostelium alters the binding of cAMP to its chemotactic receptor, J. Sci., in press.
17. De Lozanne, A. & Spudich, J., 1987, Disruption of the Dictyostelium myosin heavy chain gene by homologous recombination, Science, 236:1086.

18. Nellen, W., Datta, S., Reymond, C. D., Sivertsen, A., Mann, S., Crowley, T. & Firtel, R. A., 1987, Molecular Biology in Dictyostelium: Tools and applications, Methods in Cell Biology, 28:67.
19. Wallraff, E., Schleicher, M., Medersitzki, M., Rieger, D., Isenberg, G. & Gerisch, G., 1986, Selection of Dictyostelium mutants defective in cytoskeletal protein: use of an antibody that binds to the ends of a-actinin rods, EMBO J., 5:61.

IDENTIFICATION OF GUANINE-NUCLEOTIDE BINDING PROTEINS IN PLANTS: STRUCTURAL ANALYSIS AND EVOLUTIONARY COMPARISON OF THE RAS-RELATED YPT-GENE FAMILY FROM ZEA MAYS

Klaus Palme[1], Thomas Diefenthal[1], Chris Sander[2], Martin Vingron[2], and Jeff Schell[1]

[1]Max Planck Institut für Züchtungsforschung D-5000 Köln 30, and
[2]European Molecular Biology Laboratory Postfach 102209, D-6900 Heidelberg, FRG

INTRODUCTION

The development of plants is regulated by five types of hormones known as auxins, cytokinins, gibberellins, abscisic acid and ethylene (Phillips, 1971). Only little is known about the mechanisms of action of these hormones at the cellular and molecular level. Genetic analysis of mutants with increased resistance to growth inhibiting concentrations of hormones argues for the presence of receptor like functions in plants (King, 1988). Biochemical evidence suggests that the action of auxins may be mediated after binding to plasmalemma located receptors (for review see: Davies, 1988). However, our current picture on other elements in the plant cell, involved in the transmission of this signal to its final biological target, is only slowly emerging. Recent evidence strengthens the importance of the calcium signal in plant cells (Marme, 1983; Hepler & Wayne, 1985). Transient changes in cytoplasmic calcium levels coupled to external stimuli such as light and phytohormones can modulate numerous developmental responses (Hepler & Wayne, 1985). Furthermore, components of the phosphoinositide signalling system, playing a pivotal role in intracellular communication in mammalian cells (Berridge, 1987), have been detected in plants as well (Boss & Massel, 1985; Heim & Wagner, 1986; Heim et al., 1987). Most interestingly, auxin specific signal reception appears to be linked to this signalling pathway. Thus, addition of auxins to susceptible plant cells leads to rapid cleavage of L-α-phosphatidylinositol-4,5-diphosphate and transient production of inositol-1,4,5-triphophate (Ettlinger & Lehle, 1987), a signal now also known in plants to stimulate calcium release from the endoplasmic reticulum and vacuoles (Schumaker & Sze, 1987; Ranjeva et al., 1988). The increase of cytosolic calcium concentration appears to transmit and amplify the hormonal signal by modulating cellular activities at least partly through a phosphorylation dependent signalling network (Palme et al., 1987; Perez et al., 1987).

Recent studies indicate that a GTP-binding protein might be involved in the control of calcium release from the endoplasmic reticulum (Gill et al. 1986; Scott et al., 1987). Although changes in the cytoplasmic calcium concentration can regulate voltage dependent ion channels in plant cells (Hedrich & Neher, 1987), the presence of regulatory GTP-binding proteins in plants has not yet been demonstrated. Only indirect studies suggest that GTP can influence a thylakoid protein kinase and that GTP-binding proteins may be involved in the stimulation of flowering (Millner, 1987; Hasunuma & Funadera, 1987).

Despite our present lack of knowledge on the identity and function of putative plant specific GTP-binding proteins, we reasoned that elements of the guanine nucleotide binding domain might be conserved in plants. Our studies show that genes encoding proteins with strongly conserved GTP-binding domains can be isolated from plants. In this report we discuss the structural analysis of maize genes related to the ras super-gene family. YPT-genes involved in both mitotic and meiotic control represent a ras-related gene family. We have isolated members of this ubiquitous gene family from corn (Zea mays L.). These YPTm gene products fit the criteria for ras-like proteins; they posess four, structurally well conserved domains necessary for GTP-binding.

IDENTIFICATION AND ANALYSIS OF YPT-GENE FAMILY FROM ZEA MAYS

Ras-genes from organisms as widely divergent as human, mouse, Drosophila, Aplysia, and Saccharomyces were used to screen for ras-related genes in plants by DNA hybridisation experiments. We were, however, not successful in isolating corresponding plant genes and therefore searched for common primary structure motifs in ras-related proteins that could be used for the construction of more specific oligonucleotide probes.

Direct sequence comparison of all recently isolated members of the ras super-gene family by multiple alignments of the amino acids of the GTP-binding domain is presented in Table 1. Comparison of amino acid sequences representing the GTP-binding domain of all presently known members of the ras super-gene family points to three consensus elements, G x x x x G K S s x l, D T A G Q E, and l x g N K x D L. As indicated by the threedimensional structure of the c-H- ras-p21 oncogen (De Vos et al., 1987), the first two domains are involved in binding the phosphate moiety of the GTP-molecule, whereas elements located further downstream of the sequence are involved in determinining guanine nucleotide specificity. More extensive amino acid sequence conservation is only detectable in subfamilies as demonstrated by distance matrix comparison (Fig. 1).

	Ha	Ki	N	Dr1	Dict	RAS1	Ras2	Dro3	Dro2	R	ral	Sp	rho	rab3	rab2	rab4	rab1	YPTy	YPTm	SEC4
Ha-ras	0	9	13	26	49	59	56	80	66	66	73	52	118	110	107	108	101	101	98	107
Ki-ras	9	0	10	25	50	58	57	80	67	66	72	57	119	111	106	109	100	102	99	108
N-ras	13	10	0	27	47	57	54	79	69	67	73	54	120	110	107	108	98	100	98	106
Dros1	26	25	27	0	54	60	57	84	67	67	74	56	115	112	106	109	105	105	105	109
Dict.	49	50	47	54	0	48	46	82	69	66	72	44	120	105	103	102	108	104	103	106
RAS1,yea	59	58	57	60	48	0	25	79	66	64	75	48	120	101	101	100	103	101	102	102
RAS2,yea	56	57	54	57	46	25	0	78	69	69	74	46	116	108	100	104	103	100	102	107
Dros3	80	80	79	84	82	79	78	0	94	89	88	84	126	115	112	119	114	111	111	108
Dros2	66	67	69	67	69	66	69	94	0	55	83	65	129	115	116	117	111	111	113	113
R-ras	66	66	67	67	66	64	69	89	55	0	78	53	122	108	103	100	106	103	101	112
ral	73	72	73	74	72	75	74	88	83	78	0	76	117	104	103	107	99	96	104	110
Spras	52	57	54	56	44	48	46	84	65	53	76	0	122	105	102	104	105	103	103	105
rho,Apl	118	119	120	115	120	120	116	126	129	122	117	122	0	125	120	121	120	122	122	125
rab3	110	111	110	112	105	101	108	115	115	108	104	105	125	0	91	96	76	80	78	74
rab2	107	106	107	106	103	101	100	112	116	103	103	102	120	91	0	64	81	86	82	82
rab4	108	109	108	109	102	100	104	119	117	100	107	104	121	96	64	0	83	79	82	87
rab1	101	100	98	105	108	103	103	114	111	106	99	105	120	76	81	83	0	33	41	67
YPT,yea	101	102	100	105	104	101	100	111	111	103	96	103	122	80	86	79	33	0	45	72
YPT,mai	98	99	98	105	103	102	102	111	113	101	104	103	122	78	82	82	41	45	0	67
SEC4,yea	107	108	106	109	106	102	107	108	113	112	110	105	125	74	82	87	67	72	67	0

Figure 1 Distance matrix comparison of the ras-related proteins shown in Table 1. Distances were calculated as counts of mismatches. A mismatch is non-identity of amino acids. Gaps are treated as mismatches.

Table 1 Amino acid sequence comparison of the conserved domains of ras-related proteins. The consensus line has entirely conserved (in 20/20) amino acids as capital and very well conserved (in 17/20 or YF in 20/20) ones as lower case letters. Secondary structure (H= helix, E = extended (beta strand)), loop names (L1-L9) and functional sites (p = involved in binding of GTP phosphate, g = interaction with GTP sugar) are taken from the report of the first three dimensional H-ras structure (de Vos et al., 1987). All sequences except for YPTm (this paper) are taken from either the PIR or the EMBL sequence database. Ha-, Ki-, and N-ras as well as R-ras are human sequences. rab1,2,3, and 4 are from rat, ral is from common marmoset, an ape. Dict. stands for dictyostelium, Apl for aplysia. Dros1,2, and 3 are from drosophila. RAS1 and 2, YPT and SEC4 are from S. cerevisiae, Spras is from Schizosaccharomyces pombe.

```
struc.      EEEEEEE     HHHHHHH           EEEEEE.EEEEE         EEEEEEEE                  EEEEE     HHH
loops              ..L1.       .....L2.....            ....L3...        ......L4.........     .L5..
func.            ppppppp          g s
cons.       K    G  gvGKs l        f      T    y                l   dTAGqE y      y r   g   v
Ha-ras      KLVVVGAGGVGKSALTIQLIQNHFVDEYDPTIEDSYR-KQVVIDG---ETCLLDILDTAGQEEYSAMRDQYMRTGEGFLCVFAINNTKSF
Ki-ras      KLVVVGAGGVGKSALTIQLIQNHFVDEYDPTIEDSYR-KQVVIDG---ETCLLDILDTAGQEEYSAMRDQYMRTGEGFLCVFAINNTKSF
N-ras       KLVVVGAGGVGKSALTIQLIQNHFVDEYDPTIEDSYR-KQVVIDG---ETCLLDILDTAGQEEYSAMRDQYMRTGEGFLCVFAINNSKSF
Dros1       KLVVVGPGGVGKSALTIQLIQNHFVDEYDPTIEDSYR-KQRFIDG---ETCLLDILDTAGQEEYSAMRDQYMRTGEGFLLVFAINSAKSF
Dict.       KLVIVGGGGVGKSALTIQLIQNHFIDEYDPTIEDSYR-KQVSIDD---ETCLLDILDTAGQEEYSAMRDQYMRTGQGFLCVYSITSRSSY
RAS1,yea    KIVVVGGGGVGKSALTIQFIQSYFVDEYDPTIEDSYR-KQVVIDD---KVSILDILDTAGQEEYSAMREQYMRTGEGFLLVYSVTSRNSF
RAS2,yea    KLVVVGGGGVGKSALTIQLTQSHFVDEYDPTIEDSYR-KQVVIDD---EVSILDILDTAGQEEYSAMREQYMRNGEGFLLVYSITSKSSL
Dros3       KIVVLGSGGVGKSALTVQFVQCIFVEKYDPTIEDSYR-KQVKVNE---RQCMLEIVNTAGTEQFTAMRNLYMKNGSDSCWSTRSRRNRRL
Dros2       KLVVVGGGGVGKSAITIQFIQSYFVTDYDPTIEDSY-TKQCNIDDIHNNLIFYLVLDTAGQEEFSAMREQYMRSGEGFLLVFALNDHSSF
R-ras       KLVVVGGGGVGKSALTIQFIQSYFVSDYDPTIEDSY-TKICSVDG---IPARLDILDTAGQEEFGAMREQYMRAGHGFLLVFAINDRQSF
ral         KVIMVGSGGVGKSALTLQFMYDEFVEDYEPTKADSYR-KKVVLDG---EEVQIDILDTAGQEDYAAIRDNYFRSGEGFLCVFSITEMESF
Spras       KLVVVGDGGVGKSALTIQLIQSHFVDEYDPTIEDSYR-KKCEIDG---EGAVLDLLDTAGQEEYSAMREQYMRTGQGFLLVYNITSRSSF
rho,Apl     KLVIVGDGACGKTCLLIVFSKDQFPEVYVPTVFENY-VADIEVDG---KQVELALWDTAGQEDYDRLRPLSYPDTDVILMCFSIDSPDSL
rab3        KILIIGNSSVGKTSFLFRYADDSFTPAFVSTVGIDFKVKTIYRND---KRIKLQIWDTAGQERYRTITTAYYRGAMGFILMYDITNEESF
rab2        KYIIIGDTGVGKSCLLLQFTDKRFQPVHDLTMGVEFGARMITIDG---KQIKLQIWDTAGQESFRSITRSYYRGAAGALLVYDITRRDTF
rab4        KFLVIGNAGTGKSCLLHQFIEKKFKDDSNHTIGVEFGQKIINVGG---KYVKLQIWDTAGQERFRSVTTSYYRGAAGALLVYDITSRETY
rab1        KLLLIGDSGVGKSCLLLRFADDTYTESYISTIGVDFKIRTIELDG---KTIKLQIWDTAGQERFRTITSSYYRGAHGIIVVYDVTDQESF
YPT,yea     KLLLIGNSGVGKSCLLLRFSDDTYTNDYISTIGVDFKIKTVELDG---KTVKLQIWDTAGQERFRTITSSYYRGSHGIIIVYDVTDQESF
YPT,mai     KLLLIGDSSVGKSCFLLRFADDSYVDSYISTIGVDFKIRTVEVEG---KTVKLQIWDTAGQERFRTITSSYYRGAHGIIIVYDITDMESF
SEC4,yea    KILLIGDSGVGKSCLLVRFVEDKFNPSFITTIGIDFKIKTVDING---KKVKLQLWDTAGQERFRTITTAYYRGAMGIILVYDVTDERTF
```

```
struc.      HHHHH.HHHHHHHHH    EEEEEEE                       HHHHHHHHHH  E.EEEEE       HHHHHHHHHHH
loops                      .L6..      ....L7..             ..          L8        ..L9..
func.                                ggg g                                       ggg
cons.                              lvgNK Dl                 V        a       f E SAk    V   F
Ha-ras      EDIHQ-YREQIKRVKDSDDVPMVLVGNKCDLAART-------------VESRQAQDLARSYGI-PYIETSAKTRQGVEDAFYTLVR
Ki-ras      EDIHH-YREQIKRVKDSEDVPMVLVGNKCDLPSRT-------------VDTKQAQDLARSYGI-PFIETSAKTRQGVDDAFYTLVR
N-ras       ADINL-YREQIKRVKDSDDVPMVLVGNKCDLPTRT-------------VDTKQAHELAKSYGI-PFIETSAKTRQGVEDAFYTLVR
Dros1       EDIGT-YREQIKHVKDAEEVPMVLAGNKCDLASWN-------------VNNEQAREVAKQYGI-PYIETSAKTRMGVDDAFYTLVR
Dict.       DEIAS-FREQILRVKDKDRVPLILVGNKADLDHERQ------------VSVNEGQELAKD-SL-SFHESSAKSRINVEEAFYSLVR
RAS1,yea    DELLS-YYQQIQRVKDSDYIPVVVVGNKLDLENERQ------------VSYEDGLRLAKQLNA-PFLETSAKQAINVDEAFYSLIR
RAS2,yea    DELMT-YYQQILRVKDTDYVPIVVVGNKSDLENEKQ------------VSYQDGLNMAKQMNA-PFLETSAKQAINVEEAFYTLAR
Dros3       -TICR-TREQILRVKDTDDVPMVLVGNKCDLEEERV------------VGKELGKNLATQFNC-AFMETSAKAKVNVNDIFYDWSG
Dros2       DEIPK-FQRQILRVKDRDEFPMLMVGNKCDLKHQQQ------------VSLEEAQNTSRNLMI-PYIECSAKLRVNVDQAFHELVR
R-ras       NEVGK-LFTQILRVKDRDDFPVVLVGNKADLESQRQ------------VPRSEASAFGASHHV-AYFEASAKLRLNVDEAFEQLVR
ral         AATAD-FREQILRVKEDENVPFLLVGNKSDLEDKRQ------------VSVEEAKNRADQWNV-NYVETSAKTRANVDKVFFDLMR
Spras       DEIST-FYQQILRVKDKDTFPVVLVANKCDLEAERV------------VSRREREQLAKSMHC-LYVETSAKLRLNVEEAFYSLVR
rho,Apl     ENIPEKWTPEVRHF--CPNVPIILVGNKKDLRNDESTKREIMKMKQEPVRPEDGRAMAEKINAYSYLECSAKTKEGVRDVFETATR
rab3        NAVQD-WSTQIKTYS-WDNAQVLLVGNKCDMEDERV------------VSSERGRQLADHLGF-EFFEASAKDNINVKQTFERLVD
rab2        NHLTT-WLEDARQHS-NSNMVIMLIGNKSDLESRRE------------VKKEEGEAFAREHGL-IFMETSAKTASNVEEAFINTAK
rab4        NALTN-WLTDARMLA-SQNIVIILCGNKKDLDADRE------------VTFLEASRFAQENEL-MFLETSALTGENVEEAFMQCAR
rab1        NNVKQ-WLQEIDRYA-SENVNKLLVGNKCDLTTKKV------------VDYTTAKEFADSLGI-PFLETSAKNEKNVEQSFMTMAA
YPT,yea     NGVKM-WLQEIDRYA-TSTVLKLLVGNKCDLKDKRV------------VEYDVAKEFADANKM-PFLETSALDSTNVEDAFLTMAR
YPT,mai     NNVKQ-WLDEIDRYA-NDSVRNVLVGNKCDLAENRA------------VDTSVAQAYAQEVGI-PFLETSAKESINVEEAFLAMSA
SEC4,yea    TNIKQ-WFKTVNEHA-NDEAQLLLVGNKSD-METRV------------VTADQGEALAKELGI-PFIESSAKNDDNVNEIFFTLAK
```

To look for plant related members of the ras super-gene family, we decided to use oligonucleotides synthesized corresponding to the amino acid sequences 5-20, 53-62, and 107-118 of the c-H-ras gene. A mixture of these oligonucleotides was used under optimum stringency conditions to screen a gt10 cDNA library. We were able to isolate 20 positive clones from a maize cDNA library giving signals with varying intensity. The recombinant phages were grouped in several classes and cDNA inserts of several of these phages were subcloned in pUC119 and nucleotide sequences were determined. As an example, the nucleotide sequence of clone YPTm1 is shown in Figure 2 (Diefenthal et al., 1988). The DNA sequence was compared to the sequences in the data base maintained by the EMBL. Interestingly this comparison revealed 54.9% similarity with the yeast YPTm1 coding sequence. 5' and 3'-untranslated regions show no significant homology. Shaded areas in Figure 2 represent the areas homologous to the oligonucleotides used for screening, having similarities of 66.6 and 77.8%. The homologies with the probes were partial, corresponding to 25 nucleotides at most. Artificial hybridisation to particularly GC rich positions was not noted in this case, but occurred in several others. The deduced amino acid sequence of YPTm1 aligned to other YPT-proteins is shown in Figure 3. The percentage of amino acids strictly conserved in YPTm1 as compared to yeast YPT1 is 60.7%. Four regions corresponding to residues 10-21, 58-69, 118-125 and 146-150 are almost identical throughout the ras-gene family. In addition, amino acids 70-80 are highly conserved within the YPT-subfamily including some of the rab proteins and distinguishes this group clearly from the ral and rho genes. As in all other ras-proteins two cysteine residues are found near the carboxyl-terminus of the maize protein. Stable association with membranes seems to be the result of postranslational modification of the cysteine marked with an asterix in Figure 3, whereby palmitic acid becomes bound via a thioester linkage. Analysis of mutants of the yeast YPT-protein has in fact demonstrated that this cysteine is required for membrane attachment and biological function (Molenaar et al., 1988). In contrast to all other known ras-related sequences, the five carboxy-terminal residues around the two cysteines are clearly different in the identified maize proteins. It should be noted that up to now all YPT-related proteins from plants have hydroxylated amino acids (Ser, Thr) at their C-terminus, which most probably determine the substrate specificity of plant fatty acid acylases. This indicates that the plant YPT-genes represent a new and separate branch of the ras super-gene family.

In order to obtain an evolutionary relationship of ras-related proteins, a multiple sequence alignment was prepared (Vingron and Argos, 1988). Distances were calculated based on these alignments by simply counting mismatches and adding penalties for insertions/deletions. Two different algorithms (Felsenstein, 1985) were applied to the data in order to obtain a phylogenetic tree. The Fitch-Margoliash approach (Fitch and Margoliash, 1967) and the parsimony approach (as implemented for proteins by J. Felsenstein) resulted in very similar trees. The hand-drawn consensus tree given in Figure 4 reflects the result of these examinations.

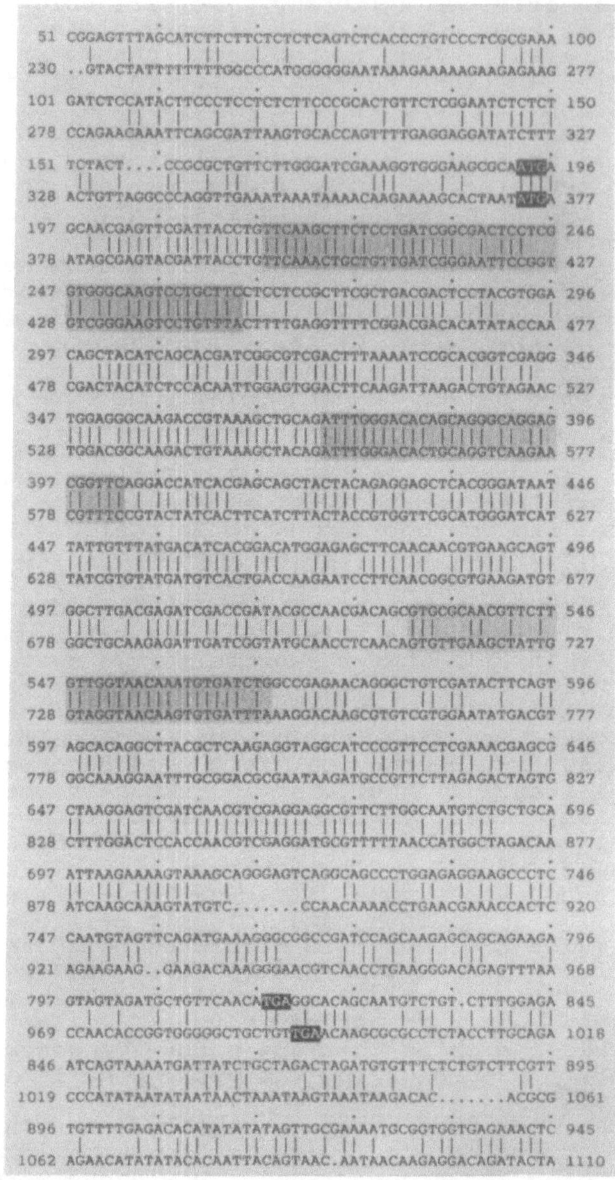

Figure 2 Alignment of the YPT-DNA sequences from Zea mays and Saccharomyces cerevisiae. Shaded areas indicate binding regions of oligonucleotides used for screening of the maize cDNA library.

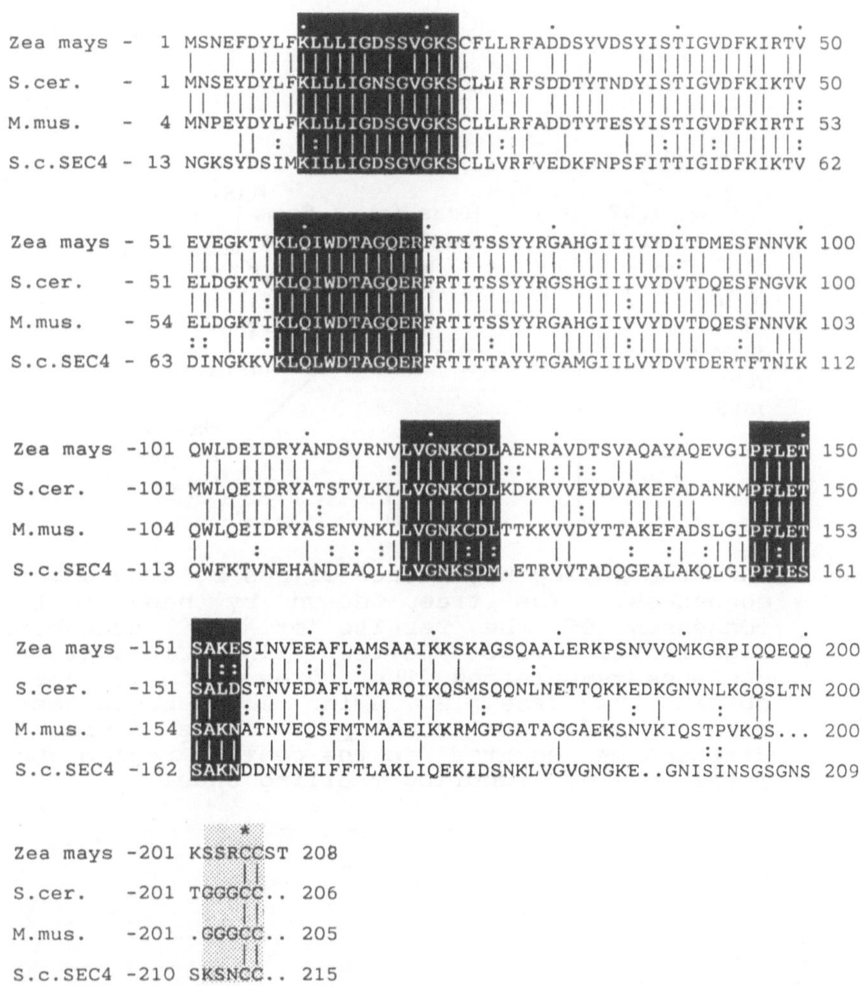

Figure 3 Alignment of YPT-related amino acid sequences from yeast(YPT1 and SEC4), mouse and maize. Boxed areas indicate conserved domains within the ras-super gene family.

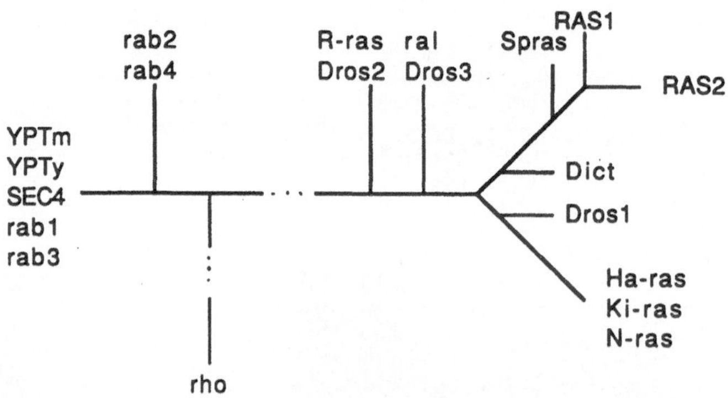

Figure 4 Unrooted phylogenetic tree of ras-related amino acid sequences. The tree, drawn by hand, reflects a consensus of the results of two tree-building algorithms. The dotted lines indicate large distances separating the three main groups. To obtain the tree, 20 amino acid sequences were aligned and, for each pair, the amino acid differences counted in the central region shown in Table 1, i.e. ignoring trailing ends.

Ras proteins roughly fall into two classes. The first class is represented by mammalian and yeast ras-proteins and includes Dictyostelium and Drosophila ras-proteins. The other is grouped around the YPT-proteins and includes SEC4 as well as the rab-proteins. Aplysia rho is clearly not in the first class and is so distant from the YPT class as well that it may belong to a third class. The positioning relative to each other of R-ras/Dros2 on the one hand and ral/Dros3 on the other hand is also not unambiguous from the calculations and the two branches might be interchanged. Further quantitative detail is in table 1, where two proteins are more likely to be in the same class the smaller their distance. As can be seen from the alignment in Table 1 the regions that are likely to be involved in GDP/GTP binding are highly conserved in these proteins. In other positions conservation within the given groupings can be seen. Thus, the grouping of the sequences is not based primarily on the phylogeny of species, but rather proposes the dominance of functional relationships within the groups as well. Yeast ras-genes seem to be involved in controlling of adenyl cyclase, whereas mammalian ras-genes act in the activation of a novel pathway for 1,2-diacylglycerol generation (Wolfman & Macara, 1987; Lacal et al., 1987).

In contrast, the YPT-gene from yeast encodes an essential function different from those of other ras-products. Loss of this function results in cytosceletal and mitotic lesions indicating a regulatory function in intracellular traffic or calcium homeostasis (Segev et al., 1988; Schmitt et al., 1988).

CONCLUSION AND PERSPECTIVES

It is now well documented that a family of guanine-nucleotide binding regulatory proteins transduces signals across cellular membranes by sequential interaction with surface receptors for various ligands and with the appropriate effector proteins. Members of a subgroup, the ras super-gene family, are likely to play a fundamental role in basic cellular functions such as control of proliferation, regulation of intracellular calcium levels and secretory control.

Up to now there was only tentative evidence indicating the presence of G-related proteins in plants (Diefenthal, 1986; Hasunuma & Funadera, 1987; Dobrak et al., 1988). The molecular cloning of cDNAs from plants encoding members of the ras super-gene family has now been achieved. Three cDNAs have been cloned from Zea mays L. representing members of the YPT-gene family. The amino acid sequence of YPTm1 is most strikingly related to the yeast and mouse YPT amino acid sequences. However, it is interesting to note that the maize proteins show slightly stronger similarities with the mouse than with the yeast YPT-proteins. Moreover, we find a similar relationship for other conserved plant proteins such as calmodulin (Palme et al., 1988c). It has often been assumed that the eucaryotic lineage arose relatively recently, 1-2 billion years ago from a procaryotic ancestry (Fox et al.,

1980). Regardless of the different rooting procedures (Fitch & Margoliash, 1967; Fox et al., 1980; Sogin et al., 1986) both multicellular plants and animals occupy a rather narrow domain within the eucaryotic line of descent. In contrast, unicellular eucaryotes including fungi and protists are located further downstream in the evolutionary timescale and span a far greater evolutionary distance. This is also supported by the molecular diversity indicated by DNA and amino acid sequence comparisons of YPT-related proteins from animals, plants and fungi. As multicelluar plants and animals evolved rather recently, it is not surprising to observe the conservation of important signalling pathways. Highly versatile signal recognition elements are not only commonly used by animals and yeast, but are expected to be used in plants as well. This prediction, made many years ago by plant physiologists, can now be confirmed by direct comparison of corresponding plant genes and their predicted protein products. Moreover, the discovery that plant-specific growth stimuli, like auxins, are linked to such pathways, will now help us in unravelling the mechanism and molecular action of this plant hormone. It can be expected that careful analysis of these elements will provide an understanding of the major distinctions between plants and animals, e.g. the ability of plants to reprogram development.

The maize YPT-proteins contain the structural motifs required for GTP-binding, GTP-hydrolase activity and the ras-specific domain (residues 53-68). YPTm1 is the most strictly conserved member of this family, whereas all others show less strict conservation patterns in the catalytic domain but keep the central feature of potential high affinity binding of GTP and GDP. The carboxy-terminal domains are highly diverged in all the maize proteins identified so far, indicating specific interactions with different plant effector proteins necessary for final developmental response.

The function of these proteins is as yet unknown. However, it was postulated that in Lemna and Neurospora GTP-binding proteins are involved in the reception of light signals (Hasunuma, 1984; Hasunuma et al., 1986; Hasunuma et al., 1987) and in the context of these observations it is interesting to detect differential expression of the YPT-genes in maize, with YPTm3 being most strongly expressed in flowers (Diefenthal et al., 1988; Palme et al., 1988b). Further analysis of these genes in both yeast and transgenic plants will lead to a more comprehensive understanding of their functions.

References

Barbacid, M. (1987) Annu. Rev. Biochem. 56, 779-827
Berridge, (1987) Annu. Rev. Biochem. 56, 159-193
Boss, W.F., Massel, M.O. (1985) Biochem. Biophys. Res. Commun. 132, 1018-1023
De Vos, A. M., Tong, L., Milburn, M.V., Matias, P.M., Jancarik, J., Noguchi, S.,S., Nishimura,S., Miura, K.,

Ohtsuka, E., Kim, S.-H. (1987) Science 239, 888-893
Davies, P.J. (1988) Plant hormones and their role in plant growth and development. Martinus Nijhoff Publishers
Diefenthal, T. (1986) Diplomarbeit, Universität Köln
Diefenthal, T., Schell, J., Palme, K. (1988) manuscript in preparation
Ettlinger, C., Lehle, L. (987) Nature 331, 176-178
Felsenstein, J. (1985) Evolution 39, 783-791
Fitch,W.M., Margoliash,E. (1967) Science 155, 279-284
Fox,G.E., Stackebrand, E., Hespel, R.B., Gibson, J., Maniloff,J., Dyer, T.A., Wolfe, W.E. Balch, Tanner, R.S., Magrum, L.J., Zablen, L.B., Blakemore, R., Gupta, R., Bonen, L., Lewis, B.J., Stahl, D.A.,Luehrsen, K.R., Chen, K.N., Woese, C.R. (1980) Science 209, 457-463
Gill, D.L., Ueda, T., Chue, S.-H., and Noel, M.W. (1986) Nature 320, 461-464
Gilman, A. (1987) Annu. Rev. Biochem. 56, 615-6 9
Hasunuma, K.(1984) Proc. Japan Acad. 60, Ser.B, 260-261
Hasunuma,K., Shinohara, Y., Funadera, K., and Watanabe, M. (1986) Abstr. of the XVI Yamada Conference; Phytochrome and Plant Photomorphogenesis, p. 155
Hasunuma, K., Funadera, K., Shinohara, Y., Furukawa, K., and Watanabe, M. (1987)
Biochem. Biophys. Res. Commun. 143, 908-912
Hedrich, R., Neher, E. (1987) Nature 329, 833-835
Heim, S., Baulecke, A., Wylegalla, C., Wagner, K.G. (1987) Plant Sci. 49, 159-165
Heim, S., Wagner, K.G (1986) Biochem. Biophys. Res. Commun. 134, 1175-1181
Hepler, P.K., R.O. Wayne (1985) Ann. Rev. Plant Physiol. 36, 397-439
King, P. (1988) Trends in Genetics 4, 157-162
La Cour, T.F.M.,Nyborg, J., Thirup, S., Clark, B.F.C. (1985) EMBO J. 4, 2385
Jurnak, F. (1985) Science 230, 32-36
Lacal, J.C., Moscat, J., Aaronson, S.A. (1987) Nature 325, 359-361
Marme,D. (1983) In: Inorganic Plant Nutrition Encyclopedia of Plant Physiology (A. Laughli, R.L. Bielski, eds.) Springer-Verlag 15B: 599-625
Millner, P.A. (1987) FEBS Lett. 226, 155-160 Molenaar, C.M.T., Prange, R., Gallwitz, D. (1988) EMBO J. 4, 971-976
Palme, K., Mayer, J., Schell, J. (1987) in: L.M.G. Heilmeyer, Jr. (ed.) Signal transduction and protein phosphorylation, 351-356, Plenum Press, N.Y.
Palme, K., Diefenthal, T., Gilles, R., Nass, N., Nitschke, K., Schell, J. (1988a) manuscript in preparation
Palme, K., Diefenthal, T., Schell, J. (1988b) manuscript in preparation
Palme, K., Nitschke, K., Schwonke, S., Schell, J. (1988c) manuscript in preparation
Perez, L., Aguilar, R., Sanchez-de-Jimenez, E., (1987) Physiol. Plantarum 69, 517-522
Phillips, I.D.J. (1971) The Biochemistry and Physiology of Plant Growth Hormones, McGraw Hill Book Company, New York.
Ranjeva, R., Boudet, A. (1987) Annu. Rev. Plant Physiol. 38, 73-93
Schmitt, H.D., Puzicha, M., Gallwitz, D. (1988) Cell 53, 635-647

Segev, N., Mulholland, J., Botstein, D. (1988) Cell, 52, 915-924
Scott, R. H., Dolphin, A.C. (1987) Nature 330, 760-762
Sogin, M.L., Elwood, H.J., Gunderson, J.H. (1986) Proc. Natl. Acad. Sci. U.S.A. 83, 1382-1387
Vingron, M., and Argos, P. (1988) Biosciences in press

DOES HIV NEF PROTEIN BELONG TO THE G-PROTEIN FAMILY?

B. Guy[1], M.P. Kieny[1], Y. Riviere[2], M. Girard[3], L. Montagnier[2] and J.P. Lecocq[1]

[1]Transgene S.A., 11 rue de Molsheim, 67000 Strasbourg, France

[2]Institut Pasteur, 28 rue du Dr. Roux, 75015 Paris, France

[3]Pasteur Vaccins, 3 avenue Pasteur, 92430 Marnes-La-Coquette France

The nef gene (previously called F, 3' orf, or orf B) is one of the regulatory genes of HIV-1. It encodes a myristilated cytoplasmic protein migrating on an SDS-PAGE gel according to a molecular weight of 25 to 27 kDa (Allan et al., 1985 ; Franchini et al., 1986). It has been demonstrated that the Nef protein down regulates viral expression (Luciw et al., 1987), and more recently, it has been shown (Ahmad et al., 1988) that a cis-regulatory element located within the LTR (long terminal repeat) of HIV is responsive to Nef. This sequence has been mapped in the Negative Regulatory Element(s) (NRE) of the LTR (Rosen et al., 1985) and its presence has a negative effect on the synthesis of proteins expressed under the control of the HIV promoter. As Nef is a cytoplasmic protein which can be membrane associated, its presence in the nucleus is dubious. We thus have to imagine that Nef does not bind directly to the NRE and that it acts by the intermediate of a cellular pathway, like other signal transducing proteins. To investigate the biologicals properties of nef, we have expressed the Nef protein both in eukaryotic cell lines using recombinant vaccinia viruses (VVTGfHIV) and in E.coli in order to obtain the recombinant Nef protein in large quantities (Guy et al., 1987).

We have first studied the modifications undergone by Nef in BHK cells infected with the recombinant virus VVTGfHIV. We have observed that Nef is a myristilated phosphorylated protein. At least two sites are phosphorylated on Nef, one site is located on Thr15 and this phosphorylation is increased by TPA, a potent activator of protein kinase C (PKC). As Thr15 is a consensus site for protein kinase C phosphorylation (Woodgett et al., 1986), and as a mutation from Thr15 to Ala prevents phosphorylation, we speculate that this site is phosphorylated by PKC. Moreover, purified Nef protein from E.coli is efficiently phosphorylated by protein kinase C in vitro on a Threonine. Additionnaly, both Thr15 and Ala15 mutants are phosphorylated on another site that has yet to be localized. It is also of interest to note that the Nef protein is secreted in the supernatant of BHK infected cells, probably through secretory vesicles (unpublished data).

Our first experiments have shown that Nef is an acylated phosphorylated protein that can be phosphorylated by PKC. It thus resembles some cellular membrane associated proteins like pp60src, C-Ki-ras (Ballester et al., 1987) and α subunits of Gi proteins (Buss et al., 1987 ; Katada et al., 1985) ; we have investigated if Nef could share properties with these proteins and we have observed that the Nef protein purified from E.coli exhibits GTP binding, GTPase and autokinase activities in vitro (Guy et al., 1987). However, a few points remain to be elucidated. First, the average kinetic values obtained in vitro with different preparations of Nef protein are relatively low compared to those published for G-proteins. This could reflect the fact that Nef is expressed in a partially denatured form from E.coli or that the purification procedure decreases its activity. Secondly, most of the GTP binding proteins posess three well identified sequences involved in the binding of GTP (Dever et al., 1987) and Nef displays only one cluster of glycines resembling the putative phosphoryl binding site, (GXXXXGK) found in the α subunits of G-proteins (Gilman, 1987). A symetric counterpart of this sequence does nevertheless exist in Nef, corresponding to a loop in the secondary structure according to Chou and Fassman. The alignment of the sequences of Nef and of the G-proteins around the putative phosphoryl binding site is shown in Fig. 1, although the significance of these homologies remains to be formaly established. Finally, though a second putative element (NKGE) is present within the Nef protein of HIV1 BRU, this peptide is not well conserved between different isolates. We have thus to consider the specificity of Nef as a GTP binding protein with some prudence.

L* L* G A G (E) S G K (S) - α subunit G proteins (Gilman 1987)

(E) K G G L (E) - G L* I* - NEF HIV1 BRU (Wain Hobson et al., 1985)

(T) R G G L (E) - G M* F* - NEF HIV2 ROD (Guyader et al., 1987)

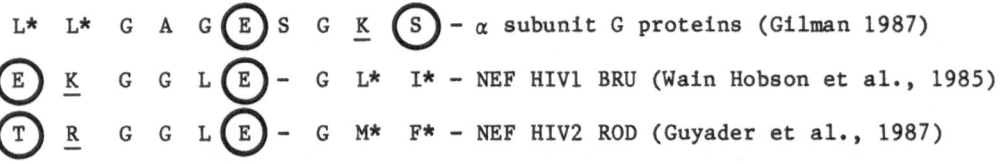

- Basic residues are underlined
- Acidic residues and residues possessing and oxygene group are encircled
- * corresponds to hydrophobic residues

Fig. 1

However, a well-known property of the G-proteins is their capability of regulating the functions of the cell (Gilman, 1987). As Nef clearly resembles proteins which regulate cellular genes, we have investigated the effect the expression of Nef on the expression of CD4 in a CEM T4$^+$ cell line, as CD4 is downregulated in HIV infected cells (Hoxie et al., 1986). The expression of Nef in VVTGfHIV-infected cells leads to a down-regulation of CD4 surface expression (40 to 60 %) (Guy et al., 1987). In our experiments, the only recombinant virus which had an effect on CD4, apart from VVTGfHIV, was the recombinant virus expressing the Env protein of HIV, in agreement with experiments showing that down regulation of CD4 was the consequence of interaction with ENV (Hoxie et al., 1986). Moreover, we have demonstrated that the expression of Nef could decrease mouse CD4 (L3T4) expression, using VV.TG.FHIV. In the same experiments, Env had no effect on mouse CD4 expression. These observations are in agreement with an indirect interaction between Nef and CD4 through an unindentified pathway. Our observation with NEF is in convergence with a report (Folks et al., 1986) showing that a CD4$^+$ cell line latently infected with HIV and which does not express the structural proteins of the virus can become CD4-negative. This phenomenon could be in that particular case the consequence of the expression of the Nef protein.

In order to map more precisely the regions of the Nef protein involved in its activities, we have mutated the nef sequence in different locations. These mutants were engineered in order to investigate the targetting of the protein and the role of the myristilation of Nef, as well as the localization of the putative phosphoryl binding site. The effect of these mutations were investigated both in vivo and in vitro (unpublished data).

First, myristilation allowing the association of Nef with the membrane is critical for phosphorylation, excretion in vesicles and activity on CD4 expression, as all mutated NEF proteins which are not myristilated lack these properties.

Secondly, mutations within the putative phosphoryl binding group (KGGLEG → KGVLEA) result in a dramatic decrease in the two phosphorylations previously described in VVTGfHIV infected cells. As the native NEF protein is phosphorylated by PKC even in the absence of TPA, the disappearance of the PKC phosphorylation in the case of the mutated NEF could have two origins : Nef could stimulate its own phosphorylation by PKC by interacting with the PKC pathway, and mutations preventing the binding of nucleotides could impair this activity. Alternatively, the second phosphorylation (autophosphorylation ?) could be necessary to stabilize PKC phosphorylation. A second observation is that the mutations on the putative phosphoryl binding site abolish the excretion of Nef in vesicles. Nef could thus, like other unidentified G-proteins, act on exocytosis (Segev et al., 1988) and promote its own excretion. The mutations could also impair the interaction of Nef with another protein involved in this process.

Surprisingly the GTP binding activity of this mutant in vitro was not dramatically decreased. GTPase activity seems to be more affected (2 or 3 times less activity in the mutant). The fact that in vivo phosphorylation is dramatically decreased could mean that another protein (like GAP for ras) (Trahey et al., 1987) could enhance the GTPase activity of the native Nef protein in vivo, but not that of the mutant.

Finally, all the mutations affecting myristilation and/or phosphorylation abolish Nef effect on CD4 expression. Membrane localization and correct folding of Nef seem thus to be critical for its activity. However, at least two domains with two different activities seem to exist on Nef. We have demonstrated that Nef mutated on the putative nucleotide binding site is no more excreted in vesicles and has no activity on CD4 expression (see before). However, we have observed that Nef from HIV2ROD (Guyader et al., 1987) and from SIVmac (BK28 clone from J. Mullins) are well excreted but have very little if any effect on CD4 expression. Both proteins are very poorly phosphorylated in the vaccinia virus expression system, and TPA has no effect on this phosphorylation. It thus appears that excretion and down regulation of CD4 expression are not linked.

In conclusion, the Nef protein belongs to a group of regulatory proteins located on the inner face of the cell membrane and which act through a cellular pathway, resulting with DNA interaction in the nucleus. Nef resembles a G protein, though not all criteria for this definition are fulfilled.Nef could be the effector itself or could interact with an inhibitor or an activator of a cellular pathway. As Nef resembles some oncogenes, it could also be a competitive inhibitor of such proteins. On the contrary of oncogenes which enhance cell growth and transformation, Nef has a negative effect on the expression of HIV. Nef could thus be an "anti-oncogene". Further mutations will allow us to better define the important domains of nef, and the consequences for the cell of the expression of Nef have to be studied in a non lytic expression system.

ACKNOWLEDGEMENTS

We thank Ch. Le Peuch from U. 249 INSERM for his help in the PKC phosphorylation experiments, A. Vidal for purifying the NEF protein, K. Dott for technical assistance, R. Drillien and D. Spehner for helpful discussions and E. Chambon for typing the manuscript.

REFERENCES

Ahmad, N. et al., 1988, in: Abstracts from the IVth International Conference on AIDS, Stockholm, Abstr. 1525.

Allan, J.S., Coligan, J.E., Lee, T.-H., McLane, M.F., Kanki, P.J., Groopman, J.E., and Essex, M., 1985, A new HTLV-III/LAV encoded antigen detected by antibodies from AIDS patients, Science, 230:810.

Ballester, R., Furth, M.E., and Rosen, O.M., 1987, Phorbol ester- and protein kinase C-mediated phosphorylation of the cellular Kirsten ras gene product, J. Biol. Chem., 262:2688.

Buss, J.E., Mumby, S.M., Casey, P.J., Gilman, A.G., and Sefton, B.M., 1987, Myristoylated α subunits of guanine nucleotide-binding regulatory proteins, Proc. Natl. Acad. Sci. USA, 84:7493.

Dever, Th.E., Glynias, M.J., and Merrick, W.C., 1987, GTP-binding domain: three consensus sequence elements with distinct spacing, Proc. Natl. Acad. Sci. USA, 84:1814.

Folks, Th., Powell, D.M., Lightfoote, M.M., Benn, S., Martin, M.A., and Fauci, A.S., 1986, Induction of HTLV-III/LAV from a nonvirus producing T-cell line:implications for latency, Science, 231:600.

Franchini, G., Robert-Guroff, M., Ghrayeb, J., Chang, N.T., and Wong-Staal, F., 1986, Cytoplasmic localization of the HTLV-III 3' orf protein in cultured T cells, Virology, 155:593.

Gilman, A.G., 1987, G-proteins: transducers of receptor-generated signals, Ann. Rev. Biochem., 56:615.

Guy, B., Kieny, M.P., Rivière, Y., Le Peuch, C., Dott, K., Girard, M., Montagnier, L., and Lecocq, J.-P., 1987, HIV F/3' orf encodes a phosphorylated GTP-binding protein resembling an oncogene product, Nature, 330:266.

Guyader, M., Emerman, M., Sonigo, P., Clavel, F., Montagnier, L., and Alizon, M., 1987, Genome organization and transactivation of the human immunodeficiency virus type 2, Nature, 326:662.

Katada, T., Gilman, A.G., Watanabe, Y., Bauer, S., and Jakobs, K.H., 1985, Protein kinase C phosphorylates the inhibitory guanine-nucleotide-binding regulatory component and apparently suppresses its function in hormonal inhibition of adenylate cyclase, Eur. J. Biochem., 151:431.

Luciw, P.A., Cheng-Mayer, C., and Levy, J.A., 1987, Mutational analysis of the human immunodeficiency virus: the orf-B region down-regulates virus replication, Proc. Natl. Acad. Sci. USA, 84:1434.

Rosen, C.A., Sodroski, J.G., and Haseltine, W.A., 1985, The location of cis-acting regulatory sequences in the human T cell lymphotropic virus type III (HTLV-III/LAV) long terminal repeat, Cell, 41:813.

Segev, N., Mulholland, J., and Botstein, D., 1988, The yeast GTP-binding YPT1 protein and a mammalian counterpart are associated with the secretion machinery, Cell, 52:915.

Trahey, M., and McCormick, F., 1987, A cytoplasmic protein stimulates normal N-ras p21 GTPase, but does not affect oncogenic mutants, Science, 238:542.

Wain-Hobson, S., Sonigo, P., Danos, O., Cole, S., and Alizon, M., 1985, Nucleotide sequence of the AIDS virus, LAV, Cell, 40:9.

STRUCTURE, FUNCTION AND GENETICS OF SIGNAL TRANSDUCING PROTEINS

STRUCTURE AND FUNCTION OF G PROTEINS FROM MAMMALIAN AND YEAST CELLS

Yoshito Kaziro

Institute of Medical Science
University of Tokyo
4-6-1, Shirokanedai, Minatoku
Tokyo 108, Japan

INTRODUCTION

A superfamily of GTP binding proteins consists of several families including (1) translational factors, (2) signal transducing G proteins, (3) protooncogenic ras proteins, and (4) other low-molecular-weight GTP binding proteins such as rho, sec (sec4, YPT1, etc.) and arf families. There are also other GTP-binding proteins such as tubulins and a group of metabolic GTP-binding proteins including succinate thiokinase and phosphoenolpyruvate carboxykinase, but they will not be discussed here.

We have spent a number of years working on the structure and function of translational factors. Then, we gradually shifted to other GTP binding proteins, i.e. signal transducing G proteins and protooncogenic ras proteins. In this article, I will concentrate mainly on the family of signal transducing G proteins, but before that, I would like to describe briefly the reaction mechanism of polypeptide chain elongation factors, from which the concept of energy-induced conformational change of proteins has been developed. The basic mechanism of the reactions catalyzed by other GTP binding proteins as well as ATP binding proteins appears to be analogous to that proposed for translational elongation factors (Kaziro, 1978).

GENERAL REACTION MECHANISMS OF GTP-BINDING PROTEINS

Figure 1 summarizes the reaction mechanism of two GTP requiring processes in polypeptide chain elongation. The first reaction, a GTP-dependent binding of aminoacyl-tRNA to the ribosomal A site is promoted by EF-Tu, a 43 kDal protein, one of the most abundant proteins in E. coli, while the second reaction, a GTP-dependent translocation of the peptidyl-tRNA-mRNA complex from the A to the P sites on ribosomes is promoted by EF-G. Only the explanation of first reaction will be described here.

EF-Tu has a tightly bound GDP (Kd=10^{-9}M) per mole of protein and is converted to EF-Tu•GTP through displacement of bound GDP with external GTP. EF-Tu•GTP is an activated form in which EF-Tu can interact with aminoacyl-tRNA and subsequently with ribosomes to form a ribosomal intermediate. Then, the GTP moiety of the complex is hydrolyzed to GDP and inorganic phosphate. The most remarkable feature of this reaction is that the reactivity as well as the conformation of the EF-Tu molecule is reversibly and qualitatively modulated through interaction with GTP or GDP, i.e. by the "phosphate-energy level" of the ligand. EF-Tu•$\overline{\text{GTP}}$ is an active form and EF-Tu•GDP is an inactive form, and the reaction is driven by the conformational change induced by the "high-energy" phosphate.

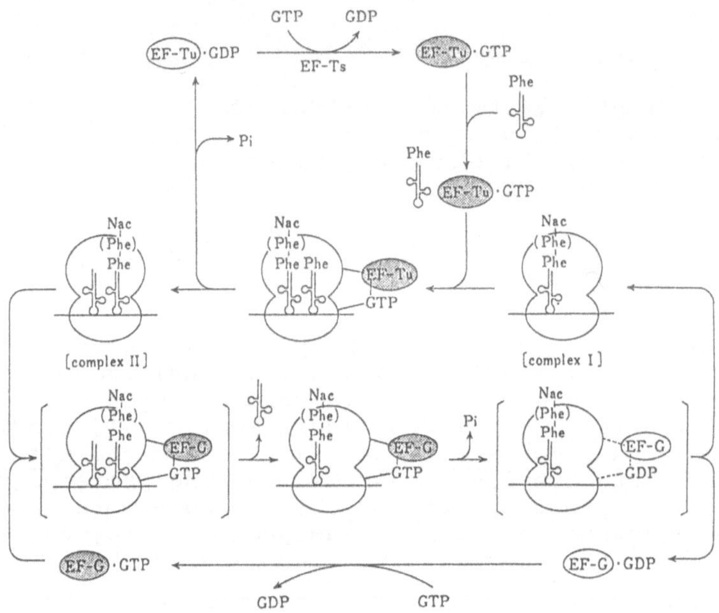

Fig. 1. Mechanism of polypeptide chain elongation. From Kaziro (1978).

If we extract only the essential part of this reaction, then the mechanism could be illustrated as shown in Fig. 2. As shown here, the protein undergoes two alternate conformational transitions induced by ligand change from GDP to GTP or *vice versa*. The former process is an activation

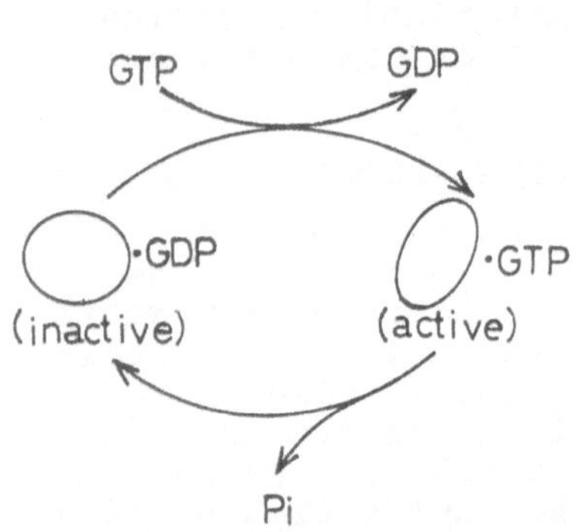

Fig. 2. Reaction mechanism of GTP binding proteins.

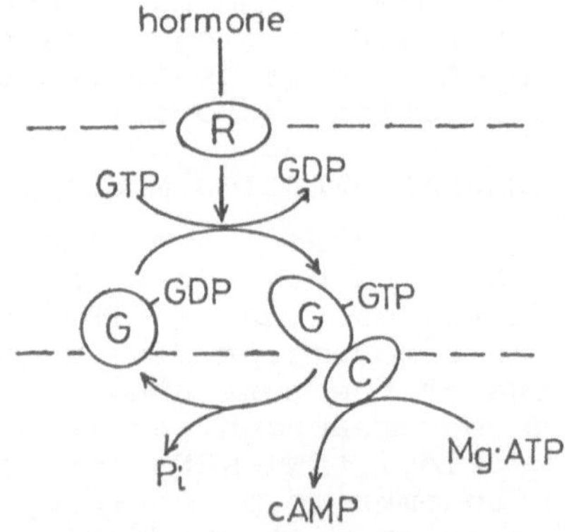

Fig. 3. Hormonal stimulation of adenylate cyclase mediated by Gsα.

```
Gsα : MGCLGNSKTEDQRNEEKAQREANKKIEKQLQKDKQVYRATHRLLLLGAGESGKSTIVKQM  60
Gi2α: MGCTVSA-------EDKAAAERSKMIDKNLREDGEKAAREVKLLLLGAGESGKSTIVKQM  53
Gi3α: MGCTLSA-------EDKAAVERSKMIDRNLREDGEKAAKEVKLLLLGAGESGKSTIVKQM  53
Gi1α: MGCTLSA-------EDKAAVERSKMIDRNLREDGEKAAREVKLLLLGAGESGKSTIVKQM  53
Goα : MGCTLSA-------EERAALERSKAIEKNLKEDGISAAKDVKLLLLGAGESGKSTIVKQM  53
Gxα : MGCRQSS-------EEKEAARRSRRIDRHLRSESQRQRREIKLLLLGTSNSGKSTIVKQM  53

Gsα : RILHVNGFNGEGGEEDPQAARSNSDGEKATKVQDIKNNLKEAIETIVAAMSNLVPPVELA 120
Gi2α: KIIHEDGYSEE---------------ECRQYRAVVYSNTIQSIMAIVKAMGNLQIDFADP  98
Gi3α: KIIHEDGYSED---------------ECKQYKVVVYSNTIQSIIAIIRAMGRLKIDFGEA  98
Gi1α: KIIHEAGYSEE---------------ECKQYKAVVYSNTIQSIIAIIRAMGRLKIDFGDA  98
Goα : KIIHEDGFSGE---------------DVKQYKPVVYSNTIQSLAAIVRAMDTLGVEYGDK  98
Gxα : KIIHSGGFNLE---------------ACKEYKPLIIYNAIDSLTRIIRALAALKIDFHNP  98

Gsα : NPENQFRVDYILSVMNVPNFDFPPEFYEHAKALWEDEGVRACYERSNEYQLIDCAQYFLD 180
Gi2α: QRADDARQLFALSCAAEEQGMLPEDLSGVIRRLWADHGVQACFGRSREYQLNDSAAYYLN 158
Gi3α: ARADDARQLFVLAGSAEE-GVMTSELAGVIKRLWRDGGVQACFSRSREYQLNDSASYYLN 157
Gi1α: ARADDARQLFVLAGAAEE-GFMTAELAGVIKRLWKDSGVQECFNRSREYQLNDSAAYYLN 157
Goα : ERKADSKMVCDVVSRMEDTEPFSAELLSAMMRLWGDSGIQECFNRSREYQLNDSAKYYLD 158
Gxα : DRAYDAVQLFALTGPAESKGEITPELLGVMRRLWADPGAQACFGRSSEYHLEDNAAYYLN 158

Gsα : KIDVIKQADYVPSDQDLLRCRVLTSGIFETKFQVDKVNFHMFDVGGQRDERRKWIQCFND 240
Gi2α: DLERIAQSDYIPTQQDVLRTRVKTTGIVETHFTFKDLHFKMFDVGGQRSERKKWIHCFEG 218
Gi3α: DLDRISQTNYIPTQQDVLRTRVKTTGIVETHFTFKELYFKMFDVGGQRSERKKWIHCFEG 217
Gi1α: DLDRIAQPNYIPTQQDVLRTRVKTTGIVETHFTFKDLHFKMFDVGGQRSERKKWIHCFEG 217
Goα : SLDRIGAADYQPTEQDILRTRVKTTGIVETHFTFKNLHFRLFDVGGQRSERKKWIHCFED 218
Gxα : DLERIAAPDYIPTVEDILRSRDMTTGIVENKFTFKELTFKMVDVGGQRSERKKWIHCFEG 218

Gsα : VTAIIFVVASSSYNMVIREDNQTNRLQEALNLFKSIWNNRWLRTISVILFLNKQDLLAEK 300
Gi2α: VTAIIFCVALSAYDLVLAEDEEMNRMHESMKLFDSICNNKWFTDTSIILFLNKKDLFEEK 278
Gi3α: VTAIIFCVALSDYDLVLAEDEEMNRMHESMKLFDSICNNKWFTDTSIILFLNKKDLFEEK 277
Gi1α: VTAIIFCVALSDYDLVLAEDEEMNRMHESMKLFDSICNNKWFTDTSIILFLNKKDLFEEK 277
Goα : VTAIIFCVALSGYDQVLHEDETTNRMHESLMLFDSICNNKFFIDTSIILFLNKKDLFGEK 278
Gxα : VTAIIFCVELSGYDLKLYEDNQTSRMAESLRLFDSICNNNWFINTSLILFLNKKDLLSEK 278

Gsα : VLAGKSKIEDYFPEFARYTTPEDATPEPGEDPRVTRAKYFIRDEFLRISTASGDGRHYCY 360
Gi2α: IT--QSPLTICFPEYTGANKYDE--------------AASYIQSKFEDLNKRKDTKEIY-- 321
Gi3α: IK--RSPLTICYPEYTGSNTYEE--------------AAAYIQCQFEDLNRRKDTKEVY-- 320
Gi1α: IK--KSPLTICYPEYAGSNTYEE--------------AAAYIQCQFEDLNKRKDTKEIY-- 320
Goα : IK--KSPLTICFPEYPGSNTYED--------------AAAYIQTQF-ESKNRSPNKEIY-- 320
Gxα : IR--RIPLSVCFPEYKGQNTYEE--------------AAVYIQRQFEDLNRNKETKEIY-- 321

Gsα : PHFTCAVDTENIRRVFNDCRDIIQRMHLRQYELL 394
Gi2α: THFTCATDTKNVQFVFDAVTDVIIKNNLKDCGLF 355
Gi3α: THFTCATDTKNVQFVFDAVTDVIIKNNLKECGLY 354
Gi1α: THFTCATDTKNVQFVFDAVTDVIIKNNLKDCGLF 354
Goα : CHMTCATDTNNIQVVFDAVTDIIIANNLRGCGLY 354
Gxα : SHFTCATDTSNIQFVFDAVTDVVIQNNLKYIGLC 355
```

Fig. 4. Deduced amino acid sequences of α subunits of rat Gs, Gi2, Gi3, Gil, Go, and Gx and bovine Gtl and Gt2. The sequences of Gsα and Gi2α are from Itoh et al. (1986); Gi3α from Itoh et al. (1988b); Gilα from Jones and Reed (1987); Goα from Itoh et al. (1986) and Jones and Reed (1987); and Gxα from Matsuoka et al. (1988); Gtαl from Tanabe et al. (1985), Medynski et al. (1985) and Yatsunami et al. (1985); and Gtα2 from Lochrie et al. (1985).

reaction with exogenous energy supply in the form of GTP, and the latter process is a relaxation reaction by hydrolysis of bound GTP to GDP and inorganic phosphate. Since the proposal of this generalized scheme by us, many examples of this type of mechanism have been demonstrated to occur in many different systems either dependent on GTP or ATP.

We can postulate an analogous mechanism for the reaction involving the signal transducing G protein. For example, the mechanism of the hormonal activation of adenylate cyclase mediated through the stimulatory G protein (Gs) may be interpreted as illustrated in Fig. 3. As shown here, interaction of hormone to receptor promotes the exchange of GDP bound to G protein with external GTP. The activated G-protein•GTP complex can now interact with adenylate cyclase and cyclic AMP is continued to be produced until G protein is converted to the G protein•GDP complex. The hydrolysis of GTP is apparently required for the shut-off of the signal transduction. When the GTPase activity is inhibited by ADP-ribosylation catalyzed by cholera toxin, this system is persistently activated and cyclic AMP is overproduced.

SIGNAL TRANSDUCING G PROTEINS

G proteins are a family of GTP-binding proteins involved in a variety of transmembrane signaling systems as transducers (for reviews see Gilman, 1987; Stryer and Bourne, 1986). Two G proteins, Gs and Gi, are involved in hormonal stimulation and inhibition, respectively, of adenylate cyclase, whereas Go (other G protein), which is present predominantly in brain tissues, may be involved in neuronal responses. Two transducins, Gt1 and Gt2, which are present in retinal rods and cones, respectively, regulate cGMP phosphodiesterase activity and mediate visual signal transduction. There is evidence suggesting the presence of additional G proteins, which may be involved in the activation of phospholipase C and phospholipase A_2, as well as the gating of K^+ and Ca^{2+} channels. G proteins consist of 3 subunits, α(39-52 kDal), β(35-36 kDal), and γ(7-10 kDal). Below, I would like to discuss the structure of cDNAs for G protein α subunits (Gαs), the organization of human genes for Gαs, and the occurrence of two Gα genes in S. cerevisiae.

ISOLATION OF cDNA CLONES FOR G PROTEIN α SUBUNITS FROM MAMMALIAN CELLS

Recently cDNA sequences for α subunits of various G proteins have been reported (see Kaziro et al., 1988). These studies have revealed that there are at least three subtypes of Giα that have closely related but distinct structures. The presence of multiple Giα subspecies had been suggested from the molecular heterogeneity, immunological distinction, and functional differences of the pertussis toxin substrates in mammalian cells (see Itoh et al., 1988a and 1988b). Transducins (Gt) also have two subtypes, Gt1α and Gt2α which are expressed in rods and cones, respectively (Lerea et al., 1986). On the other hand, four different Gsα cDNAs are generated by alternative splicing as described below.

Studies on the cDNA cloning of G protein α subunits from rat C6 glioma cells carried out in our laboratory (Itoh et al., 1986; Itoh et al., 1988b) revealed the structure of Gsα, Gi2α, Gi3α, and Goα, consisting of 394, 355, 354, and 354 amino acid residues, respectively, with molecular weights of

45,663, 40,499, 40,522, and 40,068. Gilα which is the most abundant among Giα subtypes in brain tissues, was not expressed in C6 glioma cells. The rat Gilα cDNA clones were later isolated from olfactory epithelium (Jones and Reed, 1987), and brain (Itoh et al., unpublished). We have recently isolated a new Gα clone (designated as Gxα) which codes for a protein apparently insensitive to pertussis toxin (Matsuoka et al., 1988). Human Gzα cDNA isolated independently by Fong et al. (1988) from retina, may be the counterpart of Gxα. Rat Gilα and Gxα code for 354 and 355 amino acid residues, respectively, with molecular weights of 40,345 and 40,879. The amino acid sequences of rat Gsα, Gi2α, Gi3α, Gilα, Goα, and Gxα deduced from the nucleotide sequences are shown in Fig. 4.

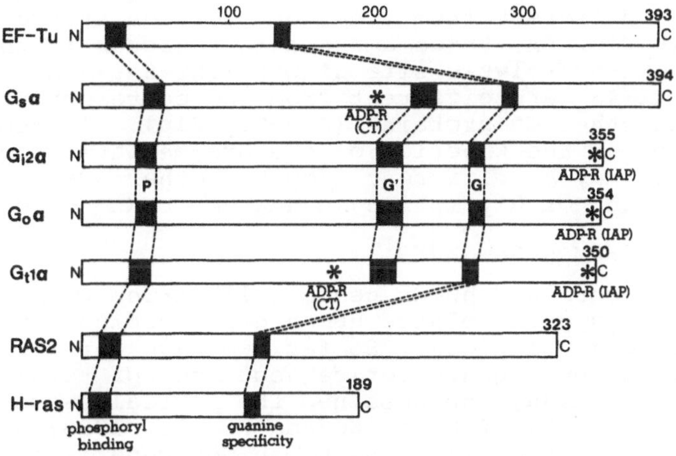

Fig. 5. Schematic representation of structures of EF-Tu, Gsα, Gi2α, Goα, Gtlα, RAS2, and H-ras p21.

Figure 5 shows a schematic representation of the structure of E. coli EF-Tu, G protein α subunits (Gα), yeast RAS2 protein, and mammalian H-ras p21 protein. A remarkable homology was found in two regions, designated as P and G sites. In EF-Tu, earlier biochemical studies indicated that the region around Cys-137 of EF-Tu (G site) is responsible for interaction with guanine nucleotides (Kaziro, 1978), and later the 4 residues Asn-Lys-Cys-Asp are found to be situated close to the guanine ring by X-ray analysis (Jurnak, 1985). On the other hand, it has been shown that the mutation of

A	Gsα:	42-57	R L L L L G A G E S G K S T I V
	Gi1α:	35-50	K L L L L G A G E S G K S T I V
	Gi2α:	35-50	K L L L L G A G E S G K S T I V
	Gi3α:	35-50	K L L L L G A G E S G K S T I V
	Goα:	35-50	K L L L L G A G E S G K S T I V
	Gxα:	35-50	K L L L L G T S N S G K S T I V
B	Gsα:	223-240	D V G G Q R D E R R K W I Q C F N D
	Gi1α:	200-217	D V G G Q R S E R K K W I H C F E G
	Gi2α:	201-218	D V G G Q R S E R K K W I H C F E G
	Gi3α:	200-217	D V G G Q R S E R K K W I H C F E G
	Goα:	201-218	D V G G Q R S E R K K W I H C F E D
	Gxα:	201-218	D V G G Q R S E R K K W I H C F E G
C	Gsα:	285-296	I S V I L F L N K Q D L
	Gi1α:	263-274	T S I I L F L N K K D L
	Gi2α:	262-273	T S I I L F L N K K D L
	Gi3α:	262-273	T S I I L F L N K K D L
	Goα:	263-274	T S I I L F L N K K D L
	Gxα:	263-274	T S L I L F L N K K D L

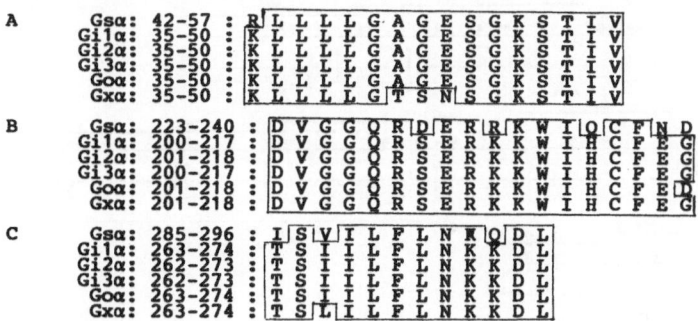

Fig. 6. Conserved sequences of G proteins. Sequences of P site (A), G' site (B) and G site (C) are shown.

amino acid residue 12 of p21 from Gly to Val decreases GTPase activity and increases transforming activity. The sequence homologous to this region (P site) was found in all GTP binding proteins. The consensus sequence, GXXXGK, was located close to phosphoryl residues of bound guanine nucleotide (Jurnak, 1985; and de Vos et al., 1988). G' site is a unique sequence which is highly conserved in all G proteins but is not as remarkable in other GTP binding proteins except for ARF proteins (see Sewell and Kahn, 1988). Only the amino terminal sequence of the G' site, Asp-Xaa-Xaa-Gly, is conserved in all GTP-binding proteins. The DXXG sequence is found in p21 at residues 57 to 60. The deduced amino acid sequences of Gαs in the P, G', and G sites are shown in Fig. 6A, 6B, and 6C, respectively.

It must be noted that the predicted amino sequence of new Gxα was different from other Gα proteins at the P site. As shown in Fig. 6A, three amino acid residues in the consensus GTP-hydrolysis site of Gxα (Thr-Ser-Asn, at positions 41-43) are different from the corresponding residues in other Gα proteins (Ala-Gly-Glu). It remains to be seen whether the kinetics of the Gxα-mediated signal transduction may be different from the other systems due to the replacement of Gly (which corresponds to Gly-12 of p21) by Ser.

ISOLATION OF HUMAN Gα GENES

We have screened human genomic libraries with the above rat cDNA clones and isolated human genes coding for Gsα, Gi1α, Gi2α, Gi3α, and Goα. So far, we have determined the gene organization and nucleotide sequences of total exons of Gsα, Gi2α, and Gi3α, and obtained the partial sequences (exons 1, 2, and 3) of Gi1α (Kozasa et al., 1988; and Itoh et al., 1988b) (see Fig. 7). The human Goα gene is a huge gene spanning at least 90 kb, however, the organization of exons is completely identical to that of Giα subfamily (T. Tsukamoto et al., unpublished).

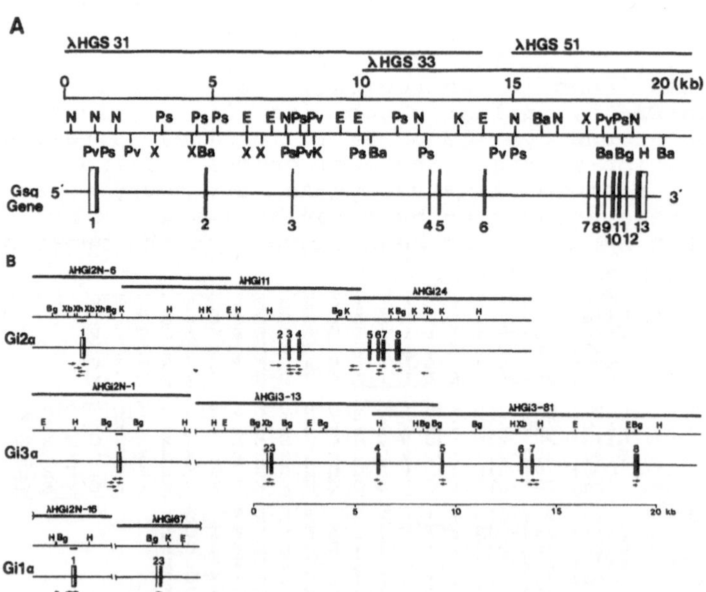

Fig. 7. Organization of human genes for Gsα (A) and three Giα subfamily (B).

STRUCTURE OF THE HUMAN Gsα GENE AND GENERATION OF 4 Gsα cDNAS BY ALTERNATIVE SPLICING

It has been known that there are two species of Gsα protein with different molecular masses (45 and 52 kDa) (Northup et al., 1980). Recently, Bray et al. (1986) isolated four different Gsα cDNAs (Gsα-1 to 4) from human brain and characterized the partial structure. Gsα-1 and Gsα-3 are identical except that Gsα-3 lacks a single stretch of 45 nucleotides. Gsα-2 and Gsα-4 have 3 additional nucleotides (CAG) to Gsα-1 and Gsα-3 3 ' to the above 45 nucleotides. Robishaw et al. (1986a) isolated two Gsα cDNAs from bovine adrenal that correspond to Gsα-1 and Gsα-4, and showed that these two cDNAs generated a 52- and a 45-kDa protein when expressed in COS-m6 cells. Mattera et al. (1986) also isolated two Gsα cDNAs from human liver that correspond to Gsα-1 and Gsα-4.

The human gene for Gsα contains 13 exons and 12 introns and spans about 20 kb of genomic DNA (Fig. 7A)(Kozasa et al., 1988). Comparison of the four types of human Gsα cDNAs reported by Bray et al. (1986) with the sequence of the human Gsα gene of Kozasa et al. (1988) suggests that four types of Gsα mRNAs may be generated from a single Gsα gene by alternative splicing as shown in Fig. 8. Gsα-1 has a sequence identical to exons 2, 3, and 4, whereas Gsα-3 lacks exon 3. Gsα-2 and Gsα-4 have 3 additional nucleotides (CAG) to Gsα-1 and Gsα-3, respectively, at the 5' end of exon 4. In the genomic sequence of the 3' splice site of intron 3, this CAG sequence is found. One additional serine residue in Gsα-2 and Gsα-4 may be the potential site for phosphorylation by protein kinase C, and the alternative use of these splice sites may confer Gsα proteins with differential regulatory properties.

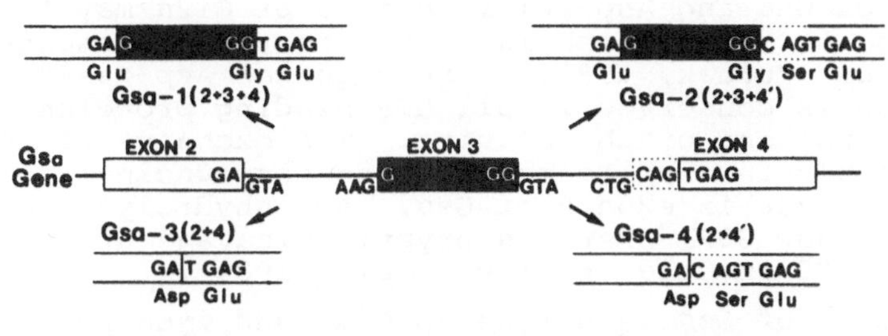

Fig. 8. Generation of four different Gsα mRNAs by alternative splicing. The Gsα gene is shown in the center. Gsα mRNAs are indicated by Gsα-1, -2, -3, and -4. For details see Kozasa et al. (1988).

HUMAN GENES FOR Giα SUBTYPES

The coding region of the human Gi2α and Gi3α genes splits into 8 exons and 7 introns (Itoh et al., 1988b)(Fig. 7B). There is an additional exon (exon 9) in the 3' non-coding region of Gi2α and Gi3α, but this is not included in the figure. For human Gi1α, we only have the sequences of exons 1 to 3 at present. All introns begin with the sequence GT at the 5' end and end at the 3' end with the sequence AG. Remarkably, the positions of the splice junctions on the sequence of cDNA for Gi2α and Gi3α were completely identical, although the length of introns are different (Itoh et al., 1988b). The same splice sites are also conserved in the partial sequence (exons 1, 2, and 3) of the human Gi1α gene as well as in the human Goα gene (T. Tsukamoto, unpublished). From the Southern blot analysis, it appears that each of the three Giα genes occurs as a single copy per haploid human genome.

ORGANIZATION OF HUMAN Gα GENES

The exon-intron organization of the Gsα, Gi2α, Gi3α, and Goα genes was compared with the predicted functional domain structure of proteins (Fig. 9). The NH_2-terminal domain encoded by exon 1 is hydrophilic and contains the site for limited tryptic digestions. Although this region may be involved in interaction with βγ subunits, its precise function has not yet been shown. Exon 2 encodes a short length region (24 and 14 amino acid residues, respectively, for Gsα and Giαs), which is the most conserved among all Gα proteins and responsible for GTP hydrolysis. Exon 3 of Gsα is unique to Gsα. The most structurally divergent domain is encoded by exons 4 to 6 of Gsα and 3 to 4 of Giα. The amino acid sequences of residues 110-140 of Gsα and 90-130 of Giα is remarkably diverse. Exon 8 of Gsα contains Arg-201, which is ADP-ribosylated in the presence of cholera toxin (Van Dop et al., 1984). ADP-ribosylation of Gsα by cholera toxin causes a decrease of affinity for βγ subunits (Kahn and Gilman, 1984). Arg-179 in exon 5 of Gi2α corresponds to this arginine residue. The domain encoded by exons 9 to 11 of Gsα, and 6 to 7 of Giα is strongly conserved among all Gα proteins. This domain is involved in formation of a core structure for GTP binding together with that coded by exon 2 (Masters et al., 1986). The sequence, Asn-Lys-Xaa-Asp, consensus to all guanine nucleotide binding proteins, occurs in exon 11 of Gsα and exon 7 of Giα. The conserved Asp-223 in exon 9 of Gsα and Asp-201 in exon 6 of Gi2α may form a salt bridge to Mg^{2+}, which is linked to the β phosphoryl group of GDP (Jurnak, 1985). The sequence Asp-Xaa-Xaa-Gly of this region is conserved in all GTP-binding proteins at the amino terminal end of the G' site. The exchange of GDP to GTP may result in displacement of the surrounding region residues 230-238 in exon 9 of Gsα. A nonhydrolyzable GTP analog, but not GDP, prevents tryptic cleavage at Lys-210 in Goα or Lys-205 in Gt1α (Hurley et al., 1984).

Exons 12 of Gsα is unique to Gsα, and exon 13 of Gsα and exon 8 of Giα encode the COOH terminus region. The domain may be involved in interaction with a receptor, since the Cys residue which is ADP-ribosylated by pertussis toxin is

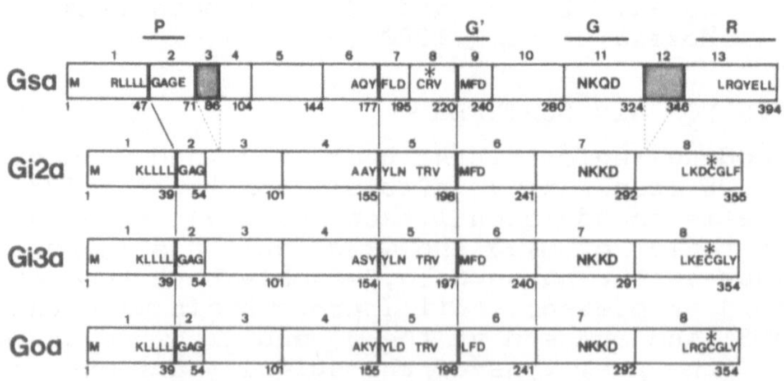

Fig. 9. Organization of the exons of mammalian G protein α subunits.

present in this region of Giα and also the structure of this region is heterogeneous. In Gxα, the Cys residue is replaced by Ile indicating that Gxα is probably refractory tomodification by pertussis toxin. Gsα which is also resistant to pertussis toxin possessed Tyr instead of Cys in this position. It was shown that the replacement of Arg to Pro at -6 position of Gsα gives rise to a mutant protein which is uncoupled with β-adrenergic receptor in S49 cells (Sullivan et al., 1987). Further studies including site-directed mutagenesis and construction of chimeric genes may throw more light on the structure-function relationship of Gα proteins.

Comparison of the exon organization of Giα subfamily and Goα with that of Gsα indicated that some of the exon junctions were conserved between Giα subfamily and Gsα. Thus, 3 out of 12 splice sites of the human Gsα gene are shared with the human Giα genes, and exon 1 and exons 7 and 8 of Gsα correspond to exon 1 and exon 5 of Giα, respectively.

CONSERVATION OF PRIMARY STRUCTURE OF EACH Gα AMONG MAMMALIAN SPECIES

Table 1 shows that, in addition to the remarkable homologies of the overall structure, there is a strong conservation of the amino acid sequence in each subtype of G protein α subunit. The amino acid sequence of Gsα is strongly conserved between human and rat; only 1 out of 394 amino acids being different. The sequence of Gi1α is completely identical between bovine and human. For Gi2α, Gi3α, Gxα, and Goα, over 98% identity of amino acid sequences is maintained among different mammalian species.

Table 1. Conservation of G protein α subunit sequences among different mammalian species

Species	Amino acid sequences	Nucleotide sequences
rGsα vs. hGsα	393/394 (99.7%)	1128/1182 (95.4%)
bGi1α vs. hGi1α	354/354 (100 %)	998/1062 (94.0%)
rGi2α vs. hGi2α	350/355 (98.6%)	985/1065 (92.4%)
rGi3α vs. hGi3α	349/354 (98.6%)	981/1062 (92.4%)
rGoα vs. bGoα	348/354 (98.3%)	992/1062 (93.4%)
rGxα vs. hGxα	349/355 (98.3%)	977/1065 (91.7%)

h, human; r, rat; b, bovine

The strong conservation of the amino acid sequence of each G protein α subunit among distant mammalian species may reflect the presence of evolutional pressure to maintain the specific physiological function of each G protein gene product. Each Gα protein may be linked to a specific receptor and thereby involved in a specific signal transducing pathway.

An evolutionary tree of G protein α subunits can be drawn based on the homologies of the predicted amino acid sequences obtained from various mammalian sources (Fig. 10). It is remarkable that the homologies among three Giα species are higher than that between rod (Gt1α) and cone (Gt2α) transducin α subunits.

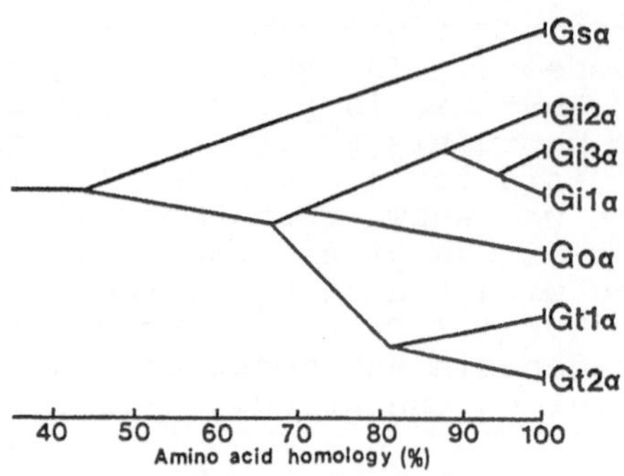

Fig. 10. Relationship among the different mammalian Gα species.

G PROTEINS FROM Saccaromyces cerevisiae

A family of GTP-binding protein, the ras family, is widely distributed among eukaryotes (see Barbacid, 1987, for a review) including yeast S. cerevisiae (DeFeo-Jones et al., 1983; Powers et al., 1984) and Schizosaccharomyces pombe (Fukui and Kaziro, 1985). It has been suggested that the RAS2 gene in S. cerevisiae is involved in the activation of adenylate cyclase (Toda et al., 1985; Broek et al., 1985), and mimics the role of mammalian Gs.

However, in view of the strong conservation of the amino acid sequences of each G protein species among different organisms (see Table 1), we speculated that G protein may occur also in yeast. We have searched for G protein homologous gene in yeast and isolated two genes GPA1 (Nakafuku et al., 1987) and GPA2 (Nakafuku et al., 1988) from S. cerevisiae, which are homologous with cDNAs for mammalian G protein α subunits.

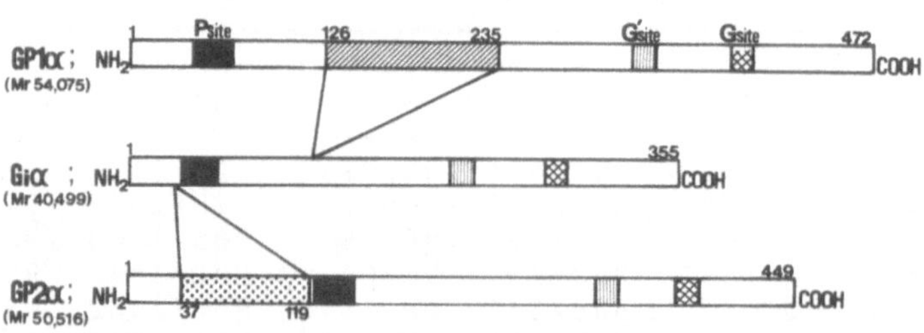

Fig. 11. Schematic representation of the structure of yeast GP1α, yeast GP2α, and mammalian Gi2α.

GPA1 and GPA2 code for the sequences of 472 and 449 amino acid residues, respectively, with calculated Mrs with 54,075 and 50,516. When aligned with the α subunit of mammalian G proteins to obtain maximal homology, GP1α (GPA1 encoded protein) and GP2α (GPA2 encoded protein) were found to contain the stretches of 110 and 83 additional amino acid residues, respectively, near the NH_2 terminus (Fig. 11).

COMPARISON OF THE AMINO ACID SEQUENCES OF YEAST GP1α AND GP2α WITH THOSE OF RAT BRAIN Giα AND Goα

The deduced amino acid sequence of yeast GP1α and GP2α is highly homologous with those of rat brain Giα and Goα. As shown in Fig. 12, the homology is most remarkable in the

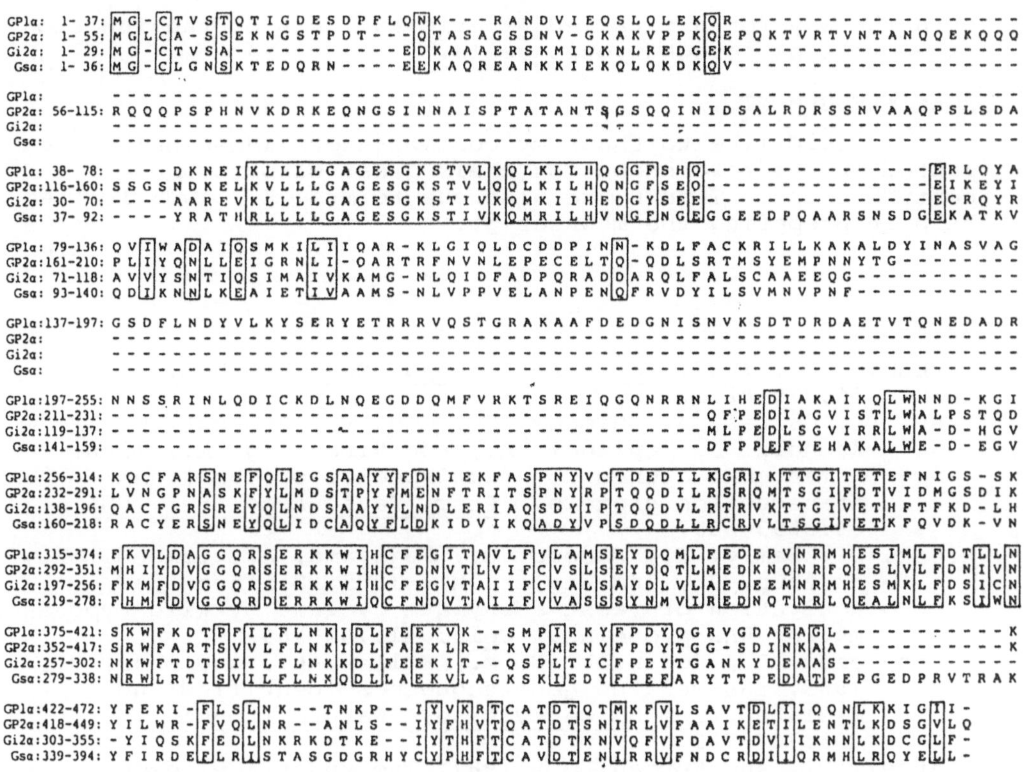

Fig. 12. Alignment of the predicted amino acid sequence of yeast GP1α and GP2α with those of rat brain Gi2α and Gsα. Identical or conservative amino acid residues are enclosed within solid lines. From Nakafuku et al. (1988).

region of GTP hydrolysis (P site)(amino acid residues 43-58 of GP1α, 125-140 of GP2α, and 35-50 of Gi2α). The region responsible for GTP binding (G site)(amino acid residues 381-397 of GP1α, 358-374 of GP2α, and 263-279 of Gi2α) was also highly homologous. Another region of homology (G' site) was found in amino acid residues 321-334 of GP1α, 298-311 of GP2α, and 203-216 of Gi2α where a sequence of 14 contiguous amino acids was completely identical in yeast GP1α, GP2α and rat Gi2α.

The overall homology in nucleotide and amino acid sequences of yeast GP1α, yeast GP2α, rat Gi2α, and rat Gsα is remarkable. Disregarding the unique sequences present in GP1α (residues 126-235) and GP2α (residues 37-119), the proteins are 60% homologous if conservative amino acid substitutions are considered to be homologous. The homology is smaller than that between rat Gi2α and Goα (85%) but is comparable to that between rat Gi2α and Gsα (60%).

As is described elsewhere in details (Miyajima et al., 1987; Dietzel and Kurjan, 1987), GPA1 is a haploid specific gene and involved in the mating factor signal transduction. On the other hands, GPA2 is expressed both in haploid and diploid cells, and may be involved in the regulation of cAMP levels in *S. cerevisiae* (Nakafuku et al., 1988). The demonstration of the occurrence of G proteins in yeast may open the way for a detailed genetic analysis of the function of G proteins in eukaryotic cells.

ACKNOWLEDGEMENT

I would like to thank my collaborators whose names are listed in the references for their great efforts and contributions. Thanks are also due to Mrs. K. Okuda for her help in preparing the manuscript.

REFERENCES

Barbacid, M., 1987, ras genes. Annu. Rev. Biochem., 56:779-827.

Bray, P., Carter, A., Simons, C., Guo, V., Puckett, C., Kamholz, J., Spiegel, A., and Nirenberg, M., 1986, Human cDNA clones for four species of Gαs signal transduction protein. Proc. Natl. Acad. Sci. USA, 83:8893-8897.

Broek, D., Samiy, N., Fasano, O., Fujiyama, A., Tamanoi, F., Northup, J., and Wigler, M., 1985, Differential activation of yeast adenylate cyclase by wild-type and mutant RAS proteins. Cell, 41:763-769.

DeFeo-Jones, D., Scolnick, E. M., Koller, R., and Dhar, R., 1983, ras-Related gene sequences identified and isolated from S. cerevisiae. Nature (London), 306:707-709.

Dietzel, D., and Kurjan, J., 1987, The yeast SCG1 gene: a Gα-like protein implicated in the a- and α-factor response pathway. Cell, 50:1001-1010.

Fong, H. K. W., Yoshimoto, K. K., Eversole-Cire, P., and Simon, M. I., 1988, Identification of a GTP-binding protein α subunit that lacks an apparent ADP-ribosylation site for pertussis toxin. Proc. Natl. Acad. Sci. USA, 85:3066-3070.

Fukui, Y., and Kaziro, Y., 1985, Molecular cloning and sequence analysis of a ras gene from Schizosaccharomyces pombe. EMBO J., 4:687-691.

Gilman, A. G., 1987, G proteins: transducers of receptor-generated signals. Annu. Rev. Biochem., 56:615-649.

Hurley, J. B., Simon, M. I., Teplow, D. B., Robishaw, J. D., and Gilman, A. G., 1984, Homologies between signal transducing G proteins and ras gene products. Science, 226:860-862.

Itoh, H., Katada, T., Ui, M., Kawasaki, H., Suzuki, K., and Kaziro, Y., 1988a, Identification of three pertussis toxin substrates (41, 40 and 39 kDa proteins) in mammalian brain. FEBS Lett., 230:85-89.

Itoh, H., Kozasa, T., Nagata, S., Nakamura, S., Katada, T., Ui, M., Iwai, S., Ohtsuka, E., Kawasaki, H., Suzuki, K., and Kaziro, Y., 1986, Molecular cloning and sequence determination of cDNAs for α subunit of the guanine nucleotide-binding proteins Gs, Gi, and Go from rat brain. Proc. Natl. Acad. Sci. USA, 83:3776-3780.

Itoh, H., Toyama, R., Kozasa, T., Tsukamoto, T., Matsuoka, M., and Kaziro, Y., 1988b, Presence of three distinct molecular species of Gi protein α subunit. J. Biol. Chem., 263:6656-6664.

Jones, D. T., and Reed, R. R., 1987, Molecular cloning of five GTP-binding protein cDNA species from rat olfactory neuroepithelium. J. Biol. Chem., 262:14241-14249.

Jurnak, F., 1985, Structure of the GDP domain of EF-Tu and location of the amino acids homologous to ras oncogene proteins. Science, 230:32-36.

Kahn, R. A., and Gilman, A. G., 1984, ADP-Ribosylation of Gs promoters the dissociation of its α and β subunits. J. Biol. Chem., 259:6235-6240.

Kaziro, Y., 1978, The role of guanosine 5'-triphosphate in polypeptide chain elongation. Biochim. Biophys. Acta, 505:95-127.

Kaziro, Y, Itoh, H., Kozasa, T., Toyama, R., Tsukamoto, T., Matsuoka, M., Nakafuku, M., Obara, T., Takagi, T., and Hernandez, R., 1988, Structure of the genes coding for G protein α subunits from mammalian and yeast cells. Cold Spring Harbor Symposia for Quantitative Biology, 53. (in press)

Kozasa, T., Itoh, H., Tsukamoto, T., and Kaziro, Y., 1988, Isolation and characterization of human Gsα gene. Proc. Natl. Acad. Sci. USA, 85:2081-2085.

Lerea, C. L., Somers, D. E., Hurley, J. B., Klock, I. B., and Bunt-Milam, A. H., 1986, Identification of specific transducin α subunits in retinal rod and cone photoreceptors. Science, 324:77-80.

Lochrie, M. A., Hurley, J. B., and Simon, M. I., 1985, Sequence of the alpha subunit of photoreceptor G protein: homologies between transducin, ras, and elongation factors. Science, 228:96-99.

Masters, S. B., Stroud, R. M., and Bourne, H. R., 1986, Family of G protein α chains: amphipathic analysis and predicted structure of functional domains. Protein Eng., 1:47-54.

Matsuoka, M., Itoh, H., Kozasa, T., and Kaziro, Y., 1988, Sequence analysis of cDNA and genomic DNA for a putative pertussis toxin-insensitive guanine nucleotide-binding regulatory protein α subunit. Proc. Natl. Acad. Sci. USA, 85:5384-5388.

Mattera, R., Codina, J., Crozat, A., Kidd, V., Woo, S. L. C., and Birnbaumer, L., 1986, Identification by molecular cloning of two forms of the α-subunit of the human liver stimulatory (Gs) regulatory component of adenylate cyclase. FEBS Lett., 206:36-41.

Medynsky, D. C., Sullivan, K., Smith, D., Van Dop, C., Chang, F.-H., Fung, B. K.-K., Seeburg, P. H., and Bourne, H. R., 1985, Amino acid sequence of the α subunit of transducin deduced from the cDNA sequence. Proc. Natl. Acad. Sci. USA, 82:4311-4315

Miyajima, I., Nakafuku, M., Nakayama, N., Brenner, C., Miyajima, A., Kaibuchi, K., Arai, K., Kaziro, Y., and Matsumoto, K., 1987, Cell, 50:1011-1019.

Nakafuku, M., Itoh, H., Nakamura S., and Kaziro, Y., 1987, Occurence in Saccharomyces cerevisiae of a gene homologous to the cDNA coding for the α subunit of mammalian G proteins. Proc. Natl. Acad. Sci. USA, 84:2140-2144.

Nakafuku, M., Obara, T., Kaibuchi, K., Miyajima, I., Miyajima, A., Itoh, H., Nakamura, S., Arai, K., Matsumoto, K., and Kaziro, Y., 1988, Isolation of a second yeast Saccharomyces cerevisiae gene (GPA2) coding for guanine nucleotide-binding regulatory protein: studies on its structure and possible functions. Proc. Natl. Acad. Sci. USA, 85:1374-1378.

Northup, J. K., Sternweise, P. C., Smigel, M. D., Shleifer, L. S., Ross, E. M., and Gilman, A. G., 1980, Purification of the regulatory component of adenylate cyclase. Proc. Natl. Acad. Sci. USA, 77:6516-6520.

Powers, S., Kataoka, T., Fasano, O., Goldfarb, M., Strathern, J., Broach, J., and Wigler, M., 1984, Genes in S. cerevisiae encoding proteins with domains homologous to the mammalian ras proteins. Cell, 36:607-612.

Robishaw, J. D., Russell, D. W., Harris, B. A., Smigel, M. D., and Gilman, A. G., 1986a, Deduced primary structure of the α subunit of the GTP-binding stimulatory protein of adenylate cyclase. Proc. Natl. Acad. Sci. USA, 83:1251-1255.

Robishaw, J. D., Smigel, M. D., and Gilman, A. G., 1986b, Molecular basis for two forms of the G protein that stimulates adenylate cyclase. J. Biol. Chem., 261:9587-9590.

Sewell, J. L., and Kahn, R. A., 1988, Sequences of the bovine and yeast ADP-ribosylation factor and comparison to other GTP-binding proteins. Proc. Natl. Acad. Sci. USA, 85:4620-4624.

Stryer, L., and Bourne, H. R., 1986, G proteins: a family of signal transducers. Annu. Rev. Cell Biol., 2:391.

Sullivan, K. A., Miller, R. T., Masters, S. B., Beiderman, B., Heideman, W., and Bourne, H. R., 1987, Identification of receptor contact site involved in receptor — G protein coupling. Nature (London), 330:758-760.

Tanabe, T. Nukada, T., Nishikawa, Y., Sugimoto, K., Suzuki, H., Takahashi, H., Noda, M., Haga, T., Ichiyama, A., Kangawa, K., Minamino, N., Matsuo, H., and Numa, S., 1985, Primary structure of the α-subunit of transducin and its relationship to ras proteins. Nature (London), 315:242-245.

Toda, T., Uno, I., Ishikawa, T., Powers, S., Kataoka, T., Broek, D., Cameron, S., Broach, J., Matsumoto K., and Wigler, M., 1985, In yeast, RAS proteins are controlling elements of adenylate cyclase. Cell, 40:27-36.

Van Dop, C., Tsubokawa, M., Bourne, H. R., and Ramachandran, J., 1984, Amino acid sequences of retinal transducin at the site ADP-ribosylated by cholera toxin. J. Biol. Chem., 259:696-698.

Yatsunami, K., and Khorana, H. G., 1985, GTPase of bovine rod outer segments: the amino acid sequence of the α subunit as derived from the cDNA sequence. Proc. Natl. Acad. Sci. USA, 82:4316-4320.

THREE FORMS OF G_i DISCRIMINATED BY SYNTHETIC PEPTIDE ANTISERA

Susanne M. Mumby

Department of Pharmacology
University of Texas Southwestern Medical Center at Dallas
Dallas, TX 75235

INTRODUCTION

Two signal transduction pathways that are regulated by guanine nucleotide-binding proteins (G proteins) have been well characterized: the light-stimulated visual cascade and the hormone-sensitive adenylyl cyclase system (for review see Stryer, 1986; Gilman, 1987). Stimulation of retinal cyclic GMP phosphodiesterase by the photoreceptor rhodopsin is mediated by the G protein G_t (also referred to as transducin). Hormone-sensitive adenylyl cyclase is under dual stimulatory and inhibitory control by two other G proteins, G_s and G_i. Receptors that stimulate adenylyl cyclase are coupled to the enzyme by G_s, whereas inhibitory receptors are coupled by G_i. These classical G proteins exhibit a characteristic heterotrimeric subunit structure. The β and γ subunits are tightly associated with each other and in some cases are functionally interchangeable between different α subunits. The α subunits display more structural heterogeneity and functional specificity, and thus they serve to distinguish the different G protein oligomers. The α subunits possess a high-affinity binding site for guanine nucleotides and an intrinsic GTPase activity that are involved in the activation and deactivation of the G protein. Toxins from *Vibrio cholerae* and *Bordetella pertussis* are capable of covalently modifying and functionally altering α subunits. Cholera toxin catalyzes the ADP-ribosylation of G_s and G_t. ADP-ribosylation by pertussis toxin blocks activation of G_t and G_i by receptors. Activation of G proteins in detergent solution (and G_t in phospholipid vesicles) by nonhydrolyzable analogs of GTP involves dissociation of the α subunit from $\beta\gamma$. It is the α subunits of G_s and G_t (α_s and α_t) that stimulate their respective effector enzymes.

It has recently become apparent that the G protein family is involved in regulating the activity of effector molecules in addition to cGMP phosphodiesterase and adenylyl cyclase. In most cases, evidence for G protein regulation includes demonstration of guanine nucleotide and/or toxin sensitivity of a biological response to receptor occupancy. Such effectors include particular classes of phospholipases and ion channels. The identity of the G proteins involved in regulation of these effectors has not yet been definitively determined.

A direct approach to defining the identity and biochemistry of the components of a signal transduction pathway is to reconstitute them functionally *in vitro*. To this end a number of closely related yet distinct α subunits have been purified from a variety of tissues, including brain, liver, retina, heart, placenta, and erythrocytes. Limited amino acid sequence has been obtained for a few of the purified proteins. The sequence has been utilized to synthesize oligonucleotide probes for the purpose of isolating cDNAs that encode for G protein subunits. Efforts in both purification of G proteins and in isolation of cDNAs for a multitude of closely related α subunits have resulted in a conceptual extension of the G protein family. The current challenge is to identify which α subunit is encoded by each cDNA and the function of each of the proteins.

Use of synthetic peptides as antigens for the production of monospecific antisera has greatly facilitated the identification of purified G protein subunits encoded by individual cDNAs. We have used this approach to discriminate between several closely related pertussis toxin substrates, including three forms of highly homologous G_i-like α subunits. Use of the specific antisera has allowed us to dismiss the possibility that α subunits, purified together with similar yet distinct apparent molecular sizes, are proteolytic products of one another; they are instead independent gene products (Mumby, et al., 1988). The three closely related forms of G_i-like α subunits do not appear to be expressed merely as cell specific isoforms, since all three are detected immunologically in a single cell type, the erythrocyte (Carty et al., 1988). It therefore seems possible that the three forms of G_i-like α subunits may respond to different receptors or serve to regulate different effectors.

PROTEINS

G_i, as purified from rabbit liver has a pertussis toxin sensitive α subunit (α_i) with an apparent molecular weight of 41,000 (Bokoch et al., 1984). In contrast to results obtained with α_s and α_t, the preparations of α_i have only a modest ability to influence effector activity. The bulk of the capacity of liver G_i to inhibit adenylyl cyclase has been associated with $\beta\gamma$ and is largely dependent on the presence of G_s (Katada et al., 1984). Inhibition of adenylyl cyclase appears to involve the association of α_s with G_i-derived $\beta\gamma$. However, recent evidence described below (under Antisera), indicates that the liver G_i preparations are composed of a heterogeneous mixture of closely related α subunits. Such heterogeneity could contribute to the difficulty in detecting an effect of α_i on adenylyl cyclase activity.

Brain is a particularly rich source of pertussis toxin-sensitive G proteins. Purification of G_i from brain requires its resolution from the more abundant G protein, G_o (for G "other") which has an α subunit of 39 kDa (Sternweis and Robishaw, 1984; Neer et al., 1984). The function of G_o is not known but it is capable of interacting with muscarinic receptors in phospholipid vesicles (Florio and Sternweis, 1985). The 39 kDa α subunit, applied intracellularly to pertussis toxin-treated neuroblastoma x glioma hybrid cells or dorsal root ganglion cells is able to restore receptor-mediated inhibition of Ca^{+2} channels (Hescheler et al., 1987; Ewald et al., 1988).

More recently, preparations of G_i α subunits from brain could be electrophoretically and chromatographically resolved into two polypeptides with approximate molecular weights of 40,000 and 41,000 (α_{40} and α_{41}) (Mumby et al., 1988; Katada et al., 1987). It was not immediately known what their structural or functional relationship might be. Ostensibly, α_{41} from brain appeared to be the same as that purified from liver. α_{40} could be a distinct subunit, a proteolytic product of α_{41}, or a precursor of α_{39}. Katada and coworkers were able to distinguish porcine α_{40} and α_{41} immunochemically with an antiserum which could recognize α_{41} but not α_{40} (Katada et al., 1987). Their results indicated α_{40} was not derived from α_{41}. We and our collaborators have been able to show by immunological methods that α_{40} and α_{41} from brain are distinct gene products by utilizing synthetic peptides as antigens (Mumby et al., 1988). The sequences of the peptides were derived from cDNAs that appeared to encode the α subunit of G_i.

COMPLIMENTARY DNA

The first two $G_{i\alpha}$ cDNAs were isolated by screening with oligonucleotide probes. The sequence of the oligonucleotides were based on amino acid sequence obtained from brain $G_{i\alpha}$. Nukada and coworkers (1986) obtained a clone from a bovine brain library (α_{i1}) that coded for an amino acid sequence approximately 89% identical to the clone isolated by Itoh and coworkers (1986) from a rat C6 glioma cell library (α_{i2}). Jones and Reed (1987) found that α_{i1} and α_{i2} were both represented in their rat olfactory mucosa library, indicating that the α_{i1} and α_{i2} clones coded for two different proteins that are 88% identical. In addition, Jones and Reed isolated a third distinct clone termed α_{i3}, that coded for another protein with an amino acid sequence 94% identical to α_{i1}. A number of additional cDNAs for α_i-like proteins have been isolated from murine and human libraries (see Table 1). With only one exception the cDNAs all fit well into either the α_{i1}, α_{i2}, or α_{i3} category. The human α_i clone isolated by Didsbury and coworkers (1987) from differentiated U937 cells is the exception. It fits into the α_{i2} category, but there are three substitutions in the region bounded by amino acid numbers 113 and 122 in this human clone compared to the rat and mouse clones (Itoh et al., 1986; Sullivan et al., 1986). The substitutions may be due to species differences between human, rat and mouse, or could be indicative of the human clone encoding a distinct protein.

Table I. α_I Proteins and cDNAs

Designation	Apparent MW[1] Purified Protein Source	Deduced Protein MW from cDNA[2]	cDNA Tissue Source
α_{i1}	41,000 brain, erythrocyte, liver, heart	40,300	Bovine brain (Nukada *et al.*, 1986) Rat olfactory epithelium (Jones & Reed, 1987) Human brain (Bray *et al.*, 1987)
α_{i2}	40,000 brain, erythrocyte, liver, neutrophil; heart	40,500	Rat C6 cells (Itoh *et al.*, 1986) Murine macrophage (Sullivan *et al.*, 1986) Rat olfactory epithelium (Jones & Reed, 1987) Human U937 cells (Didsbury *et al.*, 1987)
α_{i3}	41,000 erythrocyte, liver, heart	40,500	Rat olfactory epithelium (Jones & Reed, 1987) Human HL-60 cells (Didsbury & Snyderman, 1987) Human liver (Suki *et al.*, 1987) Rat C6 cells (Itoh *et al.*, 1988)

1. Apparent MW as defined by relative migration of purified proteins on SDS-polyacrylamide gels. Assignment of purified protein as α_{i1}, α_{i2}, or α_{i3} based on reactivity with monospecific antisera directed against peptides whose sequence was deduced from α_{i1} α_{i2}, and α_{i3} cDNAs.

2. cDNA sequences from rat olfactory epithelium library used to deduce molecular weight (Jones & Reed, 1987).

ANTISERA

For the purpose of generating subunit specific antisera, we have utilized the small region of sequence diversity for the three α_i subunits shown in Figure 1. Specificity of the antisera for the particular α subunit encoded for by the cDNA was demonstrated by immunoblotting purified α subunits from *E. coli* expressing protein coded for by α_{i1}, α_{i2} or α_{i3} cDNA (Linder and Gilman unpublished data). The antiserum from rabbit I-355 injected with the α_{i1} peptide linked to hemocyanin reacts only with protein from *E. coli* expressing α_{i1} cDNA, but not α_{i2} or α_{i3} cDNA. The α_{i2} peptide antigen elicited antibodies from rabbit J-883 that were specific for the α_{i2} protein. Antiserum K-887 produced with the α_{i3} peptide reacted with the α_{i3} protein but not with α_{i1}. A trace cross reaction of the α_{i3} peptide antiserum K-887 with α_{i2} protein can be detected.

Antiserum Code Number	Protein Name	Deduced Amino Acid Sequence	Amino Acid Number
	αs	L S V M N V P D F D F P P E F Y E H - A	150
I-355	αi1	L A G A A - E E G F M T A E - L A G V I	127
J-883	αi2	L S C A A - E E Q G M L P E D L S G V I	128
K-887	αi3	L A G S A - E E G V M T S E - L A G V I	127
	αo	V V S R M - E D T E P F S A E L L S A M	128
	αt	H M A D T I E E G T M P K E M S D I - I	123

Figure 1. Amino acid sequences utilized as antigens for generation of αi-specific antisera. Sequences of αi peptides, deduced from cDNAs referenced below, are outlined. Corresponding sequences of αs, αo, and αt are included for comparison in this region of relative α subunit sequence diversity. The αi1 and αi3 peptides were synthesized with an additional cysteine at the amino terminus to facilitate coupling to the carrier protein hemocyanin. References: αs; Robishaw et al., 1986: αi1; Nukada et al., 1986; αi2; Itoh et al., 1986: αi3; Jones & Reed, 1987: αo; Itoh et al., 1986: αt; Yatsunami & Khorana, 1985.

The identity of proteins purified from tissues can be determined by immunoblotting with the peptide antisera, specific for particular sequences. α_{41} resolved chromatographically from α_{39} and α_{40} from bovine brain reacts with the α_{i1} antiserum I-355 but not the α_{i2} or α_{i3} specific antisera (Mumby et al., 1988). α_{40} from brain reacts with the α_{i2} specific antiserum J-883 but not the α_{i1} or α_{i3} specific antisera. Although the α_{i3} specific antiserum K-887 does not react with α_{40} or α_{41} from bovine brain it does react with α_{41} from rabbit liver (unpublished observation). We conclude from these and further results with additional peptide antisera to sequences shared by multiple forms of α subunit, that the α_{41} purified from bovine brain is coded for by the α_{i1} cDNA and that α_{40} corresponds to the α_{i2} cDNA (Mumby et al., 1988). None of the α_i peptide antisera react with the α_{39} from brain (α_o) or retina (α_t). A peptide antiserum specific for α_o (Mumby et al., 1986) does not react with α_{40} or α_{41}, indicating that neither protein is a proteolytic precursor of α_{39} (Mumby et al., 1988). Similar immunological methods have been utilized by Goldsmith and coworkers (1987) to discriminate between α_{40} from neutrophils and α_{39} and α_{41} from brain.

The antisera specific for the three forms of α_i were utilized to determine the composition of G_i purified from liver, heart, and erythrocytes. The various forms of α_i have not been chromatographically resolved from these tissues as they have been from brain. Comparison of several preparations of G_i from rabbit liver by immunoblotting indicates they each include the α_{i3} form of α_{41}. In addition, α_{i1} and/or α_{i2} are represented in the preparations but in less quantity and with more variability from one preparation to another. A single preparation of G_i from porcine atria contained all three forms of α_i (Mumby, Tota, Peterson and Schimmerlik unpublished observation). Three independent chromatographic pools of proteins composed of G_s and G_i from human erythrocytes possess all three forms of α_i (Carty et al., 1988). Since all three proteins are expressed in a single cell type it seems likely they serve different purposes and are not cell specific forms of a protein that serves one function.

Tissue distribution of α_{i1} and α_{i2} has been determined in a cursory manner by immunoblot analysis of crude membrane preparations from rat (Mumby et al., 1988). α_{i2} is detected by antiserum J-883 in every tissue tested, including brain cortex, spleen, lung, liver, testis, erythrocyte, cardiac antrium, and cardiac ventricle. In contrast, α_{i1}, appears to have a more limited distribution. Antiserum I-355, which is specific for α_{i1}, reacted prominently only with the brain membrane preparation. Upon longer exposure of film to the immunoblot, α_{i1} was detected in lung. The quality of the α_{i3} antiserum K-887 is suitable for detection of α_{i3} in purified G protein preparations but not crude membrane preparations (unpublished observation). Future attempts to affinity purify the antibodies may allow α_{i3} to be detected unambiguously in membrane preparations.

FUNCTION

The term G_i, originally coined to designate the inhibitory guanine nucleotide-binding protein that regulates adenylyl cyclase, has become a generic term for the highly related G proteins with α subunits that are pertussis toxin substrates and that differ from α_o and α_t. A current challenge is to define the functions of these multiple species. It is possible that all three of the G_i-like proteins could inhibit adenylyl cyclase through dissociation of common $\beta\gamma$ subunits (Gilman, 1987). Poor inhibition of adenylyl cyclase by purified α_i from rabbit liver (Katada et al., 1984), could be due to heterogeneity of or exclusion of the active species in α_i preparations. The type specific antisera detect both α_{i1} and α_{i3} (but not α_{i2}) in two such preparations of α_i chromatographically resolved from $\beta\gamma$ (unpublished observation). Future experiments utilizing homogeneous preparations of the individual α subunits may identify one particular form of α capable of inhibiting adenylyl cyclase in the absence of $\beta\gamma$.

If we assume the three forms of α_i possess unique functions, extensive and careful experimentation will be required to determine which signal transduction pathways are regulated. The α subunits may interact with different receptors and/or effectors. Possible functions include regulation of ion channels and phospholipases. Evidence for regulation of atrial K^+ channels and neural Ca^{+2} channels by G proteins is compelling, but the exact identity of the G proteins responsible is not known. K^+ channels in pertussis toxin-treated atrial membranes can be activated by addition of nanomolar concentrations of $\beta\gamma$ (Logothetis et al., 1987; Cerbai et al., 1988) or picomolar concentrations of α_i subunits (referred to as α_k) purified from erythrocytes (Yatani et al., 1987; Cerbai et al., 1988). The exact composition of the active α_k preparations needs to be determined rigorously. The major component appears to be α_{i3} (Codina et al., 1988), but it is not yet certain that α_{i3} is the only active component.

Hescheler and coworkers (1987) have demonstrated that α_o is ten fold more potent than G_i oligomer in the ability to mediate receptor-promoted inhibition of Ca^+ channels in neuroblastoma x glioma hybrid cells. However, the exact composition of the G_i oligomer utilized is not known. Ewald and coworkers (1988) have found α_o purified from bovine brain is able to restore neuropeptide Y inhibition of Ca^{+2} channels in dorsal root ganglion cells that had been pre-treated with pertussis toxin. Subsequent experiments demonstrated that purified α_{i1} and α_{i2} were much less effective. In contrast, each of the three α subunits are effective in restoring bradykinin-mediated inhibition of Ca^{+2} current in the same cells (Ewald, Pang, Sternweis, and Miller unpublished data). It is not yet known if the three different G protein species interact with the same or different classes of channels to regulate Ca^{+2} current.

G proteins are implicated in the regulation of phospholipases C and A_2. In some (but not all) systems, pertussis toxin blocks the ability of hormones to activate phospholipase C, the phosphodiesterase that cleaves phosphotidylinositol 4, 5-diphosphate (for review, see Cockcroft, 1987). Two second messengers are produced, inositol triphosphate and diacylglycerol. Phospholipase A_2 catalyzes the cleavage of phospholipids, producing equimolar amounts of lysophospholipids and free fatty acids. The release of arachidonic acid leads to the generation of inflammatory lipid mediators. In FRTL5 rat thyroid cells, pertussis toxin inhibits norepinephrine-stimulated arachidonic acid release by phospholipase A_2 (Burch et al., 1986). Perhaps one or more of the pertussis toxin-sensitive forms of G_i participate(s) in the regulation of phospholipases C and/or A_2.

It has only recently come to light that the G proteins constitute a family of greater than ten highly related α subunits. Homogeneous G protein preparations are required for biochemical reconstitution experiments, but we are faced with heterogeneity in G protein preparations from tissues. Some family members such as α_{i1} and α_{i3} comigrate on high resolution acrylamide gels. Only by immunochemical methods have we been able to determine that these two different proteins can be present in a single preparation of G_i from a tissue such as liver. It is very difficult to obtain the required quantities of homogeneous α_i preparations from tissue sources to use in rigorous reconstitution experiments designed to test the ability of a G protein species to regulate a particular effector.. An approach being tested in our laboratory is expression of individual cDNAs for the various forms of α_i in *E. coli*. This approach has been successful for the expression and purification of two functionally active forms of α_s (Graziano et al., 1987; 1988). Similar methods of purification are being currently utilized to obtain homogeneous preparations of α_{i1}, α_{i2}, α_{i3} and α_o (Linder and Gilman unpublished data). The homogeneous α subunit preparations will be compared directly for their ability to interact with a variety of receptors and effectors in reconstitution experiments.

Acknowledgments

I would like to thank Alfred G. Gilman who provided guidance, support and laboratory facilities, Linda Hannigan for skillful technical assistance in all phases of the antibody work, and Carla Murray for excellent editorial assistance. I would also like to extend my gratitude to colleagues who allowed me to describe their unpublished data in this paper. Financial support was provided by United States Public Health Service Grant GM34497 and American Cancer Society Grant BC555I.

REFERENCES

Bokoch, G. M., T. Katada, J. K. Northup, M. Ui, and A. Gilman, Purification and properties of the inhibitory guanine nucleotide-binding regulatory component of adenylate cyclase, *J. Biol. Chem.* 259:3560-3567 (1984).

Bray, P., A. Carter, V. Guo, C. Puckett, J. Kamholz, A. Spiegel, and M. Nirenberg, Human cDNA clones for an α subunit of G_i signal-transduction protein, *Proc. Natl. Acad. Sci. USA* 84:5115-5119 (1987).

Burch, R. M., A. Luini, and J. Axelrod, Phospholipase A_2 and phospholipase C are activated by distinct GTP-binding proteins in response to α_1-adrenergic stimulation in FRTL5 thyroid cells, *Proc. Natl. Acad. Sci. USA* 83: 7201-7205 (1986).

Carty, D. J., S. M. Mumby, E. Padrell, J. Codina, R. Graf, L. Birnbaumer, A. G. Gilman, and R. Iyengar, The GTP-binding regulatory proteins of human erythrocytes: Evidence for three distinct pertussis toxin substrates in the 40-41 kDalton range. Submitted for publication.

Cerbai, E., U. Klockner, and G. Isenberg, The α subunit of the GTP-binding protein activates muscarinic potassium channels of the atrium, *Science* 240:1782-1783 (1988).

Cockcroft, S. 1987. Polyphosphoinositide phosphodiesterase: regulation by a novel guanine nucleotide binding protein, G_p, *Trends Biochem. Sci.* 12:75-78 (1987).

Codina, J., J. Olate, J. Abramowitz, R. Mattera, R. G. Cook, and L. Birnbaumer, α_i-3 cDNA encodes the α subunit of G_k, the stimulatory G protein of receptor-regulated K^+ channels, *J. Biol. Chem.* 263:6746-6750 (1988).

Didsbury, J. R., Y. Ho, and R. Snyderman, Human G_i protein α-subunit: deduction of amino acid structure from a cloned cDNA, *FEBS Lett.* 211:160-164 (1987).

Didsbury, J. R., and R. Snyderman, Molecular cloning of a new human G protein; evidence for two $G_{i\alpha}$-like protein families, *FEBS Lett.* 219:259-263 (1987).

Ewald, D. A., P. C. Sternweis, and R. J. Miller, G_o induced coupling of NPY receptors to calcium channels in sensory neurons, *Proc. Natl. Acad. Sci. USA* In press. (1988)

Florio, V. A., and P. C. Sternweis, Reconstitution of resolved muscarinic cholinergic receptors with purified GTP-binding proteins, *J. Biol. Chem.* 260:3477-3483 (1985).

Goldsmith, P., P. Gierschik, G. Milligan, C. G. Unson, R. Vinitisky, H. L. Malech, and A. Spiegal, Antibodies directed against synthetic peptides distinguish between GTP-binding proteins in neutrophil and brain, *J. Biol. Chem.* 262:14683-14688 (1987).

Gilman, A. G., G proteins: transducers of receptor-generated signals, *Ann. Rev. Biochem.* 56:615-649 (1987).

Graziano, M. P., P. J. Casey, and A. G. Gilman, Expression of cDNAs for G proteins in *Escherichia coli*: two forms of $G_{s\alpha}$ stimulate adenylyl cyclase, *J. Biol. Chem.* 262:11375-11381 (1987).
Graziano, M. P., M. Freissmuth, and A. G. Gilman, Expression of $G_{s\alpha}$ in *Escherichia coli:* Purification and properties of two forms of the protein, *J. Biol. Chem.* In press.
Hescheler, J., W. Rosenthal, W. Trautwein, and G. Schultz, The GTP-binding protein, G_o, regulates neuronal calcium channels, *Nature* 325:445-447 (1987).
Itoh, H., T. Kozasa, S. Nagata, S. Nakamura, T. Katada, M. Ui, S. Iwai, E. Ohtsuka, H. Kawasaki, K. Suzuki, and Y. Kaziro, Molecular cloning and sequence determination of cDNAs for α subunits of the guanine nucleotide-binding proteins G_s, G_i, and G_o from rat brain, *Proc. Natl. Acad. Sci. USA* 83:3776-3780 (1986).
Itoh, H., R. Toyama, T. Kozasa, T. Tusukamoto, M. Matsomka, and Y. Kaziro, Presence of three distinct molecular species of G_i protein α subunit, *J. Biol. Chem.* 263:6656-6664 (1988).
Katada, T., G. M. Bokoch, M. D. Smigel, M. Ui, and A. G. Gilman, The inhibitory guanine nucleotide-binding regulatory component of adenylate cyclase. Subunit dissociation and the inhibition of adenylate cyclase in S49 lymphoma cyc^- and wild type membranes, *J. Biol. Chem.* 259:3586-3595 (1984).
Katada, T., J. K. Northup, G. M. Bokoch, M. Ui, and A. G. Gilman, The inhibitory guanine nucleotide-binding regulatory component of adenylate cyclase. Subunit dissociation and guanine nucleotide-dependent hormonal inhibition. *J. Biol. Chem.*, 259:3578-3585 (1984).
Katada, T., M. Oinuma, K. Kusakabe, and M. Ui, A new GTP-binding protein in brain tissues serving as the specific substrate of islet-activating protein, pertussis toxin, *FEBS Lett.* 213:353-358 (1987).
Logothetis, D. E., Y. Kurachi, J. Galper, E. J. Neer, and D. E. Clapham, The $\beta\gamma$ subunits of GTP-binding proteins activate the muscarinic K^+ channel in heart, *Nature* 325:321-326 (1987).
Mumby, S. M., I. Pang, A. G. Gilman, and P. C. Sternweis, Chromatographic resolution and immunologic identification of the α_{40} and α_{41} subunits of guanine nucleotide-binding regulatory proteins from bovine brain, *J. Biol. Chem.* 263:2020-2026 (1988).

Mumby, S. M., R. A. Kahn, D. R. Manning, and A. G. Gilman, Antisera of designed specificity for subunits of guanine nucleotide-binding regulatory proteins. *Proc. Natl. Acad. Sci. USA* 83:265-269 (1986).
Nukada, T., T. Tanabe, H. Takahashi, M. Noda, K. Haga, T. Haga, A. Ichiyama, K. Kangawa, M. Hiranaga, H. Matsuo, and S. Numa, Primary structure of the α-subunit of bovine adenylate cyclase-inhibiting G-protein deduced from the cDNA sequence, *FEBS Lett.* 197:305-310 (1986).
Sternweis, P. C., and J. D. Robishaw, Isolation of two proteins with high affinity for guanine nucleotides from membranes of bovine brain, *J. Biol. Chem.* 259:13806-13813 (1984).
Stryer, L., Cyclic GMP cascade of vision, *Ann. Rev. Neurosci.* 9:87-119 (1986).
Suki, W. N., J. Abramowitz, R. Mattera, J. Codina, and L. Birnbaumer, The human genome encodes at least three non-allellic G proteins with α_i-type subunits, *FEBS Lett.* 220:187-192 (1987).
Sullivan, K. A., Y. Liao, A. Alborzi, B. Biederman, F. Chang, S. B. Masters, A. D. Levinson, and H. R. Bourne, Inhibitory and stimulatory G proteins of adenylate cyclase: cDNA and amino acid sequences of the α chains, *Proc. Natl. Acad. Sci. U.S.A.* 83:6687-6691 (1986).
Yatani, A., J. Codina, A. M. Brown, and L. Birnbaumer, Direct activation of mammalian atrial muscarinic potassium channel by GTP regulatory protein G_K, *Science* 235:207-211 (1987).

IDENTIFICATION OF FUNCTIONAL DOMAINS IN G PROTEIN α CHAINS

Susan B. Masters, R. Tyler Miller, Kathleen A. Sullivan, and Henry R. Bourne

Departments of Pharmacology and Medicine and
the Cardiovascular Research Institute
University of California
San Francisco, CA 94143-0450

INTRODUCTION

In many signaling pathways, G proteins transduce extracellular signals (hormones or sensory stimuli), detected by receptors on the surface of cells, into regulation of effector proteins that control accumulation of intracellular second messengers (for recent reviews, see Stryer and Bourne, 1986; Gilman, 1987). The G proteins are a highly conserved family of heterotrimeric guanine nucleotide binding proteins. The α subunit, the key component of the G proteins, binds and hydrolyzes guanine nucleotides and interacts with βγ subunits and specific sets of receptor and effector proteins. Signal transduction by the G proteins is driven by a cycle of GTP-dependent conformational changes in the α chain. This guanine nucleotide-dependent molecular machinery is conserved within and beyond the G protein family.

To understand in molecular detail the mechanism of signaling by G proteins, we need to relate the function of these proteins to their structure. The present review will summarize recent progress toward identification of functional domains in the G protein α chains. We will begin with a brief description of the function of G protein α chains and features of a structural model of a composite α chain, α_{avg} (Masters et al., 1986). We will then describe phenotypes produced by mutations in the α chain of G_s (termed α_s) and inferences that can be drawn from them.

FUNCTION OF α CHAINS

Fig. 1 illustrates the reactions that make up the cycle of activation and inactivation of a G protein α chain (numbers refer to reactions shown in the diagram):

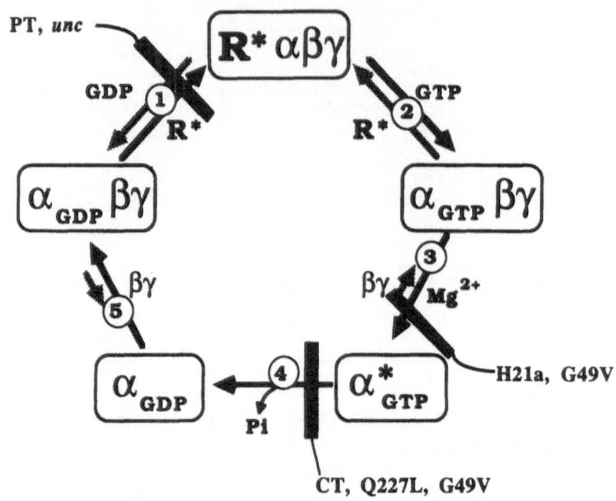

Fig. 1. The GTP-driven cycle of G protein α subunits in transmembrane signaling. See text for details. α and βγ represent the α subunit and the βγ complex of the G protein. **R*** represents the stimulated form of the receptor. Cross-hatched rectangles indicate sites at which the cycle is interrupted by mutations (***unc*, H21a, G49V and Q227L**) or by ADP-ribosylation catalyzed by cholera (**CT**) and pertussis (**PT**) toxins.

(1) In its inactive GDP-bound state, the α chain binds with high affinity to the βγ complex . When receptors become activated by external stimuli, the activated (i.e., hormone-bound) receptor (R*) binds to the $\alpha_{GDP}\beta\gamma$ complex. Release of GDP from the αβγ complex causes the complex to bind much more tightly to R*.

(2) Binding of GTP to the R*αβγ complex causes dissociation of R* and formation of a (transient) $\alpha_{GTP}\beta\gamma$ complex. (Note that binding of GDP would also cause dissociation of R*, as in reaction 1.)

(3) In the presence of Mg^{2+} ion, α_{GTP} has a very low affinity for βγ. Activated α^*_{GTP}, released from βγ, now can activate the effector molecule (e.g., adenylyl cyclase, cGMP phosphodiesterase, K^+ channel, etc.).

(4) The intrinsic GTPase activity of the α^*_{GTP} complex hydrolyzes the bound nucleotide to form a (transient) inactive α_{GDP} state.

(5) The α_{GDP} quickly associates with βγ, thus bringing the cycle back to its initiation point. The α chain continues cycling in this manner, from an inactive to an active form, for as long as the R* species is present.

One reaction in this cycle deserves special emphasis: The formation of α^*_{GTP} (3 in the diagram). The binding of GTP to α promotes a conformational change that results in (a) a marked decrease in affinity for binding βγ, (b) ability to bind and activate the effector, and (c) ability to hydrolyze GTP. These properties differ profoundly from those of α_{GDP}. How does the presence of the γ-phosphoryl on the bound guanine nucleotide produce these changes in α chains? Although poorly understood, this GTP-induced conformational change is a key element of the molecular mechanism common to all GTP binding proteins, and probably occurs in essentially the same way in each of them.

MODEL OF α CHAIN STRUCTURE

Definitive knowledge of the three-dimensional structure of the α chains must await a crystallographic structure. In the interim, we have constructed a structural model of a composite G protein α chain, α_{avg} (Masters et al., 1986). The model centers upon identification of the guanine nucleotide binding domain. Halliday (1984) identified four stretches of primary sequence that are conserved among the GTP-binding proteins of ribosomal protein synthesis (e.g., EF-Tu) and the 21,000 dalton products ($p21^{ras}$) of the *ras* oncogenes. Halliday's prediction that these conserved sequences would play important roles in the structures of guanine nucleotide binding sites has been confirmed by the crystal structures of EF-Tu and $p21^{ras}$ (Jurnak, 1985; La Cour et al., 1985; de Vos et al., 1988).

Identification of similar stretches of sequence in the α chains of the G proteins allowed us to predict the regions of sequence that compose the guanine nucleotide binding site. Fig. 2 shows the locations of the two stretches of amino acid residues that make up the putative guanine nucleotide binding site in a linear representation of α_{avg}. The other portions of α_{avg} are divided by the guanine nucleotide binding domain into three putative "domains" (Fig. 2) — the small (~40 amino acids) amino terminal domain, domain I; a large (~ 150 amino acids) internal domain, domain II; the carboxy terminal domain (~ 100 amino acids), domain III.

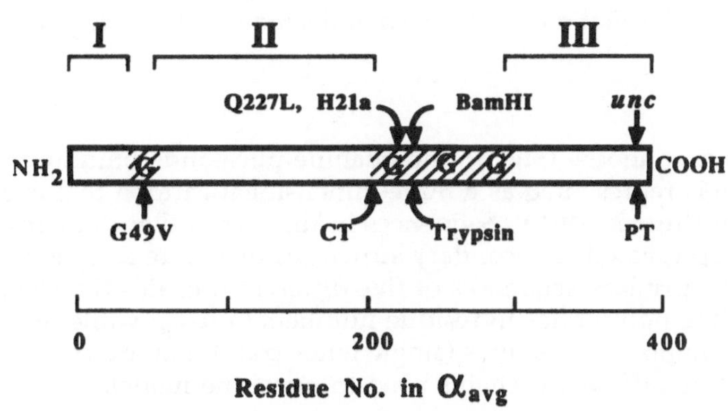

Fig. 2. Linear map of α_{avg}. Letters and arrows indicate the sites ADP-ribosylated by cholera (CT) and pertussis (PT) toxins, the locations of four regions that contribute to the GDP-binding domain (G), and three additional presumed domains (I, II, III) of the α chains. Also shown are the sites of mutations in α_s (*unc*, H21a, G49V, and Q227L), the tryptic cleavage site that is protected in the GTP-bound active conformation of α chains, and position of the BamHI restriction site used for construction of $\alpha_{i/s}$ and $\alpha_{t/s}$ chimeras.

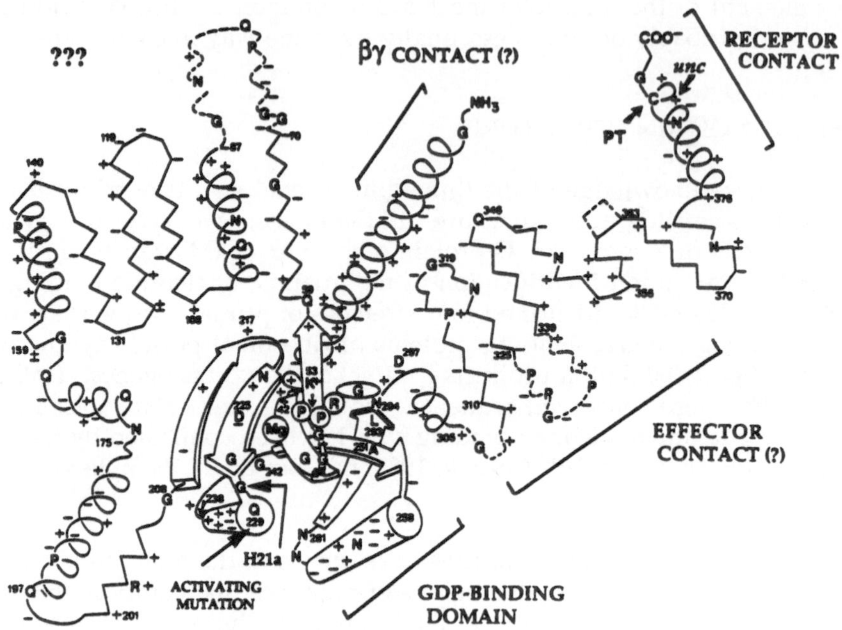

Fig. 3. Structural model of a composite G protein α chain, α_{avg} (Masters et al. 1988). The GDP binding domain (three-dimensional cartoon at the center of the model) is adapted from that of EF-Tu (Jurnak 1985). In this domain the GDP molecule (G, guanine ring; R, ribose; P, phosphoryl) nestles in a pocket formed by turns between β strands and α helices. The rest of the model depicts predicted secondary structure (β strands, α helices, turns). Residues conserved among different α chains are depicted in the single letter amino acid code, and numbers refer to positions of residues in α_{avg} (Masters et.al. 1986). Sites of α_s mutations and the pertussis toxin ADP-ribosylation site are also indicated (see Fig. 2 legend).

In the α_{avg} model (Fig. 3), the guanine nucleotide binding region (center of the figure) is represented as a three-dimensional cartoon that is based upon that of EF-Tu (Jurnak, 1985). Sequences in the three other domains of the protein are represented as secondary structures that were assigned based on analysis of the primary sequences of five different α chains (Masters et al., 1986). Numbers in the figure refer to residue numbers in α_{avg}, while letters designate some of the amino acid residues (single letter code) that are conserved among three of the five different α chains used to make the model.

Comparison of the predicted amino acid sequences of the nine G protein α chains so far identified in mammals reveals a consistent pattern of conserved and divergent sequences (Masters et al., 1986): All are highly conserved in their presumptive guanine nucleotide binding regions, and diverge widely in domains I, II, and III. Conserved α chain functions — guanine nucleotide binding and hydrolysis, and GTP-dependent conformational changes — probably are performed by residues in the highly conserved guanine nucleotide binding domain. The observation that βγ complexes can be functionally interchanged among several G proteins (Kanaho et at., 1984) suggests that the parts of α chains that interact with $\beta\gamma$ should also be conserved. Interactions that require specificity (i.e., with different sets of receptors and effectors) are probably mediated by divergent stretches of amino acids.

We have recently embarked on a series of molecular genetic experiments aimed at assigning functions to the structural elements defined in the α chain model. We have used as a model system the hormone sensitive adenylyl cyclase of S49 mouse lymphoma cells. In wild type S49 cells, G_s couples β-adrenergic and PGE_1 receptors to stimulation of adenylyl cyclase, while a different G protein, G_i, couples somatostatin receptors to inhibition of adenylyl cyclase. The propensity of S49 cells to die upon prolonged elevation of intracellular cAMP has allowed selection of mutant strains with instructive defects in signaling through the cAMP pathway. Of these, one of the most useful is the S49 *cyc*⁻ cell, which is genetically deficient in transcription and translation products of the gene that encodes α_s. S49 *cyc*⁻ cells provide a unique "null" background in which to assess signaling phenotypes produced by expression of normal and genetically altered α_s chains*. Details of the expression system, which utilizes a retroviral vector, have been reported (Sullivan et al., 1987). Here we will briefly describe the phenotypes produced by mutated α_s in S49 cells and sketch their relation to our developing model of α chain structure and function.

GUANINE NUCLEOTIDE BINDING DOMAIN

If the model correctly identifies amino acid sequences that make up the guanine nucleotide binding region of α_{avg}, mutations that alter these sequences should alter binding or hydrolysis of GTP. A large number of mutations in the guanine nucleotide binding region of $p21^{ras}$ provide useful precedents. These mutations create a constitutively active $p21^{ras}$ by decreasing its intrinsic GTPase activity. So far we have assessed function of two α_s proteins with amino acid substitutions that correspond to mutation-induced substitutions that activate $p21^{ras}$. Although both mutant α_s chains appear to exhibit decreased GTPase activity, only one of them shows the expected "activated" phenotype.

In EF-Tu, the second Halliday region is composed of a β strand connected by a loop to an amphipathic α helix. An aspartate residue in the β strand (Asp^{225} in α_{avg}, Asp^{58} in $p21^{ras}$) interacts with a Mg^{2+} ion that in turn is coordinated with the phosphoryl groups of GDP (Jurnak, 1985). Activating mutations in $p21^{ras}$ have been identified at several positions in this region (for review, see Barbacid, 1987). For example, replacement of Gln^{61} by Leu in the Ha-*ras* protein activates the protein's capacity for malignant transformation and diminishes its GTPase activity (Barbacid, 1987).

Mutational replacement of the corresponding residue in α_s (Gln^{227} in α_s, Gln^{229} in α_{avg}, see Fig. 3) produces a constitutively activated phenotype (Masters et al., unpublished) quite similar to that of the $Gln^{61} \rightarrow Leu$ mutation in $p21^{ras}$.

---

* S49 *cyc*⁻ cells, like their wild type parents, die upon continuous exposure to endogenous cAMP. Since it was important to prevent cAMP-induced death of cells transformed with potentially "activated" α_s chains, mutant α_s cDNAs were expressed in an S49 cell bearing, in addition to the *cyc*⁻ mutation, a mutation (*kin*⁻) that causes inactivation of cAMP-dependent protein kinase.

Adenylyl cyclase is constitutively elevated, even in the absence of hormonal agonist, in *cyc*⁻ cells expressing the Gln^{227}→Leu α_s. cAMP is markedly elevated in intact cells expressing this mutant α chain, and the adenylyl cyclase activity of membrane preparations is high in the absence of exogenous stimulators. This activation is nearly equivalent to that produced in wild type S49 membranes upon stimulation by hormonal agonist (e.g., the β-adrenergic receptor agonist isoproterenol) plus a hydrolysis-resistant GTP analog, such as GTPγS.

The activated phenotype of the Gln^{227}→Leu α_s appears to result from decreased GTPase activity. The rate of GTP hydrolysis by α_s can be estimated indirectly by measuring the decay of agonist-stimulated adenylyl cyclase activity following addition of a saturating concentration of antagonist (Cassel et al., 1979). Such estimates suggest that the wild type α_s hydrolyzes GTP at a rate that is at least 20-fold higher than that of Gln^{227}→Leu α_s. (Masters et al., unpublished) The phenotype of this mutant α_s constitutes strong evidence that this region of the guanine nucleotide binding domain of G protein α chains resembles those of $p21^{ras}$ and EF-Tu.

Near the amino terminus of α_{avg}, the first Halliday region forms a loop, located near the phosphoryl groups of GDP, that is thought to play a role in regulating and perhaps in catalyzing hydrolysis of bound GTP. In $p21^{ras}$, mutational replacement of Gly^{12} (corresponding to Gly^{49} in both α_s and α_{avg}; see Fig. 3) by valine (or almost any other amino acid) produces a protein with enhanced capacity to cause malignant transformation of fibroblasts (Seeburg et al., 1984). The same mutations also decrease the *ras* protein's ability to hydrolyze bound GTP (Barbacid, 1987; Trahey and McCormick, 1987).

Mutational replacement of Gly^{49} by valine has a different effect in α_s (Masters et al., unpublished). Adenylyl cyclase in membranes from *cyc*⁻ cells expressing the Gly^{49}→Val α_s responds quite poorly to stimulation by agents that act through α_s, including β-adrenergic receptor agonists, GTP analogs, and AlF_4^-. Indirect estimates of the GTPase activity of Gly^{49}→Val α_s suggest that this mutant hydrolyzes GTP at a rate that is 5- to 10-fold slower than that of wild type α_s (Masters et al., unpublished). The decrease in GTPase activity, an effect that would be expected to cause constitutive activation of adenylyl cyclase, appears to be over-shadowed by an accompanying deleterious effect on the activation of the Gly^{49}→Val α_s. Preliminary results suggest that this mutant α_s chain cannot fully undergo the GTP-induced conformation change required for activation of adenylyl cyclase (Masters et al., unpublished).

The last two Halliday sequences together form a pocket for the guanine ring, including a conserved aspartate (Asp^{297} in α_{avg}; see Fig. 3) that interacts with the 2-amino group of the guanine. The strong conservation of amino acid sequence in this region among the G proteins, EF-Tu, and $p21^{ras}$ led to the prediction (Masters et al., 1986) that the guanine ring of the bound nucleotide is located in a pocket defined by corresponding amino acid residues in all three sets of proteins. The crystal structure of a $p21^{ras}$ protein (de Vos et al., 1988) confirmed this prediction, at least with respect to EF-Tu and $p21^{ras}$. We are presently constructing mutant α_s DNAs designed to test functional similarities of these other portions of the putative GDP binding region of α chains to the corresponding sequences in other GTP binding proteins.

GTP INDUCED CONFORMATIONAL CHANGE

The crystal structures of p21^{ras} and EF-Tu provide only a static picture of the GDP-bound conformation of these proteins. Key questions, both for these proteins and for the G proteins, center on the change in conformation that occurs when bound GDP is replaced by GTP. The G proteins are converted to the "active" conformation upon binding GTP, synthetic GTP analogs, or GDP plus AlF_4^- ion. G protein activation is most easily measured by assessing the protein's ability to regulate activity of effectors, such as adenylyl cyclase. Activation is accompanied by changes in many other properties, including: Altered binding affinity for receptors and $\beta\gamma$, as well as effectors (Gilman, 1987), enhanced intrinsic tryptophan fluorescence of α chains (Higashijima et al., 1987a,b), altered reactivity of cysteine sulfhydryls (Winslow et al., 1987), susceptibility to ADP-ribosylation by cholera and pertussis toxins (Van Dop et al., 1984), and sensitivity to cleavage by trypsin (Fung and Nash, 1983; Hurley et al., 1984).

Useful clues regarding the GTP-dependent conformational change have come from studies of the phenotype of the S49 mutant cell line, H21a. The H21a phenotype (Bourne et al., 1981) results from replacement of a glycine by an alanine residue in α_s, at a position corresponding to Gly^{228} of α_{avg} (Miller et al., 1988). This residue is located at the $\beta \rightarrow \alpha$ turn in the third Halliday region (see Fig. 3). Although G_s in H21a can bind GTP (as shown by the ability of GTP analogs to regulate the affinity of β-adrenergic receptors for agonists in H21a membranes), the mutant G_s cannot stimulate adenylyl cyclase in response to GTP analogs, AlF_4^-, hormones, or cholera toxin. In addition, the GTP- and AlF_4^- -induced dissociation of α_s from $\beta\gamma$, detected by assessing altered migration of normal (S49 wild type) α_s in sucrose density gradients, does not occur with G_s from H21a (Miller et al., 1988).

Studies of α_s cleavage by trypsin provide further evidence that G_s in H21a cannot undergo the conformational change that is normally induced by GTP. Activation of transducin and of (normal) G_s by GTP analogs protects the α chains of these proteins from cleavage by trypsin at a conserved site, corresponding to Arg^{233} of α_{avg}. In contrast to the α_s of wild type S49 cells, the mutant α_s of H21a is *not* protected from proteolysis by GTP analogs (Miller et al., 1988). Thus the H21a mutation appears to have created an α chain that knows it has bound guanine nucleotide but that cannot tell GTP from GDP.

The functional defect of α_s in H21a can be precisely located in the GTP-driven activation cycle (see Fig. 1): The $\alpha\beta\gamma$ complex of H21a apparently is able to assume the "empty site" conformation (step 1), which binds tightly to the activated receptor and enhances the affinity of receptor for binding agonists. Furthermore, $\alpha\beta\gamma$ in H21a can be heterotropically displaced from association with the receptor upon binding guanine nucleotide (step 2), as shown by the ability of guanine nucleotide to lower receptor affinity for agonists in H21a membranes (Miller et al., 1988). The H21a mutation exerts its effect at the next step (step 3 in Fig. 1): Because the mutation prevents α_s from taking on its "active" conformation in the GTP-bound state, α_s does not dissociate from $\beta\gamma$ and cannot interact with the effector, adenylyl cyclase.

It is clear that the GTP induced change from an inactive to an active conformation is common to all G proteins. We would like to know how the active

GTP-bound conformation differs, in three dimensions, from the inactive GDP-bound conformation. If the guanine nucleotide binding pocket of G protein α chains resembles those of EF-Tu and $p21^{ras}$, it is obvious that the γ phosphoryl of GTP must occupy space that is not available in the GDP-bound protein. One attractive possibility is that replacement of GDP by GTP somehow displaces the β strand and induces flexion of the β → α turn in the third Halliday region. In EF-Tu this turn begins at a conserved glycine residue that corresponds to the mutated residue in the H21a α chain; this glycine residue may serve as a hinge (see Fig. 3) that allows relative movement of separate domains of the protein. The Gly → Ala substitution in H21a could decrease flexibility of the presumed hinge in this region and thereby prevent a critical conformational change.

RECEPTOR BINDING DOMAIN

Several lines of evidence point to a carboxy terminal region of the α chains (in domain III; see Fig. 2) as one of the sites involved in receptor interaction. The pertussis toxin catalyzed ADP-ribosylation of a cysteine four residues from the carboxy terminus prevents G_i, G_o, and transducin from responding to stimulation by their respective receptors (Ui et al., 1984). Furthermore, pertussis toxin-catalyzed ADP-ribosylation prevents transducin from binding to rhodopsin (Van Dop et al., 1984). The *unc* mutation in α_s creates a similar phenotype — uncoupling of G_s from interactions with the receptors that stimulate adenylyl cyclase (Haga et al., 1977). The *unc* mutation causes replacement by proline of an arginine located six residues from the carboxy terminus of α_s (Sullivan et al., 1987; Rall and Harris, 1987). Both the *unc* mutation of α_s and the cysteines modified by pertussis toxin in α_i, α_o, and α_t are located in a predicted α helix at the carboxy terminus of the composite G protein α chain, α_{avg} (Masters et al., 1986).

A retinal protein, arrestin, provides a third line of evidence implicating the predicted α helix at the carboxy terminus as a contact site of α chains with receptors. Arrestin competes with transducin for binding to the phosphorylated form of photorhodopsin. Because arrestin contains a stretch of internal sequence very similar to that of the carboxy terminus of α_t (Wistow et al., 1986), it is reasonable to propose that these conserved regions are both involved in rhodopsin binding.

Further evidence has come from phenotypes produced by recombinant chimeric α chains. Two such chimeras are composed of the amino terminal 60% of either α_i (mouse α_{i2}) or α_t (the α chain of bovine transducin) and the carboxy terminal 40% of α_s; the chimeras were constructed by splicing together the corresponding portions of the cDNAs at a conserved BamHI site near the middle of the coding region (see Fig. 2). As a result, the amino acid sequences of domains I and II in the resulting chimeras differ markedly from corresponding sequences of α_s. Expression of either the α_{i2}/α_s chimera (Masters et al. 1988) or of the α_t/α_s chimera (Masters et al., unpublished) in S49 *cyc*⁻ cells produces cells in which β-adrenergic receptors stimulate adenylyl cyclase. Other experiments have shown that the α_i/α_s chimera does not interact with somatostatin receptors, which are coupled to G_i in S49 cells (Masters et al., 1988). These results imply that the carboxy terminal portion of the chimera, which was contributed by α_s, confers on the α chain its specificity for interacting with receptors.

EFFECTOR BINDING DOMAIN

The location of the effector binding region of G protein α chains is not known. Masters et al. (1986) proposed that domain II (Fig. 2) mediates interactions with effectors. This proposal was based in part on the heterogeneity of amino acid sequence in this domain among α chains that regulate different effectors, and in part on reports suggesting that the (much shorter) cognate regions of EF-Tu (Laursen et al., 1981) and $p21^{ras}$ (Sigal et al., 1986) interact with proteins analogous to the effectors of G protein-mediated signaling systems.

The 60:40 α_i/α_s and α_t/α_s chimeras described above allowed a simple test of the notion that domain II mediates effector interactions. If this notion were correct, the chimeras would not be expected to mediate stimulation of adenylyl cyclase, because domain II in the chimeras is derived from α_{i2} or α_t, rather than from α_s. Unexpectedly, the chimeric α chains behave like α_s, in that they respond to the β-adrenergic receptor agonist isoproterenol with *stimulation* of adenylyl cyclase activity (Masters et al., 1988). Thus the phenotypes produced by expression of these chimeras make it unlikely that domain II specifies stimulation of adenylyl cyclase. Instead, they suggest that domain III, and not domain II, contains the amino acid sequences that are critical for a specific stimulatory interaction with adenylyl cyclase (and, by implication, the sequences in other α chains that specify regulation of their effectors). Further experiments may show that this model is too simple — i.e., it is quite possible that sequences in "domains" I and II interact intimately with sequences in domain III, and that sequences from more than one domain may be required for interaction with effectors — or indeed with any other protein, including receptors and the βγ complex as well.

CONTACT WITH THE βγ COMPLEX

Proteolytic removal of the amino terminal 21 residues of the α chain of rod cell transducin destroys its ability to bind to βγ (Navon and Fung, 1987). The inability to interact with βγ prevents the truncated α_t chain from binding to rhodopsin and from undergoing light-triggered exchange of guanine nucleotide. The simplest (but not the only) interpretation of these results is that βγ binds directly to the extreme amino terminus of the α chain. Although observations that βγ complexes can be functionally interchanged among several G proteins (Kanaho et al., 1984) suggest that the parts of α chains that interact with βγ should be conserved, the primary structure of the α chain amino terminus (domain I) is *not* highly conserved. Masters et al. (1986) predicted a conserved α helical secondary structure for the amino terminus (domain I) in all G protein α chains (see Fig. 3). Perhaps some feature of this conserved α helix, not dependent upon strict conservation of a long stretch of amino acid sequence, specifies interaction with βγ. Alternatively, this region may be involved in regulating interactions with βγ, without contacting the complex directly.

SUMMARY

The molecular genetic approach has just begun to provide hints about the functional domains of the G protein α chains. We have some idea of which portions of the α_s interact with receptors and effectors, and guess that the same is true of corresponding regions of other α chains. Tantalizing hints also point to a key region of α_s that is necessary for GTPase activity and for the conformational change induced by binding GTP. With this information in hand, we can pose more precise questions that will begin to provide more detailed understanding of the structure and function of these signal transducing G proteins.

ACKNOWLEDGEMENTS

Work from this laboratory is supported in part by research grants from the National Institutes of Health and from the March of Dimes.

REFERENCES

Barbacid, M., 1987, *ras* Genes, Ann. Rev. Biochem., 56:779.

Bourne, H. R., Kaslow, D., Kaslow, H. R., Salomon, M., and Licko, V., 1981, Hormone sensitive adenylate cyclase: Mutant phenotype with normally regulated beta-adrenergic receptors uncoupled from catalytic adenylate cyclase, Mol. Pharmacol. 20:435.

Cassel, D., Eckstein, F., Lowe, M., and Selinger, Z., 1979, Determination of the turn-off reaction for the hormone-activated adenylate cyclase, J. Biol. Chem., 254:9835

de Vos, A. M., Tong, L., Milburn, M. V., Matias, P. M., Jancarik, J., Miura, K., Ohtsuka, E., Noguchi, S., Nishimura, S., and Kim, S.-H., 1988, Three-dimensional structure of an oncogene protein: Catalytic domain of human c-H-*ras* p21, Science 239:888.

Fung, B. K.-K., and Nash, C. R., 1983, Characterization of transducin from bovine retinal rod outer segments, J. Biol. Chem. 258:10503.

Gilman, A. G, 1987, G Proteins: Transducers of receptor-generated signals, Ann. Rev. Biochem. 56:615.

Haga, T., Ross, E. M., Anderson, H. J., and Gilman, A. G., 1977, Adenylate cyclase permanently uncoupled from hormone receptors in a novel variant of S49 mouse lymphoma cells, Proc. Natl. Acad. Sci. USA 74:2016.

Halliday, K., 1984, Regional homology in GTP-binding proto-oncogene products and elongation factors, J. Cyclic Nucl. Res. 9:435-448.

Higashijima, T., Ferguson, K. M., Sternweis, P. C., Ross, E. M., Smigel, M. D., and Gilman, A. G., 1987a, The effect of activating ligands on the intrinsic fluorescence of guanine nucleotide-binding regulatory proteins, J. Biol. Chem. 262:752.

Higashijima, T., Ferguson, K. M., Smigel, M. D., and Gilman, A. G., 1987b, The effect of GTP and Mg^{2+} on the GTPase activity and the fluorescent properties of G_o, J. Biol. Chem. 262:757.

Hurley, J. B., Simon, M. I., Teplow, D. B., Robishaw, J. D., and Gilman, A. G., 1984, Homologies between signal transducing G proteins and *ras* gene products, Science 226:860.

Jurnak, F, 1985, Structure of the GDP domain of EF-Tu and location of the amino acids homologous to *ras* oncogene proteins, Science 230:32.

Kanaho, Y., Tsai, S. C., Adamik, R., Hewlett, E. L., Moss, J., and Vaughan, M., 1984, Rhodopsin-enhanced GTPase activity of the inhibitory GTP-binding protein of adenylate cyclase, J. Biol. Chem. 259:7378.

La Cour, T. F. M., Nyborg, J., Thirup, S., and Clark, B. F. C., 1985, Structural details of the binding of guanosine diphosphate to elongation factor Tu from *E. coli* as studied by X-ray crystallography, EMBO J. 4:2385.

Laursen, R. A., L'Italien, J. J., Nagarkatti, S., and Miller, D. L., 1981, The amino acid sequence of elongation factor Tu of *Escherichia coli*. The complete sequence, J. Biol. Chem. 256:8102.

Masters, S. B., Stroud, R. M., and Bourne, H. R., 1986, Family of G protein α chains: amphipathic analysis and predicted structure of functional domains, Protein Engineering 1:47.

Masters, S. B., Sullivan, K. A., Miller, R. T., Beiderman, B., Lopez, N. G., Ramachandran, J., and Bourne, H. R., 1988, Carboxy terminal domain of $G_{s\alpha}$ specifies coupling of receptors to stimulation of adenylyl cyclase, Science, 241:448.

Miller, R. T., Masters, S. B., Sullivan, K. A., Beiderman, B., and Bourne, H. R., 1988, A mutation that prevents GTP-dependent activation of the α chain of G_s, Nature, in press.

Navon, S. E. and Fung, B. K.-K., 1987, Characterization of transducin from bovine retinal rod outer segments. Participation of the animo-terminal region of T_α in subunit interaction, J. Biol. Chem. 262:15746.

Rall, T. and Harris, B. A., 1987, Identification of the lesion in the stimulatory GTP-binding protein of the uncoupled S49 lymphoma, FEBS Lett. 224:365.

Seeburg, P. H., Colby, W. W., Capon, D. J., Goeddel, D. V., and Levinson, A. D., 1984, Biological properties of human c-HA-*ras*1 genes mutated at codon 12, Nature 312:71.

Sigal, I. S., Gibbs, J. B., D'Alonzo, J. S., and Scolnick, E. M., 1986b, Identification of effector residues and a neutralizing epitope of Ha *ras* p21, Proc. Natl. Acad. Sci. USA 83:4725.

Stryer, L. and Bourne, H. R., 1986, G proteins: A family of signal transducers, Ann. Rev. Cell Biol. 2:391.

Sullivan, K. A., Miller, R. T., Masters, S. B., Beiderman, B., Heideman,W., and Bourne, H. R., 1987, Identification of receptor contact site involved in receptor-G protein coupling, Nature 330:758.

Trahey, M., and McCormick, F., 1987, A cytoplasmic protein stimulates normal N-*ras* p21 GTPase, but does not affect oncogenic mutants, Science 238:542.

Ui, M., Katada, T., Murayama, T., Kurose, H., Yajima, M., Tamura, M., Nakamura, T., and Nogimori, K., 1984, Islet-activating protein, pertussis toxin: A specific uncoupler of receptor-mediated inhibition of adenylate cyclase, Advances in Cyclic Nucleotide and Protein Phosphorylation Research 17:145.

Van Dop, C., Yamanaka, G., Steinberg, F., Sekura, R. D., Manclark, C. R., Stryer, L., and Bourne, H. R., 1984, ADP-ribosylation of transducin by pertussis toxin blocks the light-stimulated hydrolysis of GTP and cGMP in retinal photoreceptors, J. Biol. Chem. 259:23.

Winslow, J. W., Bradley, J. D., Smith, J. A., and Neer, E. J., 1987, Reactive sulfhydryl groups of α_{39}, a guanine nucleotide-binding protein from brain. Location and function, J. Biol. Chem. 262:4501.

Wistow, G. J., Katial, A., Craft, C., and Shinohara, T., 1986, Sequence analysis of bovine retinal S-antigen. Relationships with α transducin and G proteins, FEBS Lett. 196:23.

BEHAVIOR OF MEMBRANE GUANINE-NUCLEOTIDE BINDING PROTEINS AS SIGNAL TRANSDUCERS

Michio Ui* and Toshiaki Katada**

*Department of Physiological Chemistry, Faculty of Pharmaceutical Sciences, University of Tokyo, Tokyo 113, Japan and **Department of Life Science, Faculty of Science Tokyo Institute of Technology, Yokohoma 227, Japan

INTRODUCTION

Most of the extracellular signal substances (neurotransmitters, hormones, autacoids, *etc*) interact with particular receptor sites on the plasma membrane of mammalian cells that produce intracellular signals (or second messengers) in response to these first messengers. The membrane receptor systems are now known to be mostly composed of three protein components, *i.e.*, receptors, transducers and the effectors that are enzymes or ion channels directly or indirectly responsible for the generation of second messengers. All the transducers so far identified are guanine-nucleotide binding proteins currently abbreviated as G proteins.

The transducer G proteins are characterized by the following three common properties. Firstly, they are $\alpha\beta\gamma$-heterotrimers. They differ from one another in their biggest α-subunits, since the accompanying $\beta\gamma$ components are usually common among the different G proteins. Secondly, their activities are regulated as a result of interconversion between the GTP-bound and the GDP-bound forms. The interconversion is accompanied by the dissociation-association cycle between the trimeric form and the α-monomeric *plus* the $\beta\gamma$-dimeric forms. Thirdly, these G proteins serve as the specific substrates of the ADP-ribosylation reaction catalyzed by cholera or pertussis toxin.

The products of c-*ras* and v-*ras* oncogenes, p21, are also GTP-binding proteins. Though this family of GTP-binding proteins are not $\alpha\beta\gamma$-trimers and not ADP-ribosylated by bacterial toxins, they appear to share essentially the same mechanism for the regulation of activities as trimeric G proteins; i.e., their activities are assumed to be controlled by the interconversion between the GTP-bound and the GDP-bound forms. A similar GTP-binding protein with a smaller molecular weight has been purified from mammalian brain membranes and characterized in comparison with trimeric G proteins (Katada *et al.*, 1988).

The purpose of the present communication is to briefly review, in particular largely based on our own experimental findings, recent progresses of the transducer G protein research (Ui, 1984; 1986; 1988a,b; Katada and Ui, 1988).

BACTERIAL TOXINS AS PROBE FOR G PROTEINS

All the transducer G proteins, or G proteins having the $\alpha\beta\gamma$-trimeric structure, so far purified are capable of being ADP-ribosylated by cholera toxin or pertussis toxin. There is a site to be ADP-ribosylated in the α-subunit of each G protein; an arginine in the middle part of the amino acid sequence and a cysteine four residues from its C-terminus are ADP-ribosylated by cholera and pertussis toxins, respectively. ADP-ribosylation does not proceed, however, on the α-subunit by itself. The coexistence of ARF (ADP-ribosylation factor), another G protein, is essential for cholera toxin to catalyze ADP-ribosylation of a G protein (Gilman, 1987). In the case of the pertussis toxin-catalyzed reaction, it never occurs unless GDP-bound α-subunits are tightly complexed with $\beta\gamma$ components (Ui, 1986).

An exceptional case is transducin-1 and -2 (or G_{t-1} and G_{t-2}) which couple light-activated rhodopsin to cGMP phosphodiesterase in the rod outer segments and play the analogous role in cones, respectively (Streyer and Bourne, 1986); transducins are ADP-ribosylated by both cholera and pertussis toxins while other G proteins are ADP-ribosylated by either toxin alone.

Cholera toxin-induced ADP-ribosylation maintains the modified G proteins in its activated GTP-bound state even without coupled receptor stimulation. This is the reason why vast amounts of cAMP accumulate in cholera toxin-treated mammalian cells in which G_s linked to adenylate cyclase activation is the selective substrate of the toxin. The effect of pertussis toxin-catalyzed ADP-ribosylation is quite different and very unique. It results in uncoupling of the thus modified G proteins from receptors in such a manner that any signal given to the receptors is never transmitted to the G proteins and hence to the subsequent effectors. Consequently, reliable evidence is provided for an involvement of pertussis toxin-sensitive G proteins, if a receptor-coupled signaling in a certain cell type is blocked by prior exposure of the cell to the toxin for a period of time long enough for the whole intracellular pool of membrane G proteins to be ADP-ribosylated.

Taking advantage of this unique property of pertussis toxin, a variety of membrane signaling systems have been reported to be mediated by the toxin-sensitive G proteins in mammalian cells; *e.g.*, adenylate cyclase inhibition, phospholipase C and A_2 activation, K^+-channel opening and Ca^{2+}-channel opening and closure. The useful preparation of pertussis toxin is still referred to as islet-activating protein (IAP), too, since it was discovered during our study as to how the toxin enhanced insulin secretory responses of rat pancreatic islets (Ui, 1984).

Table 1. a-Subunits of various G proteins.

G		Molecular wt (K) from cDNA	Molecular wt (K) SDS-PAGE	Source	Effector	Toxin
G_s		44.5–46	45–52		Ad Case ↑	CT
G_i	1	40.4	41	neutrophil platelet	Ad Case ↓	IAP
	2	40.5	40		P-lipase C ↑(↓)	
	3	40.4	?		P-lipase A_2 ↑	
					K-channel ↑	
G_o		39.9	39	neural cell	Ca-channel ↓↑	
G_t	1	40.0	39	rod	cGMP PDE ↑	CT,IAP
	2	40.4	39	cone		

Abbreviations: SDS-PAGE, sodium dodecyl sulfate-polyacrylamide gel electrophoresis; Ad Case, adenylate cyclase; P-lipase, phospholipase; PDE, phosphodiesterase; CT, cholera toxin; IAP, pertussis toxin.

DIVERSITY OF TOXIN-SUBSTRATE G PROTEINS

Various G proteins have been purified from mammalian tissues. Complementary DNAs (and genomic DNAs in most cases) have been cloned and amino acid sequences have been deduced for the α-subunits of these G proteins. Their molecular weights calculated from the deduced amino acid sequences, together with those actually observed with purified proteins, are listed in Table 1.

G_s is the only G protein that stimulates adenylate cyclase (subscript $_s$ stands for stimulation) and its α-subunit is encoded by a single genomic gene. There are, however, two to four different α-subunits for G_s, because of alternative splicing of internal exons. G_i was first purified as the G protein serving as a substrate of the pertussis toxin-catalyzed ADP-ribosylation that causes reversal of receptor-coupled inhibition of adenylate cyclase (the subscript $_i$ stands for inhibition). Its α-subunit was a protein with a molecular weight of 41,000 on SDS-polyacrylamide gel electrophoresis. This is now referred to as G_{i-1} because two other cDNAs were found to code for similar G proteins. G_{i-2} encoded by one of these cDNAs (Itoh *et al.*, 1986; 1988) has recently been purified as a minor component of G_i families from mammalian brains (Katada *et al.*, 1987a) and as the major component from neutrophils (Oinuma *et al.*, 1987) and platelets (Banno *et al.*, 1987). G_k from human erythrocytes is probably G_{i-3} encoded by the third cDNA (Birnbaumer, personal communication).

The physiological function of G_o contained in large amounts in brain tissues is still unknown. G_o is encoded by its own genomic DNA, and exhibited relatively higher activity than G_i *in vitro* (Katada *et al*, 1986a,b). Immunohistochemical studies with the use of the antibody raised against purified G_o revealed that the G protein is distributed widely in peripheral nerve terminals and pancreatic islets as well as in the central nervous system and retina (Terashima *et al.*, 1987a,b,c; 1988).

G_o was coupled to purified muscarinic receptors more efficiently than G_i when these proteins were reconstituted into phospholipid vesicles (Kurose *et al.*, 1986). Certain receptor agonists including a chemotactic peptide (fMLP) stimulate phospholipase C in neutrophils. The stimulation is mediated by pertussis toxin-sensitive G proteins in some cases; the receptor-coupled stimulation of phospholipase C was abolished when endogenous G proteins were ADP-ribosylated by exposure of neutrophils (Okajima and Ui, 1984; Okajima *et al.*, 1985; Ohta *et al.*, 1985) or mast cells (Nakamura and Ui, 1985) to pertussis toxin. G_o was as effective as G_i in restoring fMLP-induced phospholipase C activation when these G proteins were reconstituted into the pertussis toxin-treated membranes of HL60 cells that had been differentiated into neutrophils (Kikuchi *et al.*, 1986). G_o could play a role similar to G_i in certain tissues.

Thus, four kinds of pertussis toxin-sensitive G proteins ($G_{i\text{-}1}$, $G_{i\text{-}2}$, $G_{i\text{-}3}$ and G_o) are candidates for the transducers that, being coupled to receptors, lead to activation or inhibition of a number of effectors with an exception of adenylate cyclase activation which is mediated by cholera toxin-sensitive G_s. The precise and differential role of these different pertussis toxin-substrate G proteins in membrane receptor-effector coupling will be one of the most important subjects of further investigations. In essentially all the reports so far published, receptor-coupled inhibition of adenylate cyclase, activation of phospholipase A_2 and regulation of K^+ and Ca^{2+} channels that occur in various cell types became unobservable any longer in the same cells whose G_i and G_o had been inactivated by cell exposure to pertussis toxin. In contrast, the activation of phospholipase C via certain receptors in certain cell types was not affected by the exposure of the cells to pertussis toxin despite the ADP-ribosylation of the apparently whole pool of endogenous G_i (and G_o) during the exposure. Cholera toxin was also ineffective in these cases. A possibility cannot be excluded, therefore, that some unidentified toxin-insensitive G proteins would communicate between certain receptors and phospholipase C in these cells.

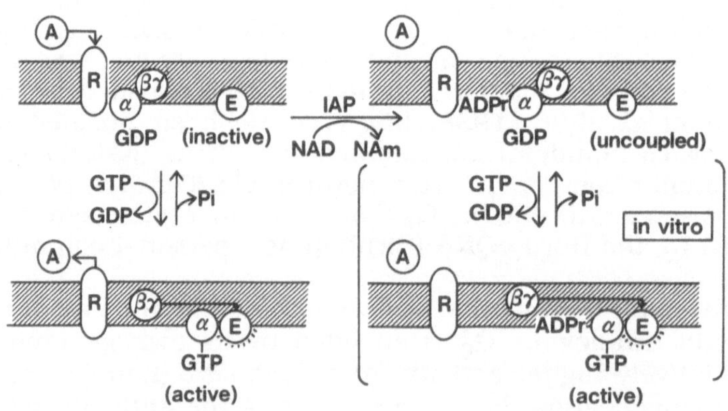

Fig. 1. A role of G proteins in signal transduction from receptors to effectors. A, agonist; R, receptors; E, effectors; ADPr, ADP-ribose; NAm, nicotinamide.

RECEPTOR-COUPLED ACTIVATION AND INHIBITION OF G PROTEINS

The guanine-nucleotide binding sites in the α-subunits of G proteins (Gα) are invariably occupied by guanine nucleotides, GTP or GDP; the sites are never emptied of guanine nucleotides unless G proteins are reversibly denatured in a concentrated $(NH_4)_2SO_4$ solution. GDP-bound Gα is tightly complexed with βγ-subunits. This GDP-bound αβγ-trimeric G protein is inactive in the sense that it does not interact with any effector protein, but it forms a complex with the receptor protein (Fig. 1, top-left). Receptor agonists bind to the receptor thus complexed with the GDP-bound G protein much more readily than do they to the uncomplexed receptor, since the affinity for agonist binding was higher with the former than with the latter.

The agonist binding to receptors gives a stimulus to cause an activation of the complexed G proteins. The activation consists of two processes; one is the release of GDP from the guanine-nucleotide binding sites on Gα in exchange for intracellular GTP (or the GTP-GDP exchange reaction at the sites) and the other is rapid resolution of the thus formed GTP-bound αβγ-complexes into GTP-bound α- and βγ-subunits. Both subunits are liberated from the receptors (see Fig. 1, bottom-left). This is the active state of G proteins. In the case of G_s, for instance, the GTP-bound Gα directly interacts with effector, the adenylate cyclase catalyst, to increase its activity. The underlying mechanism is more complicated, however, for the inhibition of adenylate cyclase. Not only α-subunits but also βγ-subunits thus resolved are responsible for the G_i (or G_o)-induced inhibition of the same cyclase catalyst (Katada *et al.*, 1986a; 1987b). In any event, the reaction involved in the formation of the active state of G proteins, *i.e.*, the GTP-GDP exchange reaction, is referred occasionally to as "turnon" reaction.

Gα has an intrinsic activity to hydrolyze bound GTP (GTPase) that is responsible for recovery of the initial inactive state as the "turnoff" reaction. The GDP-bound inactive αβγ-trimer is the real substrate of pertussis toxin-catalyzed ADP-ribosylation. The ADP-ribosylated G_i or G_o is not capable of forming a complex with receptor protein, and is hence uncoupled from receptors (Fig. 1, top-right). Stimulation of otherwise coupled receptors does not increase activities (such as those to hydrolyze GTP or release GDP in exchange for GTP) of G proteins ADP-ribosylated by pertussis toxin. The spontaneous activities of G proteins observable without receptor stimulation, however, were not affected by pertussis toxin-induced ADP-ribosylation at all (Katada *et al.*, 1986b) (Fig. 1, bottom-right).

The ADP-ribosylation of the cysteine residue of a G protein is not the only means to uncouple the G protein from receptors. Alkylation of the same cysteine by N-ethylmaleimide (NEM) proceeds very rapidly, in preference to other cysteine residues in the same peptide, at a temperature as low as 4°C. The G proteins thus alkylated at cysteines four residues from the C-termini were, just like the ADP-ribosylated G proteins, uncoupled from receptors (Katada *et al.*, 1986c; Ui and Katada, 1987).

An alteration of the affinity for agonist binding to receptors provides a good criterion for coupling of the receptors to G proteins. The addition of GTPγS (a non-hydrolyzable GTP analogue) to membranes lowered the affinity for agonist binding to the membrane receptors coupled to G proteins, reflecting the conversion of the inactive state (Fig. 1, top-left) to the active state (Fig. 1, bottom-left). ADP-ribosylation of membrane G proteins by pertussis toxin was also effective in lowering the affinity of agonist binding, and the reconstitution of purified G_i or G_o into the pertussis toxin-treated membranes could restore the original high affinity binding. Taking advantage of this approach, we have shown that G_i was more effective than G_o in the coupling to α_2-adrenergic, adenosine and opiate receptors, and *vice versa* in the coupling to muscarinic cholinergic receptors, in membrane preparations from rat whole brain (Katada *et al.*, 1986c; Ui and Katada, 1987).

Essentially all of the G protein-coupled receptors so far studied have been shown to promote the GTP-GDP exchange reaction occurring on the coupled G proteins, *i.e.*, to activate the coupled G proteins. An exceptional unique inhibition of a G protein by receptor has recently been reported by Ueda *et al.* (1988), who found that the stimulation of opiate receptors of the κ-subtype in guinea pig cerebellar membranes caused inhibition, rather than activation, of the membrane GTPase activity. The inhibition was totally reversed by pertussis toxin-induced ADP-ribosylation or the alkylation by NEM of endogenous G proteins. G_i purified from rat brain membranes, but not G_o, restored the opiate receptor-coupled inhibition of GTPase upon being reconstituted into the NEM-treated membranes. When G_i possessing ^{3}H-labeled GDP in the guanine-nucleotide binding sites were reconstituted into the NEM-treated membranes, the stimulation of opiate receptors by specific agonists lowered the initial rate of GTP-induced ^{3}H-GDP release from the membranes. Thus, this is the first to report that the GTP-GDP exchange turnon reaction of G_i is not activated but inhibited by coupled receptors.

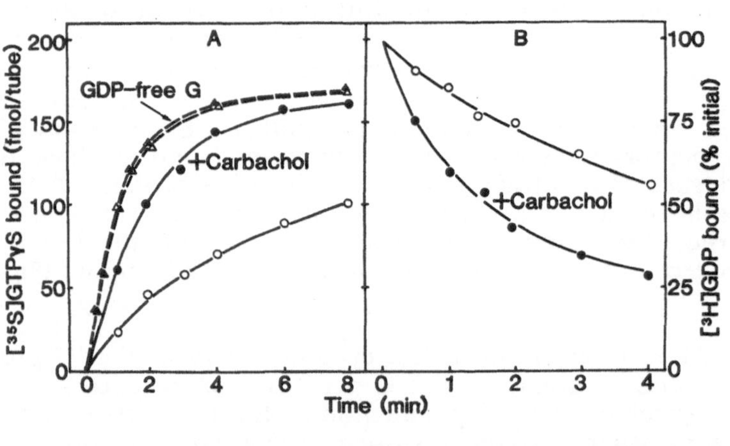

Fig. 2. GTPγS-GDP exchange reaction occurring in phospholipid vesicles into which purified G_o and muscarinic receptors had been reconstituted. The incorporation of radioactive GTPγS into GDP-containing G_o in vesicles (A) or the release of prebound radioactive GDP in exchange for GTPγS in the medium (B) was followed in the absence (open symbols) or presence (solid symbols) of carbachol. In A, GDP-free G_o had also been reconstituted into the vesicles as shown by triangles.

The crucial step of receptor-coupled activation (or inhibition) of G proteins is, as described above, the GTP-GDP exchange reaction on Gα that is accompanied by the subunit resolution of Gαβγ into Gα *plus* βγ-subunits. The interrelationship between the exchange reaction and the subunit resolution is an important mechanism involved in receptor-G protein coupling. This problem has recently been studied as follows (Figs. 2 and 3).

Guanine nucleotide-binding sites on the G_o purified from rat brain membranes were occupied by GDP (see above). This G_o preparation was reconstituted into phospholipid vesicles containing muscarinic receptors purified from bovine brain (Fig. 2). Incubation of these vesicles with [^{35}S]GTPγS resulted in progressive incorporation of the radioactive GTPγS into the vesicles (Fig. 2A; open circles). The incorporation was accompanied by the release of prelabeled GDP (Fig. 2B; open circles), clearly indicating that added GTPγS bound to G_o in exchange for pre-bound GDP spontaneously. As expected, this spontaneous exchange reaction was accelerated by the addition of carbachol, as a reflection of functional coupling between G_o and carbachol-stimulated muscarinic receptors in the vesicles (Fig. 2, A and B; solid circles).

GDP can be released from purified G_o preparations during incubation in a $(NH_4)_2SO_4$ solution, thereby rendering the guanine-nucleotide binding sites on the G_o totally freed of GDP or other guanine nucleotides (see above). This GDP-free G_o preparation was similarly reconstituted into the muscarinic receptor-containing vesicles. The radioactive GTPγS

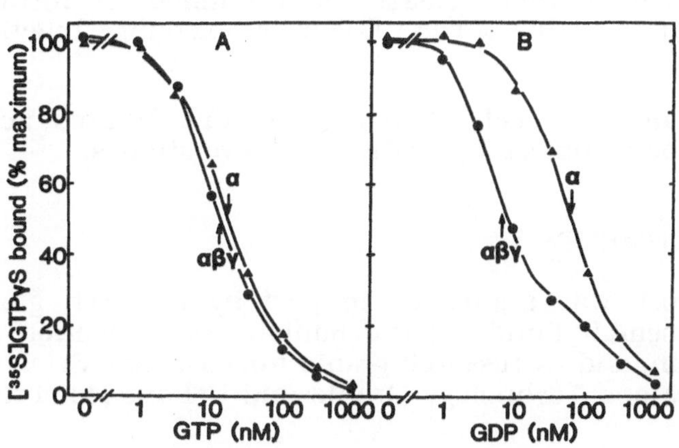

Fig. 3. Displacement curves for radioactive GTPγS in competition with GTP or GDP for the binding sites on $G_o\alpha$ or $G_o\alpha\beta\gamma$. $G_o\alpha\beta\gamma$ (solid circles) or $G_o\alpha$ (solid trigangles) was incubated with radioactive GTPγS and increasing concentrations of GTP (A) or GDP (B). The radioactivity bound to the protein was measured after incubation and plotted against GTP or GDP concentrations.

was incorporated into the GDP-free G_o much more rapidly than was it into the GDP-containing G_o (Fig. 2A; open triangles). Of much interest was the finding that the incorporation of GTPγS into the GDP-free G_o, unlike the incorporation into the GDP-containing one, was not promoted by stimulation of the coexisting muscarinic receptors (Fig. 2A; solid triangles). Thus, the results in Fig. 2 strongly suggest that a process directly susceptible to coupled receptor stimulation is the release of GDP from the guanine-nucleotide binding sites on G proteins rather than the binding of GTP to the same sites.

The affinity for GTP and GDP was then compared between Gαβγ and Gα (Fig. 3). $G_o\alpha$ used in Fig. 3 was prepared by incubating $G_o\alpha\beta\gamma$ with AMF (the mixture of 20 μM $AlCl_3$, 6 mM $MgCl_2$ and 10 mM NaF) which was as effective as GTPγS in resolving a G trimer into an α-monomer *plus* a βγ-dimer. The Gα thus resolved still possessed GDP in its guanine-nucleotide binding sites. The GDP-bound $G_o\alpha$ was then separated from βγ-subunits by the aid of a gel filtration column that was equilibrated and eluted with AMF (Katada *et al.*, 1986b). The affinity for GTP or GDP was measured by competition with [^{35}S]GTPγS for the guanine-nucleotide binding sites on the α-subunits of G_o. The incorporation of [^{35}S]GTPγS to Gαβγ or Gα was competitively antagonized by GTP or GDP added at increasing concentrations. There was no difference between Gαβγ and Gα in GTP-induced GTPγS displacement curves (Fig. 3A), but marked difference observed in the GDP-induced displacement curves (Fig. 3B). The affinity of GDP to the guanine-nucleotide binding sites on the α-subunits was thus lowered by one order of magnitude, when Gαβγ was resolved into Gα.

Thus, the results in Figs. 2 and 3 may lend a strong support to the idea that the direct effect of receptor stimulation by an agonist is exerted to resolution of coupled G proteins into the α- and βγ-subunits. Since the affinity of GDP for the binding sites on the α-subunits is lowered by the resolution, GDP is then released in exchange for intracellular GTP, thereby resulting in the turnon GTP-GDP exchange reaction on the resolved Gα.

This tentative mechanism of receptor-mediated activation of G proteins will be a subject for further extensive studies.

ACKNOWLEDGMENTS

This work was supported in part by research grants from the Scientific Research Funds of the Ministry of Education, Science, and Culture, Japan, and by research grants from Yamada Science Foundation, Toray Science and Technology Grants and Uehara Momorial Foundation in Japan.

REFERENCES

Banno, Y., Nagao, S., Katada, T., Nagata, K., Ui, M., and Nozawa, Y., 1987, Stimulation by GTP-binding proteins (G_i, G_o) of partially purified phospholipase C activity from human platelet membranes, *Biochem. Biophys. Res. Commun.*, 146: 861-869.

Gilman, A. G., 1987, G proteins: transducers of recepor-generated signals, *Ann. Rev. Biochem.*, 56: 615-649.
Itoh, H., Kozasa, T., Nagata, S., Nakamura, S., Katada, T., Ui, M., Iwai, S., Ohtsuka, E., Kawasaki, H., Suzuki, K., and Kaziro, Y., 1986, Molecular cloning and sequence determination of cDNAs for α-subunits of the guanine nucleotide-binding proteins G_s, G_i and G_o from rat brain, *Proc. Natl. Acad. Sci. USA*, 83: 3776-3780.
Itoh, H., Katada, T., Ui, M., Kawasaki, H., Suzuki, K., and Kaziro, Y., 1988, Identification of three pertussis toxin substrates (41, 40 and 39 kDa proteins) in mammalian brain. Comparison of predicted amino acid sequences from G-protein α-subunit genes and cDNAs with partial amino acid sequences from purified proteins, *FEBS Lett.*, 230: 85-89.
Katada, T., Oinuma, M., and Ui, M., 1986a, Mechanisms for inhibition of the catalytic activity of adenylate cyclase by the guanine nucleotide-binding proteins serving as the substrate of islet-activating protein, pertussis toxin, *J. Biol. Chem.*, 261: 5215-5221.
Katada, T., Oinuma, M., and Ui, M., 1986b, Two guanine nucleotide-binding proteins in rat brain serving as the specific substrate of islet-activating protein, pertussis toxin. Interaction of the α-subunits with βγ-subunits in development of their biological activities, *J. Biol. Chem.*, 261: 8182-8191.
Katada, T., Kurose, H., Oinuma, M., Hoshino, S., Shinoda, M., Amanuma, S., and Ui, M., 1986c, Role of GTP-binding proteins in coupling of receptors and adenylate cyclase, *in*: "Gunma Symposium on Endocrinology," vol. 23, pp. 45-67, Center for Academic Publications Japan, Tokyo, VNU Science Press BV, Utrecht.
Katada, T., Oinuma, M., Kusakabe, K., and Ui, M., 1987a, A new GTP-binding protein in brain tissues serving as the specific substrate of islet-activating protein, pertussis toxin, *FEBS Lett.*, 213: 353-358.
Katada, T., Kusakabe, K., Oinuma, M., and Ui, M., 1987b, A novel mechanism for the inhibition of adenylate cyclase via inhibitory GTP-binding proteins. Calmodulin-dependent inhibition of the cyclase catalyst by the βγ-subunits of GTP-binding proteins. *J. Biol. Chem.*, 262: 11897-11900.
Katada, T., and Ui, M., 1988, Unique properties of a new GTP-binding protein with molecular mass of 24,000 daltons purified from porcine brain membranes, *in*: "The Molecular Biology of Signal Transduction," Symposia on Quantitative Biology, vol. 53, in press, Cold Spring Harbor Laboratory, New York.
Katada, T., Imai, S., Tohkin, M., and Ui, M., 1988, Purification and characterization of a new GTP-binding protein with the molecular mass of 24,000 daltons from porcine brain membranes, *J. Biol. Chem.*, in press.
Kikuchi, A., Kozawa, O., Kaibuchi, K., Katada, T., Ui, M., and Takai, Y., 1986, Direct evidence for involvement of a guanine nucleotide-binding protein in chemotactic peptide-stimulated formation of inositol bisphosphate and trisphosphate in differentiated human leukemic (HL-60) cells. Reconstitution with G_i or G_o of the plasma membranes ADP-ribosylated by pertussis toxin, *J. Biol. Chem.*, 261: 11558-11562.
Kurose, H., Katada, T., Haga, T., Haga, K., Ichiyama, A., and Ui, M., 1986, Functional interaction of purified muscarinic receptors with purified inhibitory guanine nucleotide regulatory proteins reconstituted in phospholipid vesicles, *J. Biol. Chem.*, 261: 6423-6428.

Nakamura, T., and Ui, M., 1985, Simultaneous inhibitions of inositol phospholipid breakdown, arachidonic acid release, and histamine secretion in mast cells by islet-activating protein, pertussis toxin. A posssible involvement of the toxin-specific substrate in the Ca^{2+}-mobilizing receptor-mediated biosignaling system, *J. Biol. Chem.*, 260: 3584-3593.

Oinuma, M., Katada, T., and Ui, M., 1987, A new GTP-binding protein in differentiated human leukemic (HL-60) cells serving as the specific substrate of islet-activating protein, pertussis toxin, *J. Biol. Chem.*, 262: 8347-8353.

Ohta, H., Okajima, F., and Ui, M., 1985, Inhibition by islet-activating protein of a chemotactic peptide-induced early breakdown of inositol phospholipids and Ca^{2+} mobilization in guinea pig neutrophils, *J. Biol. Chem.*, 260: 15771-15780.

Okajima, F., and Ui, M., 1984, ADP-ribosylation of the specific membrane protein by islet-activating protein, pertussis toxin, associated with inhibition of a chemotactic peptide-induced arachidonate release in neutrophils. A possible role of the toxin substrate in Ca^{2+}-mobilizing biosignaling. *J. Biol. Chem.*, 259: 13863-13871.

Okajima, F., Katada, T., and Ui, M., 1985, Coupling of the guanine nucleotide regulatory protein to chemotactic peptide receptors in neutrophil membranes and its uncoupling by islet-activating protein, pertussis toxin. A possible role of the toxin substrate in Ca^{2+}-mobilizing receptor-mediated signal transduction, *J. Biol. Chem.*, 260: 6761-6768.

Streyer, L., and Bourne, H. R., 1986, G proteins: a family of signal transducers, *Ann. Rev. Cell Biol.*, 2: 391-419.

Terashima, T., Katada, T., Oinuma, M., Inoue, Y., and Ui, M., 1987a, Immunohistochemical localization of guanine nucleotide-binding protein in rat retina, *Brain Res.*, 410: 97-100.

Terashima, T., Katada, T., Oinuma, M., Inoue, Y., and Ui, M., 1987b, Endocrine cells in pancreatic islets of Langerhans are immunoreactive to antibody against guanine nucleotide-binding protein (G_o) purified from rat brain, *Brain Res.*, 417: 190-194.

Terashima, T., Katada, T., Okada, E., Ui, M., and Inoue, Y., 1987c, Light microscopy of GTP-binding protein (G_o) immunoreactivity within the retina of different vertebrates, *Brain Res.*, 436: 384-389.

Terashima, T., Katada, T., Oinuma, M., Inoue, Y., and Ui, M., 1988, Immunohistochemical analysis of the localization of guanine nucleotide-binding protein in the mouse brain, *Brain Res.* 442: 305-311.

Ueda, H., Misawa, H., Katada, T., Ui, M., Satoh, M., and Takagi, H., 1988, A novel mechanism for coupling receptors to an islet-activating protein-sensitive G-protein: inhibition of G-protein activities by κ-opioid receptor agonists in guinea pig cerebellar membranes, *J. Biol. Chem.*, in press.

Ui, M., 1984, Islet-activating protein, pertussis toxin: a probe for functions of the inhibitory regulatory component of adenylate cyclase, *Trends in Pharmacol. Sci.*, 5: 277-279.

Ui, M., 1986, Pertussis toxin as a probe of receptor coupling to inositol lipid metabolism, *in*: "Phosphoinositides and Receptor Mechanisms," pp. 163-195, J. W., Putney, ed., Alan R. Liss, Inc., New York.

Ui, M., 1988a, The multiple biological activities of pertussis toxin, *in*: "Pathogenesis and Immunity in Pertussis," pp. 121-145, A. C. Wardlaw and R. Paton, eds., John Wiley & Sons Ltd., London.

Ui, M., 1988b, G proteins identified as pertussis toxin substrates, *in*: "G Proteins and Calcium Mobilization," P. H. Naccache, ed., in press, CRC Press, Boca Raton.

Ui, M., and Katada, T., 1987, Differential roles of G_i and G_o in multiple receptor coupling in brain, *in*: "Current Communications in Molecular Biology: Inositol Lipids in Cellular Signaling," pp. 59-63, R. H. Michell & J. W. Putney, eds., Cold Spring Harbor Laboratory, New York.

TRANSDUCIN: THE MOLECULAR SWITCH IN VISUAL EXCITATION AND A MODEL FOR BIOLOGICAL COUPLING ENZYMES

Yee-Kin Ho and Vijay N. Hingorani

Department of Biological Chemistry
University of Illinois at Chicago
Chicago, Illinois 60612, U.S.A.

TRANSDUCIN, THE RETINAL cGMP CASCADE AND VISUAL EXCITATION

Visual excitation in vertebrate rod photoreceptor cells involves a light-activated cGMP cascade (for a review see Liebman, et al., 1987). Photoexcitation of rhodopsin leads to the activation of a latent cGMP phosphodiesterase (PDE) in the rod outer segments and results in the rapid hydrolysis of cGMP to 5'-GMP. The transient decrease of cGMP concentration causes the closure of the cation channels within the plasma membrane and the subsequent hyperpolarization of the photoreceptor cell. Transducin, a GTP-binding protein which is composed of three polypeptides (T_α, Mr 40,000 and $T_{\beta\gamma}$, Mr 37,000 and 8,000), has been shown to mediate the light activation signal from photolyzed rhodopsin to the PDE. The excitation occurs in a two stage amplification cascade. In the dark-adapted state, transducin exists in its latent form where T_α-GDP is associated with $T_{\beta\gamma}$. Photolyzed rhodopsin catalyzes the exchange of bound GDP for GTP in hundreds of transducin molecules. The T_α-GTP and $T_{\beta\gamma}$ subunits of the activated transducin then dissociate from the rod outer segment membrane. The T_α-GTP activates the latent PDE complex ($P_{\alpha\beta}$, Mr 88,000 and 84,000 and P_γ, Mr 14,000) by removing the inhibitory constraints imposed by P_γ upon the $P_{\alpha\beta}$ catalytic sites. The cascade is shut off via the hydrolysis of T_α-bound GTP and the phosphorylation of the photolyzed rhodopsin by rhodopsin kinase. The T_α-GDP then reassociates with $T_{\beta\gamma}$ for another activation cycle.

The protein components of the cascade including rhodopsin, the subunits of transducin (T_α and $T_{\beta\gamma}$), the T_α-Gpp(NH)p complex and the phosphodiesterase ($P_{\alpha\beta}$ and P_γ) have been purified (Fung, 1983). Specific antibodies against these purified proteins have been raised. Reconstitution assays have been established to examine each step of the cGMP cascade cycle. The overall activation of transducin by photolyzed rhodopsin can be assayed by the [^{3}H]Gpp(NH)p binding and GTPase activities. The release of the T_α-Gpp(NH)p and $T_{\beta\gamma}$ subunits from disk membrane can be assayed by the centrifugation method which monitors the solubilization of the transducin subunits. The interaction between T_α and $T_{\beta\gamma}$ can be indirectly assayed in the absence of rhodopsin by the $T_{\beta\gamma}$-enhancement of the pertussis-toxin catalyzed ADP-ribosylation of T_α. Since the purified T_α-Gpp(NH)p complex can activate the PDE in the absence of rhodopsin and $T_{\beta\gamma}$, the second stage of cascade can be examined independent of the rhodopsin catalyzed nucleotide exchange. The amino acid

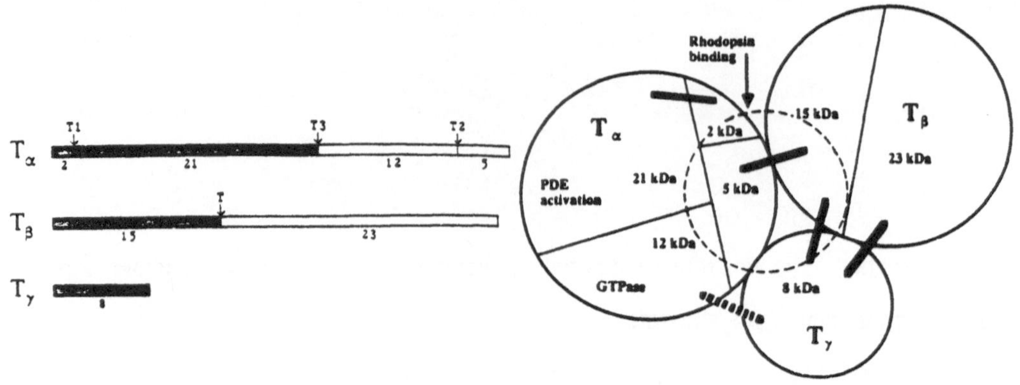

Fig. 1. Linear tryptic peptide map (left) and the schematic diagram of topological map of the cross-linked products (right) of transducin subunits. In the linear peptide map, the arrows indicate the sites of tryptic cleavage. In the topological map, the heavy bars indicate the observed cross-linking between various fragments of the subunits.

sequences of the subunits of transducin have been obtained via molecular cloning of the genes (Medynski, et al., 1985; Hurley, et al., 1984; Fong, et al., 1986) and the linear tryptic peptide map determined from proteolysis analysis as shown in Figure 1 (Fung & Nash, 1983). This information has set the ground work for our structural and functional characterization of transducin.

SUBUNIT ORGANIZATION OF TRANSDUCIN

Chemical cross-linking was applied to analyze the subunit organization of transducin in solution and the changes which occur upon activation by bound Gpp(NH)p (Hingorani, et al., 1988a). The approach was to first cross-link the transducin subunits with a bifunctional reagent (para-phenyldimaleimide or maleimidobenzyoyl n-hydroxysuccinimide ester) and then treat the cross-linked products with trypsin to generate covalently linked peptide fragments. The cross-linked subunits as well as the peptides were identified by Western immunoblotting with anti-T_α and anti-$T_{\beta\gamma}$ antisera. The molecular weights of the cross-linked products, the effect of bound Gpp(NH)p on cross-linking and the antigenicity of the peptide fragments of T_α and $T_{\beta\gamma}$ toward the antisera, allowed us to generate a topological model for the transducin subunits (Figure 1).

In addition to the expected $T_{\alpha\beta}$ and $T_{\beta\gamma}$ cross-linked products, several oligomeric products as large as $(T_{\alpha\beta\gamma})_3$ were observed and their existence diminished after the incorporation of Gpp(NH)p. A $(T_{\alpha\gamma})_2$ structure was also detected. The close proximity of T_α and T_γ suggests that T_γ may play a role in conferring the specificity of the interaction between T_α and rhodopsin. These observations could imply that transducin may form an oligomeric structure such as a tetramer of $(T_\alpha T_{\beta\gamma})_4$ in solution. We have furthered our study using sedimentation and laser dynamic light scattering methods (Goldin, S., Mazar, A., & Ho, Y.-K., unpublished result). Results obtained from both methods suggest that transducin exists in an equilibrium of monomers and tetramers. No dimers, trimers or oligomers larger than the tetramer were detected. The physiological role of the transducin tetramer remains to be elucidated.

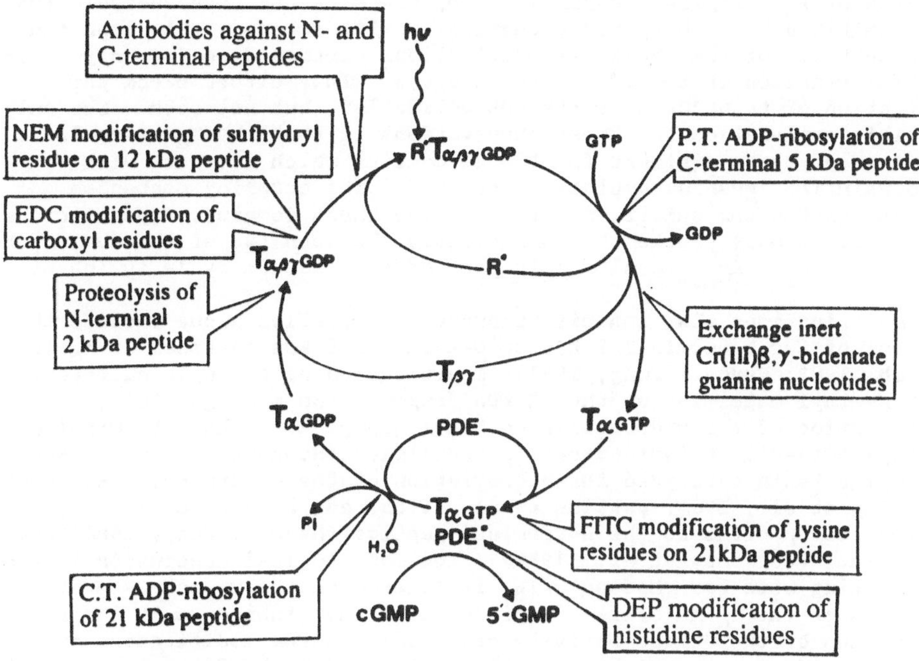

Fig. 2. Summary of specific inhibitions of the cGMP cascade by limited proteolysis, ADP-ribosylation and chemical modifications. R* represents photoexcited rhodopsin. NEM, DEP, FITC, C.T. and P.T. are abbreviations for N-ethylmaleimide, diethyl pyrocarbonate, fluorescein-5'-isothiocyanate, cholera toxin and pertussis toxin, respectively.

However, its existence is consistent with results from other studies. The four-fold internal homology seen in the T_β molecule may be essential in providing the symmetry needed for oligomer formation (Fong, et al., 1986). Kinetic studies have suggested the possibility that a single rhodopsin may interact with four transducin has been reported (Bennett & Dupont, 1985).

One may speculate on the advantage of transducin as a tetramer in facilitating the turnover of the cascade. Rhodopsin, an integral membrane protein and the transducin monomer ($T_\alpha T_{\beta\gamma}$) may each have a single active site for interacting with each other. In order to interact productively transducin must bind to the disk membrane in a precise manner. Transducin as a tetramer $(T_\alpha T_{\beta\gamma})_4$ with four active sites facing different directions would increase the productive binding to the rhodopsin membrane. A single photolyzed rhodopsin may activate four ($T_\alpha T_{\beta\gamma}$) in a single encounter thus providing the needed speed and amplification for the activation of the cascade. In the returning step of the cycle the soluble T_α-GDP and $T_{\beta\gamma}$ recombine. If the formation of the tetramer is under allosteric control, the recruiting of soluble transducin subunits could also be greatly facilitated.

IDENTIFICATION OF FUNCTIONAL DOMAINS OF T_α SUBUNIT

The T_α subunit must contain binding sites for $T_{\beta\gamma}$, rhodopsin, guanine nucleotide and PDE in order to carry out its role in mediating the signal. Using the linear tryptic peptide map for structural identification and the reconstitution assays for functional analysis, chemical modification has been extensively used to study the organization of the functional domains of T_α. Figure 2 summarizes the steps of the cascade cycle that are inhibited by each specific chemical modification. It is important to note that all of the chemical modifications performed so far only inhibit a partial reaction of the cGMP cascade cycle. They either block the activation of transducin or the PDE activation, but not both. The only possible explanation for these observations is that the T_α molecule is composed of three distinct functional domains which interact with rhodopsin/$T_{\beta\gamma}$, guanine nucleotide and PDE. The coupling mechanism depends on the communication between these three domains. Specific chemical modification only blocks the communication between two of the domains but not all three. As a result, only part of the cascade cycle is inhibited.

Modifications that inhibit transducin activation alone include the removal of the N-terminal 2 kDa or C-terminal 5 kDa peptides of T_α (Fung & Nash, 1983; Navon & Fung, 1987), modification of a single sulfhydryl group with N-ethyl maleimide on the 12 kDa fragment (Ho & Fung, 1984), modification of the hydrophilic carboxyl groups near the N-terminal with 1-ethyl-3-(3-dimethylaminopropyl carbodiimide (Hingorani et al., 1988), pertussis toxin catalyzed ADP-ribosylation of the C-terminal 5 kDa peptide (Van Dop et al., 1984; Watkins et al., 1985) and binding of monoclonal antibodies specific to the N-terminal peptide (Navon & Fung, 1988) and the C-terminal (Cerione, et al., 1988). The most logical conclusion is that the interacting site for rhodopsin/$T_{\beta\gamma}$ is located on the N- and C-terminal peptides of the T_α molecule. Modifications that inhibit only the PDE activation or the GTPase activity of T_α include the modification of a lysine residue on the 21 kDa tryptic peptide of T_α by fluorescein 5'-isothiocyanate (Hingorani & Ho, 1987a) and ADP-ribosylation of the 21 kDa peptide catalyzed by cholera toxin (Navon & Fung, 1984). Modification of histidine residue of the T_α-Gpp(NH)p complex with diethyl pyrocarbonate specifically inhibits the activation of the latent PDE (Tobias, D., Chang, L., & Ho., Y.-K., unpublished result). The guanine nucleotide binding site has been identified by comparison with other GTP-binding proteins (Halliday, 1984) and affinity analogue labeling (Hingorani, 1988). These studies indicate that the guanine binding site is on the 12 kDa tryptic fragment and phosphate binding site is associated with the N-terminal side of the 21 kDa fragment. The remainder of the 21 kDa fragment which is sandwiched between the guanine and phosphate binding sites should be the PDE activation domain.

STRUCTURAL MODELING OF T_α SUBUNIT

The homology between the GTP-binding site of T_α and those of other signal transducing G-proteins, elongation factor Tu (EF-Tu) and the ras-p21 protein has been identified. Recently, the tertiary structures of the GTP-binding domains of EF-Tu and ras-p21 were solved by X-ray crystallography (laCour 1985; Jurnak 1985; de Vos, et al., 1988). By integrating the structural homology and the biochemical characteristics of T_α, we have proposed a three-dimensional model for the T_α molecule (Hingorani & Ho, 1987b).

First, the secondary structure and hydropathy of the 350 amino acid residues of the T_α molecule was predicted from the primary sequence using the existing algorithms which provided a foundation for the folding efforts. Then, the information from functional domain analyses was incorporated into the linear tryptic peptide map (Figure 3A). Domain 1 (D1) consists of the nucleotide binding site; domain 2 (D2) is responsible for the interaction with PDE; and domain 3 (D3), consisting of the N- and C-terminal peptides binds with the $T_{\beta\gamma}$ and rhodopsin. In general, the link between functional domains is composed of flexible regions which act as movable hinges which allow a conformational change in one domain to be conveyed to another domain. A careful inspection of the primary sequence of

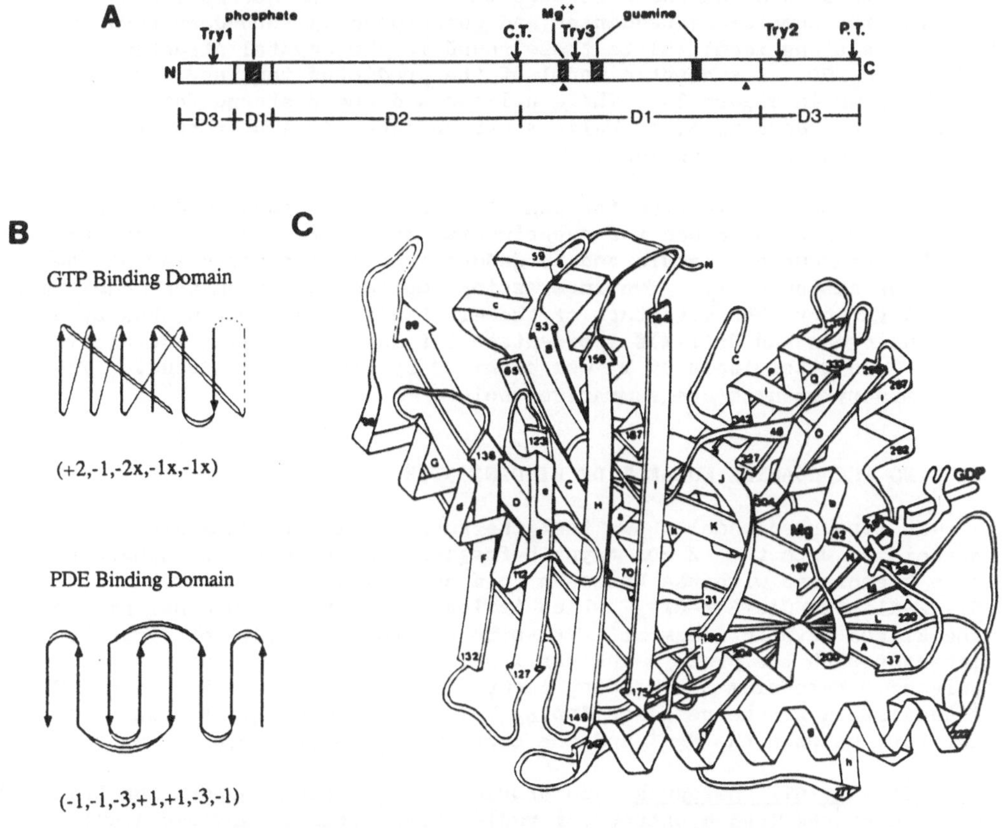

Fig. 3. Structural Model for T_α. (A) The linear peptide map of T_α with proposed functional domains. (B) Schematic diagram of the topology of the ß-pleated sheets for the two major functional domains of T_α. The arrows correspond to individual ß-strands. The topology of the proposed nucleotide binding domain of T_α and the PDE interacting domain were deduced from the comparison with that of EF-Tu and Cu,Zn superoxide dismutase respectively. In T_α the arrows, from left to right, represent ß-Strands N,M,L,A,K,J in the nucleotide binding site and ß-Strands B,C,D,G,F,E,H,I and the PDE site. (C) Proposed Tertiary Structure of the T_α Molecule. The numbers represent the number of amino acid residues from the primary sequence.

T_α reveals two such regions (Gly_{198}-Gly_{199} and Gly_{288}-Pro_{289}-Asn_{290}, both of which are located between the proposed functional domains of the T_α molecule (Figure 3A, black triangles).

The possible folding pattern for each domain of T_α was determined separately and then combined to build a complete three-dimensional model. We propose that the GTP-binding site of T_α, like that of EF-Tu, has a ß sheet topology of (+2,-1,-2x,-1x,-1x) according to the nomenclature of Richardson (1977). An examination of the secondary structure within the PDE activation domain of T_α shows several ß-strands in tand m which are separated by turns and coils. We propose that this domain, consists of a twisted ß-barrel such as in Cu, Zn superoxide dismutase (SOD) (Richardson et al., 1975) and is shown in Figure 3B. The proposed topology has allowed us to translate the primary and secondary structure of each functional domain of T_α into a three-dimensional structure. This was accomplished by placing the homologous sequences and corresponding secondary structures of T_α in positions identical to those found in the crystal structures of EF-Tu and SOD. A schematic model of the predicted tertiary structure of T_α is shown in Figure 3C. The α helices and the β strands of the molecule are designated as Helices a-i and Strands A-Q, respectively, starting from the N-terminus.

The primary motivation for constructing a structural model for T_α was to provide new clues for experimental design for future characterization of the transducin molecule and to deduce a molecular mechanism of the coupling action of T_α. Two interesting implications from the model were tested further. The structure of the GTP binding site was probed by affinity GTP analogues and Cr(III)β,γ-bidentate GTP analogues. The interaction of the N- and C-terminals of T_α was examined by fluorescence study. The preliminary results are summarized below.

PROBING THE GUANINE NUCLEOTIDE BINDING SITE

The proposed model shows that the guanine ring binding site is associated with the 12 kDa tryptic fragment, whereas the phosphate-binding site associates with the Rossmann fold containing the residues Gly_{36}-Ala-Gly-Glu-Ser-Gly_{41} which is located in the N-terminal side of the 21 kDa tryptic fragment. The ribose structure does not directly interact with the protein and has its 2',3' hydroxyl groups exposed to the solvent. Magnesium ion binding occurs in a loop between Strand K and Helix f consisting of the residues Asp_{196}-Val-Gly-Gly-Gly_{200}. A salt bridge may form between the Mg^{2+} and the carboxyl group of Asp_{196}.

Affinity GTP analogues. To provide experimental support for this model we have synthesized a battery of radioactive affinity analogues with reactive groups on different parts of the GTP molecule. By analyzing the labeling site of these analogues, the GTP binding site of transducin can be mapped out in molecular detail (Hingorani, 1988). 8-azido-GTP was used to determine the regions of transducin which are in close proximity to the purine ring. The sites which interact with the phosphate moiety were determined by using 4-(azidoanilido)-P^1-5'- GTP(AA-GTP) and 5'-p-fluorosulfonylbenzoyl guanosine (FSBG). The ribose binding regions were examined with 4-(azidoanilido)-P^1-5'-GTP(AA-GTP) and 2',3' dialdehyde derivative of GTP (oGTP). A bifunctional cross-linking analogue, 8-azido-P^3-4-(azidoanilido)-P^1-5'GTP(8-azido-AA-GTP), was used to probe the possible involvement of subunit interaction in nucleotide binding.

The effectiveness of these analogues in interacting with the GTP-binding site of transducin was examined by their ability to serve as cofactors to activate the PDE and to act as inhibitors of the GTPase activity of transducin. With the exception of FSBG which does not activate PDE nor inhibit the GTPase activity, all the analogues were found to be effective with the following order: AA-GTP>oGTP>ANPAP-GTP=8-azido-GTP. The labeled transducin was subjected to trypsin proteolysis for peptide mapping. oGTP and ANPAP-GTP were unable to covalently modify transducin. Since both modifications are at the ribose ring on the 2' and 3' hydroxy groups, these negative results indicate that this region of the GTP molecule is exposed to the solvent and is away from the protein side chains. 8-Azido-GTP and AA-GTP both labeled transducin but on different regions of the protein. AA-GTP only labeled the T_α subunit and was specific for the 21 kDa fragment. This indicates that the phosphate moiety of the bound GTP is in direct contact with this region. The phosphates have no contact with the T_β subunit. On the other hand 8-Azido-GTP labeled both the T_α and the T_β subunit. The labeling of T_α was on the 12 kDa fragment suggesting that the guanine ring interacts with this region of the protein. T_β was labeled on the 23 kDa fragment. This suggest that either T_β has a distinct nucleotide binding region or that when GTP is bound to T_α, the guanine ring is sufficiently exposed to also interact with the $T_{\beta\gamma}$ subunit. A bifunctional analogue, 8-azido-AA-GTP, was synthesized to test these possibilities. In addition to the labeled T_α and $T_{\beta\gamma}$, higher molecular weight bands were also seen which correspond to the cross-linked products of T_α and $T_{\beta\gamma}$. This result implies that the exposed guanine ring of GTP may in close proximity with the $T_{\beta\gamma}$ subunit. The overall result is in complete agreement with the proposed model.

Cr(III)β,γ-bidentate guanine nucleotides. The proposed T_α model shows that Mg^{2+} has its own binding site and does not form a tight complex with GTP. We have synthesized the exchange-inert Cr(III) β,γ-bidentate guanine nucleotide complexes as probes for GTP binding to transducin (Frey, et al., 1988). Some interesting results related to the activation mechanism of transducin were observed. (1) Cr(III)Gpp(NH)p activated the cGMP phosphodiesterase of photolyzed rod outer segment membranes up to 75 percent of the Mg(II)Gpp(NH)p level, but lacked the ability to dissociate the transducin subunits from the rod outer segment membrane. Since the Cr(III) complex tightly associates with the phosphates of GTP and practically eliminates their interaction with the Mg^{2+}, the Mg^{2+} interaction with the β-γ-phosphate of GTP may be important for the dissociation of transducin subunits. The activation of the PDE may be independent of the Mg^{2+} interaction and may be due to the steric effect of the GTP at the site. (2) Both the Δ and Λ screw sense stereoisomers of Cr(III)Gpp(NH)p were capable of activating the cGMP cascade with no apparent stereoselectivity. Indeed, the proposed GTP-binding site is quite exposed to the surface of the protein and has sufficient room to accommodate the different configurations of the Cr(III) GTP steroisomers. The lack of selectivity of stereoisomers of Cr(III)GTP may also imply that Mg^{2+} and GTP do not form a β,γ-bidentate complex at the binding site of T_α, but the interaction is monodentate in nature as seen for EF-Tu (Eccleston, et al., 1981).

INTERACTION OF THE N- AND C-TERMINII OF T_α

We propose that the rhodopsin/$T_{\beta\gamma}$ binding site of T_α is located in the two terminal regions of the peptide. Biochemical assays cannot clearly distinguish the rhodopsin binding from the $T_{\beta\gamma}$ binding, thus limiting the ability to distinguish the functions of these two sites on T_α. It is likely that they may have multiple contact points to confer the specificity of their interaction. If the N- and C-terminal regions of T_α form a distinct functional domain which is separated from the GTP-binding and PDE activating domains, they must interact closely. We have designed experiments to explore this possibility. We found that nicotinamide 1,N^6-ethenoadenine dinucleotide (ε-NAD), a fluorescent analogue of NAD, was able to serve as a substrate for the pertussis toxin catalyzed ADP-ribosylation of transducin. As a result, a fluorescent probe was incorporated into the ADP-ribosylation site of T_α at the C-terminal peptide. Removal of the N-terminal peptide of ε-ADP-ribosylated T_α with either trypsin or S. aureus V-8 protease resulted in a concommitant decrease of the fluorescence intensity. This observation suggests that the N- and C-terminal peptides of the T_α molecule may interact with each other. This interaction is unlikely to be mediated through the central portion of the peptide chain which forms two other distinct functional domains for GTP and PDE binding.

THE MOLECULAR SWITCHING MECHANISM

The goal of constructing a structural model for T_α was to provide a molecular mechanism for its coupling action. From the proposed model one can envision the rhodopsin binding site of transducin as being composed of regions from both the T_α and $T_{\beta\gamma}$ subunits. Helix a, at the amino terminal of T_α has been suggested to be essential for the interaction with rhodopsin/$T_{\beta\gamma}$. As $T_{\beta\gamma}$ associates with the carboxyl terminal region of T_α, Helix a may be anchored with the $T_{\beta\gamma}$ subunit to form the rhodopsin binding site. Under these conditions, the transducin molecule is associated with the rhodopsin membrane and is available for activation by photolyzed rhodopsin. In the absence of photolyzed rhodopsin, the guanine nucleotide binding site is in a closed conformation. Interaction with photolyzed rhodopsin opens up the nucleotide binding site and allows rapid GTP/GDP exchange. Such an interaction must involve the transfer of information between the receptor binding domain and the nucleotide binding domain. In the closed conformation, Helix i located near the opening of the guanine nucleotide binding pocket, is in a position to sterically hinder the exchange of the bound nucleotide. Hence, it provides tight binding for the bound nucleotide with a dissociation constant smaller than 10^{-7} to 10^{-8} M. The light activated signal from the binding of photolyzed rhodopsin can be transmitted to the nucleotide binding site via a conformational change through Helix i. A slight tilting of Helix i results in opening the nucleotide binding pocket through a flexible hinge region (Gly_{288}-Pro_{289}-Asn_{290}). This open conformation enables nucleotide exchange to occur.

Upon binding of GTP two major changes occur in the T_α-GTP complex. First, it dissociates from photolyzed rhodopsin and the $T_{\beta\gamma}$ subunit. Second, the PDE activation site of the T_α-GTP complex is exposed for interaction with the latent PDE. Based on the proposed model, when GTP binds to the guanine nucleotide binding pocket additional space is needed to accommodate the γ-phosphate of GTP. Strand K and Helix f are pushed away from Strand A and as a result, the groove between Strand K and the nucleotide binding domain is widened. The shift of the position of Strand K also includes similar movement on the adjacent Strand J. Such a spatial rearrangement of Strands K and J provides the molecular basis for the T_α coupling function. As can be seen in Figure 3C, one end of Strand K is linked directly to a flexible hinge region (Gly_{198}-Gly_{199}) which is directly attached to Helix F. The movement of Strand K mechanically triggers the movement of Helix f toward the PDE binding domain. Hence, the conformational changes originating at the guanine nucleotide binding domain are now transmitted to the PDE binding site and cause it to be exposed or assembled for PDE activation. The other end of Strand K and J is directed toward the rhodopsin/$T_{\beta\gamma}$ binding domain. The GTP-induced movement could disrupt the rhodopsin/$T_{\beta\gamma}$ binding site that is formed by the amino terminal Helix a and the carboxyl terminal peptide which may lead to the dissociation of both the rhodopsin and the $T_{\beta\gamma}$ subunit from the T_α-GTP complex. The interaction of Mg^{2+} with the γ-phosphate of the bound GTP may stablize the dissociated conformation. The flow of information between the three functional domains of T_α can be accomplished by shifting the spatial arrangement of a few ß-strands and α-helixes located in the interface between the three domains.

TRANSDUCIN AS MODEL FOR BIOLOGICAL COUPLING ENZYMES

Coupling phenomena in biological systems exist in many forms. In general they can be divided into four categories: (1) mechanical movement where two fixed points are linked or pulled closer together via an assembly of proteins such as microtubules and actin-myosin, (2) informational transfer in which information stored in one biopolymer is translated into another form of biopolymer to carry out different functions such as protein synthesis, RNA transcription and DNA replication, and selective degradation of biopolymers, (3) signal transduction such as cellular sensing and communication which include chemotaxis, phototaxis, hormonal regulation, phototransduction, olfaction, and (4) energy coupling which involves utilization of an electro-chemical gradient for ATP synthesis. The enzymes responsible for the coupling functions include the elongation factors in protein synthesis, transducin in visual excitation, the G-proteins in hormonal regulation of adenylyl cyclase, tubulin, the ras-p21 oncogene product, the actin-myosin ATPase, the rec-A protein, DNA polymerase, ATPase in RNA splicing system, the La protease and the FoF1 proton ATP synthetase in mitochondria. A detailed comparison of the regulatory mechanism of these systems has revealed a common motif. All of the coupling enzymes contain tightly bound nucleotides (GTP/ATP) which can be exchanged rapidly upon activation. They are either a GTPase or ATPase. Their coupling functions are generally carried out via a protein subunit association/dissociation cycle which is regulated by the binding and hydrolysis of nucleotides. This striking homology suggests that these enzymes may have had a common evolutionary path.

The well-characterized coupling cycle of elongation factors (EF-Tu and EF-Ts) (Kaziro, 1978) and transducin (T_{α} and $T_{\beta\gamma}$) can serve as models to illustrate the common regulatory principles behind all biological coupling enzymes. One can generalize the two mechanisms by grouping T_{α} and Tu as the activators, $T_{\beta\gamma}$ and Ts as the modulators for the nucleotide exchange reaction, and the cGMP phosphodiesterase and the aminoacyl tRNA as the effector molecules. In the latent form the activator contains GDP and is associated with the modulator. The interaction between the activator/modulator with an active receptor facilitates the nucleotide exchange where GTP is incorporated into the activator. The activator-GTP complex dissociates from the modulator and associates with the effector molecule.

A careful survey in the literature has led us to the conclusion that not only is the basic organization of the ATP regulated enzymes the same, but their nucleotide exchange and hydrolysis cycle is practically identical. One good example is the actin-myosin ATPase in muscle contraction which can be viewed as mechanical coupling. When ADP is bound to myosin (the activator), the myosin-ADP associates with the actin filament (the modulator). Their association promotes the nucleotide exchange reaction where ATP is exchanged for the bound ADP in myosin. The myosin-ATP complex dissociates from actin. After the hydrolysis of the bound ATP, the myosin-ADP reassociates with another actin molecule along the actin filament. Another system is the rec A protein which is involved in gene repair and recombination events. The rec A protein (the activator) in its latent form contains bound ADP. When the double stranded DNA is damaged to generate single strand portions, it can be viewed as a signal that is equivalent to the photolyzed rhodopsin. Single stranded DNA binding protein (the modulator) will associate with the SS-DNA to facilitate the ATP-ADP exchange reaction of rec A. The formation of the rec A-ATP complex leads the dissociation of the rec A-ATP from SS-DNA site. The rec A-ATP then binds to its target, the undamaged double strand DNA portion. As many exchange cycles occur, the double strand is wrapped around by the rec A-ATP complexes. The assembly of the rec A-ATP filament provides the site for the homologous double stranded DNA to associate with the damaged one and allows repair to occur through the actions of nuclease and DNA-polymerase. Similar mechanisms can be deduced for other coupling enzymes including DNA polymerase, rho factor, DNA gyrase, La protease, dynein ATPase and FoF1-ATPase.

The diversity of the coupling actions carried out by these enzymes make it certain that comparing sequence homologies will not be adequate to elucidate their relationship. We have taken two different approaches to correlate these enzymes. If the nucleotide binding sites of these enzymes were originated from a single nucleotide binding protein, their steric requirement toward nucleotide binding as well the hydrolysis mechanism should be conserved. Our preliminary findings are summarized below:

(1) All coupling enzymes share a similar mechanism of hydrolysis for the bound nucleotide. Results from previous studies indicated that the GTP or ATP hydrolysis by these classes of coupling enzymes follows a direct hydrolysis mechanism, i.e. there is no phosphorylated intermediate of the enzymes such as in Na^+/K^+ ATPase or Ca^{++}-ATPase. This feature has been demonstrated in EF-Tu, actin-myosin ATPase, FoF1-ATPase etc.

The kinetic analyses of the nucleotide hydrolysis reaction indicate that the bound GTP or ATP is hydrolyzed rapidly at the site which is then followed by a slow step of releasing the GDP/ADP and Pi. The activator with tightly bound GDP/ADP and Pi remains in its active form and is capable of interacting with the effector or stays dissociated from the receptor/modulator. This mechanism can be illustrated in a reaction scheme for transducin shown below.

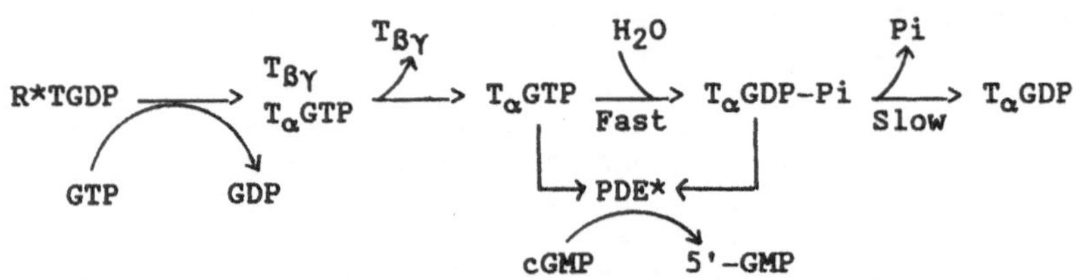

Experimental support for this mechanism comes from studies on the pre-steady state kinetics and the deuterium solvent isotope effect of the GTPase activity of transducin. Three lines of evidences were obtained (i) An initial "burst" of the Pi formation due to the rapid hydrolysis of the T_α bound GTP is detected. After 3-6 sec, the Pi is released from the site and system turns to a steady-state condition. (ii) The half life of Pi release or the turnover of the activated transducin is approximately 20 sec. (iii) No solvent deuterium isotope effect on the steady state GTPase activity of transducin was observed (Tsai, S.M., Ting, T.D., & Ho, Y.-K., unpublished result). A similar mechanism has been suggested for the nucleotide hydrolysis reaction of actin-myosin ATPase, dynein ATPase and tubulin.

(2) Their hydrolytic activity is controlled by the activator which dissociates from the modulator when triphosphate nucleotide is incorporated. The modulator only facilitates the binding of the activator to the receptor and the nucleotide exchange reaction.

(3) The binding of triphosphate nucleotide onto the site of the activator is for conformational control and is not an energy source. Therefore all of the coupling enzyme systems can be either permanently activated or carry out a single activation cycle with the binding of non-hydrolyzable nucleotide analogues such as GTPγS or App(NH)p.

(4) The divalent cation, Mg^{2+}, does not associate with the nucleotide as a bidentate ligand and serve as a substrate for the activator. Mg^{2+} has a separate binding site and may form a monodentate ligand with the β or γ phosphate of the nucleotide at the site. As a result, there is no

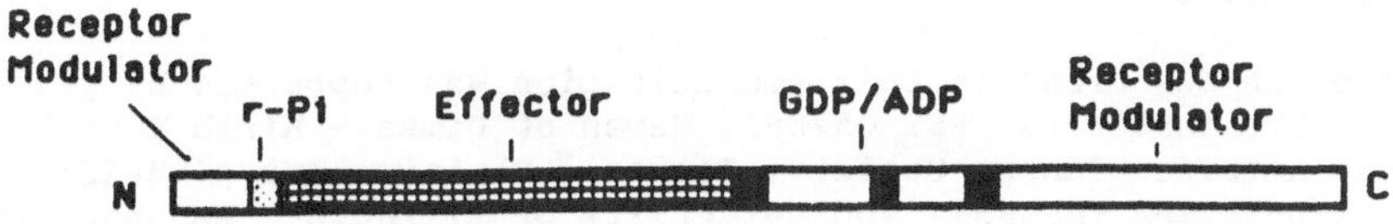

Coupling Enzymes	Receptor	Effector	Nucleotides	Modulator
Transducin	rhodopsin	cGMP PDE	GDP	$T_{\beta\gamma}$
EF-Tu	-	aminoacyl-tRNA	GDP	Ts
Gi/Gs	receptor	adenylyl cyclase	GDP	$G_{\beta\gamma}$
ras-P21	-	transformation	GDP	GAP
Myosin	-	-	ADP	Actin
Dynein	-	-	ADP	Microtubule
rec A	SS-DNA	DS-DNA	ADP	SS-DNA binding protein

Fig. 4. Arrangement of functional domains of some coupling enzymes.

stereo-selectivity towards the isomers of Cr(III)β,γ-bidentate nucleotide complexes. This is different from the nucleotide binding site of kinases or phosphoryl transferases. This feature has been shown in EF-Tu, transducin, actin-myosin ATPase.

(5) All of the coupling enzymes examined thus far are sensitive to sulfhydryl group modification. Moreover, the inhibition by sulfhydryl modification is protected when nucleotide is bound to the activator site.

If the role of nucleotide binding is to control the communication of different functional domains of the enzymes to elicit its coupling function, the functional domain arrangement with respect to receptor, nucleotide, effector binding should be organized in a similar manner in all these enzymes. The size of the activator proteins as well as their coupling function varies considerably among these enzymes. The entire coupling system may have evolved by divergent evolution from a simple nucleotide binding protein which may have existed in two different conformations upon binding of diphosphate and triphosphate nucleotides and was controlled via the hydrolysis reaction. The coupling functions evolved mainly by intron-exon suffling. Functional domains for the modulator, the receptor and the effector enzyme were spliced into the proteins. The communication among these domains is synchronized with the nucleotide exchange and hydrolysis reaction. If this hypothesis is correct, the relative positions of the receptor, nucleotide and effector binding sites on a linear peptide map of these proteins should show a high degree of similarity. A careful comparison of the available data showed the presence of such a homology (Fig. 4). The GTP- or ATP-binding domain is split into two regions with a small region closer to the N-terminal responsible for γ-phosphate binding and nucleotide hydrolysis. The second region forms the purine ring binding site and is located closer to the C-terminal side. The receptor/modulator domain is composed of the C- and N-terminal peptides. This analysis clearly suggests that all coupling enzymes have evolved through divergent evolution. Many other cellular coupling functions involve nucleotide binding proteins. Their coupling action could possibly be formulated in a similar manner as transducin in the visual excitation system.

ACKNOWLEDGMENTS

Research described in this communication was supported by grants from National Eye Institutes (EY 05788), March of Dimes - Birth Defect Foundation (5-508), American Cancer Society, Illinois Division Inc. (88-22), American Heart Association (86-833) and University of Illinois. We thank Ms. Donna Lattyak for preparing the manuscript.

REFERENCES

Bennett, N., and Dupont, Y., 1985, J. Biol. Chem. 260: 4156-4168.
Cerione, R.A., Kroll, S., Rajaram, R., Unson, C., Goldsmith, P., and Spiegel, A.M., 1988, J. Biol. Chem. 263: 9345-9352.
deVos, A. M., Tong, L., Milburn, M. V., Matias, P. M., Jancarik, J., Noguchi, S., Nishimura, S., Miura, K., Ohtsuka, E., and Kim, S.-H., Science 239: 837-952.
Eccleston, J.F., Webb, M.R., Ash, D.E., and Reed, G.H., 1981, J. Biol. Chem. 256: 10774-10777.
Frey, S. E., Hingorani, V. N., Su-Tsai, S.-M., and Ho, Y.-K., 1988, Biochemistry (in press).
Fong, H. K. W., Hurley, J. B., Hopkins, R. S., Miake-Lye, R., Johnson, M. S., Doolittle, R. F., and Simon, M. I., 1986, Proc. Natl. Acad. Sci. U.S.A. 83: 2162-2166.
Fung, B. K.-K., 1983, J. Biol. Chem. 258: 10495-10502.
Fung, B. K.-K., and Nash, C. R., 1983, J. Biol. Chem. 258: 10503-10510.
Halliday, K. R., 1984. J. Cyc. Nuc. Prot. Phos. Res. 9: 435-448.
Hingorani, V.N., 1988, Ph.D. Thesis. Department of Biological Chemistry, University of Illinois at Chicago.
Hingorani, V. N., and Ho, Y.-K., 1987a, Biochemistry 26: 1633-1639.
Hingorani, V. N., and Ho, Y.-K., 1987b, FEBS Lett. 220(1): 15-22.
Hingorani, V. N., Tobias, D. T., Henderson, J. T., and Ho, Y.-K., 1988a, J. Biol. Chem. 263: 6916-6926.
Hingorani, V.N., Tobias, D.T., and Ho, Y.-K., 1988b, Submitted for publication.
Ho, Y.-K., and Fung, B.K.K., 1984, J. Biol. Chem. 259, 6694-6699.
Hurley, J. B., Fong, H. K. W., Teplow, D. B., Dreyer, W. J., and Simon, M. I., 1984, Proc. Natl. Acad. Sci. U.S.A. 81: 6948-6952.
Jurnak, F., 1985, Science 230: 32-36.
Kaziro, Y., 1978, Biochim. Biophys. Acta. 505: 95-127.
La Cour, T.F.M., Nyborg, J., Thirup, S. and Clark, B.F.C., 1985, EMBO J. 4, 2385-2388.
Liebman, P. A., Parker, K. R., and Dratz, E. A., 1987, Ann. Rev. Physiol. 49: 765-791.
Medynski, D. C., Sullivan, K., Smith, D., Van Dop, C., Chang, F. H., Fung, B.K.-K., Seeburg, P. H., and Bourne, H. R., 1985, Proc. Natl. Acad. Sci. U.S.A. 82: 4311-4315.
Navon, S. E., and Fung, B. K.-K., 1988, J. Biol. Chem. 262: 15746-15751.
Navon, S. E., and Fung, B. K.-K., 1984, J. Biol. Chem. 259: 6686-6693.
Van Dop, C., Tsubokawas, M., Bourne, H. R., and Ramachandran, J., 1984, J. Biol. Chem. 259: 696-698.
Watkins, P. A., Burns, D.L., Kanaho, Y., Liu, T.-Y., Hewlett, E.L. and Moss, J., 1985, J. Biol. Chem. 260: 13478-13482.

A NOVEL CYTOSOLIC GTP-BINDING PROTEIN WITH PHOSPHOLIPID STIMULATED GTP-BINDING AND GTPase ACTIVITY

Ronit Sagi-Eisenberg, Linton M. Traub
Galia Gat-Yablonski and Meir Aridor

Department of Chemical Immunology
The Weizmann Institute of Science
Rehovot 76100, Israel

INTRODUCTION

The signal transducing GTP-binding proteins (G-proteins) isolated thus far are membrane associated. However, rapidly accumulating data suggest that regulatory G-proteins may also be cytosolic. This conclusion is based mainly on indirect observations such as the ability of cytosol to restore adenylate cyclase activity of cyc^- membranes (1), the presence of cholera or pertussis toxins substrates in the cytosol (2,3) and the activation of a soluble phosphoinositide-hydrolyzing PLC by GTP-γ-S (4). The findings that the purified α-subunits of Go and Gi are water soluble (5), that α-subunits of Gs are released from the membrane following activation (6) and that about 80% of the GTP-binding protein ARF is located in the cytosolic fraction (7) further support his notion. Indeed, it has been recently argued that the critical interactions of the signal transducing G-proteins occur in the cytoplasm rather than in the membrane (8). In addition, several reports (9-11) have implicated GTP-binding proteins as regulatory elements in the transport and secretion of proteins. These data again imply that the distribution and function of G-proteins should be extended from the surface of the cell to the interior (12).

RESULTS

In view of the data implying the existence of regulatory GTP-binding proteins in the cytosol, we were prompted to investigate in detail the presence of regulatory G-proteins in the cytosol. For that purpose, the cytosolic fraction (100,000 g supernatant) derived from rat basophilic leukemia cells, RBL-2H3, a histamine secreting cell line (13) grown either in culture (14) or as solid tumors in rats (15), was subjected to chromatography on DEAE-cellulose. The fractions eluted by a linear gradient of NaCl were screened for the presence of putative G-proteins.

Two parameters were used as assays to identify potential G-proteins:

1. GTP-γ-S binding using low GTP-γ-S concentrations in conjunction with a filtration method thereby restricting the assay to the detection of proteins which bind GTP with a high affinity (16).

2. GTPase activity which is a characteristic of all known G-proteins (17). Indeed, as shown in Fig. 1, activities of both GTP-γ-S binding and GTPase co-eluted from the column as a sharp single peak at approximately 120 mM NaCl. The peak fraction was further subjected to gel filtration on Ultrogel AcA-34. Again, a major peak of GTP-γ-S binding, which also exhibited GTPase activity, eluted from the column (Fig. 2).

The peak fractions exhibiting GTP-γ-S binding were next pooled and characterized. GTP-γ-S binding was dose-dependent and of high affinity. Scatchard analysis revealed a homogenous population of the GTP-binding protein with an apparent dissociation constant of 10 nM. GTP-γ-^{35}S binding was specific and could be competed with GTP-γ-S or GDP-β-S, half maximal inhibition occurring at a concentration of 20 nM GTP-γ-S or 90 nM GDP-β-S. In contrast, ATP-γ-S had no effect on GTP-γ-S binding even at a concentration of 100 μM. Binding was strictly dependent on Mg^{2+}. Maximal binding occurred at 1 mM Mg^{2+} while higher concentrations inhibited binding. It displayed fast kinetics reaching maximal binding within 5 min and was heat sensitive. Incubation for 2 min at 56°C resulted in a complete loss of the GTP-γ-S binding. GTPase activity was relatively slow and linear for at least 60 min.

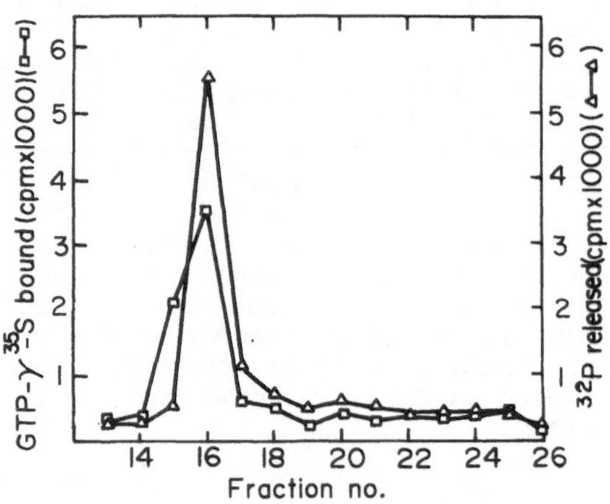

Fig. 1. Chromatography on DEAE-cellulose. The cytosolic fraction of RBL cells (100,000xg supernatant) was chromatographed on DE52. Aliquots (20μl) of the indicated fractions were incubated for 1 h at 22°C with either 2 nM GTP-[γ-^{35}S] (10^5 cpm) or with 62 nM [γ-^{32}P]-GTP, and GTP-γ-S binding (-) and GTPase activity (Δ-Δ) were determined.

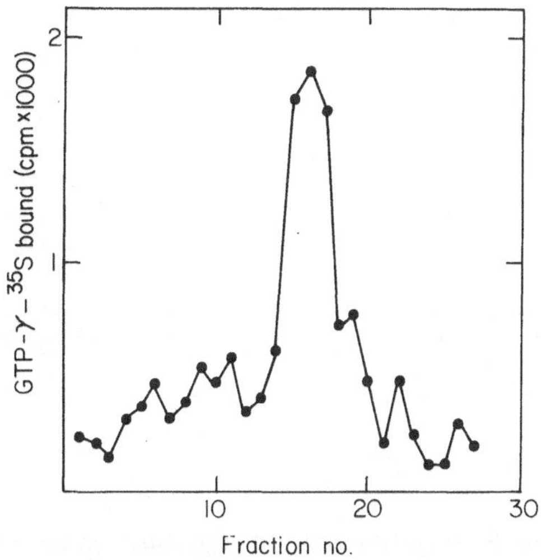

Fig. 2. Chromatography on Ultrogel AcA 34. Aliquots (20 μl) from each fraction were incubated with 2 nM GTP-[γ-^{35}S] (50,000 cpm) and GTP-[γ-^{35}S] binding assessed as described under Fig. 1.

These results demonstrate that the cytosolic fraction of the RBL cells contains a protein that binds specifically guanine nucleotides. The affinity of this protein for GTP-γ-S is similar to that reported for the membrane bound signal transducing G-proteins (18).

The soluble G-protein isolated from RBL cells failed to restore activity under experimental conditions that Gs increased cyc$^-$ cyclase activity by 10 fold. In addition, a GTP-binding protein with properties similar to those of the G-protein isolated from RBL cytosol was also found to be present in cyc$^-$ cells. As shown in Fig. 3, the protein isolated from cyc$^-$ cytosol bound specifically [α-^{32}P]-GTP on a dot blot. Furthermore, no quantitative or qualitative differences could be found when the same procedure was repeated for the S49 wild type cells or the leukemic HL-60 and K562 cells and the hepatoma FaO cells. These findings suggest that the soluble G-protein, we termed Gc, is different from Gs. They also reveal that Gc is an ubiquitous protein present in the cytosolic fractions of several cell types.

Both GTP-γ-S binding and GTPase activity (not shown) of Gc were affected by the presence of phospholipids. Phosphatidylserine (PS) and phosphatidylinositol (PI) significantly stimulated GTP-γ-S binding and GTPase activity while phosphatidylethanolamine (PE) had no effect on both (Fig. 4). Hence, Gc appears to include within its structure a regulatory domain, which following its interaction with phospholipids, activates the catalytic domain. This finding implies that Gc could associate with membrane phospholipids and may be found both in cytosolic and particulate fractions.

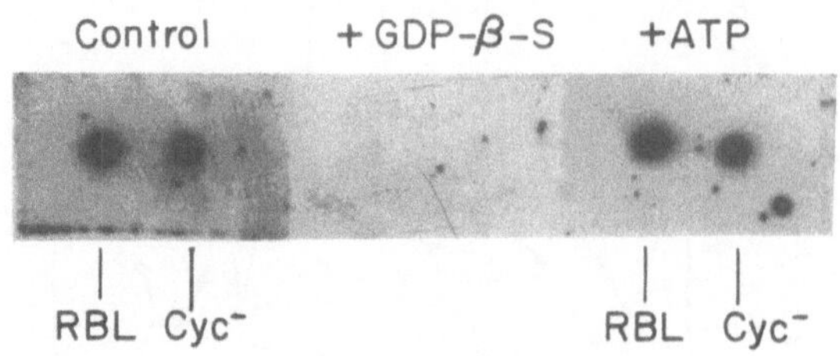

Fig. 3. [α-^{32}P]-GTP binding to Gc derived from RBL and cyc$^-$ cells. Cytosolic fractions derived from 1x10^8 RBL or cyc$^-$ cells were chromatographed on DE52 columns and fractions eluted by a linear gradient of NaCl were assayed for GTP-[γ-^{35}S] binding. Aliquots from the peak fractions derived from each cell preparation were tested for [α-^{32}P]-GTP binding on dot blots.

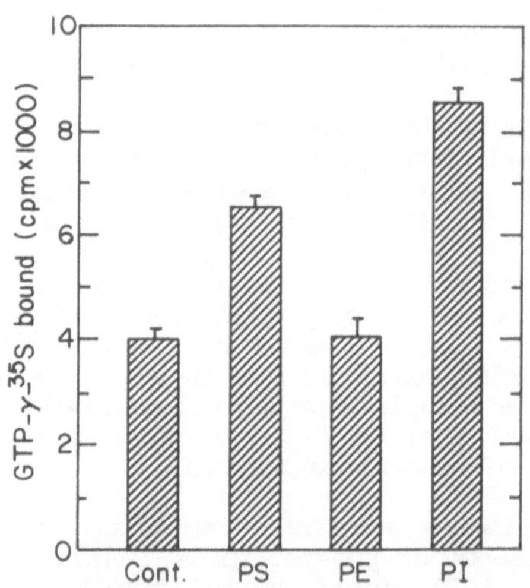

Fig. 4. Effect of phospholipids on GTP-γ-S binding. Twenty μl aliquots of Gc were incubated at 22°C for 5 min with 2 nM GTP-[γ-^{35}S] (100,000 cpm) in the absence or presence of the indicated phospholipids.

To assess whether Gc shares structural similarity with known signal transducing G-proteins, rabbit antisera raised against a synthetic decapeptide (Lys-Glu-Asn-Leu-Lys-Asp-Cys-Gly-Leu-Phe), corresponding to the carboxyl-terminal end of the α-subunit of transducin (TD), was used as an anti α-subunit probe. These antibodies have previously been shown to cross-react with the α-subunit of Gi (19).

An immunoblot analysis, using these antibodies, is depicted in Fig. 5. The anti α antisera decorated a 40 kDa protein present in the plasma membranes of the RBL cells (Fig. 5, PM). This band probably represents the α-subunit of Gi present in this preparation. This region also aligned well with the 39 kDa α-subunit of transducin in rod outer segments (Fig. 5, ROS). Gc, however, contained a major immunoreactive band at approximately 100 kDa (Fig. 5, Gc). Similar immunoblotting experiments with an antipeptide antiserum raised against the N-terminal decapeptide of the β-subunit of transducin (MS/1) revealed activity with the 36 kDa β-subunits present in the membrane preparation (Fig. 6, PM) and the rod outer segments (Fig. 6, ROS). Gc exhibited immunoreactivity again in the 100 kDa region (Fig. 6, Gc). Thus, the cytosolic G-protein, Gc, exerts immunoreactivity with both anti TD-α and β subunit antibodies. These observations indicate the presence of both an α and β-like structure within a single polypeptide chain of 100 kDa.

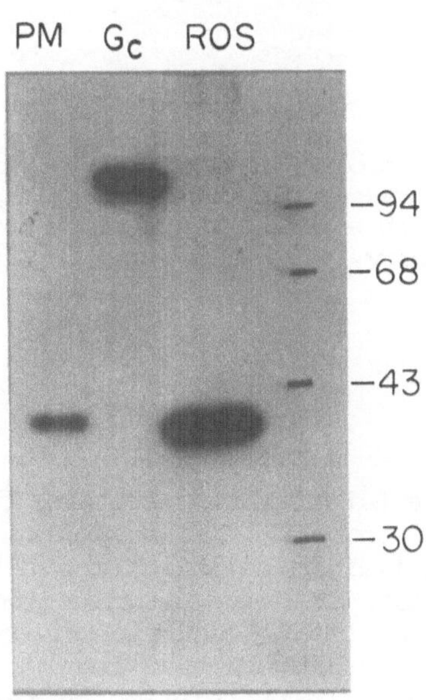

Fig. 5. Immunological cross-reactivity of Gc with anti G-α antibodies. RBL tumor plasma membranes (PM), Gc enriched DE52 fraction (Gc) and rod outer segment membranes (ROS) were analysed by the immunoblotting. Samples were visualized by autoradiography after reaction with 10 μg/ml anti-α-subunit antibodies followed by incubation with ^{125}I-donkey anti-rabbit IgG (0.2μCi/ml).

To ascertain that the anti G-protein immunoreactivity and the GTP-γ-S binding present in this preparation do indeed reside in the same functional entity, photoaffinity labeling was undertaken. As shown in Fig. 7, in the presence of 100 μM ATP-γ-S and 30 nM azido - [α-^{32}P]-GTP, the α-subunit of TD in rod outer segments was intensely labeled (lane b). This labeling was completely abolished by 100 μM GTP-γ-S and significantly inhibited by 100 μM GDP-β-S (not shown). Under similar conditions, a major band of 100 kDa was labeled in the Gc preparation (lane a). The labeling was abolished by both 100 μM GTP-γ-S and GDP-β-S (not shown).

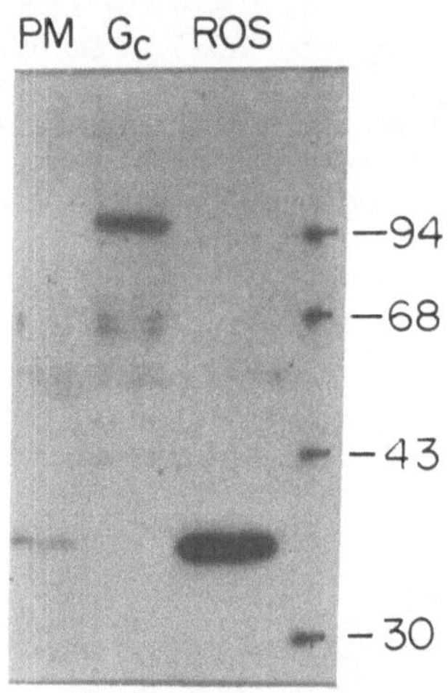

Fig. 6. Immunological cross-reactivity of Gc with anti G-β antibodies. Immunoblotting was as described under Fig. 5 except that the samples reacted with a 1:500 dilution of anti-β subunit antisera as the first antibody.

Of all the soluble GTP-utilizing proteins described thus far, the eukaryotic elongation factor EF-2 has a similar molecular weight (20). Therefore, it was essential to establish the relationship between EF-2 and Gc. For that purpose, EF-2 was isolated from rat liver cytosol and purified to homogenity by the procedure described in Ref. 21. By several criteria Gc could be distinguished from EF-2: 1. EF-2 but not Gc served as a substrate for phosphorylation by rat liver Ca^{2+}-calmodulin kinase III; 2. In contrast to Gc, EF-2 revealed a total lack of immunological cross-reactivity with the anti TD-α antibodies; 3. EF-2 but not Gc served as a substrate for ADP-ribosylation by diphteria toxin.

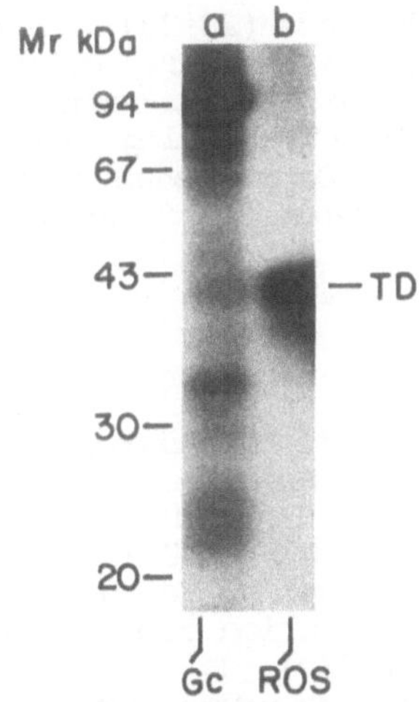

Fig. 7. Photoaffinity labelling of ROS and Gc with azido-[α-^{32}P]-GTP. 20 μg Gc enriched DE52 fractions (lane a) or ROS (lane b) were incubated with 28 nM azido-[α-^{32}P]-GTP in the presence of 100 μM ATP-γ-S and 200μg/ml PS. Samples were irradiated with a mercury vapor lamp with a 400nm cut off filter for 3 min. Labelled proteins were visualized by autoradiography.

SUMMARY AND CONCLUSIONS

In summary, a 100 kDa protein with several GTP-binding protein-like attributes is present in the cytosolic fraction of a number of cells. The protein, isolated from rat basophilic leukemia cell cytosol, is dissimilar to the previously described heterotrimeric G-proteins and the elongation factor EF-2. The protein, termed Gc, binds guanine nucleotides specifically and with high affinity and it exhibits a GTPase activity. Both GTP binding and GTPase activity are stimulated by the acidic phospholipids, phosphatidylserine and phosphatidylinositol but not by phosphatidylethanolamine. Thus, in spite of its cytosolic localization, Gc is capable of interacting with phospholipids present in the inner leaflet of the cell membrane.

Gc immunoreacts with antipeptides antisera raised against the C-terminal end of TD-α and the N-terminal end of TD-β. Both immunoreactive domains of Gc as well as its GTP-binding site reside on the same 100 kDa polypeptide chain as is evidenced by Western blots and photoaffinity labeling.

Based on our findings, we would like to suggest that Gc may represent a novel member of the G-protein family that, as was recently proposed (8), serve as cytoplasmic shuttles that are localized on the membrane surface like the cytoskeleton. Moreover, their subcellular localization may be determined by modifications such as phosphorylation turning them into programmable messengers (22). Indeed, we and others have shown that G-proteins such as Gi, TD and the ras proteins undergo phosphorylation by protein kinase C and by the insulin and insulin like growth factor I receptor kinases (23-26). As such, Gc could be involved in the activation of phospholipase C (4) or in the control of exocytosis (27). Furthermore, the indicated presence of both an α and β-like structure within a single GTP-binding polypeptide chain raises the intriguing possibility that Gc represents an ancestor which is evolutionarily related to the heterotrimeric signal transducing G-proteins Gi and TD.

REFERENCES

1. M.K. Bhat, R. Iyengar, J. Abramowitz, E.M. Bordelon-Riser, and L. Birnbaumer, Naturally soluble component(s) that confer(s) guanine nucleotide and fluoride sensitivity to adenylate cyclase, Proc. Natl. Acad. Sci. USA 77:3836 (1980).
2. P.A. Watkins, J. Moss, and M. Vaughan, Effects of GTP on choleragen-catalyzed ADP-ribosylation of membrane and soluble proteins, J. Biol. Chem. 255:3959 (1980).
3. T. Nakamura, and M. Ui, Simultaneous inhibitions of inositol phospholipid breakdown, arachidonic acid release, and histamine secretion in mast cells by islet-activating protein, pertussis toxin, J. Biol. Chem. 260:3584 (1985).
4. H. Deckmyn, S-M. Tu, and P.W. Majerus, Guanine nucleotide stimulated soluble phosphoinositide-specific phospholipase C in the absence of membranes, J. Biol. Chem. 261:16553 (1986).
5. Sternweis, P.C., The purified α subunits of Go and Gi from bovine brain require βγ for association with phospholipid vesicles, J. Biol. Chem. 261:631 (1986).
6. Nielsen, T.B., P.M. Lad, M.S. Preston, and M. Rodbell, Characteristics of the guanine nucleotide regulatory components of adenylate cyclase in human erythrocyte membranes, Biochim. Biophys. Acta. 629:143 (1980).
7. R.A. Kahn, C. Goddard, and M. Newkirk, Chemical and immunological characterization of the 21-kDa ADP-ribosylation factor of adenylate cyclase, J. Biol. Chem. 263:282 (1988).
8. Chabre, M., The G protein connection: is it in the membrane or the cytoplasm, Trends. Biochem. Sci. 12:213 (1987).
9. Melancon, P., B.S. Glick, V. Malhotra, P.J. Weidman, T. Serafini, M.L. Gleason, L. Orci, and J.E. Rothman, Involvement of GTP-binding "G" proteins in transport through the golgi stack, Cell 51:1053 (1987).
10. N. Segev, J. Mulholland, and D. Botstein, The yeast GTP-binding γPT-1 protein and a mammalian counterpart are associated with secretion machinery, Cell 52:915 (1988).
11. B. Goud, A. Salminen, N.C. Walworth, and P.J. Novic, A GTP-binding protein required for secretion rapidly associates with secretory vesicles and the plasma membrane in yeast, Cell 53:753 (1988).
12. H.R. Bourne, Do GTP-ases direct membrane traffic in secretion, Cell 53:669 (1988).
13. D.C. Seldin, S. Adelman, K.R. Austen, R.L. Stevens, Homology of the rat basophilic leukemia cell and the rat mucosal mast cell, A. Hein, J.P. Caulfield, and R.G. Woodbury, Proc. Natl. Acad. Sci. USA 82:3871 (1985).
14. Sagi-Eisenberg, R., H. Lieman, and I. Pecht, Protein kinase C regulation of the receptor-coupled calcium signal in histamine-secreting rat basophbilic leukemia cells, Nature 313:59 (1985).

15. Eccleston, E., B.J. Leonard, J.S. Lowe, and H.J. Welford, Basophilic leukaemia in the albino rat and a demonstration of the basopoietin, Nature New Biol. 244:73 (1973).
16. Sternweis, P.C., and J.D. Robishaw, Isolation of two proteins with high affinity for guanine nucleotides from membranes of bovine brain, J. Biol. Chem. 259:13806 (1984).
17. Cassel, D., and Z. Selinger, Catecholamine-stimulated GTPase activity in turkey erythrocyte membranes, Biochim. Biophys. Acta 452, 538 (1976).
18. R.M. Huff, and E.J. Neer, Antibodies directed against synthetic peptides distinguish between GTP-binding proteins in Neutrophil and brain, J. Biol. Chem. 261:1105 (1986).
19. Goldsmith, P., P. Gierschik, G. Milligan, C.G. Unson, Subunit interactions of native and ADP-ribosylated α_{39} and α_{41}, two guanine nucleotide-binding proteins from bovine cerebral cortex, R. Vinitsky, H.L. Malech, and A.M. Spiegel, J. Biol. Chem. 262:14683 (1987).
20. Nairn, A.C., B. Bhagat, and C.H. Palfrey, Identification of calmodulin-dependent protein kinase III and its major Mr 100,000 substrate in mammalian tissues, Proc. Natl. Acad. Sci. USA 82:7939 (1985).
21. Nairn, A.C., and C.H. Palfrey, Identification of the major Mr 100,000 substrate for calmodulin-dependent protein kinase III in mammalian cells as Elongation factor-2, J. Biol. Chem. 262:17299 (1987).
22. Rodbell, M., Programmable messengers: a new theory of hormone action, Trends. Biochem. Sci. 10:461 (1985).
23. Katada, T., A.G. Gilman, Y. Watanabe, S. Bauer, and K.H. Jakobs, Protein kinase C phosphorylates the inhibitory guanine-nucleotide-binding regulatory component and apparently suppresses its function in hormonal inhibition of adenylate cyclase, Eur. J. Biochem. 151:431 (1986).
24. Zick, Y., R. Sagi-Eisenberg, M. Pines, P. Gierschik, and A.M. Spiegel, Multisite phosphorylation of the α subunit of transducin by the insulin receptor kinase and protein kinase C, Proc. Natl. Acad. Sci. USA, 83:9294 (1986).
25. Zick, Y., A.M. Spiegel, and R. Sagi-Eisenberg, R., Insulin-like growth factor I receptors in retinal rod outer segments, J. Biol. Chem. 262:10259 (1987).
26. Ballester, R., M.E. Furth, and O.M. Rosen, Phorbol ester- and protein kinase C- mediated phosphorylation of the cellular kirsten ras gene product, J. Biol. Chem. 262:2688 (1987).
27. Cockcroft, S., T.W. Howell, and B.D. Gomperts, Two G-proteins act in series to control stimulus-secretion coupling in mast cells: Use of neomycin to distinguish between G-proteins controlling polyphosphoinositide phosphodiesterase and exocytosis, J. Cell Biol. 105:2745 (1987).

ROLE OF A G PROTEIN HOMOLOG IN YEAST PHEROMONE RESPONSE

Janet Kurjan

Department of Biological Sciences
Columbia University
New York, New York 10027

INTRODUCTION

The recent discovery of a G protein homolog involved in a yeast signal transduction system indicates that G proteins are conserved over a wide evolutionary distance (Dietzel and Kurjan, 1987; Nakafuku et al., 1987; Miyajima et al., 1987; Jahng et al., 1988). The extensive genetic characterization possible in yeast allows a type of approach to the study of G proteins not possible in vertebrate systems. Using such a genetic approach, we isolated the SCG1 gene, which encodes a homolog to the α subunits of G proteins. Genetic results have also provided evidence that SCG1 is involved in pheromone response in yeast. Recently, putative homologs to the vertebrate β and γ subunits have also been identified (M. Whiteway, personal communication). In this paper, after presenting background information on pheromone response, I will describe the isolation and characterization of SCG1 and will discuss our currently favored model of the mechanism of action of SCG1 in the pheromone response pathway.

PHEROMONE RESPONSE IN YEAST

Haploid cells of the yeast *Saccharomyces cerevisiae* secrete peptide pheromones; cells of a mating type secrete a-factor, and cells of α mating type secrete α-factor (reviewed in Sprague et al., 1983). Haploid cells are able to respond to the pheromone produced by the opposite mating type. Response to pheromone results in arrest of cell growth in the G1 phase of the cell cycle, cell wall changes, and morphological changes often called "shmooing". In addition, a number of genes that play a role in mating and/or pheromone response are induced by exposure to pheromone (Shimoda et al., 1976; Betz et al., 1978; Terrance and Lipke, 1981; Manney, 1983; Strazdis and MacKay, 1983; Hagen and Sprague, 1984; Nakayama et al., 1985; Hartig et al., 1986). Mutants defective in pheromone production and/or pheromone response are unable to mate to form the third cell type, the a/α diploid (MacKay and Manney, 1974a; Manney and Woods, 1976; Hartwell, 1980; Kurjan, 1985; Michaelis and Herskowitz, 1988). Pheromone production and response, therefore, plays a critical role in mating.

Each of the haploid mating types expresses a receptor for the pheromone produced by the opposite mating type. The α-factor receptor is encoded by the STE2 gene, and the a-factor receptor is encoded by the STE3 gene (MacKay and Manney, 1974b; Jenness et al., 1983; Hagen et al., 1986). The STE2 and STE3 sequences (as determined by DNA sequencing) show similarities to the receptors for vertebrate G protein-mediated signal transduction systems, i.e. they contain seven putative membrane spanning domains and a hydrophilic carboxy terminus with a high proportion of serine and threonine residues (Burkholder and Hartwell, 1985; Nakayama et al., 1985; Hagen et al., 1986; Sibley et al., 1987). The STE2 and STE3 products, however, are not homologous to one another or to vertebrate receptors. The a- and α-factor receptor-pheromone interactions are interchangeable, indicating that the pheromone response pathways in the two mating types converge at a point after the pheromone-receptor interactions (Bender and Sprague, 1986; Nakayama et al., 1987).

IDENTIFICATION OF THE SCG1 GENE

The identification of the SCG1 gene utilized a mutant (sst2) that shows an alteration in response to pheromone in both a and α cells; a sst2 cells are supersensitive to α-factor, and α sst2 cells are supersensitive to a-factor (Chan and Otte, 1982a; Chan and Otte, 1982b). The sst2 mutation results in response to very low levels of pheromone and a defect in recovery from pheromone arrest. The SCG1 gene was identified by isolation of a multicopy plasmid (pC3) from a yeast library that is able to suppress the supersensitivity of sst2 cells to pheromone (Dietzel and Kurjan, 1987). Plasmid pC3 was able to suppress the sst2 mutation in both a and α strains. Sequencing of a 1.9 kb fragment that was sufficient for suppression of the sst2 mutation identified an open reading frame of 472 amino acids. This open reading frame is homologous to the α subunits of mammalian G proteins that are involved in a number of different signal transduction systems (Figure 1; Gilman, 1987). There are close matches to the consensus sequences for the regions of the G_α subunits, ras proteins, and EF-Tu that have been implicated in guanine nucleotide binding and GTPase activity. We named this gene SCG1, for sst2-complementing gene or Saccharomyces cerevisiae G protein. Another group identified the SCG1 gene (and named it GPA1) by virtue of its homology to a G_α cDNAs (Nakafuku et al., 1987). The genes were isolated from different strains and show five amino acid polymorphisms.

NULL scg1 MUTATIONS RESULT IN CONSTITUTIVE PHEROMONE RESPONSE

The ability of SCG1 to suppress the sst2 mutation when present on a multicopy plasmid suggested that it might play a role in pheromone response. To determine whether this possibility was correct, disruption mutations of scg1 were constructed and tested for a phenotype (Dietzel and Kurjan, 1987). The disruption mutations were made by inserting DNA fragments containing either the LEU2 or URA3 gene into the SCG1 open reading frame. The wild-type gene was then replaced with the mutant gene in a diploid (using the technique of Rothstein, 1983), to produce a/α SCG1/scg1 diploids. Haploid scg1 spores were then obtained by tetrad analysis. The scg1 spores arrest growth as very small colonies containing very large and abnormally shaped cells.

The morphology of the scg1 cells showed a resemblance to haploid cells exposed to the opposite pheromone, which arrest growth and continue to grow to form enlarged cells. This resemblance suggested that the scg1 phenotype might represent constitutive expression of the pheromone response pathway. An alternative explanation for the scg1 phenotype is that SCG1 is required for growth, i.e. transition through the Start phase of the cell cycle, and is not involved in pheromone response. In this case, the ability of high levels of SCG1 to suppress the sst2 defect might occur by competition of the

SCG1 product with a component of the pheromone response pathway, possibly another G protein. If the first hypothesis is correct, an a/α scg1/scg1 diploid should not show the scg1 haploid phenotype, because a/α diploids do not respond to pheromone. If the second hypothesis is correct, an a/α scg1/scg1 diploid would be likely to have a phenotype similar to the scg1 haploids. An a/α scg1/scg1 diploid was constructed, and its growth and cellular morphology was shown to be identical to wild-type strains (Dietzel and Kurjan, 1987; Miyajima et al., 1987). In addition, SCG1 expression is haploid-specific, i.e., it is not expressed in a/α diploids. Based on these results, we proposed that SCG1 is a component of the pheromone response pathway, and that disruption of scg1 function results in constitutive expression of this pathway.

Results of Jahng et al. (1988) provide additional evidence that the scg1 null phenotype represents constitutive expression of the pheromone response pathway. These experiments involved the isolation and analysis of temperature-sensitive scg1 mutants. At the restrictive temperature, these mutants arrest in the G1 phase of the cell cycle with a morphology that resembles wild-type cells after exposure to pheromone. Double mutants containing a temperature-sensitive scg1 mutation and a deletion of the pheromone receptor are able to mate when transferred to the restrictive temperature. In addition, transcripts induced by exposure to pheromone in a

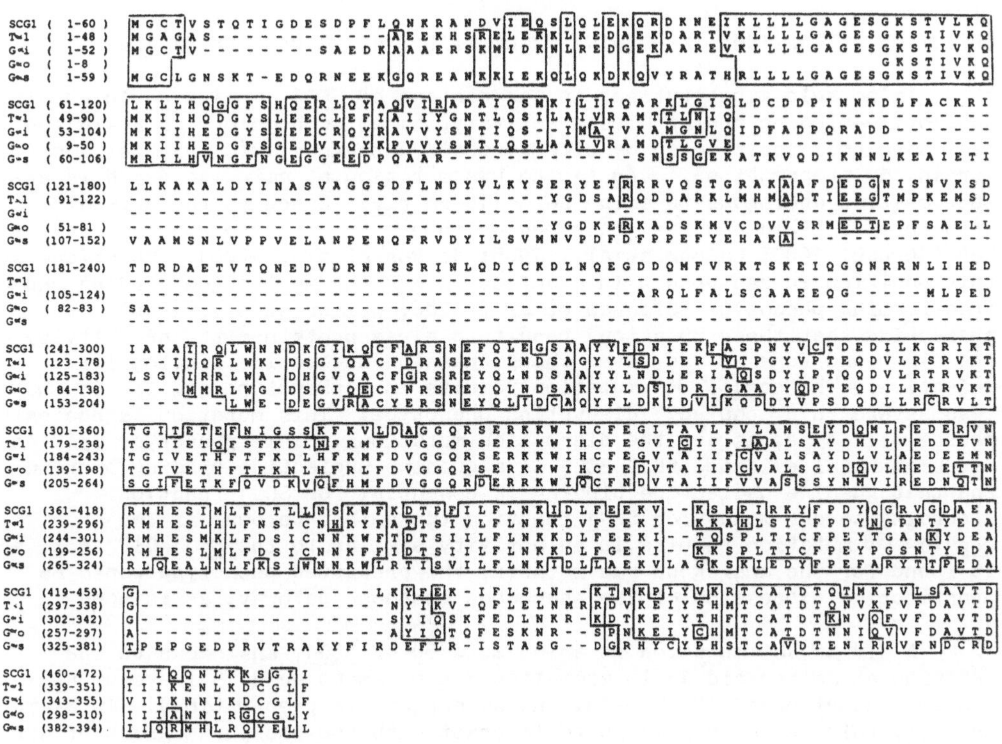

Fig. 1. Homology of SCG1 to vertebrate α subunits. The amino acid sequence of the putative SCG1 protein is compared with the sequences of the α subunits of the regulatory G proteins, transducin, G_i, G_o, and G_s (Tanabe et al., 1985; Bray et al., 1986; Itoh et al., 1986, Robishaw et al., 1986). Identical amino acids and conservative amino acid substitutions are boxed. Conservative amino acids are grouped as follows: N, D, E, Q; S, T, A, G, P; R, H, K; I, L, V, M; and Y, F, W. Reprinted by permission of Cell; first published in Dietzel and Kurjan (1987).

wild-type strain are induced by transfer of the temperature-sensitive mutants to the restrictive temperature. These results indicate that inactivation of SCG1 elicits several aspects of pheromone response independent of pheromone receptor and are consistent with the hypothesis that the scg1 null phenotype represents constitutive activation of the pheromone response pathway.

COMPLEMENTATION BY RAT HOMOLOGS

The similarity in amino acid sequence of SCG1 and mammalian G protein α subunits is extensive. Expression of rat α_s under the control of a yeast promoter (construct provided by D. Ecker and J. Stadel) result in complementation of the growth and morphological defects of haploid scg1 strains (Dietzel and Kurjan, 1987), indicating that there is functional as well as sequence conservation. The scg1 strains expressing α_s are sterile, however, indicating that α_s cannot totally substitute for SCG1. According to the model described below, this sterility suggests that α_s can interact with yeast β/γ, but not with the pheromone receptors. High level expression of rat α_i allows only slight growth of scg1 mutants, and α_o has not shown complementation of scg1 (D. Tipper, J. Stadel, and J. Kurjan, in preparation).

EFFECT OF ALTERATIONS IN GUANINE NUCLEOTIDE BINDING REGIONS OF SCG1

Amino acid substitutions in the regions of SCG1 implicated in guanine nucleotide-binding and/or GTPase activity, based on homology to EF-Tu and ras, have been constructed by site-directed mutagenesis (C. Dietzel and J. Kurjan, in preparation). An Asn to Lys mutation at amino acid 388 is at the position analogous to amino acid 116 in ras. Based on the crystal structure of EF-Tu and ras, this region interacts with the guanine ring of GDP (Jurnak, 1985; LaCour et al., 1985; de Vos et al., 1988). Mutations at this position in ras result in a defect in guanine nucleotide-binding and a transforming phenotype (Clanton et al., 1986; Walter et al., 1986), suggesting that these mutations result in a ras protein with the activity of the GTP-bound form of ras, even though guanine nucleotide binding is defective. The $scg1^{Lys388}$ mutation results in a phenotype similar to, but less severe than, the scg1 disruption phenotype. This mutation is recessive to the wild-type, as are the disruption mutations. This phenotype provides circumstantial evidence that SCG1 is a guanine nucleotide-binding protein and that guanine nucleotide binding is essential for SCG1 function. According to our model for SCG1 function (described below), if the $scg1^{Lys388}$ mutation results in an activity similar to the SCG1-GTP form, as is found for the analogous ras mutants, the predicted phenotype would be constitutive expression of the pheromone response pathway, as is seen.

A Gly to Val mutation at amino acid 50 of SCG1, analogous to the mutation at amino acid 12 in mammalian ras or amino acid 19 in yeast RAS, was also constructed (C. Dietzel and J. Kurjan, in preparation). In EF-Tu and ras, this region was shown to interact with the phosphate group of GDP (Jurnak, 1985; LaCour et al., 1985; de Vos et al., 1988). In mammalian ras, this mutation leads to a GTPase defect and a transforming phenotype (McGrath et al., 1984; Seeburg et al., 1984; Sweet et al., 1984). In yeast, RAS is involved in activation of adenylate cyclase, and this mutation leads to hyperactivation of adenylate cyclase (Toda et al., 1985). The $SCG1^{Val50}$ mutation results in two phenotypes, a slight morphological change (somewhat enlarged cells) and a significant decrease in mating and pheromone response. The mating defect is partially dominant to the wild-type. A possible explanation for the mating defect is that the $SCG1^{Val50}$ protein is defective in guanine nucleotide exchange, thus preventing activation of pheromone response and a defect in mating. Biochemical characterization of these mutants and testing of additional mutants should provide more information on the role of guanine nucleotide binding in the mechanism of SCG1 action.

EFFECT OF ALTERATIONS OF THE CARBOXY TERMINUS OF SCG1

Alterations in the carboxy terminus of SCG1 have been made (C. Dietzel and J. Kurjan, in preparation) based on results suggesting that the carboxy termini of G_α subunits are involved in interactions with the corresponding receptors (Masters et al., 1986; Sullivan et al., 1987). A five amino acid carboxy terminal truncation of SCG1 results in a defect in pheromone response and mating. The mating defect is dominant to the wild-type. Another mutation, a Ser to Cys change four amino acids from the carboxy terminus, gives rise to the sequence Cys-A-A-X (where A is an aliphatic amino acid) similar to the sequence present at the carboxy terminus of ras and some of the G_α subunits. In ras, this sequence is involved in palmitylation and membrane localization (Chen et al., 1985; Fujiyama and Tamanoi, 1986). This mutation in SCG1 results in a slight decrease in mating and is dominant. The sterility of the carboxy terminal SCG1 mutants is consistent with an interaction between the carboxy terminus of SCG1 and the pheromone receptors. Alternatively, SCG1 could interact with an intermediary that interacts with the receptors.

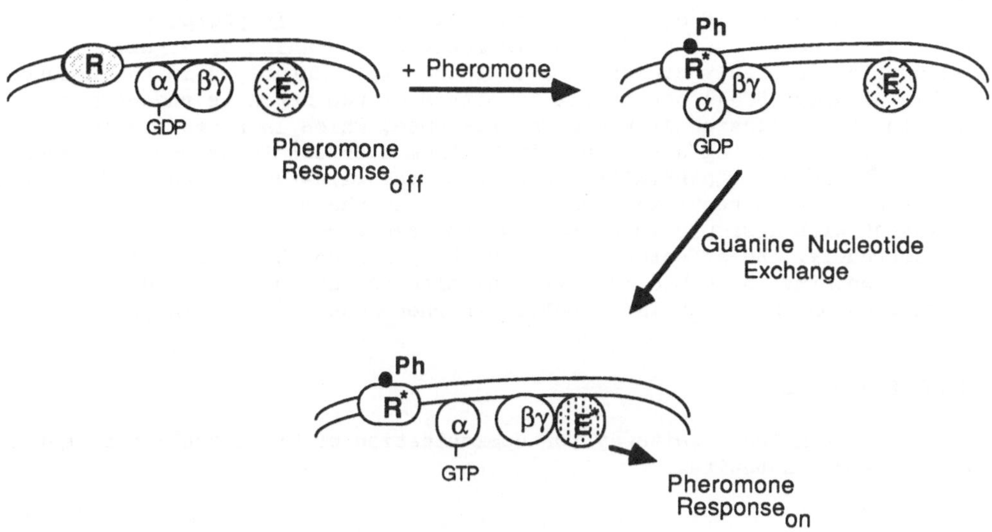

Fig. 2. Model for pheromone response pathway. A model that fits the current data for the action of the G protein subunits in the pheromone response pathway is shown and described further in the text. This model was described in Dietzel and Kurjan (1987) as Model II. The steps after the pheromone-receptor interactions are assumed to be the same in a and α cells. R-receptor; R^*-activated receptor; P-pheromone; E-inactive effector; E^*-active effector. α, β, and γ represent G protein subunits; α is the SCG1 product. The effector is currently unidentified.

MODEL FOR THE PHEROMONE RESPONSE PATHWAY

A model for the role of SCG1 in the pheromone response pathway must accomodate recent results suggest that STE4 and STE18 encode β and γ subunits, respectively, that are involved in pheromone response (M.Whiteway, personal communication). Disruption mutations in either of these genes result in a sterile phenotype. ste4 and ste18 mutations are epistatic to scg1 null mutations, i.e scg1 ste4 and scg1 ste18 double mutants show normal growth and morphology, but are sterile. These epistatic relationships indicate that STE4 and STE18 act downstream of SCG1 in the pheromone response pathway; i.e. in the scg1 ste4 and scg1 ste18 mutants, the scg1 mutation activates the pathway, but a mutation in a downstream component (ste4 or ste18) inactivates the pathway. A model for the pheromone response pathway, therefore, must have β and γ acting downstream of α.

The model shown in Figure 2 (as proposed by Dietzel and Kurjan, 1987) is consistent with these epistatic relationships and with the scg1, ste4/ste18, and double mutant phenotypes. In this model, in the absence of pheromone α-GDP (SCG1-GDP) would be bound to β/γ (STE4/STE18). After guanine nucleotide exchange β/γ would be freed and would activate the effector. In an scg1 null mutant, the effector and therefore the pheromone response pathway, would be constitutively active. In ste4 or ste18 mutants (whether SCG1 is active or not), the pheromone response pathway would be inactivated, leading to sterility. This model, therefore, is consistent with all of the results that have been obtained at this point. The effector in the pathway has not been identified.

CONCLUSIONS

The discovery of α, β, and γ homologs involved in pheromone response in yeast indicates that the G protein-mediated mechanism of signal transduction is highly conserved evolutionarily. Mutations in the α subunit of this pathway (SCG1) can result in either of two opposite phenotypes, constitutive expression of pheromone response, which is recessive to the wild-type, or a mating defect resulting from a defect in pheromone response, which is dominant or partially dominant to the wild-type. These phenotypes and the phenotypes resulting from mutations in the β and γ subunits are consistent with a simple model for the pheromone response pathway (Figure 2). Currently, the effector in this pathway has not been identified. Further experiments should identify the effector and provide additional information on the mechanism involved in pheromone response in yeast.

ACKNOWLEDGEMENTS

I thank Malcolm Whiteway for communication of his unpublished results on the β and γ subunits.

REFERENCES

Bender, A., and Sprague, G.F., Jr., 1986, Yeast peptide pheromones, a-factor and α-factor, activate a common response mechanism in their target cells, Cell, 47:929.

Betz, R., Duntze, W., and Manney, T.R., 1978, Mating factor-mediated sexual agglutination in Saccharomyces cerevisiae, FEMS Letters, 4:107.

Bray, P., Carter, C., Simons, C., Guo, V., Puckett, C., Kamholz, J., Spiegel, A., and Nirenberg, M. 1986, Human cDNA clones for four species of $G_{\alpha s}$ signal transduction protein, Proc. Natl. Acad. Sci. USA, 83:8893.

Burkholder, A.C., and Hartwell, L.H., 1985, The yeast α-factor receptor: structural properties deduced from the sequence of the STE2 gene, Nucl. Acids Res., 13:8463.

Chan, R.K., and Otte, C.A., 1982a, Physiological characterization of Saccharomyces cerevisiae mutants supersensitive to G1 arrest by a-factor and α-factor pheromones, Mol. Cell. Biol., 2:21.

Chan, R.K., and Otte, C.A., 1982b, Isolation and genetic analysis of Saccharomyces cerevisiae mutants supersensitive to G1 arrest by a-factor and α-factor pheromones, Mol. Cell Biol, 2:11.

Chen, Z.Q., Ulsh, L.S., DuBois, G., and Shih, T.Y., 1985, Posttranslational processing of p21ras proteins involves palmitylation of the C-terminal tetrapeptide containing cysteine-186, J. Virol., 56:607.

Clanton, D.J., Hattori, S., and Shih, T.Y., 1986, Mutations of the ras gene product p21 that abolish guanine nucleotide binding, Proc. Natl. Acad. Sci. USA, 83:5076.

De Vos, A.M., Tong, L., Milburn, M.V., Matias, P.M., Jancarik, J., Noguchi, S., Nishimura, S., Miura, K., Ihtsuka, E., and Kim, S.-H., 1988, Three-dimensional structure of an oncogene protein: catalytic domain of human c-H-ras-p21, Science, 239:888.

Dietzel, C., and Kurjan, J., 1987, The yeast SCG1 gene: a G_α-like protein implicated in the a- and α-factor response pathway, Cell, 50:1001.

Fujiyama, A., and Tamanoi, F., 1986, Processing and fatty acid acylation of RAS1 and RAS2 proteins in Saccharomyces cerevisiae, Proc. Natl. Acad. Sci. USA, 83:1266.

Gilman, A.G., 1987, G proteins: transducers of receptor-generated signals, Ann. Rev. Biochem, 56:615.

Hagen, D.C., and Sprague, G.F., Jr., 1984, Induction of the yeast α-specific STE3 gene by the peptide pheromone a-factor, J. Mol. Biol., 178:835.

Hagen, D.C., McCaffrey, G., and Sprague, G.F., Jr., 1986, Evidence the yeast STE3 gene encodes a receptor for the peptide pheromone a factor: gene sequence and implications for the structure of the presumed receptor, Proc. Natl. Acad. Sci. USA, 83:1418.

Hartig, A., Holly, J., Saari, G., and MacKay, V.L., 1986, Multiple regulation of STE2, a mating-type-specific gene of Saccharomyces cerevisiae, Mol. Cell. Biol., 6:2106.

Hartwell, L.H., 1980, Mutants of Saccharomyces cerevisiae unresponsive to cell division control by polypeptide mating hormone, J. Cell Biol., 85:811.

Itoh, H., Kozasa, T., Nagata, S., Nakamura, S., Katada, T., Ui, M., Iwai, S., Ohtsuka, E., Kawasaki, H., Suzuki, K., and Kaziro, Y., 1986, Molecular cloning and sequence determination of cDNAs for α subunits of the guanine nucleotide-binding proteins G_s, G_i, and G_o from rat brain, Proc. Natl. Acad. Sci. USA, 83:3776.

Jahng, K.-Y., Ferguson, J., and Reed, S.I., 1988, Mutations in a gene encoding the α subunit of a Saccharomyces cerevisiae G protein indicate a role in mating pheromone signaling, Mol. Cell. Biol., 8:2484.

Jenness, D.D., Burkholder, A.C., and Hartwell, L.H., 1983, Binding of α-factor pheromone to yeast a cells: chemical and genetic evidence for an α-factor receptor, Cell, 35:521.

Jurnak, F., 1985, Structure of the GDP domain of EF-Tu and location of the amino acids homologous to ras oncogene proteins, Science, 230:32.

Kurjan, J., 1985, α-Factor structural gene mutations in Saccharomyces cerevisiae: effects on α-factor production and mating, Mol. Cell. Biol., 5:787.

La Cour, T.F.M., Nyborg, J., Thirup, S., and Clark, B.F.C., 1985, Structural details of the binding of guanosine diphosphate to elongation factor Tu from E. coli. as studied by X-ray crystallography, EMBO J., 4:2385.

MacKay, V.L., and Manney, T.R., 1974a, Mutations affecting sexual conjugation and related processes in Saccharomyces cerevisiae. I.

Isolation and phenotypic characterization of nonmating mutants, Genetics, 76:255.

MacKay, V., and Manney, T.R., 1974b, Mutations affecting sexual conjugation and related processes in Saccharomyces cerevisiae. II. Genetic analysis of nonmating mutants, Genetics, 76:273.

Manney, T., 1983, Expression of the BAR1 gene in Saccharomyces cerevisiae: induction by the α mating pheromone of an activity associated with a secreted protein, J. Bacteriol., 155:291.

Manney, T.R., and Woods, V., 1976, Mutants of Saccharomyces cerevisiae resistant to the α mating-type factor, Genetics, 82:639.

Masters, S.B., Stroud, R.M., and Bourne, H.R., 1986, Family of G protein α chains: amphipathic analysis and predicted structure of functional domains, Prot. Eng., 1:47.

McGrath, J.P., Capon, D.J., Goedell, D.V., and Levinson, A.D., 1984, Comparative biochemical properties of normal and activated human ras p21 protein, Nature, 310:644.

Michaelis, S., and Herskowitz, I., 1988, The a-factor pheromone of Saccharomyces cerevisiae is essential for mating, Mol. Cell. Biol., 8:1309.

Miyajima, I., Nakafuku, M., Nakayama, N., Brenner, C., Miyajima, A., Kaibuchi, K., Arai, K., Kaziro, Y., and Matsumoto, K., 1987, GPA1, a haploid-specific essential gene, encodes a yeast homolog of mammalian G protein which may be involved in mating factor signal transduction, Cell, 50:1011.

Nakafuku, M., Itoh, H., Nakamura, S., and Kaziro, Y., 1987, Occurrence in Saccharomyces cerevisiae of a gene homologous to the cDNA coding for the α subunit of mammalian G proteins, Proc. Natl. Acad. Sci. USA, 84:2140.

Nakayama, N., Miyajima, A., and Arai, K., 1985, Nucleotide sequences of STE2 and STE3, cell type-specific sterile genes from Saccharomyces cerevisiae, EMBO J., 4:2643.

Nakayama, N., Miyajima, A., and Arai, K., 1987, Common signal transduction system shared by STE2 and STE3 in haploid cells of Saccharomyces cerevisiae: autocrine cell-cycle arrest results from forced expression of STE2, EMBO J., 6:249.

Robishaw, J.D., Russell, D.W., Harris, B.A., Smigel, M.D., and Gilman, A.G., 1986. Deduced primary structure of the α subunit of the GTP-binding stimulatory protein of adenylate cyclase, Proc. Natl. Acad. Sci. USA, 83:1251.

Rothstein, R.J., 1983, One-step gene disruption in yeast, Meth. Enzym., 101:202.

Seeburg, D.H., Colby, W.W., Capon, D.J., Goedell, D.V., and Levinson, A.D., 1984, Biological properties of human c-Ha-ras1 genes mutated at codon 12, Nature, 312:71.

Shimoda, C., Yanagishima, N., Sakurai, A., and Tamura, S., 1976, Mating reaction in Saccharomyces cerevisiae. IX. Regulation of sexual cell agglutinability of a-type cells by a sex factor produced by alpha type cells, Arch. Microbiol., 108:27.

Sibley, D.R., Benovic, J.L., Caron, M.G., and Lefkowitz, R. J., 1987, Regulation of transmembrane signaling by receptor phosphorylation, Cell, 48:913.

Sprague, G.F., Jr., Blair, L.C., and Thorner, J., 1983, Cell interactions and regulation of cell type in the yeast Saccharomyces cerevisiae. Ann. Rev. Microbiol., 37:623.

Strazdis, J.R., and MacKay, V.L., 1983, Induction of yeast mating pheromone a-factor by α cells, Nature, 305:543.

Sullivan, K.A., Miller, R.T., Masters, S.B., Biederman, B., Heidman, W., and Bourne, H.R., 1987, Identification of a receptor contact site involved in receptor-G protein coupling, Nature, 330:758.

Sweet, R.W., Yokoyama, S., Kamata, T., Feramisco, J.R., Rosenberg, M., and Gross, M., 1984, The product of ras is a GTPase and the T24 oncogenic mutant is deficient in this activity, Nature, 311:273.
Tanabe, T., Nukada, T., Nishikawa, Y., Sugimoto, K., Suzuki, H., Takahashi, H., Noda, M., Haga, T., Ichiyama, A., Kangawa, K., Minamino, N., Matsuo, H., and Numa, S., 1985, Primary structure of the α-subunit of transducin and its relationship to ras proteins, Nature, 315:242.
Terrance, K., and Lipke, P.N., 1981, Sexual agglutination in Saccharomyces cerevisiae, J. Bacteriol., 148:889.
Toda, T., Uno, I., Ishikawa, T., Powers, S., Kataoka, T., Broek, D., Cameron, S., Broach, J., Matsumoto, K., and Wigler, M., 1985, In yeast, RAS proteins are controlling elements of adenylate cyclase, Cell, 40:27.
Walter, M., Clark, S.G., and Levinson, A.D., 1986, The oncogenic activation of human $p21^{ras}$ by a novel mechanism, Science, 233:649.

SIGNAL-TRANSDUCING G-PROTEINS IN *DICTYOSTELIUM DISCOIDEUM*

Anthony A. Bominaar, B. Ewa Snaar-Jagalska, Fanja Kesbeke and Peter J. M. Van Haastert

Cell biology and Genetics Unit, Zoological Laboratory, PO Box 9516, 2300 RA Leiden, The Netherlands

INTRODUCTION

Most of the studies on signaltransduction and the role of G-proteins therein have been done on mammalian cells. This brings about several disadvantages, such as laborious growth conditions non-homogeneous material and the use of highly specialized cells. The use of tumor cell lines brings the risk of using cells in which especially the signaltransduction pathways are somehow impaired. Using micro-organisms can overcome many of these disadvantages. In our studies we use the cellular slime mold *Dictyostelium discoideum* which is an easy to grow haploid organism with a simple life cycle. The fact that it is haploid makes it easier to perform both classical as well as molecular genetics. *Dictyostelium discoideum* is a soil living organism which feeds on bacteria during its vegetative stage. Upon food deprivation it enters a social stage in which cells aggregate into multicellular structures in response to excreted cAMP. Subsequently differentiation to two different cell types (i.e. spores and stalk cells) takes place (Schaap 1986) . In the aggregative stage cAMP elicits several cellular responses which are thought to be mediated by two types of cAMP surface receptors, the A sites which exist in a low (A^L) and a high (A^H) affinity form and the B sites of which three kinetic forms can be distinguished. A fast dissociating form (B^F) a slower dissociating form (B^S) and a very slowly dissociating form (B^{SS}) (Van Haastert, et al. 1986) . Some of the responses upon stimulation by cAMP are: a transient elevation of the intracellular levels of cAMP and cGMP, activation of phospholipase C and excretion of cAMP (for review see Devreotes 1983, Janssens and Van Haastert 1987) . In many ways the responses shown by *Dictyostelium discoideum* resemble those observed in the well known adenylate cyclase system for adrenergic receptors (Levitzki 1988) . Hence it seems likely that the signal transduction pathway shows some resemblance as well. One of the crucial points in signal transduction in the adrenergic system is the presence of stimulating and inhibiting G-proteins. As far as *Dictyostelium discoideum* is concerned many clues have been presented for G-proteins functioning in the cAMP dependent signal transduction.

In this paper we will use the criteria for G-protein involvement as they were defined by Gilman (Gilman 1987) to determine whether there are G-proteins present and functioning in *Dictyostelium discoideum*. In the second part of this paper we will discuss the fact that *Dictyostelium discoideum* is 2 billion years away from mammals on the evolutionary timescale and what the implications are for G-protein conservation.

THE GILMAN CRITERIA

In 1987 A.G Gilman defined seven criteria for G-protein involvement in signal transduction (Gilman 1987) . Some of these criteria, we think, are more strict than others. In short the seven criteria are the following.

1) An appropriate ligand and GTP are both required for a response. This is a strict criterium, falsification means no signal transduction G-protein is involved.
2) The response can be provoked independently of the receptor by non-hydrolyzable analogues or AlF_4^-. This criterium is strict as well, falsification indicates that GTP hydrolysis is not involved in signal termination. This of course is only true if association of the analogue or AlF_4^- happens in the first place.
3) There is a negative heterotropic interaction between the binding of guanine nucleotide to a G-protein and the binding of an agonist to a G-protein linked receptor. Falsification of this criterium would imply that there is no interaction between the receptor and the G-protein thus resulting in an inactive system.
4) Toxins have effects in many known G-proteins and can be used with intact cells or purified components. Since it has not been shown that being sensitive to toxins is required for proper functioning, we do not consider this criterium a strict one; on the other hand, being sensitive to toxins such as cholera toxin and pertussis toxin makes it more likely that a G-protein is involved.
5) Mutants can be useful in the definition of G-protein regulated function. The value of this criterium is of course solely dependent on the quality of the mutants used.
6) Antibodies can be used for establishing the role of a G-protein. As with the foregoing, the value depends on the quality. Not being able to show a G-protein or its function using an antibody does not imply that the G-protein is not present. Being able to show a protein or its function using an anti G-protein antibody makes it likely that a G-protein is involved.
7) Purification and reconstitution of the individual components. This is considered the ultimate criterium by Gilman. Not being able to fulfill this criterium however does not exclude the role of G-proteins.

G-PROTEINS IN *DICTYOSTELIUM DISCOIDEUM*

As mentioned before there are several lines of evidence suggesting the presence of G-proteins in *Dictyostelium*. On the receptor level it has been shown that GTP and its analogues induce a shift from B^S and B^{SS} to B^F for the B-sites and a shift from A^H to A^L for the A-sites of the cAMP surface receptor (Van Haastert, et al. 1987) . Inaddition cAMP has been shown to stimulate high affinity GTPase by decreasing the K_m from 6.5 to 4.5 µM resulting in a stimulation of GTP-ase activity up to 65% in a dose dependent manner (Snaar-Jagalska, et al. 1988b). GTP binding is increased by cAMP due to an increase in both the affinity (the K_d goes from 0.22 to 0.16 µM) as well as the number of sites (31.45 to 55.90 nM) (Snaar-Jagalska, et al. 1988a) . Both stimulation of GTP-ase activity and GTP binding show a specificity pattern for cAMP analogues which suggest surface receptor involvement (Snaar-Jagalska, et al. 1988b) . On the effector site it has been shown that GTP or its non-hydrolysable analogues Gpp(NH)p or GTPγS can stimulate or inhibit (depending on the protocol used) adenylate cyclase in vitro (Van Haastert, et al. 1987) and a stimulation of the IP_3 production has been observed as well (Europe-Finner and Newell 1987) .

Cholera toxin has been shown to ADP-ribosylate a 42 kDa protein which can be photo-affinity labeled using 8-N_3-GTP. This protein co-migrates with the 42 kDa Gsα subunit of mammalian cells on two dimensional gel electroforesis (Leichtling, et al. 1981) . Pertussis toxin inhibits desensitization of adenylate cyclase in vivo (Snaar-Jagalska and Van Haastert 1988) and was shown to inhibit both GTP-binding (Snaar-Jagalska, et al. 1988a) and high affinity GTP-ase activity in membranes of *Dictyostelium;* however, no detectable ADP-ribosylation could be observed. The latter seems to be due to the presence of an ADP-ribosylase inhibitor in preparations of *Dictyostelium* since the addition of such a preparation to purified transducin inhibited pertussis toxin dependent ADP-ribosylation in this system as well (Snaar-Jagalska and Van Haastert 1988) .

Using antibodies against α-subunits of G-proteins (the common antibody A569 was used (Mumby, et al. 1986)) two proteins were detected specifically with molecular weights of 40 and 52 kDa respectively (Snaar-Jagalska, et al. 1988c) .

cAMP SIGNALTRANSDUCTION MUTANTS

Until now two types of mutants have been identified in which the signal transduction is impaired at, or close to the G-protein level. The first is the synag 7 mutant in which extracellular cAMP is unable to induce activation of adenylate cyclase (Franz 1980) . The surface receptor for cAMP however is present and GTP still modulates the number and affinity of this receptor, indicating a functional coupling between the receptor and the putative G-protein. However GTP or its non-hydrolizable analogues are not able to induce an activation of adenylate cyclase suggesting that the G-protein is no longer able to interact properly with the enzyme (Snaar-Jagalska and Van Haastert 1988) . Recent work showed that wild type responses can be restored in this mutant by the addition of a high speed supernatans from wild type cytoplasm. Preliminary results indicate that a high molecular weight, heat labile factor is responsible for restoring the wild type phenotype in synag 7 cells (Van Haastert, et al. 1987) . Thus it seems likely that in *Dictyostelium discoideum* an additional cytosolic factor is involved in the activation of adenylate cyclase by G-proteins.

A second type of mutants which are interesting in view of G-protein functioning are the mutants of the frigid A class (*fgd* A) (Coukell, et al. 1983) . In these mutants signal transduction in response to cAMP is completely absent, neither adenylate cyclase activation nor IP_3 formation are observed in vivo. In vitro however both stimulation and inhibition of adenylate cyclase by GTP analogues is possible, cAMP binding and its inhibition by GTP and GDP are reduced , GTPγS binding is present but not stimulated by cAMP. GTPase activity is reduced but still present and can be stimulated by cAMP. cAMP stimulation of adenylate cyclase is still absent as are the cAMP or GTP induced IP_3 formation. Basal levels of both cAMP and IP_3 are normal indicating that the effector enzymes are present. Thus it looks as if in the adenylate cyclase pathway all the required components are present but that somehow the interaction is blocked between the G-protein and the effector. As far as the IP_3 route is concerned there is a complete blockage of interaction which could be due to a missing G-protein (Kesbeke, et al. 1988) . This latter view is supported by the finding that on western blots using the common Gα anti body A569 the 40 kDa antigen is missing in the *fgd* A strain HC 213 (Snaar-Jagalska, et al. 1988c) .

A clue for the absence of adenylate cyclase activation in these mutants is provided by the experiment in which was shown that stimulation of saponin permeabilized cells with cAMP in the presence of IP_3 did result in a stimulation of adenylate cyclase. Thus it appears that there is some level of cross talk between the IP_3 route and the adenylate cyclase route (Van Haastert, et al. 1988) . The properties of both synag 7 and the *fgd* A mutants are summarized in table 1.

Tabel 1. In vivo and in vitro properties of the signaltransduction mutants synag 7 and *fgd* A (strain HC 85).

Properties	Synag 7	HC 85	References
in vivo			
cAMP binding	normal	reduced	(Kesbeke et.al. 1988) (Snaar-Jagalska and Van Haastert 1988)
cAMP production	absent	absent	(Kesbeke et.al. 1988) (Snaar-Jagalska and Van Haastert 1988)
cGMP production	normal	absent	(Kesbeke et.al. 1988)
IP_3 production	normal	absent	(Van Haastert et.al. 1988)
basal cAMP level	normal	normal	(Kesbeke et.al. 1988)
basal IP_3 level	normal	normal	(Van Haastert et.al. 1988)
in vitro			
cAMP binding	normal	reduced	(Snaar-Jagalska and Van Haastert 1988) (Kesbeke et.al. 1988)
GTP effects	normal	reduced	(Snaar-Jagalska and Van Haastert 1988) (Kesbeke et.al. 1988)
GTP binding	reduced	normal	(Snaar-Jagalska and Van Haastert 1988) (Kesbeke et.al. 1988)
cAMP stimulation	normal	absent	(Snaar-Jagalska and Van Haastert 1988) (Kesbeke et.al. 1988)
GTP-ase activity	reduced	reduced	(Snaar-Jagalska and Van Haastert 1988) (Kesbeke et.al. 1988)
cAMP stimulation	normal	normal	(Snaar-Jagalska and Van Haastert 1988) (Kesbeke et.al. 1988)
Adenylate cyclase			
cAMP stimulation	absent	absent	(Kesbeke et.al. 1988) (Van Haastert et.al 1986)
GTP stimulation	absent	normal	(Kesbeke et.al. 1988) (Van Haastert et.al 1986)
GTP inhibition	normal	normal	(Van Haastert et.al 1986)
Western blot analysis			
40 kDa protein	normal	absent	(Snaar-Jagalska et.al. 1988c)
52 kDa protein	normal	reduced	(Snaar-Jagalska et.al. 1988c)

Normal: the effect is comparable to the effect in wild type, Reduced: the effect is significantly reduced as compared to wild type, Absent: the effect could not be observed. *: wild type phenotype is restored by the supernatant factor.

Summarizing these two mutants show that although GTP (or its analogues) are a prerequisite for receptor-effector coupling at least in the case of activation of adenylate cyclase other factors are involved as well. Whether the cytosolic factor in synag 7 and the IP_3 requirement are parts of the same mechanism remains to be elucidated.

ARE G-PROTEINS PRESENT IN *DICTYOSTELIUM DISCOIDEUM*

Returning to the Gilman criteria we can conclude that most of these are at least partially fulfilled and none of them has been falsified in *Dictyostelium*.

-It has been shown that there is a negative heterotropic relationship between agonist binding to the receptor and GTP binding to the putative G-protein.

-Non-hydrolyzable GTP analogues are capable of inducing responses independent from the receptor, AlF_4^- however is not.

-Cholera toxin was shown to ADP-ribosylate a Gs_α like protein and pertussis toxin inhibited several processes both in vivo and in vitro.

-Common Gα antibodies are cross reactive with two *Dictyostelium* proteins.

-Strain HC 213 from the *fgd* A class lacks one of the above mentioned proteins and is severely impaired in its signal transduction.

-A strict requirement for GTP in the signal transduction has not been shown but in case of in vitro stimulation of adenylate cyclase GTP has an additional stimulating effect over the effect of cAMP alone. That stimulation is possible without the addition of GTP is probably due to the fact that in these experiments crude homogenates were used in which the GTP content is high enough to allow G-protein activation.

-Isolation and reconstitution of the individual components has never been achieved. Considering the instability of the individual components, especially adenylate cyclase and the receptor, it is not very likely to be performed on a short timescale.

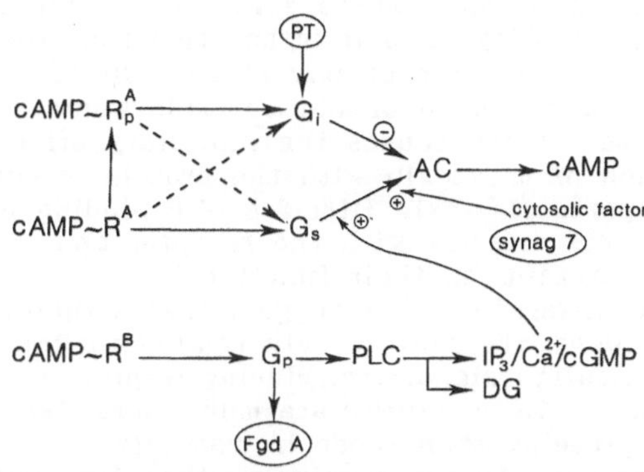

Fig 1. Schematic representation of the signal transduction in *Dictyostelium discoideum*. R: receptor, G: G-protein, AC: adenylate cyclase, PLC: phospholipase C, PT: pertussis toxin, arrows indicate interactions.

Concluding we think that it is justified to state that in *Dictyostelium discoideum* at least one but more likely three G-proteins are present and functioning in the signal transduction from cAMP to both adenylate cyclase and phospholipase C. A possible mechanism for the interaction with the receptor, the effectors, the cytosolic factor and IP_3 is given in figure 1.

DISCUSSION

The fact that using criteria, based on mammalian data, it is still possible to determine whether G-proteins are present in an organism which is 2 billion years apart on the evolutionary timescale indicates that either all the criteria refer to the most basic properties of the proteins in question or that most properties are highly conserved throughout the evolution.

There are differences however between mammalian G-proteins and those of *Dictyostelium discoideum*. One of the most striking features in *Dictyostelium* signal transduction is the observed cross-talk between to different pathways. This type of cross-talk has not been observed in mammalian systems and its existence there is unlikely when we consider the reconstitution experiments performed with purified compounds. The same holds for the requirement of the cytosolic factor for the activation of adenylate cyclase making it tempting to assume that in *Dictyostelium* this cytosolic factor is the molecule involved in cross-talking. If this were true it would mean that in *Dictyostelium* there is an additional system monitoring and integrating G-protein activation directly at the G-protein-effector interaction level. Until now *Dictyostelium discoideum* is the only organism known, in which this cross-talk occurs; before we can decide whether this is a so called primitive property or not it should be established if in other lower eukaryotes the same system is observed or not.

An other difference between *Dictyostelium* G-proteins and those of mammals is that the *Dictyostelium* G-proteins can not be activated using phosphate analogues like ALF_4^-. Such an activation has only been reported for the signal transducing G-proteins of mammalian systems and not (yet) for other GTP binding proteins like elongation factors or RAS p21 hence this could be a property of signal transducing G-proteins obtained later during the evolution.

A third difference is that ADP-ribosylation by pertussis toxin seems to be strongly inhibited in membranes whereas functional inhibition by pertussis toxin still occurs. In mammalian systems it is assumed that the ADP-ribosylation and the effect of pertussis toxin are closely related. In *Dictyostelium* it appears that these two processes can be uncoupled. This would mean that it is not the ADP-ribosylation itself which causes the inhibitory effects on G-proteins but that some interaction of the toxin with the protein is sufficient. Another possibility is that pertussis toxin (itself a GTP binding protein (Mattera et al. 1987)) interacts directly with the receptor and the effector thereby blocking cellular G-proteins in their function.

The latter two differences may suggest that structurally the G-proteins of *Dictyostelium discoideum* are more closely related to RAS p21 and elongation factors than to mammalian signal transducing G-proteins. And this would mean that the G-proteins of *Dictyostelium* are much more like the archetypal GTP-binding regulatory proteins than those in mammals.

Summarizing we come to the conclusion that G-proteins are present in *Dictyostelium discoideum* and that the study of these proteins in this organism can teach us more about the general properties of these proteins and the systems they are functioning in.

LITERATURE CITED

- Coukell, M. B., S. Lappano, and A. M. Cameron. 1983.Isolation and characterization of cAMP unresponsive (frigid) aggregation deficient mutants of *Dictyostelium discoideum* . Dev.Genet. 3:283-297
- Devreotes, P. N. 1983. Cyclic nucleotides and cell-cell communication in *Dictyostelium discoideum.* Adv.Cyclic Nucleotide Res. 15:55-96.
- Europe-Finner, G. N. and P. C. Newell. 1987. GTP-analogues stimulate inositol trisphosphate formation transiently in *Dictyostelium.* J.Cell Sci. 87:513-518.
- Franz, C. E. 1980. Phenotype analysis of aggregation mutants of *Dictyostelium discoideum.* PhD Thesis, University of Chicago, Chicago.
- Gilman, A. G. 1987. G-proteins: transducers of receptor generated signals. Ann.Rev.Biochem. 56:615-649.

- Janssens, P. M. W. and P. J. M. Van Haastert. 1987. Molecular basis of transmembrane signal transduction in *Dictyostelium discoideum*. Microbiol.Rev. 51:396-418.
- Kesbeke, F., B. E. Snaar-Jagalska, and P. J. M. Van Haastert. 1988. Signal-transduction in *Dictyostelium fgd* A mutants with a defective interaction between surface cAMP receptor and a GTP binding regulatory protein. J.Cell.Biol. 107:521-528
- Leichtling, B. H., D. S. Coffman, E. S. Yaeger, H. V. Rickenberg, W. al-Jumaliy, and B. E. Haley. 1981. Occurence of the adenylate cyclase "G-protein" in membranes of *Dictyostelium discoideum*. Biochem.Biophys.Res.Comm. 102:1187-1195.
- Levitzki, A. 1988. Signal transduction in hormone dependent adenylate cyclase. Cell.Biophys. 12:133-143.
- Mattera, R., J. Codina, R. D. Sekura, and L. Birnbaumer. 1986. The interaction of nucleotides with pertussis toxin. J.Biol.Chem. 261:11173-11179.
- Mumby, S. M., R. A. Kahn, D. R. Manning, and A. G. Gilman. 1986. Antisera of designed specificity for subunits of guanine nucleotide binding regulatory proteins. Proc.Natl.Acad.Sci.USA 83:265-269.
- Schaap, P. 1986. Regulation of size and pattern in the cellular slime molds. Diff. 33:1-16.
- Snaar-Jagalska, B. E. and P. J. M. Van Haastert. 1988. Pertussis toxin inhibits cAMP-induced desensitization of adenylate cyclase in *Dictyostelium discoideum* submitted.
- Snaar-Jagalska, B. E. and P. J. M. Van Haastert. 1988. *Dictyostelium discoideum* mutant synag 7 with altered adenylate cyclase interaction. J.Cell.Sci. in press.
- Snaar-Jagalska, B. E., R. J. W. De Wit, and P. J. M. Van Haastert. 1988a. Pertussis toxin inhibits cAMP surface receptor stimulated binding of 35S GTPγS to *Dictyostelium discoideum* membranes. FEBS Lett. 232:148-152.
- Snaar-Jagalska, B. E., K. H. Jakobs, and P. J. M. Van Haastert. 1988b. Agonist stimulated high affinity GTP-ase in *Dictyostelium discoideum*. FEBS Lett. 236:139-144.
- Snaar-Jagalska, B. E., F. Kesbeke, M. Pupillo, and P. J. M. Van Haastert. 1988c. Immunological detection of G-protein alpha subunits in *Dictyostelium discoideum*. Biochim.Biophys.Rec.Comm. in press.
- Van Haastert, P. J. M., R. J. W. De Wit, P. M. W. Janssens, F. Kesbeke, and J. DeGoede. 1986. G-protein mediated interconversions of cell surface cAMP receptors and their involvement in excitation and desensitization of guanylate cyclase in *Dictyostelium discoideum*. J.Biol.Chem. 261:6904-6911.
- Van Haastert, P. J. M., L. C. Penning, and F. Kesbeke. 1988. *Dictyostelium* mutants with malfunctioning G-proteins reveal cross talk of signal transduction pathways. submitted.
- Van Haastert, P. J. M., B. E. Snaar-jagalska, and P. M. W. Janssens. 1987. The regulation of adenylate cyclase by guanine nucleotides in *Dictyostelium discoideum* membranes. Eur.J.Biochem. 162:251-258

STRUCTURE AND FUNCTION OF TUBULINS

TUBULIN AS A G-PROTEIN : REGULATION OF TUBULIN-TUBULIN INTERACTIONS BY GTP HYDROLYSIS

Marie-France Carlier and Dominique Pantaloni

Laboratoire d'Enzymologie
Centre National de la Recherche Scientifique
91198 Gif-sur-Yvette, France

As proposed by Hughes in 1983[1], tubulin belongs to the large family of GTP-binding proteins, which contains several other subclasses : the elongation and initiation factors of protein synthesis, and proteins involved in signal transduction such as G_s, G_i, transducin and the Ras p21 proteins ; recently this family has been extended to include low molecular weight YPT1 and SEC4 proteins involved in the regulation of secretion (for a review see reference 2).

All these GTP-binding proteins are characterized by the fact that the hydrolysis of GTP regulates the interaction with the effector : in the case of EF Tu, GTP hydrolysis causes the dissociation of EF Tu from the ternary complex formed with the aminoacyl tRNA and the ribosome ; in the case of G-proteins, GTP hydrolysis disrupts the interaction of the G-protein with the effector. In all these cases, the conformation of the GDP and of the GTP-protein are drastically different. In the case of tubulin, GTP hydrolysis plays the same regulatory role, as emphasized below :

The β-subunit of the αβ heterodimer of tubulin binds GTP tightly in an exchangeable manner. The GTP-tubulin complex is the functional unit able to interact with another GTP-tubulin unit and undergo self-assembly into microtubules. GDP-tubulin in contrast does not polymerize. Hydrolysis of GTP occurs on polymerized tubulin as a single turnover reaction. Following Pi release into the medium, GDP remains tightly bound to tubulin subunits in the microtubules that spontaneously disassemble following hydrolysis of GTP[3]. In other words, a conformation change is linked to GTP hydrolysis on microtubules, that destabilizes the tubulin-tubulin interactions in the polymer. The above features are therefore in perfect agreement with the main functional properties of GTP-binding proteins : GTP hydrolysis is associated to protein-protein interactions ; GTP hydrolysis is the conformational switch that induces the destabilization of this interaction.

Structural analogies between the GTP-binding sequences of G-proteins and tubulin

In view of the functional analogies between tubulin and the GTP-binding proteins, attempts have been made to correlate sequence elements of tubulin with the known consensus sequences of the GTP-binding domain of G-proteins, in order to propose a structure of the GTP binding site on β-tubulin. Data derived from sequence data base and reported by Dever et al.[4] and Sternlicht et al.[5] concerning different classes of G-proteins and tubulin are displayed in Table I. From this comparison it appears that peptides 111-105 and 205-208 of the N-terminal domain of β-tubulin do match the consensus sequences I and II attributed to the phosphoryl binding site, while peptide 300-297 matches the sequence Asn-Lys-X-Asp characteristic of the guanine base binding site. In addition however, a glycyl-rich region I_A formed by peptide 143-148, mostly found in nucleotidases but not in G proteins, must be considered as a highly flexible region controlling the access of GTP to its site. In addition, too, a hydrophobic peptide 64-70 (upstream of regions I, I_A, II and III) is predicted to be part of the guanine binding site too.

In spite of these plausible sequence analogies, a problem remains : the spacing between regions I and II is conserved and equal either to 40-70, or 150-170 for different classes of G-proteins, while it is 94 for tubulin ; a better fit to the standard spacing is obtained when considering the spacing between I_A and II, which is 57 residues. A large discrepancy also exists when comparing the spacing between regions II and III which is conservatively 50-70 in all G-proteins, but 89 in β-tubulin. Also, the hydrophobic 64-70 peptide is located upstream of the I, II, III sequences.

All these data indicate that the possible similarity in the predicted 3-dimensional structures of the GTP-binding sites of G-proteins and tubulin is somewhat elusive.

Table I

Consensus sequences of the GTP binding site of G-proteins and tubulin (adapted from Dever et al., 1987 PNAS 84, 1814 and Sternlicht et al. 1987, FEBS Lett. 214, 226).

Proteins	P-binding I	P-binding II	G-binding III	I-II spacing	II-III spacing
Consensus sequence	GXXXXGK	DXXG	NKXD		
EF and IF factors	GHVDHGK A A	DCPG TA	NKCD M I	40-70	50-60
Ras-proteins	GGGGVGK A	DTAG	NKSD L C	41	56
Signal Transducing proteins	GAGESGK	DVGG	NKKD	150-170	66
Tubulin (β)	GETYHGK 111 105 I GGTGSG 143 148 IA	DNEA 205 208	NKAD 300 297 RAILVDL 64 70 hydrophobic	94 57	89

Experiments have been done to identify peptides involved in GTP binding, using affinity labeling. Results are summarized in Table II. Except for the labeled 64-70 region found by Kim et al.[7] using 8-azido-GTP as a photo-affinity probe, no clear correlation appears between the distribution of label and the predicted sequences of the GTP binding region. Tubulin, in some respects, appears closer to ATPases than to G-proteins.

I will therefore rather focus on the mechanistic aspects of GTP hydrolysis in the regulation of microtubule dynamics, and try to derive general conclusions concerning the role of GTP hydrolysis in the regulatory function of G-proteins.

Table II

Localization of GTP E-site in β-tubulin N-terminal domain using affinity labeling.

Probe	Peptide identified on β	References
8-azido GTP	64-70	Kim, Ponstingl and Haley[7]
GTP_{ox} and 3'-p-azidobenzoyl GTP	73-147 on β 37-154 on α	Maccioni and Seeds[8]
GTP (UV irrad.)	155-174	Hesse et al.[9]

GTP hydrolysis in microtubule assembly : Regulation of dynamic instability of microtubules by Pi release

Microtubules grow by endwise association of GTP-tubulin subunits. GTP hydrolysis occurs subsequently on the polymerized subunits. The kinetic study of the rate of polymer growth as a function of tubulin dimer concentration leads to the determination of the rate constants for tubulin association to and dissociation from microtubule ends). It was found that in a regime of growth terminal subunits dissociate from microtubules at a 2 orders of magnitude slower rate than internal GDP-subunits[10]. In other words, tubulin subunits newly incorporated in the microtubule and located at the ends are not GDP-subunits. Such a difference between polymer subunits is not observed when microtubules are assembled in the presence of a non hydrolyzable analog of GTP. The terminal subunits were first thought to be GTP subunits forming a stabilizing cap at the tip of microtubules and preventing the depolymerization of the unstable GDP core.

The loss and rebuilding of the protective "GTP cap" via statistic fluctuations in polymer length was proposed by Mitchison and Kirschner[11] and theoretically developed by Hill[12,13] as a possible explanation for the dynamic instability of microtubules (i.e. the property of microtubules to undergo phase transitions between states of slow growth (with a GTP cap) and catastrophic depolymerization (following the loss of the cap).

In order to understand the molecular mechanism by which GTP hydrolysis is linked to the switch between states of strong and weak tubulin-tubulin interactions in the microtubule, it is necessary to know the dynamic properties of the transient species involved in the polymerization process. The following kinetic scheme is proposed :

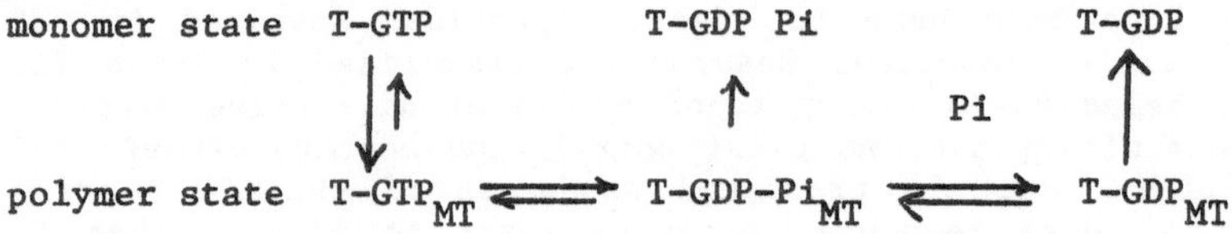

According to this scheme GTP is hydrolyzed on microtubules only, within two elementary steps : cleavage of the γ-phosphate, followed by the release of inorganic phosphate in the medium. In the central T-GDP-Pi_{MT} complex, Pi is non-covalently but tightly bound to the nucleotide site. Although no evidence has been obtained for a long-lived T-GDP-Pi_{MT} intermediate, this complex can be reconstituted by adding Pi back to GDP-microtubules. Microtubule GDP-subunits bind Pi with a low affinity (25 mM) and the reconstituted GDP-Pi-polymer is very stable[14] : GDP-Pi subunits dissociate from microtubule ends at a 2 orders of magnitude slower rate than GDP-subunits.In addition, beryllium fluoride (BeF_3, H_2O) and aluminum fluoride AlF_4^-, recently proposed by Bigay et al.[15] to act as structural analogs of Pi binding to the site of the γ-phosphate of GTP on GDP-transducin, also bind mole to mole to GDP-tubulin in microtubules in competition with Pi but with a 3 orders of magnitude higher affinity, and stabilize microtubules in a Pi-like fashion[14]. Therefore the T-GDP-$BeF_{3\,MT}$ complex mimics the T-GDP-Pi_{MT} transient state in microtubule assembly. These results reinforce the view that GTP hydrolysis is involved in microtubule dynamic instability. In addition, the main conclusion is that the elementary step that triggers the destabilization of tubulin-tubulin interactions in microtubules is not the cleavage of the γ-phosphate of GTP, but Pi release. The liberation of Pi therefore appears as the key reaction that controls dynamic instability. The following model accounts for the data :

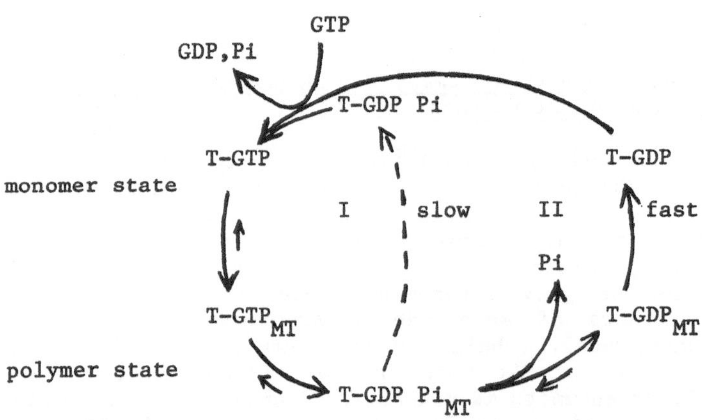

According to this model, tubulin cycles between the monomer and polymer states through two essential routes I (slow turnover) and II (rapid turnover). The extent of dynamic instability depends on the balance between these two cycles. At steady- state, microtubules are stabilized when capped by slowly dissociating GDP-Pi terminal subunits ; the phase transition to the state of catastrophic depolymerization is materialized by the reaction of Pi liberation.

It is interesting to note that while Pi exhibits low affinity rapid equilibrium binding to GDP-subunits in microtubules, the Pi analog BeF_3^- binds with a much higher affinity and is in slow association-dissociation with microtubules[16] ; both features are characteristic of analogs of the transition state in enzymatic catalysis[17]. These observations therefore suggest that the T-GDP-BeF^-_{3MT} complex may mimic the conformation of the T-GDP-P^*_{MT} transition state which would be energetically closer to the T-GTP_{MT} state than to the T-GDP-Pi state, assuming the following sequence of elementary reactions on the microtubule :

$$\text{T-GTP}_{MT} \rightleftharpoons \text{T-GDP-P*} \rightleftharpoons \text{T-GTP-Pi} \rightleftharpoons \text{T-GDP} + \text{Pi}$$

The same considerations apply to the case of other G-proteins on which (BeF_3^-, H_2O) and AlF_4^- have been shown to induce the effects of GTP-γS upon binding to the GDP species[18].

It is important to note that Pi and BeF_3^- or AlF_4^- bind only to the conformation of tubulin that has a GTPase activity, i.e. only to the microtubular GDP-tubulin subunit. Dimeric GDP-tubulin does not bind Pi nor its analogs[16]. In a parallel fashion, we may expect that G-proteins, which are GTPases only when interacting with the effector E, will bind BeF_3^- only in the E-G-GDP conformation, while free G-GDP should not bind BeF_3^-.

Analogy with other nucleotidases

The finding that Pi release, and not cleavage of the γ-phosphate of the nucleotide, is the regulatory switch involved in a vectorial process is not new and other examples can be found : in actomyosin ATPase, Pi release is coupled to force generation[19]. In the case of actin, which hydrolyzes ATP upon polymerization, we have shown that Pi release is also associated to the destabilization of actin-actin interactions in the filament[20-22]. We derive the same conclusion from the observed effect of BeF_3^- on the activity of recA protein, an eukaryotic protein involved in genetic recombination, which hydrolyzes ATP when bound to ssDNA : in this case the release of Pi is associated to the destabilization of the interaction of recA with ssDNA[23]. Despite the fact that all these proteins are ATPases, the coupling between ATP hydrolysis and the regulation of the interaction with an effector is exerted via the same mechanism as G-proteins.

We therefore suggest that G-proteins are part of a wider family of nucleotidases specialized in transduction of information via interaction with an effector. A different class of ATPases is represented by the ion motive ATPases (for a review see ref. 24).

REFERENCES

1. Hughes, S.M. (1983) FEBS Lett. 164, 1-8.
2. Bourne, H.R. (1988) Cell 53, 669-671
3. Carlier, M.F. and Pantaloni, D. (1978) Biochemistry 17, 1908-1915.
4. Dever, T.E., Glynias, M.J. and Merrick, W.C. (1987) Proc. Natl. Acad. Sci. USA 84, 1814-1818
5. Sternlicht, H., Yaffe, M.B. and Farr, G.W. (1987) FEBS Lett. 214, 226-235.
6. McCormick, F., Clark, B.F.C., La Cour, T.F.M., Kjeldgaard, M., Norskov-Lauritsen, L. and Nyborg, J. (1985) Science 230, 78-82.
7. Kim, H., Ponstingl, H. and Haley, B.E. (1987) Fed. Proc. 46, 2229.
8. Maccioni, R.B. and Seeds, N.W. (1983) Biochemistry 22, 1572-1579.
9. Ponstingl, (1987) J. Biol. Chem.
10. Carlier, M.F., Hill, T.L. and Chen, Y. (1984) Proc. Nat. Acad. Sci. USA 81, 771-775.
11. Mitchison, T. and Kirschner, M.W. (1984) Nature (London) 312, 237-242.
12. Hill, T.L. and Chen, Y. (1984) Proc. Nat. Acad. Sci. USA 81, 5772-5776.
13. Chen, Y. and Hill, T.L. (1985) Proc. Natl. Acad. Sci. USA 82, 4127-4131.
14. Carlier, M.F., Didry, D., Melki, R. and Pantaloni, D. (1988) Biochemistry 27, 3555-3558.

15. Bigay, J., Deterre, P., Pfister, C. and Chabre, M. (1985) FEBS Lett. 191, 181-185.
16. Carlier, M.F., Didry, D., Simon, C. and Pantaloni, D., submitted.
17. Lienhard, G.E., Secemski, I.I., Koehler, K.A. and Lindquist, R.N. (1971) Cold Spring Harbor Symp. Quant. Biol. 36, 45-51.
18. Bigay, J., Deterre, P., Pfister, C. and Chabre, M. (1987) EMBO J. 6, 2907-2913.
19. Hibberd, M.G., Danzig, J.A., Trentham, D.R. and Goldman, Y.E. (1985) Science 228, 1317-1319.
20. Korn, E.D., Carlier, M.F. and Pantaloni, D. (1987) Science 238, 638-644.
21. Carlier, M.F. and Pantaloni, D. (1988) J. Biol. Chem. 263, 817-825.
22. Combeau, C. and Carlier, M.F. (1988) J. Biol. Chem., in press.
23. Moreau, P. and Carlier, M.F. submitted.
24. Pedersen, P.L. and Carafoli, E. (1987) TIBS 12, 146-150 ; 186-189.

TUBULIN STRUCTURE AND NUCLEOTIDE BINDING

Eva-Maria Mandelkow, Klaus Linse and Eckhard Mandelkow
Max-Planck-Unit for Structural Molecular Biology
c/o DESY, Notkestraße 85, D-2000 Hamburg 52, F.R.G.

ABSTRACT

In this report we summarize our recent studies to identify the GTP binding site of tubulin, the subunit of microtubules, using the method of direct photoaffinity labeling. We compare the results with predictions derived from a comparison of tubulin with other nucleotide binding proteins. The binding site of the base (peptide 63-77) is consistent with that of other GTP binding proteins; however, its position relative to other presumptive sites (e.g. the phosphate binding loop) differs from the consensus pattern of G-proteins. This means that tubulin represents a distinct class of nucleotide binding proteins.

INTRODUCTION

Microtubules are ubiquitous cytoskeletal fibers of eukaryotic cells. They are responsible for cell shape, transport processes (e.g. axonal transport), mitosis, motility (e.g. flagellar beating), and others. Their core consists of the subunit protein, tubulin, to which a variety of microtubule-associated proteins (MAPs) are attached. In terms of microtubule assembly the functional subunit is a heterodimer of α- and β-tubulin which contains two nucleotide binding sites. One of them binds GTP non-exchangeably (GTP_n site); the other binds GTP or GDP exchangeably (GTP_e site). GTP binding at this site activates tubulin for assembly, and during this process GTP is hydrolyzed to GDP. This site is therefore responsible for the microtubule dynamics of living cells. Because of this many authors have investigated the GTP binding properties of tubulin, often with conflicting results. One problem is that tubulin has not yet been crystallized so that the GTP binding cannot be related to a three-dimensional structure of the molecule.

Our approach to the problem was triggered by three developments. One was the growing body of sequence data on nucleotide binding proteins which suggested that new information might be derived from homologies with known proteins (Wierenga & Hol, 1983; Leberman & Egner, 1984; Sternberg & Taylor, 1984; Halliday, 1984; Mandelkow et al., 1985).

Secondly, the three-dimensional structures of some GTP binding proteins became available (La Cour et al., 1985; Jurnak, 1985; De Vos et al., 1988). Finally, the method of direct photoaffinity labeling (Maruta & Korn 1981) allowed one to circumvent the ambiguities associated with the use of derivatized nucleotide analogs.

When comparing α- and β-tubulin to other nucleotide-binding proteins, one can detect four regions of homology which are termed I, II, III, and IV in Fig. 1 (cf. Mandelkow et al., 1985). Region I is a glycine cluster that could be involved in a loop connecting a β-sheet and an α-helix near the phosphate-binding site; region II may be involved in the binding of the ribose, and region III is thought to be near the base of the nucleotide. These predictions are based on homologies with ATP and dinucleotide binding proteins (e.g. dehydrogenases, $pp60^{src}$; Sternberg & Taylor, 1984). The region IV is homologous to GTP binding proteins (e.g. EF-Tu or $p21^{ras}$; Leberman & Egner, 1984); it is thought to be involved in the binding of the base. Thus, tubulin shows characteristics of two different classes of nucleotide binding proteins, resulting in seemingly conflicting predictions. This prompted our attempt to distinguish between the possibilities, as summarized below (see Linse & Mandelkow, 1988).

MATERIALS AND METHODS

Most procedures are described in detail elsewhere (Linse & Mandelkow, 1988; Linse, 1988). Briefly, microtubule protein from porcine brain and phosphocellulose-purified tubulin were prepared as described (Mandelkow et al., 1985), using a reassembly buffer of 0.1 M PIPES pH 6.9, with 1 mM each of $MgSO_4$, EGTA, GTP, and DTT. Direct photoaffinity labelling of tubulin by [α-^{32}P]GTP (Amersham, 1 mCi/ml, 3000 Ci/mmol) was done at protein concentrations of 50-100 μM and concentrations of [α-^{32}P]GTP around 0.1 μM, supplemented by equimolar concentrations of cold GTP. After pre-incubation for 5 min at 4°C the solution was irradiated by UV light of 254 nm. Limited proteolysis of tubulin was done using α-chymotrypsin (Sigma), Staphylococcus aureus V8 protease (Miles), subtilisin type VIII (Boehringer Mannheim), and clostripain (Sigma). The GTP-binding peptide was isolated by the following steps: (1) limited digestion of β-tubulin into its two major domains; (2) separation of the fragments by FPLC using a Mono Q column (Pharmacia); (3) total tryptic digestion of the radioactive fragment, (4) separation of the tryptic peptides by reversed phase HPLC (Beckman). The radioactive tryptic peptide was then sequenced in a gas phase sequencer by Dr. Wittmann-Liebold, MPI Berlin.

RESULTS

The analysis proceeded in several successive steps as follows:

(1) Fig. 2 shows a comparison of an SDS gel and an autoradiogram of tubulin and its chymotryptic fragments.

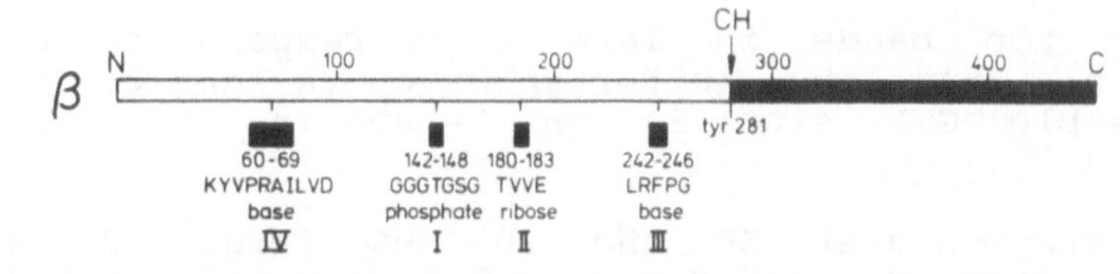

Fig. 1. Diagram of β-tubulin, its domain structure, and homologies with nucleotide binding proteins. The protein contains 445 amino acid residues (Krauhs et al., 1981). CH = site of limited proteolysis by chymotrypsin at Tyr 281 (Mandelkow et al., 1985). The regions labelled I, II, and III correspond to the regions of homology with ATP- and dinucleotide binding proteins (Sternberg & Taylor, 1984); they would be predicted sites of phosphate, ribose, and base binding, respectively. Region IV is homologous to a region in elongation factor EF-Tu (Leberman & Egner, 1984) that is involved in base binding. The numbering follows that of Krauhs et al. (1981) after alignment with α-tubulin.

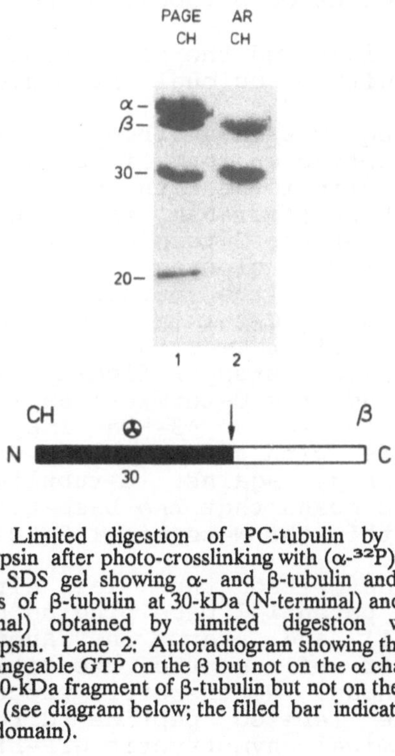

Fig. 2. Limited digestion of PC-tubulin by chymotrypsin after photo-crosslinking with (α-^{32}P)GTP. Lane 1: SDS gel showing α- and β-tubulin and two fragments of β-tubulin at 30-kDa (N-terminal) and 20-kDa (C-terminal) obtained by limited digestion with chymotrypsin. Lane 2: Autoradiogram showing the label of the exchangeable GTP on the β but not on the α chain, and on the 30-kDa fragment of β-tubulin but not on the 20 kDa fragment (see diagram below; the filled bar indicates the labelled domain).

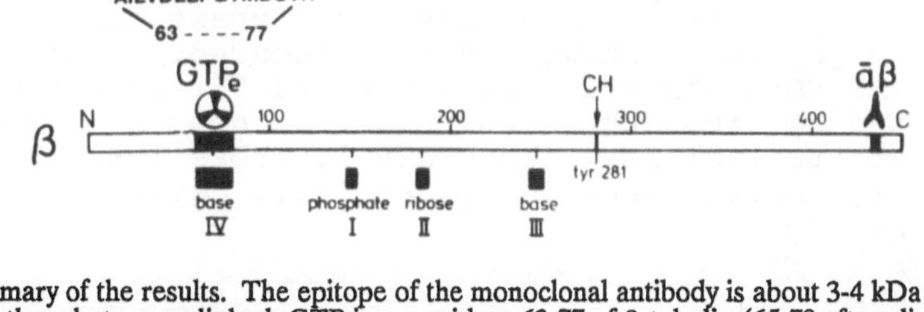

Fig. 3. Summary of the results. The epitope of the monoclonal antibody is about 3-4 kDa from the C-terminus, the photo-crosslinked GTP is on residues 63-77 of β-tubulin (65-79 after alignment with α-tubulin). This is consistent with region IV (Fig. 1, homology with other GTP-binding proteins). All Figures reprinted from Linse & Mandelkow, 1988.

The two top bands in lane 1 correspond to α- and β-tubulin; only the latter shows up in lane 2. Thus the exchangeable GTP site is on β-tubulin, but not on α-tubulin.

(2) The fragments at 30 and 20 kDa (Fig. 2, lane 1) represent the N- and C-terminal domains of β-tubulin, as shown previously. Comparison with lane 2 shows that the GTP_e site is on the N-terminal domain of β-tubulin, but not on the C-terminal domain. This is consistent with the predictions since all homologies are found in the N-terminal domain (Fig. 1).

(3) The GTP_e is within 18 kDa of the N-terminus since it binds to an 18 kDa fragment obtained from V8 protease that contains the N-terminus of β-tubulin, (not shown).

(4) The next steps involved the identification of peptides by end-labeling with monoclonal antibodies (see Matsudaira et al., 1985).
(4.1) The method requires an antibody label for one of the ends of the polypeptide chain. Since the region of highest antigenicity of tubulin is near the C-terminus we chose a β-monoclonal antibody (Amersham) whose epitope was located between 3 and 5 kDa of the C-terminus by a combination of proteases with different cleavage sites near the C-terminus (data not shown). For the purpose of this study the antibody served as a C-terminal end-label to identify tubulin peptides (as diagrammed in Fig. 3).
(4.2) The next step consisted of finding large proteolytic fragments that bind the C-terminal antibody label but not GTP. One such peptide is a 43-kDa fragment produced by limited proteolysis with clostripain. This fragment binds the monoclonal antibody against β-tubulin, but not the labeled GTP. This means that the base-binding site must be located roughly within the N-terminal 8 kDa.

(5) The results described thus far already distinguish between the two potential base binding sites (regions III and IV in Fig. 1) and argue in favor of region IV. However, a further refinement of the site was not possible with the above methods. We therefore proceeded to isolate and identify the labeled peptide directly. We first isolated the N-terminal chymotryptic GTP-binding fragment of β-tubulin and used this for the identification of a tryptic peptide carrying the nucleotide. The chymotryptic fragment was obtained by FPLC using an ion exchange column and subjected to total tryptic digestion. The peptides were separated by HPLC on a reversed phase column. The radioactive peptide peak was then sequenced, yielding the sequence AILVDLEPGTMDSVR (residues 63-77 of β-tubulin). The GTP-binding residue itself could not be identified because of its lability. Thus GTP is crosslinked in region IV (Fig. 1) where tubulin shows homologies with other GTP-binding proteins.

DISCUSSION

The binding site of the exchangeable GTP on tubulin has been investigated by a number of authors, resulting in several different proposals (for example, the GTP$_e$ site was reported to be on α-tubulin, β-tubulin, or both; the label was found near the N-terminus, near the C-terminus, or in between). This prompted us to re-investigate the matter using the method of direct photoaffinity labeling with GTP which is less susceptible to artefacts than photosensitive GTP analogues (Maruta & Korn, 1981). Using seval methods to narrow down the GTP$_e$ site we find (a) that it is on β-tubulin (in agreement with Geahlen & Haley, 1977; Hesse et al., 1985), (b) on the N-terminal domain defined by chymotryptic cleavage (Nath & Himes, 1986; Linse & Mandelkow, 1986), (c) within the N-terminal 80 residues (defined by clostripain cleavage and antibody end labeling), (d) finally, on the tryptic peptide 63-77 (Linse & Mandelkow, 1988), consistent with the homology region IV of Fig. 1.

This can be compared with the two classes of nucleotide-binding proteins that contain homologies with tubulin (Mandelkow et al., 1985). The first class contains several dehydrogenases and other nucleotide- and dinucleotide-binding proteins, including proteins related to pp60^{src} (Wierenga & Hol, 1983; Sternberg and Taylor, 1984). The second class contains GTP binding proteins such as elongation factors, G-proteins and proteins related to p21^{ras} (Leberman and Egner 1984; Halliday, 1984). The experiments support the homologies with GTP-binding proteins.

Since tubulin binds GTP and not ATP this result might have been expected. However, several problems remain which make a direct comparison of tubulin with GTP-binding proteins of known structure difficult. Direct photolabeling is thought to link the base to the protein (cf. Carroll et al., 1985). Consistent with this, the labeled peptide is in a region homologous with the base binding site of GTP binding proteins. However, these proteins contain a consensus sequence NKXD (Dever et al., 1987) which is absent from tubulin. Judging from the surrounding residues the corresponding stretch of tubulin would be DLEP. This tetrapeptide would have the necessary hydrogen bonding capacity to interact with the base (as is postulated for NKXD); but in any case the lack of correspondence makes tubulin distinct from the other known GTP binding proteins.

Another problem concerns the putative phosphate binding region. There is no direct evidence on this, but there is a glycine-rich region (residues 142-148 of β-tubulin, region I of Fig. 1) whose homology to many nucleotide binding proteins is suggestive (Krauhs et al., 1981). If this is indeed the phosphate binding loop then it would be downstream from the base binding region. This would be in contrast to the other GTP or ATP binding proteins: Their homologies are in the order phosphate-ribose-base binding region (from N to C terminus, see Sternberg & Taylor, 1984; Dever et al., 1987; this sequence is obeyed by regions I-III of Fig. 1 predicted from ATP binding proteins). The discrepancey could be interpreted in two ways: (1) The glycine-rich region is not involved in phosphate binding. This would mean that the position of the phosphate loop is not known; it may actually be upstream from the base binding region but in this case it would not conform to the consensus sequence (since there is no other glycine cluster). (2) Alternatively, residues 142-148 in fact constitute the phosphate binding loop; in this case the folding of the nucleotide binding domain of tubulin is topologically different from that of other proteins. Whatever the choice, it means that tubulin cannot be related directly to other nucleotide binding proteins.

Finally, we note that comparisons of protein sequence are generally based on certain accepted procedures which include, among others, that sequences are read with defined orientations (i.e. from N- to C-terminus). Using these criteria, Dever et al. (1987) concluded that tubulin differed from other GTP-binding proteins since certain consensus elements were lacking (e.g. NKXD, DXXG), as discussed above. By contrast, Sternlicht et al. (1987) proposed that the consensus sequences of Dever et al. actually occurred in tubulin, but this was achieved only by reading the sequence both forwards and backwards and by combining sequence stretches with different orientations. Whether this approach is adequate remains to be shown.

Acknowledgement: We are grateful to Prof. B. Wittmann-Liebold (MPI Berlin) for help with the sequencing work.

REFERENCES

Carroll,S., McCloskey,J., Crain,P., Oppenheimer,N., Marschner,T. and Collier,R.(1985). Proc. Natl. Acad. Sci. U.S.A. 82, 7237-7241.
Dever, T.E., Glynias, M.J. and Merrick, W.C. (1987). Proc. Natl. Acad. Sci. U.S.A. 84, 1814-1818.
De Vos, A., Tong, L., Milburn, M., Matias, P., Jancarik, J., Noguchi, S., Nishimura, S., Miura, K., Ohtsuka, E. and Kim, S.-H. (1988). Science 239, 888-893.
Geahlen, R.L. and Haley, B.E. (1977). Proc. Natl. Acad. Sci. U.S.A. 74, 4375-4377.
Halliday,K.(1984). J. Cyc. Nucl. and Prot. Phosphor. Res. 9, 435-448.
Jurnak, F. (1985). Science 230, 32-36.
Krauhs,E., Little,M., Kempf,T., Hofer-Warbinek,R., Ade,W. and Ponstingl,H.(1981). Proc. Natl. Acad. Sci. U.S.A. 78, 4156-4160.
La Cour, T.F.M., Nyborg, J., Thirup, S. and Clark, B.F.C. (1985). EMBO J. 4, 2385-2388.
Leberman, R. and Egner, U. (1984). EMBO J. 3, 339-341.
Linse, K. and Mandelkow, E.-M. (1986). J. Cell Biol. 103, 545a.
Linse, K. and Mandelkow, E.-M. (1988). J. Biol. Chem. 263.
Linse, K. (1988). Ph.D. Thesis, University of Hamburg.
Mandelkow, E.-M., Herrmann, M. and Rühl, U. (1985). J. Mol. Biol. 185, 311-327.
Maruta, H. and Korn, E.D. (1981). J. Biol. Chem. 256, 499-502.
Matsudaira, P., Jakes, R., Cameron, L. and Atherton, E. (1985). Proc. Natl. Acad. Sci. U.S.A. 82, 6788-6792.
Nath, J.P. and Himes, R. H. (1986). Biochem. Biophys. Res. Comm. 135, 1135-1143.
Sternberg,M. and Taylor,W. (1984). FEBS Letters 175, 387-392.
Sternlicht,H., Yaffe,M. and Farr, G.(1987). FEBS Letters 214, 226-235.
Wierenga, R.K. and Hol, W.G.J. (1983). Nature 302, 842-844.

TUBULIN AS A G PROTEIN?

Mark M. Rasenick, Yan Kun and Nan Wang

Dept. of Physiology and Biophysics and the Committee on Neuroscience Univ. of Illinois College of Medicine Chicago, IL 60680 USA

GTP binding proteins in the synaptic membrane appear to mediate several receptor induced processes. These processes and the G-proteins mediating them include the stimulation [Gs] and inhibition [Gi] of adenylate cyclase, the gating of K^+ [Gi] and Ca^{++} channels and the activation of phospholipase c [G?]. Several other processes might be G protein mediated as well, but the identity of the specific G proteins involved has not been determined. For many of these G proteins, the amino acid sequence and the nature of gene coding for that protein have been established. Furthermore, the tertiary structure of some is known or surmised and specific molecular domains have been suggested to participate in various aspects of G protein function.

Activation of G proteins

A myriad of studies using purified G proteins (either in detergent solutions or in phospholipid vesicles) have led to the formulation of hypotheses for the mechanism of G protein activation and inactivation as well as the mechanism whereby G proteins activate or inhibit effector systems (see Neer and Clapham, 1988 for review). In broad terms, the $G\alpha$ protein (i.e. the specific protein possessing an exchangeable GTP binding site) exists in the inactive form with GDP bound and in complex with one or more other proteins. When a signal (e.g. occupation of a receptor by agonist; activation of rhodopsin by light; or the binding of aminoacyl tRNA) occurs, GTP binds to the protein and the active, GTP bound subunit dissociates from the complex, interacting with an effector system. [see figure 1 for a description of receptor-effector coupling in the adenylate cyclase system].

The G protein environment

Despite the obvious sophistication of the afore mentioned reconstitution studies, correlation of their conclusions with hormone or neurotransmitter responsiveness is not always possible. One reason for this is that, whereas soluble G proteins have been observed

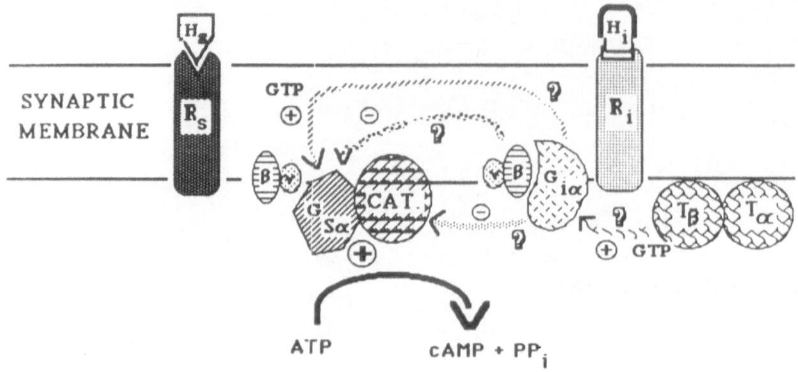

Figure 1. Regulation of adenylate cyclase.

This cartoon represents the identified components of the adenylate cyclase system and possible schemes for the activation and inhibition of that enzyme. Subsequent to the binding of an agonist (Hs or Hi) to a receptor effecting the stimulation (Rs) or inhibition (Ri) of adenylate cyclase, GTP is bound to the α subunit of a G protein associated with that receptor (Gs or Gi respectively). Gs or Gi subsequently dissociate from Rs or Ri (decreasing the affinity of agonist binding) and β and γ subunits dissociate from the α subunit. The free Gs α activates the catalytic moiety (CAT) of the enzyme and the β and γ subunits liberated from Giα may associate with Gsα inhibiting the stimulation of adenylate cyclase. It is likely that Giα inhibits the catalytic moiety directly as well. We have observed transfer of hydroslysis-resistant GTP analogs from Giα (or the closely related Goα) to Gsα. We have also observed the transfer of these analogs from tubulin to Giα or Goα (Rasenick and Wang, 1988). Tubulin is an α-β heterodimer (although the monomers have significant sequence homology) and both monomers bind GTP. The exchangeable binding site is on β rather than α tubulin. Note that although experimental evidence is consistent with the above schemes, the actual mechanism of adenylate cyclase regulation awaits further clarification.

(Sayhoun et al., 1981; Krishna-Bhat et al., 1980; Rasenick et al., 1984), there is little evidence that they participate in signal transduction when dissociated from the membrane. Further, the plasma membrane in which receptors, G proteins and effectors reside is comprised of a variety of lipids and proteins and it is likely that many of these membrane elements interact with components of the cytoskeleton.

Considerable evidence exists that alteration of membrane lipids or disruption of microtubules will modify the relationship between receptors and G proteins in the adenylate cyclase system (see reviews by Hanski et al., 1979; Zor, 1983 and Rasenick et al., 1985). This relationship between receptors, G proteins and effector molecules (known as coupling) is modified in neuronal systems by the disruption of cells and this modification is not due to the loss of cellular components. Rather, it appears that the biological coupling of receptors and effectors to G proteins requires the ordered juxtaposition of those components, and it is the membrane and cytoskeleton which provide such order.

This review will attempt to provide a mechanistic explanation for the regulation of G protein function by cytoskeletal components. As the adenylate cyclase and rod outer segment systems have provided paradigms for G protein mediated signal transduction systems, they will also serve as the framework for this chapter.

Cytoskeletal regulation of adenylate cyclase

Numerous studies (see Zor, 1983; and Rasenick et al., 1985 for reviews) have demonstrated the existence of some interaction between cytoskeletal elements and the adenylate cyclase system. Data from most of those studies pointed to microtubules and G proteins as the locii for cytoskeletal control. In leukocytes, perturbation of microtubules with colchicine or vinblastine caused an increase in adenylate cyclase activity, but membranes from those cells did not respond to those agents (Kennedy and Insel, 1979; Rudolph et al., 1979). Membranes from various regions of the rat brain showed a potent augmentation of adenylate cyclase after treatment with microtubule disrupting agents (Rasenick et al., 1981) however, microscopic examination of those membranes revealed no formed microtubules. This observation led to the suggestion that tubulin, (perhaps membrane tubulin) rather than microtubules, was interacting with G proteins to modify adenylate cyclase activity.

Similarities between tubulin and other G proteins

Several similarities between tubulin and "traditional" G proteins exist. The most obvious is that all of these proteins bind and hydrolyze GTP. As in the case with G proteins, tubulin is a substrate for ADP-ribosylation by cholera and pertussis toxins (Amir Zaltsman et al., 1982; Lim et al., 1985). Further, both tubulin and the traditional G proteins form complexes with other proteins and the trigger for the formation and/or dissociation of the complex involves the exchange of GDP for GTP. The active conformation in the case of the initiation of microtubule polymerization or in the continuation of the signal transduction cascade is the GTP bound form (although the GDP-Pi bound form may also be active).

Nucleotide binding proteins have been examined with respect to amino acid sequence homologies, and two major groups have been sorted out. One group contains the G proteins involved in signal transduction as well as elongation factors and oncogene products such as $p21^{ras}$ (Leberman and Egner, 1984; Halliday, 1984). The second group contains several dehydrogenases and other nucleotide (often ATP) binding proteins (Sternberg and Taylor, 1984). Tubulin appears to share features with both groups (Sternlicht et al., 1987; Linse and Mandlekow, in press). These homologous regions are illustrated in figure 2. Both α and β tubulin are highly conserved in these regions and the regions are nearly identical between α and β tubulin. The base binding region (region A) homologous with other G proteins appears to be correct not only from sequence but from direct photoaffinity studies (Linse and Mandlekow, in press). Those studies gave

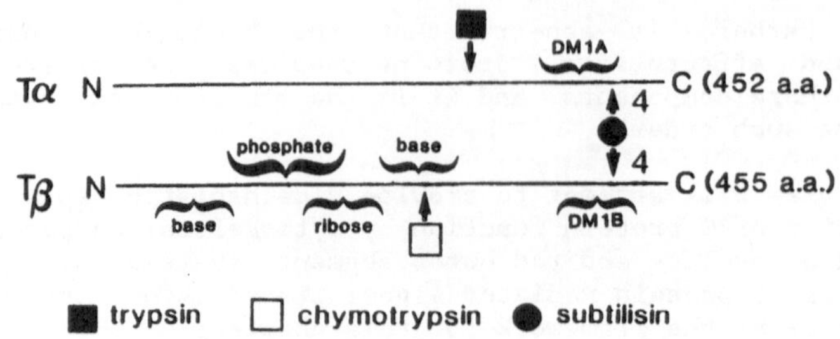

DM1A (415–443) EGEFSEREDMAALEKDYEEVGVDSVEGE

DM1B (420–437) NMNDLVSEYQQYQDATAD

Figure 2. Schematic model of tubulin.

Proteolytic cleavage sites on tubulin are illustrated along with the regions thought to be involved with exchangeable GTP binding on β tubulin. The base binding region nearest the amino terminal bears greatest homology to G proteins (Linse and Mandlekow, in press). Monoclonal antibody DMIA and DMIB binding sites are indicated on α and β tubulin and the sequences of regions likely to include the α and β specific epitopes are noted (de la Vina et al., 1988).

further credence to β being the site of exchangeable nucleotide binding, as discerned from photolabelling studies with azido GTP analogs (Geahlen and Haley, 1980; Rasenick and Wang, 1988).

Despite the similarities listed above, it is clear that several distinctions between tubulin and "traditional" G proteins exist. The four G protein consensus sequences described by Halliday (1984) can be observed in tubulin only when certain domains are inverted (Sternlicht et al., 1988; Linse and Mandlekow, in press.) The spacing between these domains does not correspond with the intra-domain spacing noted in other G proteins. Although α–β tubulin heterodimers complex with other heterodimers, α and β tubulin are bound tightly to one another and disociated only by denaturation. The thesis of this chapter concerns the possibility that tubulin acts not as a surrogate G protein but rather as a protein capable of complexing (and exchanging GTP with) Gs and/or Gi; thus modifying neuronal adenylate cyclase or other G protein mediated signal transduction systems.

Nucleotide exchange among G proteins

The direct interaction between α subunits of signal transducing G proteins has been proposed by this (Hatta et al., 1986; Rasenick et al., 1987) and other laboratories (Marbach et al., 1988; Ho and Hingorani, this volume). This laboratory has utilized the hydrolysis-resistant photoaffinity GTP analog, P^3-(4-azidoanilido)-P^1 5′ GTP [AAGTP] (see figure 3) to probe the behavior of synaptic membrane G proteins without removing those proteins from their natural milieu.

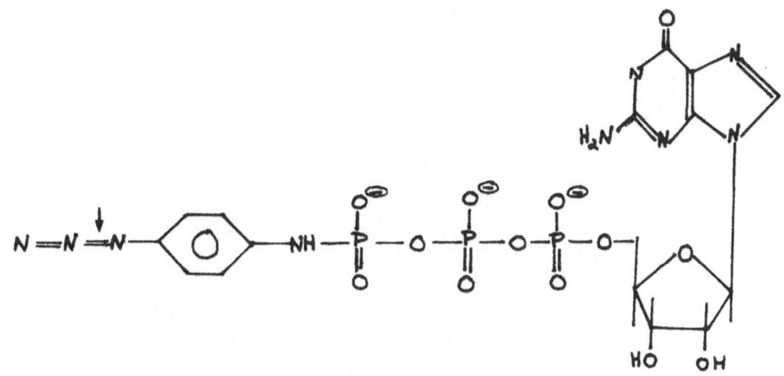

P3-(4-AZIDOANILIDO)-P1-5'GTP

Figure 3. P^3-(4-azidoanilido)-P^1-5'GTP

Nitrene free radicals are formed at the site indicated by the arrow subsequent to UV irradiation of this compound. Most experiments employ the α[^{32}P] version of this compound although [^{3}H] AAGTP has been used as well. AAGTP is resistant to hydrolysis and does not appear to be a substrate for protein phosphorylation.

AAGTP, which binds to synaptic membrane Gi/o with a Kd of 2.12 μM and Gs with a Kd of 4.87 μM (Gordon and Rasenick, 1988) is a potent and stable stimulator or inhibitor of adenylate cyclase. When synaptic membranes are incubated in a buffer containing 1 mM MgCl and 10^{-4}M AAGTP at 23°C for 3 minutes and subsequently washed, adenylate cyclase is inhibited to about 50% of the control level. Subsequent incubation of membranes with GppNHp or any other hydrolysis-resistant GTP analog) reverses adenylate cyclase inhibition. When [^{32}P] AAGTP is used in parallel experiments, and SDS PAGE and radioautography are performed, Gi/o binds 5-8 fold more AAGTP than Gs. If membranes are incubated with GppNHp (or other GTP analogs) prior to photolysis, a decrease in AAGTP bound to Gi/o and a parallel increase in AAGTP bound to Gs are seen (figure 4). In each case, the total AAGTP bound to Gi/o plus Gs are constant (Hatta et al., 1986). Curiously, the increasing concentrations of GTP analog added to the membranes causes an apparent increase in the binding capacity of Gs and a concomitant decrease in that of Gi/o. This is revealed in experiments where [^{3}H] AAGTP is used in the second incubation. In such experiments (Hatta and Rasenick, unpublished observations), the distribution of [^{3}H] and [^{32}P] AAGTP is comparable. Although such a result would be consistent with positive cooperativity in the binding of AAGTP by Gs, the Hill coefficients calculated for each of the G proteins are not significantly different from 1 (Gordon and Rasenick, 1988).

Complexes of G protein α subunits

Thus, a complex series of interactions among the α subunits of the synaptic membrane G proteins appears to exist. We have proposed that one of these interactions involves a direct transfer of nucleotide from Gi to Gs. Such a transfer is postulated because the shift of [^{32}P] AAGTP occurs after the membranes have been washed (the free [AAGTP] is < 10^{-13}M) and the K_{off} for AAGTP from Gi/o has a $t_{1/2}$ of

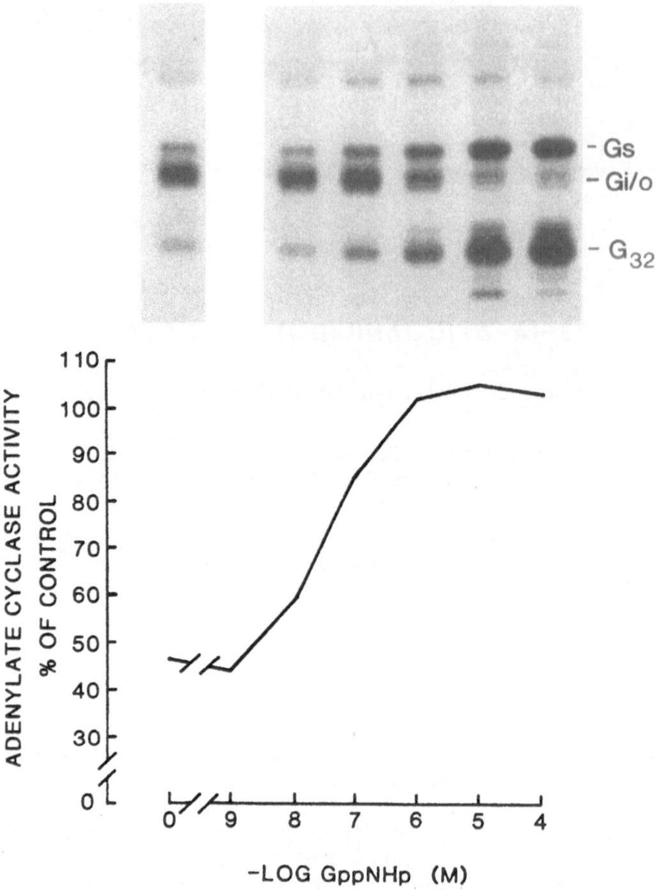

Figure 4. Correlation of nucleotide transfer from Gi/o to Gs with the GppNHp-induced override of adenylate cyclase inhibition.

(Upper) SDS/PAGE analysis of AAGTP-labeled synaptic membranes.

Synaptic membranes were incubated with 1.2×10^{-7} M [^{32}P]AAGTP and washed as described below. After the 2nd incubation with GppNHp, reactions were subjected to 5 min of UV photolysis on ice. The membranes were then washed and submitted to SDS/PAGE and autoradiography. The sum of [^{32}P] cpm in the 40 and 42 kDa bands was constant (2047 ± 192 in the illustrated experiment).

(Lower) Adenylate cyclase activity of synaptic membranes incubated with AAGTP.

Synaptic membranes prepared from rat cerebral cortex were incubated with AAGTP (1.2×10^{-4}M) for 3 min at 23°C, subsequently washed and assayed for adenylate cyclase activity with the indicated concentration of GppNHp. Adenylate cyclase activity is expressed as a percentage of the control activity of the membranes which were not incubated with AAGTP. Values are means of 3 experiments. Basal adenylate cyclase activity in control membranes was 50.4 pmol/mg protein/min. Persistent inhibition of adenylate cyclase was dose-dependent between 10^{-8}-10^{-4}M.

about 25 minutes. As the time required for equilibration of the nucleotide exchange process is 5 minutes, release and rebinding of nucleotide are unlikely. Further, the spatial juxtaposition of G proteins referred to above might allow a direct transfer of nucleotides. It is noteworthy that, other nucleotide binding proteins (the mitochondrial hydrolases) which are present in high local concentration (and perhaps an ordered configuration) in the mitochondrial have been observed to transfer NAD directly from one protein to the next (Srivastava and Bernhard, 1987). If G protein α subunits behave in a similar manner, it is likely that these proteins exist as a complex on the synaptic membrane.

Tubulin-G Protein nucleotide exchange

Recently, we have observed that tubulin, polymerized with hydrolysis-resistant GTP analogs promotes inhibition of adenylate cyclase which persists subsequent to washing. This inhibition is similar to that caused by hydrolysis-resistant GTP analogs alone, and tubulin stripped of nucleotide by pretreatment with charcoal has no effect on adenylate cyclase. If tubulin is polymerized with [^{32}P] AAGTP and added to synaptic membranes (in experiments parallel to those described with AAGTP alone), washing of the membranes followed by UV photolysis and SDS PAGE reveals that AAGTP binds primarily to Gi. To distinguish between the possibilities that nucleotide (AAGTP or GppNHp) is released from tubulin and bound by Gi, it was demonstrate that after binding .8 mol of AAGTP/mol tubulin that about half of the nucleotide dissociates from tubulin after three cycles of desalting P6-dg chromatography. The remaining AAGTP (.42-.46 mol/mol tubulin) remains bound for greater than 100 minutes under the experimental conditions. Although in the presence of synaptic membranes, nucleotide is released from tubulin and bound by Gi, saturation of the synaptic membrane G proteins with GppNHp prevents the release of nucleotide from tubulin. Such a result is consistent with the hypothesis that tubulin transfers nucleotide directly to synaptic membrane G proteins (figure 5). Further, although the binding of AAGTP to G proteins occurs rapidly at ice temperature, the transfer of nucleotide between Gi and Gs or between tubulin and Gi shows a t 1/2 of 90 seconds at 23°C (Rasenick and Wang, 1988).

Although incubation of tubulin-GppNHp with cerebral cortex membranes causes stable inhibition of adenylate cyclase, similar experiments performed with C6 glioma membranes result in a stable stimulation of that enzyme (Kun and Rasenick, unpublished observations). The simplest explanation for these apparently discrepant data is the lack of Gi_1 in C6. Gi_1 is the predominant in Gi brain and occurs in significant excess over Gs in that tissue. Tubulin has no effect upon the adenylate cyclase catalytic moiety isolated from either brain or C6 membranes, requiring the presence of a "genuine" G protein in order to influence adenylate cyclase.

Tubulin-G protein complexes

The formation of complexes among GTP-binding subunits is well established for tubulin. The suggestion that G proteins can form complexes with tubulin has been made as well (Higyashi and Ishibashi, 1987; Rasenick et al., 1985; Sayhoun et al., 1981, 1988). Further, microtubule preparations polymerize in association with components of the adenylate cyclase system (Margolis and Wilson, 1979). Preliminary data (Wang and Rasenick unpublished) indicate that ^{125}I tubulin binds

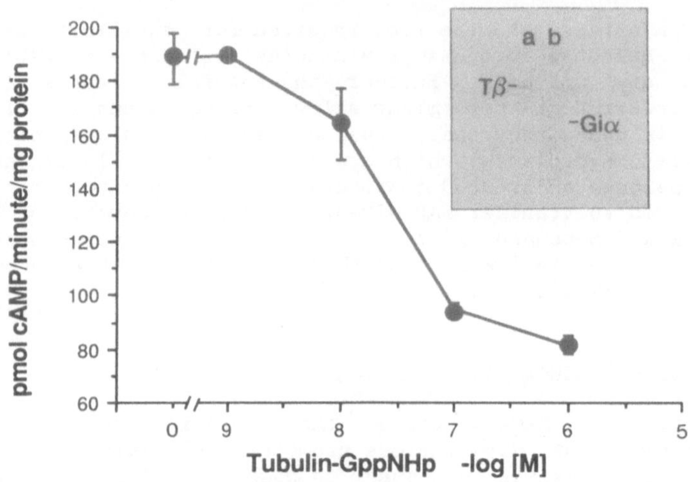

Figure 5 Incubation of tubulin-GppNHp (or tubulin-AAGTP) with synaptic membranes results in a dose-dependent inhibition of adenylate cyclase which persists after membranes are washed (similar to that observed with GppNHp - see fig. 4). Less than 5% of the added tubulin-AAGTP remains bound to the membranes under these conditions, however. (see Rasenick and Wang, 1988, for experimental details)

The inset shows the autoradiographs of SDS-PAGE patterns of tubulin-AAGTP and synaptic membrane 1.5 x 10^{-7}M tubulin-AAGTP. UV photolysis of membranes was carried out subsequent to washing of the membrane. AAGTP, initially bound to tubulin is transferred to Gi. If tubulin-AAGTP is UV irradiated prior to incubation with membranes, no nucleotide is transferred to Gi. If membrane G proteins are saturated with a GTP analog and washed prior to incubation with tubulin-AAGTP, no AAGTP is transferred to Gi (the AAGTP remains bound to tubulin).

to purified Gs or Gi. The specific Gi subtype has not yet been established. Go and transducin prepared from bovine ROS bind tubulin with an affinity about 10^3x lower than Gs or Gi. Other data indicate that monoclonal antibodies with the epitopes noted in figure 2 block the transfer of binding of tubulin to that G protein (Wang and Rasenick, unpublished). Thus, it is likely that a multifaceted interaction between tubulin and G proteins is required for the transfer of GTP between those molecular species.

Considerable data are available for the rate and nature of tubulin polymerization, but the domains of the molecule involved in binding to MAPs or to other tubulin dimers are as yet unidentified. It is assumed, however, that during the assembly into a microtubule several points of contact must be made. Specifically, α subunits "lock into" some interaction with β subunits from another heterodimer. Finally, as the microtubule coils around, a vertical association among α and β heterodimers might occur (see figure 6).

Such multiple association sites on tubulin evoke the possibility that similar sites exist on Giα and Gsα, and that the Gα subunits "polymerize" either with tubulin or with one another. Although there is no evidence that large complexes of G proteins form, it has been suggested that the photoreceptor G protein (transducin) exists as a complex. Equilibrium binding studies with transducin have evoked the possibility that oligomeric assemblies of transducin molecules account for the observed cooperative binding behavior of transducin to rhodopsin (Wessling-Resnick and Johnson 1988). Further, data obtained from chemical crosslinking as well as light scattering studies indicate the existence of transducin tetramers in significant quantity (see Ho and Hingorani, this volume).

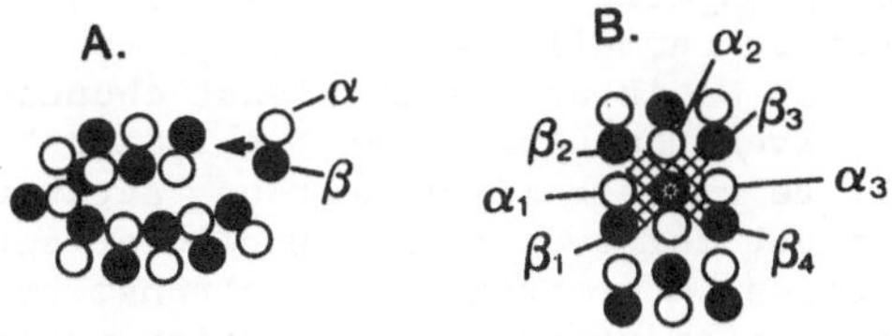

Figure 6. Dimer-Dimer interactions in the formed and forming microtubule.

A. During microtubule assembly, α–β dimers bind GTP and associate with an assembling protofilament. Thirteen dimers make up 1 "turn" of a microtubule and the dimers continue to build in a spiral fashion.

B. In a formed microtubule, a given β subunit (indicated with the asterisk) is likely to associate (bind?) to specific regions of the α tubulin on either side in the protofilament (α_1 and α_3) as well as the α tubulin in the protofilament "above" α_2). Further the noted β tubulin has specific interaction sites with the adjacent (β_1 and β_4) and above (β_2, β_3) β tubulins. Mandlekow (1985) have speculated on the existence of such sites, however, the nature of these domains and their location on the tubulin molecule is unknown. It is hypothesized that one or more of these sites participates in tubulin Gα-binding.

Tubulin and the regulation of neuronal signal transduction

Rat synaptic membranes have been shown to contain (perhaps intrinsically) tubulin dimers (Zisapel et al., 1980). Although apparently polymerization competent (Battachyrria and Wolf, 1976), no assembled microtubules appear associated with the membrane. If these dimers contain GTP rather than GDP, then tubulin complexed with G proteins might transfer GTP to those proteins. Exchange or release of GTP from dimeric tubulin has been demonstrated (Croom et al., 1985) and tubulin and G_i have comparable affinity [$K_D \simeq 2 \times 10^{-8}$ M] for GTP (Zeeburg and Caplow, 1979; Bokoch et al., 1984). As it has been observed that GTP analogs are not released from tubulin into the medium but are capable of transfering from tubulin to G_i (Rasenick and Wang, 1988), it is hypothesized that such a process takes place on the synaptic membrane. Clearly under physiologic conditions GTP rather than an hydrolysis resistant GTP-analog is bound to tubulin or an "activated" G protein. The GTP-hydrolysis time appears sufficiently slow, however, that GTP exchange could occur prior to hydrolysis of that nucleotide.

Intracellular regulation of adenylate cyclase

The occurrence of such a GTP transfer process might provide intraneuronal regulation of adenylate cyclase. Certain features of the mammalian control nervous system might be difficult to coordinate with the mechanism of adenylate cyclase regulation presented in figure 1. The overwhelming majority (quantitatively) of neurotransmitter molecules which act on mammalian brain adenylate cyclase, inhibit that enzyme. Further, Gs is one of the least abundant synaptic membrane G proteins. Given the surfeit of "inhibitors", it is hard to explain how adenylate cyclase might ever be activated. G protein complexes could account for such an intercellular activation via the exchange of nucleotide. As such a neurotransmitter and receptor coupled to Gi could have an effect without prior presence of a neurotransmitter linked to the stimulation of adenylate cyclase.

Calcium, G protein and neurotransmitter interactions

The transfer of nucleotide from tubulin to G_i (or Gs) might be regulated by the reversible association of membrane tubulin with MAP_2. The latter is a substrate for calcium-calmodulin kinase (the major postsynaptic density protein) (Kelley et al, 1984) and phosphorylated MAP_2 does not bind tubulin (Burns and Islam, 1986). Under such a scenario, transmitters which promote increases in intracellular [Ca^{++}] could alter adenylate cyclase via the mobilization of tubulin and subsequent transfer of nucleotide to G_i or G_s. It is noteworthy in this regard that phosphatidyl inositol appears to be a potent inhibitor of microtubule polymerization (Lee et al., 1986). The mechanism of this inhibition is unknown, but it is intriguing to speculate that the neurotransmitter mediated hydrolysis of phosphatidyl inositol might increase the availability of free tubulin dimers, associated with the synaptic membrane. thus increasing the likelihood of tubulin-G protein association and subsequent nucleotide exchange.

Conclusions

Clearly, much of the above discussion is highly speculative. It is also clear, however, that the regulation of G-protein mediated signal transduction is a complex process which may be even more complex in neural systems. This review poses substantially more questions than answers, but the ideas that tubulin may share functional as well as structural features with G proteins are still quite new. With the addition of further information, it is hoped that a mechanistic understanding will polymerize.

Acknowledgment

The authors thank Dr. Hiroki Ozawa for helpful discussions and Ms. Janice Gentry for preparing this manuscript. The research discussed above was supported by US Public Health Service Grants MH 39595 and MH 00699 as well as grant BNS 87-19758 from the U.S. National Science Foundation. MMR is a recipient of a Research Scientist Development Award from the U.S. National Institute of Mental Health.

REFERENCES

Amir-Zaltsman, Y., Ezra, Z., Scherson, T., Littauer, U. and Salomon, Y., 1982, ADP-ribosylation of microtubule proteins as catalyzed by cholera toxin, EMBO J., 1:181-186.

Battachyrria, B., and Woff, J., 1976, Polymerisation of membrane tubulin, Nature, 264:576-577.

Bokoch, G., Katada, T., Northup, J., Ui, M. and Gilman, A., 1984, Purification and properties of the inhibitory guanine nucleotide-binding regulatroy component of adenylate cyclase, J. Biol. Chem. 259:3560-3567.

Croom, H., Correia, J., Baty, L. and Williams, R.J., 1985, Release of exchangeably bound guanine nucleotides from tubulin in a magnesium-free buffer. Biochem., 24:768-775.

de la Vina, S., Andreu, D., Medrano, F.J., Nieto, J.M., and Andreu, J.M., 1988, Tubulin structure probed with Antibodies to Synthetic peptides. Mapping of three major types of limited proteolysis fragments, Biochemistry, 27:5352-5365.

Geahlen, R.L. and Haley, B.E., 1979, Use of GTP photoaffinity probe to resolve aspects of the mechanism of tubulin polymerization, J. Biol. Chem., 254:11982-11987.

Gordon, J.H., and Rasenick, M.M., 1988, In situ binding of aphotoaffinity GTP analog to synaptic membrane G proteins, FEBS Lett., 235:201-206.

Halliday, K., 1984, Regional homology in GTP-binding protoncogene products and elongation factors, J. Cyclic Nucleotide Protein Phosphor. Res., 9:435-466.

Hanski, E., Rimon, G., and Levisky, A., 1979, Adenylate cyclase activation by β-adrenergic receptors as a diffusion controlled process, Biochemistry, 18:846-853.

Higashi, K. and Ishibashi, S., 1985, Specific binding of tubulin to a guanine nucleotide-binding inhibitory regulatory protein in the adenylate cyclase system, N_i, Biochem. Biophys. Res. Comm., 132:193-197.

Ho, Y.K., and Hingorani, V.N., Transducin: the molecular switch in visual excitation and a model for biological coupling enzymes, in: this volume.

Kennedy, M. and Insel, P., 1979, Inhibitor of microtubule assembly enhance beta-adrenergic and prostaglandin E_1-stimulated cyclic AMP accumulation in S49 lymphoma cells, Molec. Pharmocol., 16:215-223.

Krishna-Bhat, M., Iyengar, R., Abromowitz, J., Bordelon-Riser, M.E., and Birnbaumer, L., 1980, Naturally soluble component(s) that confer(s) quanine nucleotide and fluoride and fluoride sensitivity to adenylate cyclase, Proc. Natl. Acad. Sci. USA, 77:3836-3840.

Leberman, R., and Egner, U., 1984, Sturctural differences between brain β1- and β2-tubulins: implications for microtubule assembly and colchicine binding, EMBO J., 4:51-56.

Lee, S-H., Flynn, G., Yamauchi, P.S. and Purich, D., 1986, Several metabolic factors governing the dynamics of microtubule assembly and disassembly, Ann. N.Y. Acad. Sci., 466:519-528.
Lim, L., Sekura, R. and Kaslow, H.J., 1985, Adenine nucleotides directly stimulate pertussis toxin, J. Biol. Chem., 260:2585-2588.
Linse, K., and Mandelkow, E.-M., The GTP-binding peptide of β-tubulin: localization by direct photoaffinity labelling and comparison with nucleotide binding proteins, J. Biol. Chem., in press
Mandlekow, E.M., Herrmann, M. and Ruhl, U., 1985, Tubulin domains probed by limited proteolysis and subunit-specific antibodies, J. Mol. Biol., 185:311-327.
Marbach, I., Shiloach, J., and Levitzki, A, 1988, Gi affects the agonist-binding properties of β-adrenoceptors in the presence of Gs, Eur. J. Biochem., 172:239-246.
Margolis, R., and Wilson, L., 1979, Regulation of the microtubule steady state in vitro by ATP, Cell, 18:673-679.
Neer, E.J., and Clapham, D.E., 1988, Roles of G protein subunits in transmembrane signalling, Nature, 333:129-134.
Rasenick, M.M., Stein, P.J. and Bitensky, M.W., 1981, Evidence that the regulatory subunit of adenylate cyclase interacts with cytoskeletal components, Nature, 294:560-562.
Rasenick, M.M., Wheeler, G.L., Bitensky, M.W., Kosack, C.M., Malina, R.L. and Stein, P.J., 1984, Photoaffinity identification of colchicine solubilized regulatory subunit from rat brain adenylate cyclase, J. Neurochem., 43:1447-1454.
Rasenick, M.M., O'Callahan, C.M., Moore, C.A. and Kaplan, R.S., 1985, GTP-binding proteins which regulate normal adenylate cyclase interact with microtubule proteins. In: Microtubules and Microtubule Inhibitors, pp. 313-323, Eds. DeBrabander, M. and DeMey, J. Elsevier - Amsterdam.
Rasenick, M.M., Marcus, M.M., Hatta, Y., DeLeon-Jones, F. and Hatta, S., 1987, Regulation of nueronal adenylate cyclase, In: Molecular mechanisms of neuronal responsiveness, pp. 123-133, Eds. Y. Erlich, R. Lenox, E. Kornecki, and W. Berry, Plenum, New York.
Rasenick, M.M., and Wang, N., 1988, Exchange of guanine nucleotides between tubulin and GTP-binding proteins that regulate adenylate cyclase: cytoskeletal modification of neuronal signal transduction, J. Neurochem., 51:300-311.
Rudolph, S.A., Hegstrand, L.R., Greengard, P. and Malawista, S., 1979, The interaction of colchicine with hormone-sensitive adenylate cyclase in human leukocytes, Mol. Pharmacol., 16:805-812.
Sahyoun, N., LeVine III, H., Davis, J., Hebdon, G. and Catrecasas, P., 1981, Molecular complexes involved in the regulation of adneylate cyclase, Proc. Natl. Acad. Sci., 78:6158-6162.
Srivastava, D. and Bernhard, S., 1986, Metabolite transfer via enzyme-enzyme complexes, Science, 234:1081-1086.
Sternberg, M.J.E., and Taylor, W.R., 1984, Modeling the ATP-binding site of oncogene products, the epidermal growth factor receptor and related proteins, FEBS Lett., 175:387-390.
Sternlicht, H., Yaffe, M. and Farr, G., 1987, A model of the nucleotide-binding site in of tubulin, FEBS Lett., 214:223-235.
Wessling-Resenick, M., and Johnson, G.L., 1987, Transducin interaction with rhodopsin, J. Biol. Chem., 262:12444-12447.
Zeeburg, B. and Caplow, M., 1979, Determination of free and bound microtubular protein and guanine nucelotide under equilbirium conditions, Biochem., 18:3880-3886.
Zisapel, N., Levi, M. and Gozes, I., 1980, Tubulin: an integral protein of mammalian synaptic vesicle membranes, J. Neurochem., 34:26-32.
Zor, V., 1983, Role of cytoskeletal organizaton in the regulation of adenylate cyclase-cyclic adenosine monophosphate by hormones, Endocrine Rev., 4:1-21.

AUTHOR INDEX

SUBJECT INDEX